MAGNESIUM DEFICIENCY IN FOREST ECOSYSTEMS

Nutrients in Ecosystems

VOLUME 1

Series Editors

Reinhard F. Hüttl
Helmut Beringer

Managing Editor

Bernd Uwe Schneider

Magnesium Deficiency in Forest Ecosystems

Editors

REINHARD F. HÜTTL

and

WOLFGANG SCHAAF

Brandenburg Technical University, Cottbus, Germany
Chair of Soil Protection and Recultivation

KLUWER ACADEMIC PUBLISHERS

DORDRECHT / BOSTON / LONDON

Library of Congress Cataloging-in-Publication Data is available.

ISBN 0-7923-4220-8

Published by Kluwer Academic Publishers,
P.O. Box 17, 3300 AA Dordrecht, The Netherlands.

Kluwer Academic Publishers incorporates
the publishing programmes of Martinus Nijhoff,
Dr W. Junk, D. Reidel, and MTP Press.

Sold and distributed in the U.S.A. and Canada
by Kluwer Academic Publishers,
101 Philip Drive, Norwell, MA 02061, U.S.A.

In all other countries, sold and distributed
by Kluwer Academic Publishers Group,
P.O. Box 322, 3300 AH Dordrecht, The Netherlands.

Printed on acid-free paper

Printed and bound in Great Britain by Hartnolls Ltd., Bodmin, Cornwall.

Contents

Part II   Magnesium in forest ecosystems

viii

List of Authors

Dr. Sabine Augustin, Niedersächsische Forstliche Versuchsanstalt, Grätzelstr. 2, D-37079 Göttingen, Germany

Dr. Hans-Peter Ende, Zentrum für Agrarlandschafts- und Landnutzungsforschung, ZALF, Institut für Rhizosphärenforschung und Pflantzenernährung, Eberswalder Str. 84, D-15374 Müncheberg, Germany

PD Dr. Karl-Heinz Feger, Institut für Bodenkunde und Waldernährungslehre, Albert-Ludwigs-Universität, Bertoldstr. 17, D-79085 Freiburg i. Br., Germany

Prof. Dr. Siegfried Fink, Institut für Forstbotanik und Baumphysiologie, Albert-Ludwigs-Universität, Bertoldstr. 17, D-79085 Freiburg i. Br., Germany

Dr. Klaus Katzensteiner, Institut für Waldökologie, Universität für Bodenkultur, Peter-Jordan-Str. 82, A-1190 Wien, Germany

Prof. Dr. Martin Kaupenjohann, Universität Hohenheim, Institut für Bodenkunde und Standortslehre, Fachgebiet Bodenchemie, D-70599 Stuttgart, Germany

Guy Landmann, Ministère del l'agriculture, de la pêche et de l'alimentation, Département de la santé des forêts, 19, avenue du Mairie, F-75732 Paris SP 07, France

Dr. Stephan Raspe, Institut für Bodenkunde und Waldernährungslehre, Albert-Ludwigs-Universität, Bertoldstr. 17, D-79085 Freiburg i. Br., Germany

Dr. Wolfgang Schaaf, Lehrstuhl Bodenschutz und Rekultivierung, Brandenburgische Technische Universität Cottbus, Postfach 101344, D-03013 Cottbus, Germany

Dr. Stefan Slovik, Julius-von-Sachs-Institut für Biowissenschaften der Universität Würzburg, Lehrstuhl für Botanik, 1, Mittlerer Dallenbergweg 64, D-97082 Würzburg, Germany

Dr. Kaisu Makkonen-Spiecker, Institut für Waldwachstum, Albert-Ludwigs-Universität, Bertoldstr. 17, D-79085 Freiburg i. Br., Germany

Series Editors' Note

Intensification of forest and agricultural land use revealed that an adequate maintenance of soil fertility is an inevitable prerequisite. To guarantee continued land use productivity the concept of sustainability was developed. On many sites optimal productivity can only be achieved by appropriate nutrient management based on profound knowledge of all relevant soil–plant relations. More recently, besides economic considerations, ecological concerns have brought about a new orientation of nutrient-related research into terrestrial ecosystems. To better understand the effects of nutrient element input over time and space applied through management, via atmospheric deposition or internal processes, an ecosystematic research onset was developed. Based on the actual measurement of input/output and internal cycling fluxes with a great emphasis on process studies, this approach allows insight into the ecological sustainability of terrestrial ecosystems. Ecosystem processes also affect hydro-geochemical properties beneath the rooting zone especially with regard to groundwater but also the atmosphere through gaseous emissions. The scope of this book series is, therefore, to show results of ecosystematic studies providing new insight into the complex interactions of mineral and organic nutrients including CO_2 and water related to the vitality, stability and/or elasticity of terrestrial ecosystems, whether managed or not. This includes ecological, ecophysiological and modeling work as well as studies into the restabilization, regeneration or restoration of terrestrial ecosystems with particular emphasis on nutrient element cycling. The book series also aims at closing the gap between well-established text book knowledge and single research results published in numerous journal contributions in this interdisciplinary field of ecological research.

Reinhard F. Hüttl
Helmut Beringer

Intended publication volumes:

Title	Guest editors
Magnesium Deficiency in Forest Ecosystems	*R. F. Hüttl and W. Schaaf*
Changes in Atmospheric Chemistry and Effects on Forest Ecosystems	*R. F. Hüttl and K. Bellmann*
Sulfur Deficiency in Agricultural Ecosystems	*H. Schnug*

Introduction

R. F. HÜTTL AND W. SCHAAF

Brandenburg Technical University Cottbus, Chair of Soil Protection and Recultivation, P.O. Box 10 13 44, 03013 Cottbus, Germany

The health status of forest trees and stands is determined by numerous site factors such as chemical, physical, and biological soil factors, water supply, climate, weather conditions, management history as well as atmospheric deposition impacts. In this context, the nutrient supply is an important evaluation parameter. Forest trees well supplied with nutrients are more resistant to stresses that affect the forest ecosystem than other trees. This is true for both biotic and abiotic influences.

Therefore the investigation of the so-called 'new type forest damage' was aimed at the exact determination of the health status of damaged trees. When considering the complete forest ecosystem, health (=vitality) means the sustainable ability to withstand negative environmental influences and still remain stable and productive. From this viewpoint, an optimal nutritional status is a prerequisite for an optimal health status.

The term 'new type forest damage' comprises a number of damage symptoms which have been observed in various tree species on very different sites since the mid-1970s, particularly in Europe and North America. However, they occurred much more intensively in the 1980s. Generally, this forest damage was thought to be related to negative impacts of air pollutants.

From intensive research into the new type forest damage within the last two decades it was learned that various damage types, for example foliar losses which have been attributed to this new problem have, in fact, been known for a long time. However, there are a number of damage types which reflect new or new type phenomena. Interestingly enough, these phenomena are generally associated with nutritional disturbances.

In this context, Mg-deficiency – a truly new phenomenon in high elevation forests – represents the most widespread disorder. To explain this far-reaching change in the nutrient supply of forest ecosystems, direct as well as indirect damage pathways have been suggested. It is now clear that anthropogenic impacts in combination with natural stress factors affect forests in different ways that are always site and stand specific. However, forest decline symptoms associated with nutritional disturbances can generally be mitigated by appropriate nutrient amendments.

Since forest ecosystems function as complex networks of their various compartments and are influenced by both biotic and abiotic site factors it is not possible to understand the role of a single element without also knowing the relevant chemical, physical, and biological processes within these systems. Also the interaction of this particular element with other elements is of importance. The knowledge of these

R. F. Hüttl & W. Schaaf (eds): Magnesium Deficiency in Forest Ecosystems, 00–00.

processes and interactions is not only a necessary prerequisite to correctly understand and eventually solve these problems but may also help to avoid or reduce such phenomena in the future, for example in many of the fast developing countries over the world.

Hence, this volume summarizes the knowledge on the role of magnesium in forest ecosystems gathered from recent studies as well as from historical findings and experiences.

Part I
Magnesium deficiency.
Symptoms and development

1
Visual magnesium deficiency symptoms (coniferous, deciduous trees) and threshold values (foliar, soil)

H.-P. ENDE and F. H. EVERS

1.1. Introduction

Before the appearance of the so-called 'new type' forest damage at the end of the 1970s, scientists had paid limited attention to Mg as a nutrient element of forest trees. Some fertilizer trials and special investigations did already exist in northern and north-eastern Germany on soils derived from acidic glacial sands. But, in most of the forested area, Mg supply had been sufficient; on the other hand, analysis and methods for the determination of Mg were complicated and not very reliable. After the development of leaf and needle analysis as a diagnostic method for the investigation of the nutritional status of forest trees, needle samples were usually drawn from the top whorl (half- to one-year-old needles), the less appropriate crown position for the detection of Mg deficiency. Today, the Mg supply of forest trees can be characterized exactly by means of AAS analysis as well as differentiated and standardized sampling methods.

A laboratory experiment creating Mg deficiency symptoms on pine plants has already been described by Möller (1904). In field trials with pine species laid out in the 1920s, deficiency experiments (optimal supply of all nutrients except for one) revealed the role of Mg in tree health: intense needle chlorosis disappeared only after fertilization with Mg-containing substances (Becker-Dillingen 1937, 1939; Brüning, 1959). In the USA and Canada, similar symptom expressions and growth reductions were discovered in coniferous and deciduous tree stands (Heiberg and White, 1951; Linteau, 1962). On these sites, the deficiency symptoms disappeared and growth came to regular levels in the same way after Mg fertilization.

This empirical–optical kind of diagnosis mainly applied to forest plantations and very young forest stands. Height growth and terminal shoot length measurements indicated acute deficiency and a growth stimulating effect of fertilization (Jover and Barneoud, 1978). In older stands, it is much more difficult and needs exact forest production measurements to determine the reaction to fertilizer treatments. A significant evaluation of the contribution of single nutrients to yield improvement is impossible without soil analyses and, above all, leaf or needle analyses.

Improved determination of the Mg supply requires not only sophisticated analytical methods, but also understanding of the methodology of sampling, e.g. concerning the organs to be sampled for analyses, appropriate crown positions and sampling times. A number of fundamental articles on this topic have been published.

R. F. Hüttl & W. Schaaf (eds): Magnesium Deficiency in Forest Ecosystems, 3–22.
© *1997 Kluwer Academic Publishers. Printed in Great Britain.*

1.2. Symptoms (leaves and needles)

Pronounced Mg deficiency is revealed by a yellow discoloration of whole leaves and needles or parts of them (chlorosis). However, not all kinds of chlorosis are due to poor Mg supply. Thus, highly differentiated symptom descriptions are needed to be able to discriminate between Mg deficiency and other nutritional disturbances or damage causing similar discoloration in the field. Hartmann *et al.* (1988), Hanisch and Kilz (1990) and Bergmann (1993) have edited excellent books in color for symptom diagnosis and damage assessment. While in forest plantations and young stands, the discoloration of the sun-exposed leaves and needles is conspicuous, symptom diagnosis can be difficult in older stands, especially of deciduous trees.

1.2.1. Leaves

The typical Mg deficiency symptom is a discoloration of the intercostal area, that is the region between the veins, mostly starting in the center of the leaf. Sometimes the leaf as a whole appears light-green to yellowish in color, but more often the green leaf-ribs and veins clearly stand out against the yellow to golden-brown intercostal fields. Similarities exist with the symptoms of potassium and manganese deficiency as well as frost damage and thawing salt damage.

1.2.2. Needles

Needles also turn yellow when Mg deficient, but the symptomatology is much more differentiated. On coniferous trees with more than one needle age class the discoloration starts in the oldest needle age class at the needle tips (tip yellowing). In a more advanced phase, all needles except for those of the current age class turn yellow, more intensely on the light-exposed parts of the branches and on the upper side of the needles. The yellow tones range from brownish-yellow (oldest needles) to light yellow (1–2-year-old needles). This is true for Norway spruce and Douglas fir mainly. On silver fir trees symptom expression is similar, but not always that marked, and needles often drop before turning yellow.

Similarities exist with potassium deficiency in the initial stage. In more advanced stages of K deficiency, the current needles also turn yellow; not so of Mg deficiency. The yellow color of K-deficient needles is not as intense as that of Mg-deficient ones.

On Scots pine the chlorosis also starts in the oldest needle age classes with golden-yellow needle tips. Later, the needles often divide into three zones of different colors: the rust-brown tips, a yellow zone in the middle and a small green base. Finally the whole needles turn yellow-brown and the same process begins in the youngest needle age class.

The needles of larches show similar zones and later on total discoloration during the vegetative period.

1.3. Sampling

The fundamental investigations on needle analysis by Tamm 1956, Wehrmann 1959, Strebel 1960, Ingestad 1962, Höhne 1963, Nebe 1963 and Reemtsma 1966 were carried out with material from the top whorl of conifers (spruce-type). This was the standard sampling position until today. For the determination of the Mg supply to conifers with more than one needle age class, however, this crown position is rather inadequate as the top of the tree obtains an optimum supply with nutrients, especially with Mg and even in a case of deficiency. In general, this is also true for the youngest compared with the oldest parts of the side branches (Hunger, 1972). That is why the investigations of Reemtsma (1966) into the nutritional status of Norway spruce stands used the nutrient contents of older needle age classes. The Mg contents of this area and the discovery of a Mg gradient from the current to the oldest needles are effective tools for the evaluation of the nutritional status of trees.

To determine such gradients or age sequences, usually the needles of the 7th whorl (from the top of the tree) of Norway spruce, silver fir and Douglas fir are collected, separating the age classes. Samples are withdrawn regularly from dominant trees, the crowns of which are completely exposed to light. Needle samples of Scots pine trees with only three or four needle age classes are usually collected at the third or fourth whorl. It is recommended to collect the samples at the beginning of the vegetative resting period (late fall to early winter).

A corresponding standard for leaf sampling from deciduous trees does not exist. Usually fully developed leaves from the upper, light-exposed crown position are collected. A more exact definition is given by Rzeznik and Nebe (1987) who collected the third and fourth leaf from the top of each of three shoots at the tops of beech trees. Leaves from the inside of the crown have been collected and analyzed by some authors, but the gain in information concerning the nutritional state by use of this method is unclear. An essential factor to be considered is the time of year of sampling. According to investigations of Guha and Mitchell (1966) and le Tacon and Toutain (1973), in the month of August the Mg contents in leaves are in balance with the other nutrients. This finding has been confirmed by Ende (1991) who collected the third to seventh leaf from the top of the terminal shoot of young beech trees.

1.4. Discussion of supply ranges and threshold values

A great number of vessel-culture and field experiments with fertilizers have been carried out to evaluate the nutritional status of forest trees (in the case of coniferous trees the current needles were analyzed). In combination with yield measurements these data allowed for the definition of ranges for optimum, normal and deficient nutrient supply. These values are still useful for a first orientation, if not applied to the wrong tree age class (cf. Höhne, 1964a,b). For a more exact assessment of the nutritional state of conifers older needle age classes are also analyzed. Ratio values are good diagnostic methods. The N/Mg ratio, for example, may reveal disharmonic nutrition even at comparatively high Mg values. If biomass data are available, it can also be helpful to relate the analysis data to the weight of 100 or 1000 needles. A comprehensive survey of nutrient contents, subdivided for tree species, has been compiled by van den Burg (1985). Ranges and threshold values for Mg contents in leaves and needles of many tree species are found in articles by Leaf (1968) and Lyr *et al.* (1992).

1.4.1. Norway spruce (Picea abies [L.] Karst.)

Regarding tree nutrition, Norway spruce is by far the most intensively investigated tree species. The first systematic work was carried out by Tamm (1956), Strebel (1960), Ingestad (1962) and Höhne (1963). In combination with forest production data, the relationships between nutrition and growth were demonstrated and nutritional states were characterized by needle analysis data of the 6-month-old needles of the top whorl. Strebel (1960) who investigated 20–60-year-old stands in Bavaria did not find any relationship between height growth and Mg content in needles. This result is not surprising as the Mg contents were at a relatively high level between 1.12 and 2.88 mg g^{-1} in dry matter. Only in one spruce plantation with reduced growth were low Mg contents of 0.3–0.5 mg g^{-1} dry weight (dw) found.

Höhne (1963) investigated spruce stands in the eastern Ore Mountains and did not find any relationship between growth and Mg nutrition; Mg contents ranged between 1.2 and 2.1 mg g^{-1} dw. In spruce stands of the Beskides with optimum growth, similar Mg values were found by Nebe and Beneš (1966). In older spruce stands of the Ore Mountains, Hunger and Fiedler (1965) found a positive correlation between height growth and Mg contents in 6-month-old needles ranging from 0.65 to 2.24 mg g^{-1}. By systematic experiments with seedlings, Ingestad (1962) defined a range for optimum Mg supply between 0.9 and 1.6 mg g^{-1} dw.

Until the beginning of the 1970s Mg contents of more than 1 mg g^{-1} dw in needles were common. Mg-deficiency symptoms had only rarely been reported, e.g. from spruce seedlings in Swedish nurseries (Ingestad, 1960) at Mg contents around 0.5 mg g^{-1} dw. Then, in more recent investigations on permanent observation plots, more and more a slow but continuous decrease in Mg values was recorded (Kenk *et al.*, 1984; Reemtsma, 1986; Hüttl, 1989; Raitio and Tikkanen, 1989; Nebe, 1991). For example, between 1962 and 1984 in the Thuringian Forest the Mg contents on acidic sites had decreased by more than 50% in current, and by more than 80% in the older needles. The N/Mg ratio in current needles had increased from 12 (1962) to 28 (1984), in three-year-old needles from 10 to 71 (Nebe, 1991). In the Black Forest, the Mg contents of current foliage had decreased from 1.4 (0.9–2.0) to 1.0 (0.2–1.7) mg g^{-1} dw between 1975 and 1983; in four-year-old needles from 0.7 (0.3–1.4) to 0.4 (0.2–1.2) mg g^{-1} dw; accordingly, N/Mg ratios in current needles had increased from 8 to 9, in four-year-old needles from 13 to 56 (Hüttl, 1989).

Even before visible Mg-deficiency symptoms had occurred, Altherr and Evers (1975) had observed growth-limiting Mg deficiency in a fertilizer trial in the Odenwald region. A re-investigation of this trial eight years later by Kenk *et al.* (1984) showed a further decrease in Mg contents (see Table 1).

Table 1. Mg contents (mg g⁻¹ dw) of Norway spruce needles from a medium-aged stand in the Odenwald region, 1974 and 1982 (from Kenk *et al.* 1984)

Treatment	Sampling position		Year	
	Whorl	Needle age class	1974	1982
Control (untreated)	I		0.77	0.50
	VII	1	0.69	0.52
		2	0.37	0.30
		3	0.31	0.20
		4	0.29	0.20
		5	0.27	0.23
Fertilized	I		1.13	0.78
	VII	1	0.91	0.81
		2	0.44	0.48
		3	0.39	0.34
		4	0.36	0.31
		5	0.31	0.30

This time the untreated spruces exhibited a distinctive chlorosis, especially in the older needle age class; a slight yellowing also occurred of the two-year-old needles. The fertilized spruces did not show any yellowing symptoms. The Mg contents of the untreated spruces indicate the threshold range below which chlorotic symptoms reveal the deficiency. This finding corresponds with those of many other authors defining the threshold range for Mg between 0.3 and 0.4 mg g⁻¹ dw (cf. Hüttl, 1991) and is valid for other spruce species, e.g. *P. engelmannii* (Tomlinson, 1985), *P. rubens* and *P. pungens* (Lowry and Avard, 1969), *P. omorica* (Hunger, 1990).

In the inverse direction, the Mg content cannot be derived from the degree of discoloration observed, as the yellowing varies considerably with light exposure and climatic as well as genetic conditions (Makkonen-Spiecker and Evers, 1993).

Referred to the analytical values of current needles, Mg contents lower than 0.7 mg g⁻¹ dw indicate the beginning of deficiency (cf. Liu and Hüttl, 1991). In Mg-deficient spruce stands of the Southern Black Forest, Liu (1988) found a close negative correlation between the Mg contents of the current needles and the estimated discoloration of the four-year-old foliage: At Mg contents lower than 0.7 mg g⁻¹ dw all the stands investigated exhibited yellowing symptoms of at least 20% and up to 100% of the four-year-old needle class (Figure 1). On the other hand, yellowing symptoms of more than 40% also occurred at Mg contents higher than 1.0 mg g⁻¹ dw in the current needles.

Good growth yield requires substantially higher Mg contents. According to Reemtsma (1986), physiologically optimum values are higher than 1.2 mg g⁻¹ dw in current needles (provided that no pronounced negative gradient to the older needle age classes exists).

The Mg supply of comparable trees can be different even though the contents of the current needles are on the same level. As Mg in a deficient tree can be retranslocated from the older to the current needles, the nutritional state can be estimated

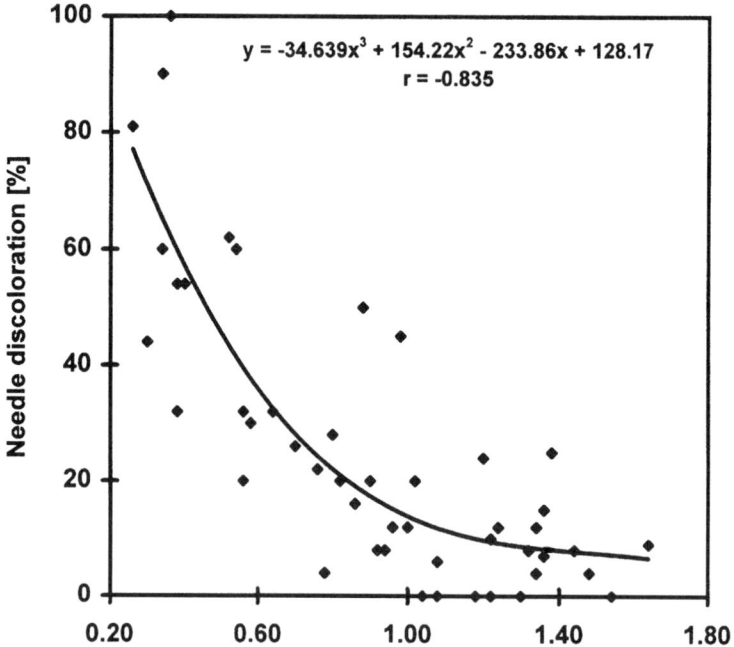

$$y = -34.639x^3 + 154.22x^2 - 233.86x + 128.17$$
$$r = -0.835$$

Figure 1. Correlation between Mg contents of current needles and the discoloration of four-year-old needles from Norway spruce (*Picea abies* [L.] Karst.) trees of the southern Black Forest, Germany; sampling time: winter 1986/1987 (modified from Liu, 1988)

from the Mg gradient along the needle age classes: the stronger the negative gradient from the current to the oldest needles, the less sufficient the supply (Reemtsma, 1986; le Goaster *et al.*, 1990/91). Frequently, only the current needles have sufficient Mg supply. Thus, for an exact estimation of the nutritional status, needle analysis data of different age classes are needed.

From a series of analytical data obtained from 200 experimental and observational spruce stands in South-Western Germany, the diagram given in Figure 2 has been constructed (Evers, 1994). Mg supply ranges are shown between the single lines of the graph. The range of strong deficiency (--) has experimentally been defined as the range of symptom expression (Kenk *et al.*, 1984).

The other ranges are demarcated based on growth conditions mainly. Growth might not be a good indicator of Mg supply, as Reemtsma (1986) points out; it is certain, however, that in the range between mediocre (n–) and deficient (–, – –, — — —), Mg supply is a growth-limiting factor.

In the ranges of normal supply (n–, n, n+), regular growth is possible. In the (n–) and (n) range, growth can still be stimulated by Mg fertilization, whereas in the ranges better than (n) no definite relationships have been found between growth and Mg nutrition.

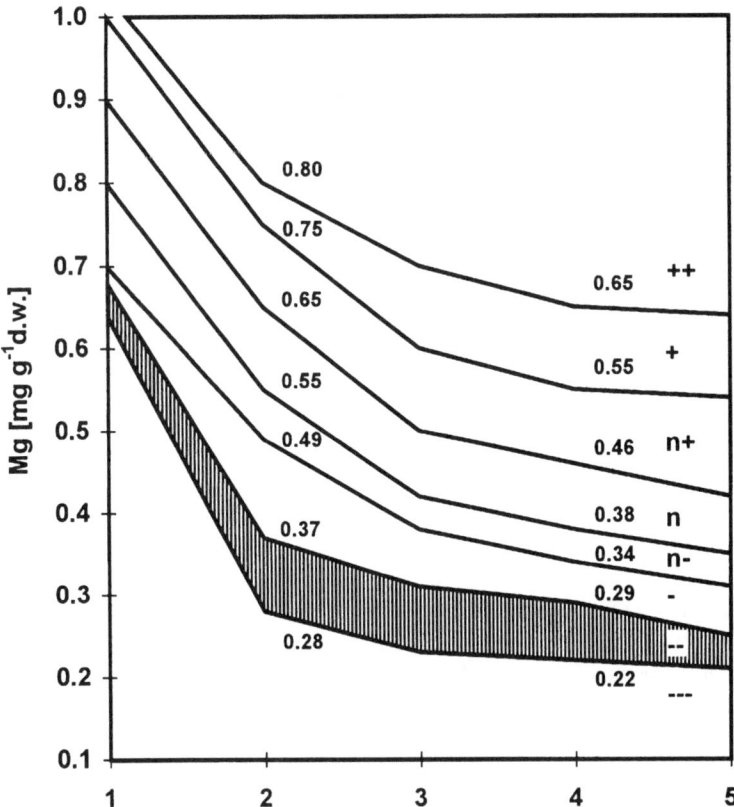

Figure 2. Ranges of Mg supply (mg g⁻¹ dw) of spruce needles in the age sequence from the first (current) to 5th needle age class of the 7th whorl (modified from Evers, 1994). Optimum supply: ++ optimum, + good; normal supply: n+ sufficient, n moderate, n– mediocre; deficient supply: – slightly deficient, – – deficient, — — — extremely deficient. Hatching: Transition zone between slight and extreme deficiency – range of symptom development

In the range of good and optimum supply, the Mg supply is seen as physiologically optimum. N/Mg ratios are around 20 in the older needles and 8–15 in the current needles. A physiologically optimum Mg supply is an important base for resistance of the trees against damaging factors, frost, drought, emissions and others.

In practice, the age gradients of the Mg contents may take different courses and cross one or two of the ranges given in Figure 2. In this case, the N/Mg ratios have to be looked at. N/Mg ratios higher than 20 (30 in older needles) indicate a critical level (Hüttl, 1991), especially with regard to the recently high N-deposition rates in forest ecosystems. With increasing N/Mg ratios, disharmonic Mg nutrition can be induced even on sites where Mg supply used to be sufficient, especially if Mg deposition rates are decreasing, e.g. as a consequence of dust emission reductions (Ende *et al.*, 1995).

1.4.2. Silver fir *(Abies alba Mill.)*

The Mg contents of needles from the top whorl of silver firs are 1.4−2 times higher than those of spruces under comparable site conditions (Rehfuess, 1967) and vary from 1.5 to 4.0 mg g^{-1} dw in Southern Germany fir stands (age 50 years and above). After a re-investigation in part of these stands in 1981, these values had decreased by more than 30% (Evers, 1984). At average Mg values of 1.38 mg g^{-1} dw, a high degree of disease and mortality was found, but the influence of Mg nutrition in this damage was not investigated. In another investigation in South-Western Germany, the current needles of healthy fir stands had Mg contents of 2.11 mg g^{-1} dw and the 5th needle age class had around 1 mg g^{-1} dw. In the needles of diseased firs, Mg values of 1.69 and 0.62, respectively, were found (Evers, 1979). Landmann *et al.* (1987) found distinctly lower Mg values in diseased firs of the Vosges mountains: in current needles 0.7 and in 4-year-old needles 0.4 mg g^{-1} dw. In the needles of healthy firs, they found 1.1 and 0.7 mg g^{-1} dw, respectively. These values are in the range of the average Mg supply given by Berchtold (1979) for diseased Bavarian fir stands (1.05 and 0.85 mg g^{-1} dw, respectively). Bonneau (1993) investigating fir (and spruce) stands of the Vosges Mountains found that needle loss as well as needle chlorosis only appeared at the end of critical climatic situations (e.g. high precipitation in spring followed by a drastic summer drought period as recorded from 1983 to 1985). Under moderate climatic conditions, the Mg contents were in the current needles below 1.3, in the four-year-old needles below 0.8 mg g^{-1} dw. At the end of the critical climatic situation, the needles turned green again (Bonneau, 1993).

In fir stands with 3−5 m tree height and pronounced needle-tip yellowing, Zech and Popp (1983) found 0.22−0.34 mg g^{-1} dw. Doubtless these are extreme values.

1.4.3. Pine *(Pinus sylvestris L.)*

On nutrient-poor glacial sands of the north-eastern lowlands of Germany and adjacent Polish areas, Mg deficiency in pine stands with pronounced needle-tip yellowing has been described for a long time (Becker-Dillingen, 1939, Brüning, 1959). In a number of plantation experiments, the growth improving effect of Mg fertilizers has been demonstrated as well. With combined NPKCaMg-fertilizers, height growth can be increased by 46% compared with controls; the same treatment without Mg resulted in 16% growth increase (Brüning *et al.*, 1970).

Slight tip yellowing of pine occurred in the lowlands of Lower Saxonia, Germany, at Mg contents of 0.73 mg g^{-1} dw in current and 0.6 mg g^{-1} dw in 2-year-old needles and at N/Mg ratios of 25−30 (Hartmann and Thomas, 1993). Krauß (1965) had defined the minimum range for Mg contents in current needles as between 0.2 and 0.8 mg g^{-1} dw where acute visible deficiency symptoms and strong fertilizer responses are to be expected. Mg contents between 0.8 and 1.3 mg g^{-1} dw were defined as the optimum range. Later on, three levels of supply were differentiated (Krauß et al., 1986, Hofmann and Krauß, 1988):

Level 1 (strong deficiency): 0.25–0.31–0.4 mg Mg g^{-1} dw
Level 2 (slight deficiency): 0.40–0.49–0.6 mg Mg g^{-1} dw
Level 3 (normal to optimal): 0.60–0.75–0.9 mg Mg g^{-1} dw

In experiments with pine seedlings, Ingestad (1960) found a Mg deficiency range between 0.2 and 0.7 mg g^{-1} dw (cf. Zöttl, 1973; Hüttl, 1991).

On glacial sand sites, Mg deficiency mainly occurs in the plantation and younger stands. In 45–70-year-old stands on typical sites of the north-eastern lowlands of Germany, Hippeli (1967) rarely found acute Mg deficiency; Mg contents varied from 0.9 to 1.7 mg g^{-1} dw. More recent investigations by Hippeli and Branse (1992), based on yearly sampling, revealed a continuous decrease of the Mg supply over the last 25 years down to an average level of 0.8 mg g^{-1} dw, approaching deficiency.

In 68 pine stands of Bavaria, Wehrmann (1959) found Mg contents between 0.75 and 1.8 mg g^{-1} dw; most of the data exceeded 1.0 mg g^{-1} dw. No relationships to growth yield were detected. Mg contents in older Bavarian pine stands analyzed by Kreutzer (1967) were in the same range (0.89–1.35 mg g^{-1} dw). On long-term observation plots in more recent investigations, no clear tendency of decreasing Mg content has been found; the average values have been around 0.7 mg g^{-1} dw since the beginning of the 1960s (Sauter, 1991).

1.4.4. Beech (Fagus sylvatica L.)

When evaluating a fertilizer trial established in 1958, Altherr and Evers (1974) found an unexpected growth response. The definite proof that this growth effect was due to Mg only was achieved by special experiments with Mg sulfate (Altherr and Evers, 1977). The growth-stimulating effect of Mg was verified by leaf analyses as well. The leaves of the unfertilized beeches exhibited intercostal chlorosis and had low Mg contents. The leaves of the fertilized trees were green and larger and had distinctly higher Mg contents. In Table 2, the data of these and other investigations are given. Looking at the values corresponding to yellowing or chlorosis, it is conspicuous that these are generally below 0.56 mg g^{-1} dw. The lowest value in green leaves is 0.73 mg g^{-1} dw (Büttner et al., 1993). In between these values (0.6–0.7 mg g^{-1} dw) the range of symptom expression can be defined. This is in accordance with Altherr and Evers (1974, 1977), van den Burg (1976), Hofmann and Krauß (1988), Zech et al. (1990/91), Ende (1991) and Krauß (1992) and corresponds well with experiments in birch seedlings, e.g. by Ericsson and Kähr (1995). The N/Mg ratios also have to be regarded: The threshold value for symptom expression is between 30 and 35; higher ratios indicate acute Mg deficiency at absolute Mg contents below 0.6 mg g^{-1} dw (Ende 1991).

Optimum Mg supply can be assumed where the Mg contents of leaves exceed 1.5 mg g^{-1} dw with Mg ratios below 15 as found in the Vosges mountains (le Tacon

Table 2. Mg contents of beech leaves (mg g⁻¹ dw) and symptom expression

Stands					Leaf analysis						Leaf color	Reference
					Mg		Mg_{min}		Mg_{max}			
Geographical situation	Number of stands	Stand age	Treatment	Sampling date	mg g⁻¹	N/Mg	mg g⁻¹	N/Mg	mg g⁻¹	N/Mg		
Vosges Mountains, France	65	>100	0	August 1969	1.60	13	0.70	25	2.50	9	nd	le Tacon and Toutain (1973)
Odenwald, Germany	5	51	0 F	July 1968	0.50 1.17	50 22					chlorot. green	Altherr and Evers (1974)
Sommeren, The Netherlands	1		0 0	August 1974	0.40 0.90	63 30					chlorot. green	van den Burg (1976)
Odenwald, Germany	3	58	0 F	August 1975	0.41 0.93	54 30					chlorot. green	Altherr and Evers (1977)
Various sites (Poland)	7	19	0	August 1984			0.80	31	1.60	11	nd	Rzeznik and Evers (1987)
Black Forest, Germany	1	12	0 F	August 1989	0.25	78	1.60	12	2.20	8	chlorot. green	Ende (1991)
Solling, Germany	1	115	0 0 F	Sept. 1993	0.28 0.73 1.55	67 33 16					chlorot. green green	Büttner et al. (1993)
Hils, Germany	1	146	0 F	Sept. 1993	0.56 0.95	42 25					yellow green	Büttner et al. (1993)

0, control; F, fertilized with Mg-containing substances; chlorot., intercostal chlorosis; nd, not documented

Table 3.   Magnesium levels and symptoms in Dutch oak stands

Symptoms on leaves	Mg content (mg g⁻¹ dw)	N/Mg ratio
Quercus robur L.		
Green (no symptoms)	1.40	19
Slight chlorosis	1.20	23
Pronounced chlorosis	1.00	24
Quercus rubra L.		
Green (no symptoms)	1.00	24
Intercostal chlorosis	0.70	35
Intercostal chlorosis and necrotic flecking	0.40	63

from van den Burg (1976)

Table 4.   Mg levels in a 60-year-old mixed hardwood stand

Species	Mg contents (mg g⁻¹ dw)	N/Mg ratio
Red oak (*Q. rubra*)	1.30	19
White oak (*Q. alba*)	0.80	31
Chestnut oak (*Q. prinus*)	1.10	23
Scarlet oak (*Q. coccinea*)	0.90	25

from Auchmoody and Hammack (1975)

and Toutain, 1973), in Poland (Rzeznik and Nebe, 1987) and in the Black Forest after fertilization with $MgSO_4$ (Ende, 1991).

1.4.5.   Oak *(Quercus species)*

In Dutch oak stands (*Quercus robur* L.), within an area of high N emission due to cattle industries, van den Burg (1976) defined the levels of leaf discoloration and Mg supply shown in Table 3. The levels for *Quercus rubra* L. on the same site are also shown. In a 60-year-old mixed hardwood stand in West Virginia, USA, Auchmoody and Hammack (1975) investigated four oak species and found the Mg values shown in Table 4.

The leaf samples were taken at the same time at the end of August. The data demonstrate that the oak species have different Mg contents (and N/Mg ratios) under the same site conditions. A comparison with the results of van den Burg shows that the Dutch *Q. rubra* has a lower Mg content at a higher N/Mg ratio. It can thus be assumed that the Mg supply was good on the North American site. However, the values for *Q. alba* (0.8 mg g⁻¹ dw, N/Mg = 31) are not far from those of the diseased Dutch *Q. rubra*. As it seems likely that the uptake of Mg in the white oak stand was not different from the other species at the same site, this example demonstrates that absolute leaf analyses values should only be applied to the species investigated.

The threshold value for Mg deficiency is about 1.20 mg g⁻¹ dw (N/Mg > 23) in the Dutch *Q. robur* and about 0.70 mg g⁻¹ dw (N/Mg > 35) in the Dutch *Q. rubra* stand. Six of 16 oak stands (*Q. robur* and *Q. sessiliflora*) investigated in Lower Saxonia had Mg contents in leaves below 1.2 mg g^{-1} dw and N/Mg ratios higher than 24 (Thomas and Büttner, 1992). It is questionable, though, if a value of 1.08 (N/Mg = 24) on a calcareous site really indicates Mg deficiency.

1.5. Discussion of soil analytical data

1.5.1. Soil sampling and analysis

Compared with foliar analysis, methods of soil sampling and soil analysis are much more differentiated and less standardized for forest nutrition purposes. The results of soil analysis depend to a high degree on the sampling method (by soil horizons or depth classes) as well as the extraction method (strong mineral acids or low concentrated neutral salt solutions). Extractions to determine the exchangeable cations, the base saturation (the proportions of K$^+$, Ca^{2+}, Mg^{2+} in all cations) and the cation exchange capacity (CEC) are common methods, making recent results of different authors comparable.

1.5.2. Relationships between soil and leaf analytical parameters and deficiency symptoms

Due to the mentioned problems with soil sampling and analysis (see Chapter 3), it would seem difficult to establish significant correlations between soil and leaf analytical data. Many investigations, however, revealed relationships between Mg contents of the soil and the foliage, especially those concerning the 'high-altitude discoloration' of spruce. From the beginning of the 1980s, a series of investigations revealed that Mg deficiency and the associated yellowing symptoms were mainly found on base-poor substrates derived from acidic crystalline or sandstone bedrock (Bosch et al., 1983; Zöttl and Mies, 1983a; Nebe et al., 1987; Raitio and Tikkanen, 1989; Ke and Skelly, 1990/91), on sites influenced by emission and/or sites subjected to extreme utilization (e.g. litter raking).

Accordingly, in a number of investigations significant correlations have been found between the exchangeable magnesium contents in the soil and the Mg contents of spruce needles. Figure 3 is taken from an investigation in the Southern Black Forest (Zöttl and Mies, 1983b) where the soil samples were withdrawn from 0–10 cm soil depth and current needles were taken from 12–20-year-old spruces. Exchangeable Mg contents below 4.5 μmol$_c$ g^{-1} soil were correlated with Mg contents in current needles less than 0.6 mg g^{-1} dw and with yellowing symptoms. Kaupenjohann et al. (1987) found similar relationships in North-Eastern Bavaria. Ke and Skelly (1990/91) verified these results for sites in the US states of Pennsylvania, West Virginia, New York and New Hampshire, where they found highly significant correlations between soil pH, exchangeable Mg, base saturation and Mg contents in needles. In diseased, chlorotic stands pH values (in H$_2$O) ranged from 3.94 to 4.53 and exchangeable Mg in the soil ranged from 0.8 to 2.4 μmol$_c$ g^{-1} soil. In healthy stands exchangeable Mg contents ranged from 2.4 to 27.9 μmol$_c$ g^{-1} soil. Nearly identical results were given

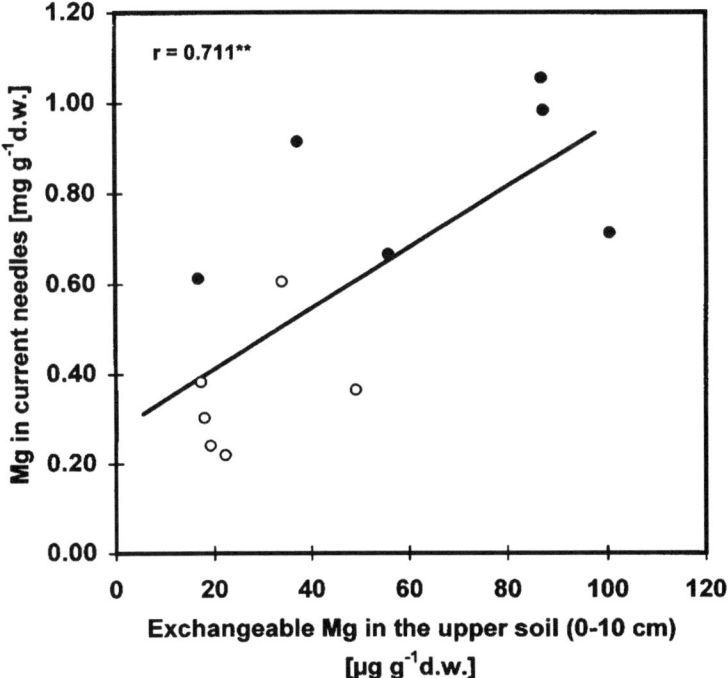

Figure 3. Relationship between exchangeable magnesium in the upper soil and the Mg contents of spruce needles; ● = healthy, ○ = diseased (from Zöttl and Mies, 1983b)

by Zöttl and Hüttl (1986) and Liu and Trüby (1989) who found a large increase in yellowing symptoms in spruce stands with exchangeable Mg contents below $2.4 \mu mol_c g^{-1}$ soil.

Highly significant correlations have also been found between pH values of the humus layer and the Mg contents in spruce needles. Heinsdorf *et al.* (1988) found extremely deficient spruces ($0.3 mg g^{-1}$ dw in the oldest needle age class) in the Thuringian Forest at a pH of 2.7 (KCl). By contrast, Ke and Skelly found a regular Mg supply ($0.9 mg g^{-1}$ dw) and green needles at a soil pH of 3.2.

The effects of fertilizer-Mg on the exchangeable Mg in the soil, the Mg contents in needles, and tree growth have been demonstrated in a fertilizer experiment by Evers *et al.* (1986).

Even though the exchangeable Mg contents in the soil seem to be related to the Mg contents of leaves or needles, there are no such relationships to the health status of the trees if expressed by the degree of discoloration. Frequently trees without any chlorotic symptoms are next to those with a strong discoloration. In such situations, Kandler and Miller (1990/91) found corresponding Mg contents in the needles, but not in the soil (exchangeable Mg^{2+}). The development of chlorotic symptoms under poor Mg supply seems to depend also on climatic and genetic factors. Systematic

Table 5. Mg contents (mg g^{-1} dw) of green and yellow needles from Mg-deficient Norway spruce plants of different clones under irrigation and under drought stress in vessel culture experiments in standardized culture substrate

Clone number	Treatment	Needle color	Needle age class	
			1991 (current)	1990 (1-year-old)
7666	Irrigation	Green	0.54	0.34
		Yellow	0.41	0.28*
7666	Drought	Green	0.87	0.29
		Yellow	0.78	0.24*
7550	Irrigation	Green	0.49	0.28
		Yellow	0.33	0.24*
7550	Drought	Green	0.57	0.25
		Yellow	0.68	0.36*
8112	Irrigation	Green	0.41	0.29
		Yellow	0.30	0.27*
8184	Irrigation	Green	0.33	0.26
		Yellow	0.22*	0.15*

*, pronounced yellowing
(From: Makkonen-Spiecker and Evers, 1993)

Mg and water deficiency experiments with cloned spruces by Makkonen-Spiecker and Evers (1993) demonstrated that yellow needles turned green if the spruces experienced good water supply after a dry period, and that needle discoloration appeared at different levels of Mg contents dependent on the genetic constitution of the spruces (see Table 5).

In this experiment the needles of clone 8184 were green at Mg contents as low as 0.26 mg g^{-1} dw (1-year-old needles). At this level, the needles of clones 7666 and 7559 already exhibited pronounced yellowing symptoms under the same substrate and water supply conditions. Moreover, in clone 7550 (drought series) the Mg contents of the yellow needles were higher than those of the green needles. These examples demonstrate the potential influence of genetic constitution on the development of yellowing symptoms. Similar relationships between needle chloroses and drought stress were found by Bonneau (1993) in fir and spruce stands of the Vosges Mountains.

An investigation carried out in the high altitudes of the Bavarian Forest by Hofmann-Schielle (1991) also indicated the role of the genetic constitution. In stands known as autochthoneous, needle discoloration and dieback were clearly less pronounced than in planted stands from unknown provenances.

1.6. Summary

Mg deficiency has appeared in forest stands on a larger scale only within the last two decades as the so-called 'high elevation yellowing' of mainly Norway spruce stands, but also on different sites and tree species. Typical symptoms of Mg deficiency are yellow to golden-yellow discolorations of leaves and needles. Conifers like *Abies*, *Picea*, *Pinus* show a tip-yellowing of the needles of the older age classes within the

Table 6. Ranges of supply and threshold values for Mg contents (mg g^{-1} dw) of the foliage of different tree species

Age class of foliage	Strong deficiency Current	3–5 y	Slight deficiency Current	3–5 y	Regular supply Current	3–5 y	Optimum supply Current	3–5 y
Norway spruce (*Picea abies*)	≤0.7	≤0.3	≤0.8	≤0.4	>0.9	≥0.5	>1.2	0.7
Silver fir (*Abies alba*)	≤0.8	≤0.4	≤0.9	≤0.5	≥1.1	≥0.7	>1.5	
Scots pine (*Pinus sylvestris*)	≤0.4		≤0.6				>0.9	
European beech (*Fagus sylvatica*)	≤0.6		≤0.7		>0.8		>1.5	
European oak (*Quercus robur*)	≤1.0				>1.2		>1.5	

Modified from Evers, 1994

light-exposed part of the tree crown. At a more advanced stage, whole needles turn yellow, often more pronounced on the upper than on the under sides. With progressing deficiency, more and more of the younger needle age classes turn yellow until only the current needles stay green and the oldest needles have turned brown or have dropped.

Fagus spp., *Quercus* spp. and other deciduous tree species indicate Mg deficiency by a typical discoloration spreading between the veins of the leaves. The symptoms first occur on the older leaves.

Even though the yellowing symptoms described are specific for Mg deficiency, it is not possible to quantify the deficiency level from the degree of discoloration. Over a wide range of insufficient Mg supply the intensity of yellowing is varied by e.g. light exposure, water supply and genetic disposition. To evaluate Mg status precisely, foliar and soil analyses are indispensable. Mg deficiency mainly appears on base-poor acidic substrates, frequently on sites intensely utilized over recent centuries. In the upper mineral soil (0–20 cm depth) of such sites, exchangeable Mg contents below 2.0 μmol (c) g^{-1} soil and a low base saturation are found.

Close correlations have been found between the Mg contents in the soil (exchangeable) and the foliage. This is generally true for whole stands but not for single trees. On the same Mg-deficient site 'green' trees can frequently be observed right beside trees with pronounced discolorations. Beside other factors, differences in the genetic constitution can be an explanation for this phenomenon.

Due to the dependence on climatic influences and genetic variation, the Mg status of trees cannot be estimated from the symptomatology only. Reliable information can only be rendered by foliar analyses. Mg contents of leaves or needles in the region of 0.3 mg g^{-1} dw reveal a strong Mg deficiency. At this level, where needles show a pronounced yellowing, a considerable depression of the photosynthetic potential can be observed (Lange *et al.*, 1987). This does not indicate a higher mortality; the needles at least of spruces may turn green after a longer period of time (Kandler and Miller, 1990/91). The vitality of such trees, however, is reduced and the trees are more susceptible to various damage factors.

Mg contents lower than $0.3 \, mg \, g^{-1}$ dw are found only under extreme conditions in the older needle age classes of conifers (Table 6). In current needles, values below $0.7 \, mg \, g^{-1}$ dw are rarely found. The Mg status of a coniferous tree can be well characterized by the Mg gradient from the oldest to the youngest age classes (Figure 2). For Norway spruce, regular Mg contents are above $0.9 \, mg \, g^{-1}$ dw in current needles; values higher than $1.2 \, mg \, g^{-1}$ dw indicate optimum supply (Figure 2). The optimum N/Mg ratio lies between 8 and 15 in current needles and around 20 in the older needle age classes. High atmospheric N input into forest ecosystems may result in elevated N/Mg ratios and disharmonic Mg nutrition.

1.7. References

Altherr E, Evers FH. 1974. Unerwarteter Düngungserfolg bei Magnesiummangel in einem jungen Buchenbestand auf mittlerem Buntsandstein des Odenwaldes. Allg. Forst- u. J. Ztg. 145, 121–125.

Altherr E, Evers F. 1975. Magnesium-Düngungseffekt in einem Fichtenbestand des Buntsandstein-Odenwaldes. Allg. Forst- u. J.Ztg. 146, 217–225.

Altherr E, Evers FH. 1977. Nachweis eines Magnesium-Düngungseffekts in einem Buchenbestand auf mittlerem Buntsandstein des Odenwaldes. Allg. Forst- u. J.Ztg. 148, 45–48.

Auchmoody LR, Hammack KP. 1975. Foliar nutrient variation in four species of upland oaks. USDA Forest Service Res. Pap. NE-331, 1–16.

Becker-Dillingen J. 1937. Die Gelbspitzigkeit der Kiefer, eine Magnesia-Mangelerscheinung. Ern. d. Pflanze. 33, 1–7.

Becker-Dillingen J. 1939. Die Ernährung des Waldes. Handbuch der Forstdüngung. Verl. Ges. f. Ackerbau, Berlin. 589 p.

Berchtold R. 1979. Beziehungen zwischen der Intensität des Tannensterbens und dem Ernährungszustand der betroffenen Bestände. Dipl. Arb. Univ. München. 110 p.

Bergmann W. 1993. Ernährungsstörungen bei Kulturpflanzen. Gustav Fischer Verlag, Jena and Stuttgart. 835 p.

Bonneau M. 1993. Fertilisation sur résineux adultes (*Picea abies* Karst et *Abies alba* Mill) dans les Vosges: composition foliaire en relation avec la défoliation et le jaunissement. Ann. Sci. Forest. 50, 159–175.

Bosch C, Pfannkuch E, Baum U, Rehfuess KE. 1983. Standorts- und ernährungskundliche Untersuchungen zu den Erkrankungen der Fichte (*Picea abies* Karst.) in den Hochlagen des Bayerischen Waldes. Forstwiss. Cbl. 102, 167–181.

Brüning D. 1959. Forstdüngung – Ergebnisse älterer und jüngerer Versuche. Neumann Verlag, Radebeul, Germany. 210 p.

Brüning D, Trillmich H-D, Uebel E. 1970. Einfluß gestaffelter Magnesiumgaben zu einer Kiefernkultur – Nadeluntersuchungen und ertragskundliche Ergebnisse. Deutsch Akad. Landwirtsch.-wiss., Berlin, GDR, Tagungsber. 103, 61–74.

Burg J van den 1976 Zink excess and magnesium deficiency of beech, pedunculate oak, red oak and Scots pine in Eastern North Brabant and Western Limburg. Rijksinstituut voor onderzoek In de bos- en landschapsbouw 'De Dorschkamp', Wageningen, Uitvoerig verslag band 14, nr. 1, 36 p.

Burg J van den. 1985. Foliar analyses for determination of tree nutrient status – a compilation of literature data. Rijksinstituut voor onderzoek in de bos- en landschapsbouw 'De Dorschkamp', Wageningen, Rapport nr. 414, 615 p.

Büttner G, Hartmann G, Thomas FM. 1993. Vorzeitige Vergilbung und Nährstoffgehalte des Buchenlaubes in Südniedersachsen. Forst u. Holz. 48, 627–630.

Ende H-P. 1991. Wirkung von Mineraldünger in Buchen- und Fichtenbeständen des Grundgebirgs-Schwarzwaldes. Freiburger Bodenkundl. Abh. 27, 98 p.

Ende H-P, Gluch W, Hüttl RF. 1995. Ernährungskundliche und morphologische Untersuchungen im

Kronenraum von *Pinus sylvestris* L. In: Atmosphärensanierung und Waldökosysteme. Eds. R.F. Hüttl, K. Bellmann, W. Seiler. pp. 112–128. Umweltwissenschaften, Eberhard-Blottner-Verlag, Taunusstein, Germany.

Ericsson T, Kähr M. 1995. Growth and nutrition of birch seedlings at varied relative addition rates of magnesium. Tree Physiol. 15, 85–93.

Evers FH. 1979. Ernährungszustand gesunder und erkrankter Tannenbestände. Forst- u. Holzwirt. 34, 366–369.

Evers FH. 1984. Läßt sich das Baumsterben durch Walddüngung oder Kalkung aufhalten? Forst- u. Holzw. 39, 75–80.

Evers FH. 1994. Magnesiummangel, eine verbreitete Erscheinung in Waldbeständen – Symptome und analytische Schwellenwerte. Mitt. Ver. Forstl. Standortskde. u. Forstpflanzenz. 37, 7–16.

Evers FH, Hildebrand EE, Kenk G, Kremer WL. 1986. Boden-, ernährungs- und ertragskundliche Untersuchungen in einem stark geschädigten Fichtenbestand des Buntsandstein-Schwarzwaldes. Mitt. Verein f. forstl. Standortskde. u. Forstpflanzenzüchtg. 32, 72–80.

Guha MM, Mitchell RL. 1966. The trace and major element composition of the leaves of some deciduous trees. II. Seasonal changes. Plant Soil. 24, 90–112.

Hanisch B, Kilz E. 1990. Waldschäden erkennen – Fichte und Kiefer. Verlag Eugen Ulmer, Stuttgart. 334 p.

Hartmann G, Thomas FM. 1993. Ernährungszustand von Kiefern mit Nadelvergilbungen im nordwestdeutschen Flachland. Forst u. Holz. 48, 667–668.

Hartmann G, Nienhaus F, Butin H. 1988. Farbatlas Waldschäden – Diagnose von Baumkrankheiten. Verl. Eugen Ulmer, Stuttgart. 256 p.

Heiberg SO, White DP. 1951. Potassium deficiency of reforested pine and spruce stands in Northern New York. Soil Sci. Soc. Am. Proc. 15, 369–376.

Heinsdorf D, Krauß H-H, Hippeli P. 1988. Ernährungs- und bodenkundliche Untersuchungen in Fichtenbeständen des mittleren Thüringer Waldes unter Berücksichtigung der in den letzten Jahren aufgetretenen Umweltbelastungen. Beitr. f. d. Forstw. 22, 160–167.

Hippeli P. 1967. Der Einfluß wiederholter NPKCaMg-Düngung auf die Ernährung mittelalter Kiefernbestände auf verbreiteten grundwasserfernen Standorten des nordostdeutschen Tieflandes. Arch. Forstwes. 16, 1073–1086.

Hippeli P, Branse C. 1992. Veränderungen der Nährelementkonzentrationen in den Nadeln mittelalter Kiefernbestände auf pleistozänen Sandstandorten Brandenburgs in den Jahren 1964 bis 1988. Forstwiss. Cbl. 111, 44–60.

Hofmann G, Krauß H-H. 1988. Die Ausscheidung von Ernährungsstufen für die Baumarten Kiefer und Buche auf der Grundlage von Nadel- und Blattanalysen und Anwendungsmöglichkeiten in der Überwachung des ökologischen Waldzustandes. Sozialist. Forstwirtsch. 38, 272–273.

Hofmann-Schielle C. 1991. Beziehungen zwischen den Böden und der Intensität der Fichtenerkrankung in den Hochlagen des Inneren Bayerischen Waldes. Forstwiss. Cbl. 110, 228–239.

Höhne H. 1963. Blattanalytische Untersuchungen an jüngeren Fichtenbeständen. Arch. f. Forstwesen. 12, 341–360.

Höhne H. 1964a. Der Einfluß des Baumalters auf das Gewicht sowie den Mineral- und Stickstoffgehalt einjähriger Fichtennadeln. Arch. f. Forstwes. 13, 153–167.

Höhne H. 1964b. Über den Einfluß des Baumalters auf das Gewicht und den Elementgehalt 1- bis 4 jähriger Nadeln der Fichte. Arch. f. Forstwes. 13, 247–265.

Hüttl RF. 1989. Liming and fertilization as mitigation tools in declining forest ecosystems. Water Air Soil Poll. 44, 93–118.

Hüttl RF. 1991. Die Nährelementversorgung geschädigter Wälder in Europa und Nordamerika. Freiburger Bodenkundl. Abh. 28, 440 p.

Hunger W. 1972. Zum Ernährungszustand älterer Fichtenbestände im Klimagefälle des Sächsischen Hügellandes. Flora. 161, 472–494.

Hunger W. 1990. Zum Elementgehalt in den Nadeln chlorosebefallener Omorikafichten (*Picea omorica* [Pancic] Purkyne). Beitr. f.d. Forstwirtsch. 24, 112–115.

Hunger W, Fiedler HJ. 1965. Düngungsdiagnosen für ältere Fichtenbestände des Erzgebirges und Vogtlandes. Arch. Forstwes. 14, 963–986.

Ingestad T. 1960. Magnesiumbrist hos gran (Mg deficiency in spruce). Svenska Skogsforen. Tidskr. 58, 69–76.

Ingestad T. 1962. Macro element nutrition of pine, spruce, and birch seedlings in nutrition solutions. Medd. Skogsforskningsinst. 51 (7), 1–150.

20

Jover J, Barneoud C. 1978. Carence magnesienne sur épicéa commun. Ann. Rech. Sylvicoles AFOCEL. 443–466.

Kandler O, Miller W. 1990/91. Dynamics of 'acute yellowing' in spruce connected with Mg deficiency. Water Air Soil Poll. 54, 21–34.

Kaupenjohann M, Zech W, Hantschel R, Horn R. 1987. Ergebnisse von Düngungsversuchen mit Magnesium an vermutlich immissionsgeschädigten Fichten (*Picea abies* [L.] Karst.) im Fichtelgebirge. Forstwiss. Cbl. 106, 78–84.

Ke J, Skelly JM. 1990/91. Foliar symptoms on Norway spruce and relationships to magnesium deficiencies. Water Air Soil Poll. 54, 75–90.

Kenk G, Unfried P, Evers FH, Hildebrand EE. 1984. Düngung zur Minderung der neuartigen Waldschäden – Auswertung eines alten Düngungsversuchs zu Fichte im Buntsandstein-Odenwald. Forstwiss. Cbl. 103, 307–320.

Krauß H-H. 1965. Untersuchungen über die Melioration degradierter Sandböden im nordostdeutschen Tiefland IV. Arch. Forstwes. 14, 499–532.

Krauß HH. 1992. Beitrag zu Ernährung und Wachstum der Buche im nordostdeutschen Tiefland. Beitr. Forstwirtsch. u. Landsch. ökol. 26, 17–23.

Krauß H-H, Heinsdorf D, Hippeli P, Tölle H. 1986. Untersuchungen zu Ernährung und Wachstum wirtschaftlich wichtiger Nadelbaumarten im Tiefland der DDR. Beitr. Forstwirtschaft. 20, 156–164.

Kreutzer K. 1967. Ernährungszustand und Volumenzuwachs von Kiefernbeständen neuer Düngungsversuche in Bayern. Forstwiss. Cbl. 86, 28–53.

Landmann G, Bonneau M, Adrian M. 1987. Le dépérissement du sapin pectiné et de l'épicéa commun dans le massif Vosgien, est-il en relation avec l'état nutritionel des peuplements? Revue Forestière Française. 39, 5–11.

Lange OL, Zellner H, Gebel J, Schramel P, Köstner B, Czygan F-C. 1987. Photosynthetic capacity, chloroplast pigments, and mineral content of the previous years spruce needles with and without the new flush: analysis of the forest-decline phenomenon of needle bleaching. Oecologia. 73, 351–357.

le Goaster S, Dambrine E, Ranger J. 1990/91. Mineral supply of healthy and declining trees of a young spruce stand. Water Air Soil Poll. 54, 269–280.

le Tacon F, Toutain F. 1973. Variations saisonnières et stationelles de la teneur en éléments minéraux des feuilles de hêtre (*Fagus sylvatica*) dans l'est de la France. Ann. Sci. Forest. 30, 1–29.

Leaf AL. 1968. K, Mg, and S Deficiencies in Forest Trees. In: Symposium on Forest Fertilization, April 1967, at Gainesville, Florida. pp. 88–122. Tennessee Valley Authority, Muscle Shoals, AB.

Linteau A. 1962. Some experiments in forest soil fertilization. In: Forest Fertilization in Canada. Ed. A Lafond. pp. 25–37. Laval University Forest Res. Foundat.

Liu JC 1988. Ernährungskundliche Auswertung von diagnostischen Düngungsversuchen in Fichtenbeständen (*Picea abies* Karst.) Südwestdeutschlands. Freiburger Bodenkundl. Abh. 21, 193.

Liu JC, Hüttl RF. 1991. Relations between damage symptoms and nutritional status of Norway spruce stands (*Picea abies* Karst.) in Southwestern Germany. Fertilizer Res. 27, 9–22.

Liu JC, Trüby P. 1989. Bodenanalytische Diagnose von K- und Mg-Mangel in Fichtenbeständen (*Picea abies* Karst.). Z. Pflanzenernähr. Bodenkde. 152, 307–311.

Lowry LG, Avard PM. 1969. Nutrient content of black spruce and check pine needles. III. Seasonal variations and recommended sampling procedures. Pulp Pap. Res. Inst. Can. 10, 1–54.

Lyr H, Fiedler H-J, Tranquillini W. 1992. Physiologie und Ökologie der Gehölze. Gustav Fischer Verl., Jena u. Stuttgart. 620 p.

Makkonen-Spiecker K, Evers FH. 1993. Untersuchungen zur Reaktionsweise junger Klonfichten (*Picea abies* [L.] Karst.) auf Trockenstreß und Magnesiummangel. Projekt Europ. Forschungszentr. f. Maßnahmen zur Luftreinhaltung (PEF), Bericht KfK-PEF. 114, 81 p.

Möller A. 1904. Karenzerscheinungen bei der Kiefer. Z. f. Forst- und Jagdwesen. 36, 745–756.

Nebe W. 1963. Über die Beurteilung der Düngebedürftigkeit von Mittelgebirgsstandorten durch Blattanalysen. Arch. f. Forstwes. 12, 1024–1052.

Nebe W. 1991. Veränderung der Stickstoff- und Magnesiumversorgung immissionsbelasteter älterer Fichtenbestände in ostdeutschen Mittelgebirgen. Forstw. Cbl. 110, 4–12.

Nebe W, Beneš. 1966. Standort, Höhenwachstum und Ernährungszustand optimal wachsender Fichtenbestände des Slowakischen Erzgebirges und der Beskiden. Arch. f. Forstwes. 15, 1225–1233.

Nebe W, Fiedler H-J, Ilgen G, Hofmann W. 1987. Immissionsbedingte Ernährungsstörungen der Fichte (*Picea abies* [L.] Karst.) in Mittelgebirgslagen. Flora. 179, 453–462.

Raitio H, Tikkanen E. 1989. Nutritional disturbances of young pines in a dry heath forest. Plant Soil. 113, 229–235.

Reemtsma JB. 1966. Untersuchungen über den Nährstoffgehalt der Nadeln verschiedenen Alters an Fichte und Nadelbaumarten. Flora. 156, 105–121.

Reemtsma JB. 1986. Der Magnesium-Gehalt von Nadeln niedersächsischer Fichtenbestände und seine Beurteilung. Allg. Forst- u. J.Ztg. 157, 196–200.

Rehfuess KE. 1967. Standort und Ernährungszustand von Tannenbeständen (*Abies alba* Mill.) in der südwestdeutschen Schichtstufen-Landschaft. Forstwiss. Cbl. 86, 321–348.

Rzeznik Z, Nebe W. 1987. Wachstum und Ernährung von Buchen-Provenienzen. Beitr. f.d. Forstwirtsch. 21, 106–111.

Sauter U. 1991. Zeitliche Variationen des Ernährungszustands nordbayerischer Kiefernbestände. Forstwiss. Cbl. 110, 13–33.

Strebel O. 1960. Mineralstoffernährung und Wuchsleistung von Fichtenbeständen (*Picea abies*) in Bayern. Forstwiss. Cbl. 79, 17–42.

Tamm CO. 1956. Studier över skogens näringsförhallanden IV. Medd. Stat. Skogsforskningsinst. 46, 1–27.

Thomas FM, Büttner G. 1992. Der Ernährungszustand von Eichen in Niedersachsen. Forst u. Holz. 47, 464–470.

Tomlinson GH. 1985. Acid deposition, nutrient imbalance and tree decline: A commentary. In: Effects of Atmospheric Pollutants on Forests, Wetlands and Agricultural Ecosystems. Eds. TC Hutchinson, KM Meema. pp. 189–199. NATO ASI Series Vol. G 16, Springer Verlag, Berlin.

Wehrmann J. 1959. Methodische Untersuchungen zur Durchführung von Nadelanalysen in Kiefernbeständen. Forstwiss. Cbl. 78, 77–97.

Zech W, Popp E. 1983. Magnesiummangel, einer der Gründe für das Fichten- und Tannensterben in NO-Bayern. Forstwiss. Cbl. 102, 50–55.

Zech W, Schneider BU, Röhle H. 1990/91. Element composition of leaves and wood of beech (*Fagus sylvatica* L.) on SO_2-polluted sites of the NE-Bavarian Mountains. Water Air Soil Poll. 54, 97–106.

Zöttl HW. 1973. Diagnosis of nutritional disturbances in forest stands. FAO-IUFRO-Symposium on forest fertilization, Paris, pp. 75–95.

Zöttl HW, Mies E. 1983a. Nährelementversorgung und Schadstoffbelastung von Fichtenökosystemen im Südschwarzwald unter Immissionseinfluß. Mitt. Dtsch. Bodenkundl. Ges. 38, 429–434.

Zöttl HW, Mies E. 1983b. Die Fichtenerkrankung in Hochlagen des Südschwarzwaldes. Allg. Forst- u. J.-Ztg. 154, 110–114.

Zöttl HW, Hüttl RF. 1986. Nutrient supply and forest decline. Water Air Soil Poll. 31, 449–462.

2

Temporal and spatial development of magnesium deficiency in forest stands in Europe, North America and New Zealand

G. LANDMANN, I. R. HUNTER and W. HENDERSHOT

2.1. Introduction

The quantification of any particular feature of forest ecosystems is faced with several problems. The unambiguous identification of any phenomenon is a prerequisite for its quantification. In this respect, and despite a rather straightforward symptomatology (see Chapter 1), Mg deficiency may remain unnoticed as long as symptoms are not acute, and confusion with K deficiency symptoms is possible in some regions. Furthermore, forest ecosystems extend over vast areas and the number of educated observers (scientists, forest managers or owners) is low compared to agricultural land, so that the 'observation pressure' is generally rather low. The quality of information, especially in earlier times, may vary depending on the geographical context.

Until recently, Mg deficiency was not a matter of great concern in Europe. Only in the early 1980s was acute Mg deficiency acknowledged as a component of the 'new type forest decline' and at least partly attributed to acid rain. The spatial distribution of Mg deficiency symptoms is reasonably well known, and the continuous monitoring of forest health since the mid-1980s allowed its temporal development to be followed.

In northeastern North America, research on the possible contribution of atmospheric pollution to forest decline led to extensive studies of forest nutrition during the 1980s. This research was primarily focused on the mechanisms by which pollution could affect forest health and only motivated in a few instances by obvious Mg deficiency symptoms.

In New Zealand, acid rain is not a matter of concern, as rainfall is clearly not acidic and only marginally loaded with anthropogenic sulfate or nitrate (Bridgman, 1989). Magnesium deficiency is considered by forest managers to be a classical nutritional deficiency triggered by local factors.

The main aim of this chapter is to describe the spatial and temporal patterns of Mg deficiency in these three areas. The information already available (e.g. Hüttl 1991, 1993; Roberts *et al.*, 1989) has been assessed, and completed. Note that all regions cited can be found in Figures 1, 6 and 7. A second aim is, as far as possible, to interpret these patterns in relation to possible causal factors. A more detailed process-oriented analysis of the causes of Mg deficiency is provided in Chapters 3 and 6.

R. F. Hüttl & W. Schaaf (eds): Magnesium Deficiency in Forest Ecosystems, 23–64.
© 1997 *Kluwer Academic Publishers. Printed in Great Britain.*

2.2. Criteria used to quantify Mg deficiency

Foliar analysis probably represents the best method for quantifying Mg deficiency, as it is objective, allows subdeficiencies to be identified (see Chapter 1 for a discussion on threshold values for visible symptoms) and is relatively inexpensive. Because of the temporal fluctuations in foliar Mg, already pointed out by Evers (1972), repeated analyses may be needed to unambiguously characterize Mg nutrition and identify any long-term trend.

Soil analysis may also be used as a diagnostic tool. Its interpretation is often more difficult as foliar chemistry is not necessarily well correlated with soil chemistry. However, a relatively close relationship between foliar Mg and exchangeable Mg was recently found for Norway spruce (*Picea abies*) in several European regions (Black Forest: Liu and Trüby, 1989; Zöttl and Mies, 1983; Vosges: Landsmann *et al.*, 1995a) (see also Chapter 1). In a survey of Norway spruce plantations in northeastern USA (West Virginia, Pennsylvania, New York, New Hampshire), Ke and Skelly (1990/91) also reported a significant relationship between foliar Mg and exchangeable Mg. Ouimet and Fortin (1989) observed a similar relationship for sugar maple in a study involving 230 sites in Quebec. Thus, any decline in Mg contents of forest soils may be considered as increasing the risk of Mg deficiency, especially if the threshold value ($2\,\mu$mol(c)$\,g^{-1}$ in the upper mineral horizon was suggested in several European studies, see Chapter 1) is reached.

Some forest stands, for example of silver fir (*Abies alba*) in the Vosges (Landmann *et al.*, 1995a) and Norway spruce in the Bavarian Forest (Hoffmann-Schielle, 1991), may grow healthily on extremely Mg-depleted soils. Specific rooting patterns, specific chemical and mineralogical soil properties, and genetic constitution (autochthoneous provenances versus introduced provenances) were suggested as possible reasons for these observations. Similar factors may explain why no relationship was found between foliar Mg in *Pinus radiata* and common soil extracts of Mg (Hunter, 1992) in New Zealand.

Extraction methods other than those used for quantifying exchangeable Mg may be used. For example, Hantschel *et al.* (1988) found a close relationship between Mg in soil percolation extracts from undisturbed soil cores from the Fichtelgebirge and the corresponding foliar Mg levels. However, this and other methods have been applied by a few research groups only, and a standard methodology is usually lacking. Little use can be made of existing data with a view to analyzing temporal and spatial patterns of Mg deficiency.

Soil solution chemistry may provide more relevant data with regard to tree nutrition. In the Fichtelgebirge, Kaupenjohann *et al.* (1987) showed a close relationship between the Mg concentrations of spruce needles and the Mg content of the soil solution. Unfortunately, data from other studies is difficult to compare because the different devices (zero-tension lysimeters, tension lysimeters, centrifugation) used may show up some variation in the solution chemistry. However, the few available long-term time series may be of interest.

Wood chemistry, i.e. the analysis of the mineral composition of tree-rings, could provide historical insight (see, for example, Zech *et al.*, 1990/91, Bondietti *et al.*, 1990; Shortle *et al.*, 1992). The interpretation of the results is not straight forward, especially for the recent tree-rings (the period of greatest interest), as translocation of mineral elements between rings is still possible. To date, this method has not been applied in large-scale surveys.

Lastly, visual assessment of discoloration has been widely used, either in large-scale ground monitoring surveys covering most of Europe (UN–ECE/EC, 1994) or at regional levels, as in North America (e.g. Millers-Weeks *et al.*, 1989). There are severe limitations in this assessment, namely its poor reliability and the lack of differentiation between the different types of discoloration. However, this relatively inexpensive method, if carried out by trained survey crews, may depict at least the major changes in crown discoloration. Provided Mg deficiency is the major cause of discoloration in the area considered, a rough estimate of temporal development of Mg deficiency may be possible.

In this chapter, we consider a combination of these methods, combined with experts' judgements.

2.3. Long-term evolution of Mg nutrition and current extent of Mg deficiency

For each of the three study areas considered in this chapter, the information available on a possible long-term trend of Mg nutrition and the current extent of Mg deficiency is examined.

2.3.1. Europe

Historical records and recent increase of Mg deficiency   Magnesium deficiency has been known for a long time by forest nutritionists. The earliest reports of Mg deficiency in Europe originate from northeastern Germany (former GDR) where Möller (1904) and Becker-Dillingen (1937) described needle tip yellowing on young Scots pine (*Pinus sylvestris*) plantations. Because symptoms are only visible for a short period of time and only occur on young trees, this disease was generally referred to as a 'Kinderkrankheit der Kiefer' (child disease of pine) (Baule and Fricker, 1967). Later in the same region, Brüning (1966) reported acute yellowing and premature loss of older needles which developed from 1957 onwards in Scots pine thickets. These observations were attributed partly to Mg deficiency, and partly to K deficiency. Indeed, a number of earlier fertilization experiments had shown that K and Mg were the limiting factors in the degraded acidic sandy diluvial soils of northern and northeastern Germany (Becker-Dillingen, 1940; Brüning, 1961). A similar situation is reported for The Netherlands where Van Goor (1970) reported Mg deficiency symptoms on Scots pine, Corsican pine, douglas fir and Norway spruce during the 1950–1970 period.

Other early indications of poor Mg nutrition were reported in the 1960s in (formerly West) Germany and France. These observations of Mg deficiency mostly related to forest stands growing on sandstone: a beech (*Fagus sylvatica*) stand in the Odenwald (northern Baden Württemberg; Altherr and Evers, 1974), a Norway spruce (*Picea abies*) stand in the northern Black Forest (Evers and Hausser, 1973), a silver fir stand (out of 5 sampled) in the nearby region of the Vosges (Vallée, 1967). In a regional survey carried out in 1969 in northeastern France, as many as 10 beech stands out of 45 sampled on acidic soils were Mg deficient according to the usual threshold values $(0.60 \, mg \, g^{-1})$ (Evers 1994; Ende and Evers, this vol). These stands were mostly located in the western Vosgian piedmont, on sandstone (Le Tacon and Toutain, 1973, and unpublished data). Strikingly, most of these earlier reports did not mention symptoms of yellowing, as one might have expected from the low foliar Mg concentrations. This was interpreted by some authors to be a consequence of the relatively low N/Mg ratio at that time (Hüttl, 1993) (see 2.5.2).

Other cases are related to base-poor granites. Several cases of young Norway spruce plantations poorly supplied with Mg, some showing yellowing symptoms, were found during the 1960s in the western Massif Central (Limousin) and the northeastern Massif Central (Morvan, Beaujolais) (Bonneau, personal communication; De Champs, personal communication). In nearly all cases, these were plantations on former agricultural land.

In the 1970s, new cases of Mg deficiency symptoms on Norway spruce were reported, especially from areas considered severely affected by Mg deficiency: Fichtelgebirge and Upper Palatinat Forest in Bavaria (Kreutzer, 1975) and the Black Forest (Zöttl et al., 1977). In the Vosges, Jover and Barneoud (1978) described the first case of acute Mg deficiency. This occurred in 1972 in a young spruce plantation set up on a former agricultural land, and recovered rapidly after Mg fertilization.

Overall, however, Mg deficiency was not a matter of great concern until the late 1970s, as damage symptoms remained localized and did not (with a few exceptions) trigger obvious damage (Zöttl and Mies, 1983). Magnesium was not generally considered a limiting nutrient. Many earlier fertilization experiments set up on acid soils did not, unlike the few examples mentioned above, include a Mg treatment. It was only during the 1980s that an unprecedented development of Mg symptoms was reported in several mid-elevation mountains dominated by acidic depleted soils.

The affected regions (see Figure 1) are (with the first citation for each):

The Harz Mountains (former FRG: Hartmann et al., 1985, 1986; former GDR: Fiedler et al., 1984).

The Thuringian Forest (Southwest of former GDR) (Fiedler et al., 1984).

The Solling (Reemstma, 1986; this area is not among the most affected areas).

The Fichtelgebirge (northeastern Bavaria) (Zech and Popp, 1983).

The Ore Mountains (South of former GDR: Nebe et al., 1987).

The Bavarian Forest (eastern Bavaria, border with Czech Republic) (Bosch et al., 1983).

The Bohemian Forest (Austria, border with Czech Republic) (Glatzel et al., 1987).

The Eggegebirge (North Rhine Westphalia) (Prinz et al., 1985).

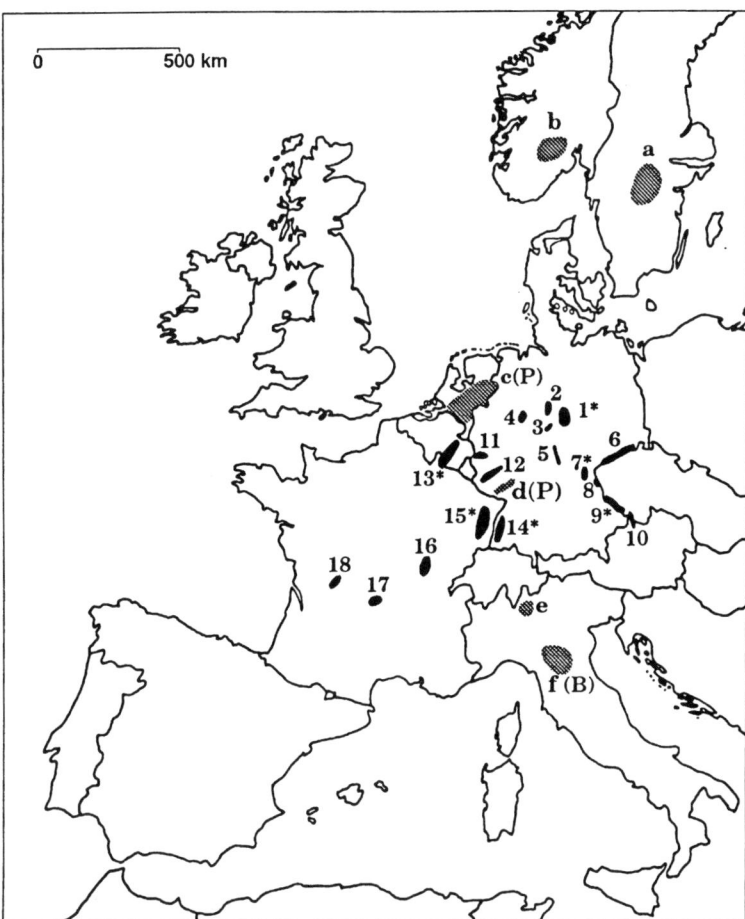

Figure 1. European areas where Mg deficiency symptoms on Norway spruce are reported to be common. Mountainous areas (black): 1. Harz, 2. Hills, 3. Solling, 4. Eggegebirge, 5. Thuringian Forest, 6. Ore Mountains, 7. Fichtelgebirge, 8. Upper Palatinat Forest, 9. Bavarian Forest, 10. Bohemian Forest, 10. Eifel, 12. Hunsrück, 13. Ardennes, 14. Black Forest, 15. Vosges massif, and Massif Central, 16. Forez, 17. Artense, 18. Plateau de Millevaches. *Indicate the main deficiency areas (see text). Cross-hatched: lowland areas and areas with scattered, episodic or subtle symptoms: a. South central Sweden, b. South central Norway, c. Eastern Netherlands, d. Palatinat Forest, e. Northern Alps, f. Tuscan Apennines. From various sources (see text) and personal communications of experts

The Ardennes (Belgium: Wiessen *et al.*, 1984; France: Nys, 1989).

The Black Forest (Zöttl and Mies, 1983).

The Vosges Massif (Bonneau, 1985).

The Massif Central (Bonneau and Landmann, 1986).

The frequency and severity of Mg symptoms may be very different among these regions but it is difficult from the available information to suggest a ranking.

Although some authors have suggested possible chronologies (for example, the Ore Mountains were reported to have been affected more recently than most of the other regions), the available information does not allow us to trace back the spread of Mg deficiency prior to 1985. On the basis of current information, a few, mostly small areas, are to be added to this list (see below).

Evidence for a long-term deterioration of Mg nutrition   Although the chronicle presented above suggests a deterioration of Mg nutrition of forests growing on acidic soils, analytical data, especially time series of Mg contents of foliage and soils, would make the picture more consistent. Two approaches may be used. The first is based on analogue comparisons, i.e. on comparisons of stands in similar site conditions at various dates. The second approach is based on repeated analyses of the same stands. Unfortunately, foliar and soils analyses have, until recently, mostly been used as diagnostic tools and rarely as part of a long-term monitoring program with standardized sampling methods, so that relatively few data are available. The most reliable data often stem from the control plots of fertilization trials.

The impression created by the data from earlier nutritional studies is that Mg levels in the foliage of conifers in the 1960s varied but was generally quite high in comparison with current values. In Norway spruce stands growing on poor sites (quartz porphyry) in the Thuringian Forest, a drastic deterioration of the Mg nutrition, from 1.2 (0.9) mg g^{-1} in 1962 to 0.6 (0.2) mg g^{-1} in 1984 in current (and 4-year-old) needles (Nebe, 1991) has been recorded. On the glacial sand sites of the lowlands of northeastern Germany, Hippeli (1967) rarely found acute Mg deficiency; Mg contents varied from 0.9 to 1.7 mg g^{-1} while recent investigations showed mostly low values, down to 0.8 mg g^{-1} (see below). In the western part of Czechoslovakia (Slavkow Forest/Kaiserwald), Materna (1989) compared the results of an extensive nutritional survey with those of a survey carried out 20 years earlier, and found a very drastic deterioration of the Mg nutrition. More than 60% of the samples had an Mg content below 0.8 mg g^{-1} compared with 1.3 mg g^{-1} in the earlier survey, and some yellowing was also visible. In the Black Forest, the Mg content of current-year needles of Norway spruce decreased from 1.4 mg g^{-1} in current needles in 1975 to 1.0 mg g^{-1} in 1983, and that in 4-year-old needles from 0.7 mg g^{-1} to 0.5 mg g^{-1} during the same period (Hüttl, 1989). Foliar analyses carried out in the 1960s/early 1970s in silver fir (*Abies alba*) growing in the Neckarland and Black Forest (southwestern Germany; Rehfuess, 1967), in northeastern Bavaria (Rehfuess, 1968), and in the Vosges massif (Chichery, 1970 in Landmann et al., 1995a) on base-poor acidic soils showed Mg foliar concentrations (1.5–2.8 mg g^{-1}) currently only reached by the best supplied stands (Hüttl and Zöttl, 1985; Landmann et al., 1995a).

These historical comparisons of stands growing in similar site conditions suggest a long-term deterioration of Mg nutrition in several areas in Europe. Although it is difficult to suspect a systematic bias in these comparisons, one should keep in mind that they do not relate to the same stands, and that they are usually based only on two or a few single years. Reliable long-term time series of Mg foliar levels would strengthen the picture. Most of the few available time series originated from control plots of fertilization trials. A few examples are listed here, following roughly a north – south gradient:

- In Sweden, Aronsson (1985) re-sampled a number of Norway spruce stands spread over Sweden which had been investigated by Tamm during the 1950s and 1960s. In the stands of southwestern Sweden, a decrease in Mg content was identified in 10 out of 15 stands but Mg concentrations did not reach critical values.

- In northeastern Germany (former GDR), Hippeli and Branse (1992) reported a dramatic decline in Mg foliar levels in 7 Scots pine stands growing in the lowlands on sandy, podzolic soils. After a dramatic drop between 1964 and 1975, the decrease in Mg levels has been slowing and Mg levels have seemingly reached an asymptote ($0.9\,mg\,g^{-1}$) (Figure 2a).

- In Lower Saxony (Solling, Eggegebirge, and districts Hohe Heide and Hohe Hessen), Reemtsma (1986) reported a strong historical deterioration of Mg nutrition between the mid/late 1960s and the late 1970s/early 1980s in a number of Norway spruce stands, some of these becoming deficient.

- In Austria, a progressive decrease was noticed in permanent Norway spruce stands between the mid 1960s and the mid 1970s, followed by a stabilization at levels above the deficiency level (Figure 2b) (Stefan, 1989).

- In the Odenwald, a distinct deterioration has been documented between 1974 and 1982 in a Norway spruce stand growing on sandstone (Kenk et al., 1984).

- In the Black Forest, Evers (1984) re-investigated a part of the silver fir stands sampled by Rehfuess (1967) and found a 30% decrease in the Mg foliar levels.

- In the Vosges, a decrease in foliar Mg was documented for a mature silver fir stand growing at a high-altitude site on a base-poor granite (Landmann et al., 1995a).

- In the western Massif Central (Plateau de Millevaches), a slight decrease in foliar Mg was found between 1968–70 and 1987 in 19 young spruce stands (Landmann et al., 1995a).

In summary, available evidence from time series of Mg foliar contents also points to a deterioration of the Mg nutrition of forest trees on acidic soils. This decrease in Mg supply is documented mostly for conifers. Data for hardwoods are very few. Van den Burg (1991) compared foliar analyses from the 1950s with analyses from the 1970s and 1980s, and found only a slight decline of foliar Mg concentrations for pedunculate oak (Quercus robur) in The Netherlands. There are a few exceptions to this overall negative trend. No such trend was found at experimental sites in north (Norrliden) and central (Lisselbo) Sweden (Tamm and Popovic, 1989), which is consistent with the acid rain hypothesis (see 2.5.3). The absence of a negative trend

a   Mg mg $^{-1}$

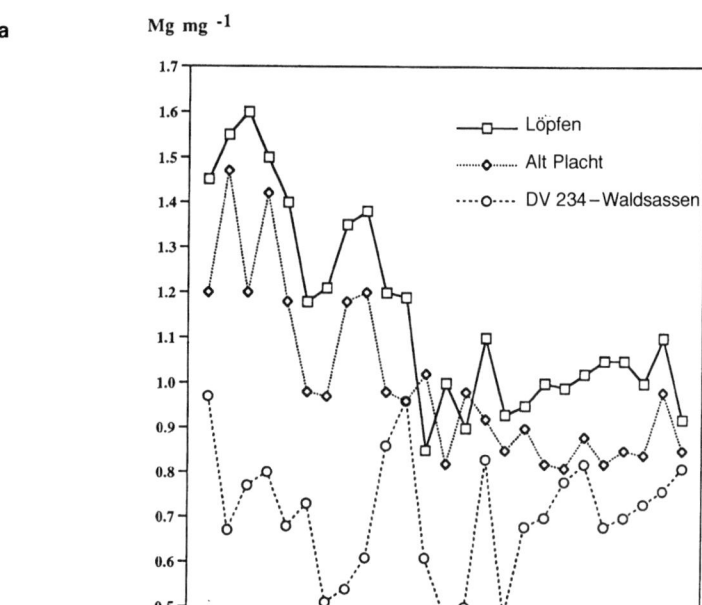

Figure 2. Examples of year-to-year variations and long-term evolution of Mg foliar concentrations in current-year needles of well-identified forest stands. **a.** Two Scots pine stands in northeastern Germany (Löpfen and Alt Placht) (Heinsdorf, 1993) and one Scots pine stand in Bavaria (DV234–Waldsassen) (Sauter, 1991). **b.** 10-year moving average values for seven Norway spruce stands in Austria (Stefan 1989)

in the ten Scots pine stands monitored in the lowlands of Bavaria (an example is given in Figure 2a) is more surprising, especially as N levels increased dramatically over the 1960–1988 period (Sauter, 1991).

A number of historical comparisons of the chemical status of forest soils have recently been reported (Johnson *et al.*, 1991). Some authors still question the reality of a general decrease in exchangeable Mg in acid forest soils (Rehfuess, 1989) or believe this conclusion needs to be founded on a sounder basis (Roberts *et al.*, 1989). Clearly, there is a lack of old soil analyses which can be considered as representative for larger regions, and of examples where both recent and older foliage and soil analyses are available. However, while each study has some weaknesses, the overall evidence is that severe Mg depletion has occurred in many cases. Concentrating on areas where Mg deficiency is currently common, one may quote the studies by Hocke (1991) in the Hesse, and Lefèvre (in Landmann, 1995) and Thimonier (1994) in the Vosges. In these studies, a mean decrease of exchangeable Mg by about 30–50% of Mg (and Ca) over a period of 20–30 years was reported. In other studies, the decrease of Mg was smaller.

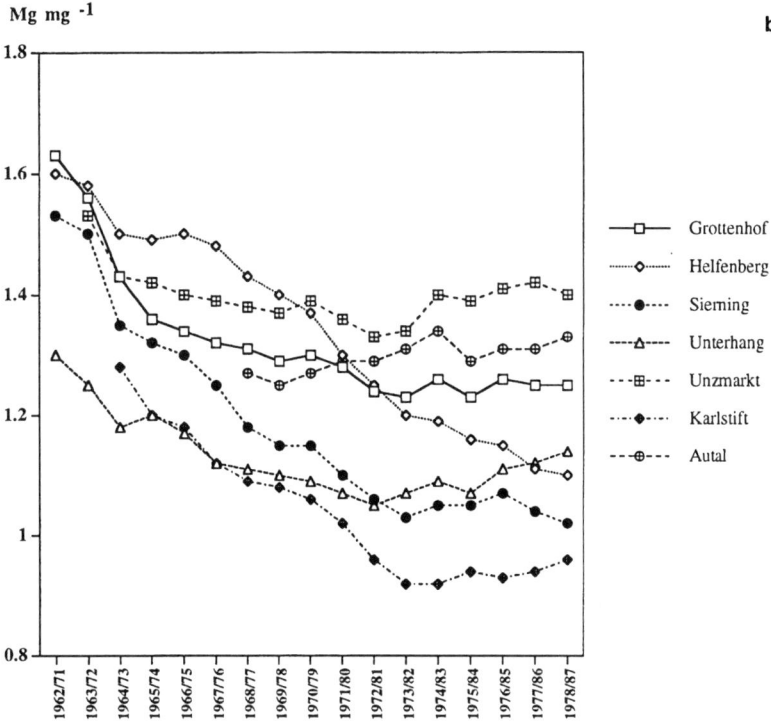

Mg mg $^{-1}$

b

Legend:
- Grottenhof
- Helfenberg
- Sieming
- Unterhang
- Unzmarkt
- Karlstift
- Autal

Figure 2. continued

Important decreases in Mg contents were also found in regions where Mg deficiency is not common, such as southwestern Sweden (Falkengren-Grerup, 1989). Although a detailed comparative evaluation of all these recent studies has yet to be made, it appears that the current levels of Mg in soils are still much higher (and were probably so originally) in southwestern Sweden for example, than in the main Mg deficiency areas (Harz, Vosges, etc.) where exchangeable Mg levels below $4\,\mu mol(c)\,g^{-1}$ in the upper mineral horizon are common. Interestingly, in the Vosgian study, the mean Mg concentration for 13 soils decreased by 48%, from $2.8\,\mu mol(c)\,g^{-1}$ to $1.5\,\mu mol(c)\,g^{-1}$, between 1969 and 1989 (Landmann, 1995), i.e. it crossed the value of $2\,\mu mol(c)\,g^{-1}$ in the upper mineral horizon (suggested as a threshold below which the risk of Mg deficiency increases sharply for Norway spruce), illustrating that in this region the risk of Mg deficiency severely increased during the two last decades. Data for other species are too few to derive threshold values (see also Chapter 1). Although field observations indicate that the most profound changes in soil chemistry have occurred in soils whose original pH and base saturation have been the highest, the hypothesis that depletion might stop before Mg deficiency appears (Rehfuess, 1989) was not verified in this case, and a similar situation has probably occurred in the other Mg deficiency areas.

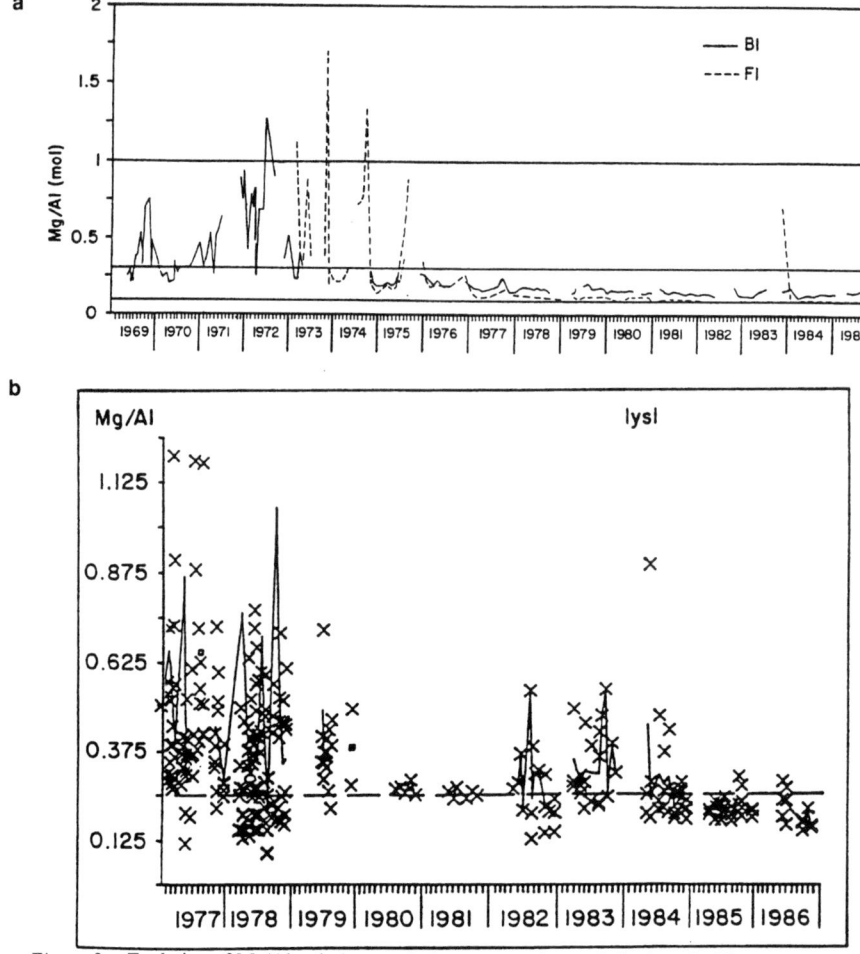

Figure 3. Evolution of Mg/Al ratio in percolation water at 1-m soil depth under Norway spruce in the Solling (**a**; from Matzner, 1987) and in the Harz (**b**; Lange Branke watershed; from Hauhs, 1985)

Long-term time series of soil solution chemistry are exceptional. Ulrich and co-workers reported for a spruce stand in the Solling (starting in 1969) and another in the Harz (Lange Branke watershed, starting in 1977) a sharp deterioration of Mg/Al ratio in the early 1970s (Solling; Figure 3) or late 1970s (Lange Branke; Ulrich, 1989). Dendrochemical studies by Zech *et al.* (1990/91) have also suggested a long-term decrease of Mg and Mg/Al in wood, but more work is needed before this approach can be used to derive soil depletion history for different regions with confidence.

In summary, although reliable time series of soils and foliar analyses from a large number of representative sites are lacking, there is little doubt that Mg nutrition has markedly decreased in several European regions characterized by originally acidic and base-poor soils. Information is still insufficient to draw clear regional patterns.

Areas severely affected by Mg deficiency   Reliable data on the frequency of Mg deficiency cannot be directly derived from the systematic surveys of forest condition because of various limitations: different grid densities are used in the different countries, and do not provide comparative data (Köhl *et al.*, 1994); the European network (16×16 km gridnet) which is the only transnational network, is insufficient to provide reliable data for the smaller regions of interest here; the visual assessment of foliar discoloration proved inconsistent between different countries (Montoya *et al.*, 1992; Ferretti *et al.*, 1995), and the various types of discoloration (yellowing, reddening) are not distinguished. Therefore, the map presented here (Figure 1) relies mostly on expert judgement, and remains qualitative. Symptoms on other species, in particular silver fir, beech, oaks (*Quercus* sp.), Scots pine, Corsican pine (*Pinus nigra* var. *maritima*), and Douglas fir (*Pseudotsuga menziesii*) are less frequent and/or more subtle. The real status of these species is therefore less well known.

A crude classification may be suggested (only a few additional relevant references are quite here):

–   Main Mg deficiency mountainous areas. These are large, forested, mid-elevation mountainous areas, with base-poor soils, where Mg deficiency is reported to be common: the Harz (Stock, 1988, 1990), the Fichtelgebirge, the Bavarian Forest, the Black Forest, the Ardennes (Weissen and Maréchal, 1991), and the Vosges Mountains. Norway spruce is the dominant or an important tree species in these areas. When other species (mostly silver fir, beech, Scots pine) are mixed, they generally show less acute damage.

–   Secondary Mg deficiency mountainous areas. These are also areas with base-poor soils, smaller in size or apparently less affected by Mg deficiency. Norway spruce is also an important species. Most of these areas are also mid-elevation mountains (reaching 600–1200 m): the Solling, the Hills, The Eggegebirge, the Thuringian Forest, the Upper Palatinate Forest, the Bohemian Forest, the Ore Mountains, the Hunsrück, the Eifel, the Palatinate Forest, the Forez (northeastern Massif Central), the Artense (central Massif Central), and the Plateau de Millevaches (western Massif Central) (Landmann *et al.*, 1995a).

–   Lowland Mg deficiency areas. These are areas with sandy soils, acidic, depleted soils (often of glacial origin) where Mg deficiency is reputed to be common, but mostly the species are other than Norway spruce, namely Scots pine, Corsican pine (*Pinus nigra* var. *maritima*) and Douglas fir (*Pseudotsuga menziesii*). These species generally cope better with poor Mg supply than Norway spruce, and show more subtle symptoms. In some of these regions, Mg deficiency may be associated with K deficiency. These areas include northeastern Germany, and part of The Netherlands, where Mg deficiency seems to be currently the most common deficiency after (in contrast to most of the above mentioned mountainous areas) phosphorus (Houdijk and Roelofs, 1993; Van den Burg, 1990, 1991).

- Marginal Mg deficiency areas where Mg deficiency appears to be rather scattered. Episodic Mg deficiency symptoms were recently found on Norway spruce at various locations in the inland of southeastern Norway (Solberg et al., 1992), while permanent, acute symptoms were only detected at one location in the Vestfold county on a poor coarse rocky base-poor soil derived from rhyolite (Solberg, personal communication). There is recent evidence for scattered, episodic Mg deficiency on Norway spruce from the inland of southern Sweden (Nihlgård, personal communication). Magnesium deficiency has also been associated with the decline of Scots pine in southern Finland (Raitio, 1993), but other stress factors seems to be prevalent in this decline.

- Areas where Mg deficiency might be expected with regard to parent material or soil type, i.e. central and northern Scandinavia, coastal areas in southern Scandinavia, parts of the United Kingdom (where a nutritional survey carried out in 1985 confirmed that the mean Mg levels were above deficiency, although some trees were in the range where deficiency symptoms might be expected, Innes, 1995), Brittany and Landes region (western France) and in northwestern Spain. The possible reasons for the absence of Mg deficiency in these areas are discussed further.

- Areas free from Mg deficiency, because soils are nutrient-richer overall. These comprise most of southern Europe (e.g. Bussotti et al., 1992), large parts of Central Europe, especially in Hungary, where Mg-rich soils predominate (Manninger, personal communication), most of the Carpathian (especially in Bulgaria; Ignatova, personal communication), the Alps (see the results of nutritional surveys in Austria: Stefan, 1991, 1992; Switzerland: Landolt et al., 1984; Wyttembach et al., 1985; and France: Landmann et al., 1995a) and the Pyrenees (Landmann et al., 1995a). A few deficient beech stands and some poorly supplied, declining, Norway spruce and silver fir have been found in the Tuscan Apennines (Ferretti, unpublished data), and in the Alps (northern French Alps: Landmann et al., 1995a; northern Italian Alps, Province of Trento: Ferretti et al., 1993; and Province of Ligura: Buffoni et al., 1990), respectively. However, such cases remain rare.

With more results of large-scale nutritional surveys becoming available, it will be possible to quantify more precisely the occurrence of Mg deficiency. Examples of such surveys include the 1983 German inventory based on the 4×4km gridnet (Knabe and Cousen, 1988; Evers and Schöpfer 1988a, 1988b), the Austrian nutritional monitoring program started in 1983, based on the yearly analysis of a systematic gridnet (ca. 300–1500 sample plots depending on the years) (Stefan, 1991, 1992), the 1983 Swiss systematic survey (8×4km) based on 840 stands (Landolt et al., 1984) and the Dutch surveys of 1051 stands in 1990 (Daamen, 1991) and of 150 stands in 1990 (Hendricks et al., 1994). Results from these surveys cannot be easily compared as they were carried out in different years, with different

Table 1. Concentrations of Mg in 3-year-old Norway spruce needles of 40–80-year-old stands (3 trees/stand, except Lower Saxony 6 and Baden Wurtemberg 2). Results of the German nutritional inventory carried out in 1982, 1983 and 1984 (depending on the regions), based on the 8×8km gridnet (4×4km in Baden Wurtemberg and partly 2×2km in Bavaria) (Knabe and Cousen, 1988)

Forest growth area (Wuchsgebiete)	Number of samples	Arithmetic mean (mg g^{-1})	CV%
Harz (Lower Saxony)	7	0.58	21.6
Northern Eifel	15	0.61	10.4
Taunus	16	0.67	17.0
Lowland eastern Lower Saxony	8	0.68	14.4
Odenwald	44	0.69	25.6
Hunsrück	28	0.71	29.7
Frankenwald and Fichtelgebirge	184	0.78	24.0
Black Forest	167	0.79	20.6
Palatinat Forest	24	0.82	15.0
Bavarian Forest	58	0.87	17.4
Bavarian Alps	93	1.06	27.3

protocols. Analytical differences are also likely. Nevertheless, it is clear from these surveys that Mg deficiency is much more common in Germany and The Netherlands than in Austria and Switzerland. However, from Table 1 it is clear that even in the main deficiency areas, the proportion of severely deficient Norway spruce stands remains relatively small. On the other hand, other deficiencies are usually more common; for example, in Austria, nitrogen deficiency is most common (about 35% of the values below 11 mg g^{-1}, for the current needle year) well ahead of Mg deficiency (2% <0.7 mg g^{-1}). Within the UN International Co-operative Programme on Assessment and Monitoring of Air Pollution Effects on forests, there is currently an attempt to synchronize the sampling.

Recent evolution of Mg deficiency Foliage discoloration assessed in ground surveys can be caused by various factors, such as mineral deficiencies, fungi, frost. However, field experience suggests that a large proportion of foliage discoloration recorded in the main Mg deficiency areas on Norway spruce and silver fir, is caused by Mg deficiency. Thus, the discoloration data from regions with acidic soils may be considered as crude indicators of the temporal variations in Mg deficiency in these species. The same does not hold true for pine species because the oldest needles turn yellow as a result of senescence before shedding and most of the discoloration in broadleaved species reflects autumnal discoloration.

The compilation of available data allows a distinct temporal pattern to emerge (Figure 4). A synchronous increase in discoloration occurred in the middle European areas with acid soils over the years 1983–85. The picture for this period is not complete, as only national networks were set up at that time. The distinct yellowing peak in 1985 was followed by a more or less synchronous decrease over the 1986–88 period. Since 1989, a second, but generally less distinct peak occurred in 1989 or 1992/93. The relative importance of the first and second yellowing peaks may

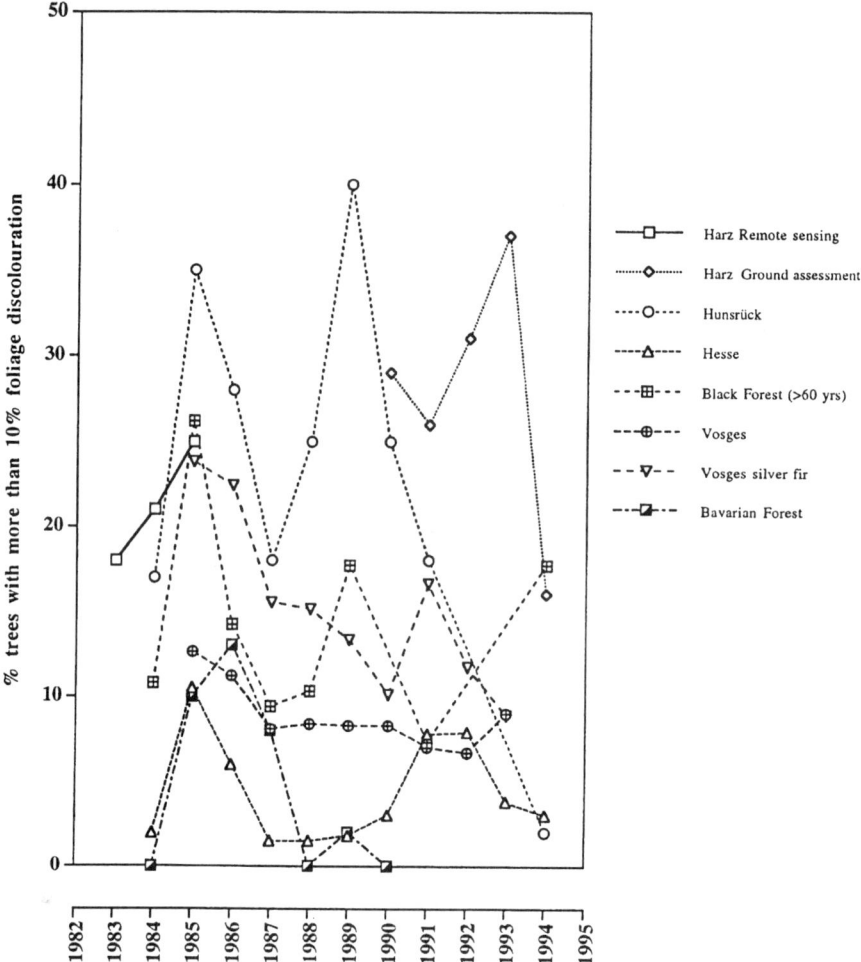

Figure 4. Evolution of discoloration of Norway spruce in a few areas where Mg deficiency is considered the main cause of foliage discoloration (data from various national authorities in charge of forest damage assessment)

vary locally as shown by the Hessian survey (Figure 5). The factors possibly causing these fluctuations are discussed in Section 2.5.4. Although Mg deficiency symptoms are believed to be more common than in the 1970s, the overall decrease since the mid-1980s was rather unexpected. Local studies of affected stands confirmed this recovery, and demonstrate that even very severely affected young trees may recover (Bonneau *et al.*, 1990/91; Kandler and Miller, 1990/1991).

It has been hypothesized that the recent decrease of Mg deficiency may be due to the effects of liming (Block *et al.*, 1991). This explanation is clearly not a general one, as a similar decrease occurred in regions (e.g. the Vosges Mountains) where

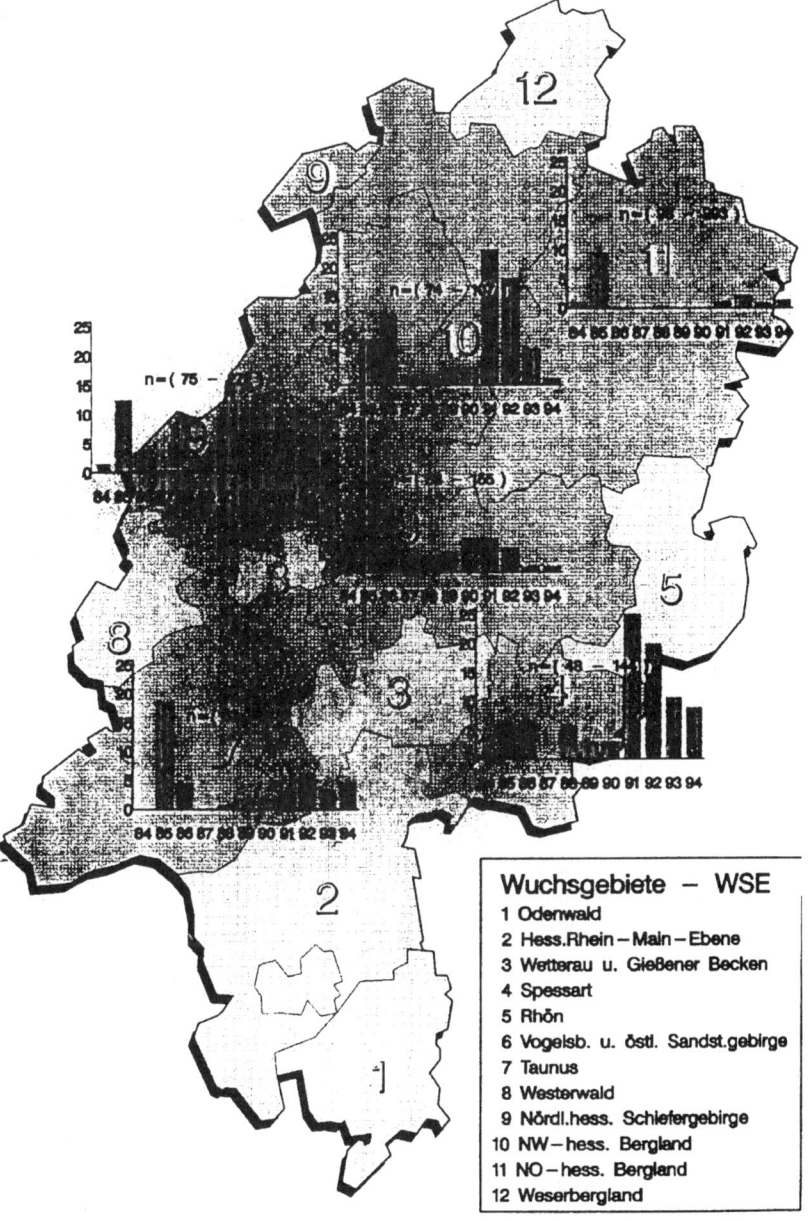

Figure 5. Evolution of discoloration (≥10% of the foliage) of Norway spruce in the Land of Hesse (Germany) (Eichhorn, unpublished)

liming has been carried out only minimally (< 1% of the forest area), and in control plots of fertilization trials, as seen above.

This short-term evolution of Mg deficiency is confirmed by foliar analyses, especially from fertilization experiments. The comparison between the evolution of yellowing symptoms and that of Mg foliar contents for the same stands shows a general agreement. However, there may be some discrepancies. For example, in the Vosges, Mg foliar contents decreased sharply in 1991, down to the 1985 level, presumably as a result of the severe drought of 1989 and 1991, but yellowing increased only slightly (Bonneau, 1993). Recent experiments have shown that other factors, especially radiation, influence the expression of the symptoms (Makkonen-Spiecker, 1995). The variations of foliage yellowing and Mg foliar content do not therefore necessarily strictly follow each other.

Small-scale patterns of Mg deficiency   In the areas affected by Mg deficiency, sharp contrasts have been observed between chlorotic and symptom-free stands. Only a few studies investigated these local variations. In the southern Black Forest, soils sampled in the chlorotic part of a 40-year-old Norway spruce stand had much lower levels of exchangeable Mg than those sampled under the healthy part of the stand (Hüttl, 1993). Also in the Harz, Blanck *et al.* (1988) found that small-scale patterns of yellowing of Norway spruce were related to soil chemistry; the Mg content in the fine soil and in the soil solution in the upper mineral soil layers was higher in the symptom-free parts of the stands. However, on sites with very low Mg supply (< 2 μeq g^{-1}), the intensity of yellowing was also influenced by the water supply: the most severe symptoms were clearly located on the drier soils.

At a smaller scale, it has also been often noticed that chlorotic trees and symptom-free trees grow close together in the same stands. Kandler and Miller (1990/91) sampled soils under individual trees, chlorotic or not, but could not find any differences between the Mg contents of the rhizospheres of yellowing and green spruce seedlings. Other studies focused on micro-scale patterns of soil chemistry in relation to tree health also found a lack of a clear relationship (see van Bredow *et al.*, 1986; Schulte-Bisping, 1989). Kandler and Miller (1990/91) argued that the lack of correlation between micro-scale spatial distribution of acute yellowing and abiotic (soil) factors suggests that biotic factors, such as microbial populations, intervene in this complex etiology. An alternative, and more-often cited explanation is that differences in rooting patterns and/or genetic variations may account for this observation. In fact, Hüttl (1991) demonstrated that different provenances of Norway spruce grown in a homogenized Mg-poor substrate contained different Mg foliar levels. These genotypes, when transplanted from the nursery to a uniform forest site displayed the same phenotypic traits. The results give some support to the hypothesis of genetic variation.

2.3.2. North America

Historical records and current distribution of Mg deficiency   Mg deficiency was identified during the middle part of the 20th century in a number of stands in eastern

Canada and the USA. These deficiencies occurred solely in plantations established on abandoned farm land (Stone, 1953; Gagnon, 1965; Leaf, 1968; Phu, 1975). There have been no published reports of declines due to Mg deficiencies in natural stands during this same time period.

Regarding the timing of the appearance of Mg deficiency symptoms recently recorded, relatively little is known. In Canada, the areas where Mg deficiency was found in the 1980s in the Lower Laurentians and to a lesser extent in the Higher Laurentians (Bernier and Brazeau, 1988; Bernier et al., 1989) are believed to have maintained healthy forest stands up to the early 1980s when the situation worsened over a period of a few years. Timing of Mg deficiency in the Lower Laurentians coincided with the development of other deficiencies, particularly in K and P, in large areas of maple forests south of the St. Lawrence River. In fact, there have been periods of decline or dieback on several occasions during the 1900s (Gagnon, 1988). One of the most widely supported theories about these earlier declines is that they were caused by fluctuations in climatic conditions. There is no evidence that these earlier declines were due to Mg deficiency and it is only during this last period that we have any real foliar data to back up hypotheses about the cause. The episode of decline during the 1980s is generally considered to be more widespread in terms of areas and species affected.

Evidence for a long-term deterioration of Mg nutrition   Overall, there is little information on the long-term evolution of mineral nutrition. Bernier et al. (1989) reported that, in comparison with foliar analyses obtained in the Higher Laurentians from 20 balsam fir (*Abies balsamea*) stands in 1967 (Brazeau and Bernier, 1973), there was a dramatic reduction (30–60%) in the concentrations of K, Ca and Mg observed in the mid-1980s. As pointed out earlier, such a historical comparison must be made with some care because of the inter-annual variability of Mg foliar contents.

Historical information on the nutritional status of forest trees in north-eastern United States is also scarce. There is no consensus as to the novelty of Mg deficiency, which can currently be seen at some high-elevation sites. In his review, Hüttl (1991) concluded that a large-scale development of the deficiency symptoms during the 1980s also occurred in North America because earlier investigations in similar site conditions did not show such low Mg levels in *Picea rubens* (see Robarge et al., 1989). Unfortunately, no time series are available, and many ecologists are sceptical about the novelty of the symptoms or consider that it is not possible to reach any conclusion because of the lack of suitable data.

As to the changes to soil chemistry, there is little evidence of a long-term decrease of exchangeable Mg in forest soils. However, the research of Bondietti *et al.* (1990) indicates that some forests underwent changes in both Ca and Mg contents in wood. These changes are believed to be related to variations in the availability in the soil. In general, there is an increase in Ca and Mg in wood formed in the mid-1900s followed by a decrease towards the end of the century. The peak corresponds with a period of rapid increase in NO_2 and SO_4 in precipitation which may have resulted in an increased mobilization of cations. The subsequent decrease is thought to correspond to a decrease in the amount of available Ca and Mg. Overall, given the scarcity of historical data, it seems difficult to reach definite conclusions on the historical development of Mg deficiency in North America. There are some indications that Mg nutrition has deteriorated in some areas, but the evidence is still weak.

Current extent of Mg deficiency in Canada and USA and recent research results on Mg deficiency   As to the current extension of Mg deficiency, a number of local or regional studies have recently been carried out. The distribution of the areas and locations where Mg deficiency has been found so far is shown in Figure 6. In Canada, the decline of sugar maple (*Acer saccharum* Marsh.) in southern Quebec stimulated intense research activity in the early 1980s to identify the cause and possible remedial procedures. Bernier *et al.* (1989) identified areas in the Lower Laurentians north and east of Quebec City that showed severe symptoms of Mg deficiency in sugar maple as well as in other plants on the same sites, including yellow birch (*Betula alleghensis*), red maple (*Acer rubrum*), balsam fir, and mountain maple (*Acer spicatum*). In other areas of the Lower Laurentians, north of Montreal, Mg was found to be deficient in declining stands of sugar maple (Côté *et al.*, 1993). Additions of Mg resulted in improved foliar Mg levels. In this study, no data were available to judge the timing of the onset of Mg deficiency, although discussion with the landowner indicated that the condition of the stand had deteriorated during the mid-1980s. Janicki and Jones (1990/91) recently identified Mg deficiencies in Norway spruce stands in southwestern Quebec.

In eastern United States, discoloration (although often only diffuse) and poor-to-deficient Mg nutrition have been documented at many high-elevation sites in the northern Appalachians, including Camels Hump (Vermont; Hoshizaki *et al.*, 1988), the Green Mountains (Vermont), the Adirondacks (New York; Friedland *et al.*, 1988) and the Whiteface Mountains (New York; Hüttl, 1989). In the southern Appalachians, Robarge *et al.* (1989) carried out a nutritional survey in 40 stands. Crown thinning, yellowing of older needles and poor Mg (and K) nutrition was reported in Fraser fir (*Abies fraseri*) and red spruce.

The available data do not allow us to assess the frequency of Mg deficiency and to compare it with the European situation. There is no indication that Mg deficiency could be the triggering factor of the severe decline of red spruce (*Picea rubens* Sarg.) which has been observed in these and other areas, especially the Green Mountains of Vermont, the White Mountains of New Hampshire, the Adirondacks of New York, and in the high-elevation forests of Virginia, North Carolina, and Tennessee (Bruck *et al.*, 1989 and references therein). Although Mg deficiencies were hypothesized as being a potentially contributing factor to the dieback of forests in the southern Appalachians, more recent evidence seems to contradict this idea. A number of experiments were set up to examine the effect of Mg nutrition, in combination with other factors, on the health of red spruce and loblolly pine (*Pinus taeda* L.) seedlings (Edwards *et al.*, 1991; McLaughlin *et al.*, 1993; van Miegroet *et*

Figure 6. Distribution of areas and locations where Mg deficiency symptoms have been observed in North America. 1. Quebec, Canada, 2. New Hampshire, USA, 3. Vermont, USA, 4. New York, USA, 5. Pennsylvania, USA, 6. Virginia, USA, 7. West Virginia, USA, 8. Tennessee, USA, 9. North Carolina, USA

al., 1993). The most convincing evidence indicates that addition of Mg fertilizer had no significant effect on physiological properties of the plants, although Ca addition did (van Miegroet *et al.*, 1993). They concluded that Mg, although low in the foliage, is not a factor in the decline of forests in this area. These results are consistent with field evidence which suggests that dieback is not closely linked to Mg deficiency in the Appalachians. Only acute yellowing significantly affects the physiological performance and health of forest trees (see Chapter 4).

Beside natural high-elevation forests, Mg deficiencies have also been reported by Ke and Skelly (1990/91) in mid-elevation-planted Norway spruce plantations that were established during the 1920s and 1930s. The study was undertaken in the Appalachians and included 12 sites in West Virginia, Pennsylvania, New York, and New Hampshire. The Mg-deficient trees showed the same types of symptoms as were observed in declining stands in Germany (crown thinning, needle discoloration, chlorosis, necrosis and loss). Yellowing trees were slightly to very Mg deficient ($0.25-0.75\,\text{mg}\,\text{g}^{-1}$ in current needles) and soil Mg levels in the deficient plantations were also found to be low ($0.8-2.4\,\mu\text{mol(c)}\,\text{g}^{-1}$). A significant negative correlation was found in this study between crown thinning and foliar Mg. Therefore, Mg deficiency may be considered the major cause of decline in these stands.

In summary, Mg deficiency does not appear to be rare in northeastern North America. It seems less widespread than in Europe, but the lack of systematic surveys of mineral nutrition and discoloration does not allow a sound comparison with the European situation to be made. There are fewer, if any, cases of very acute yellowing in conifers similar to those observed in the mid-1980s in Europe. With the exception of some mid-elevation stands in the Appalachians, Mg deficiency does not seem to cause major health problems to the affected forests. The relative importance of nutrient deficiencies in the recent dieback of sugar maple is still unclear. Bernier and Brazeau (1989) stated that nutrient deficiencies have not been reported in previous cases of forest decline which were attributed to insect defoliations and (or) adverse climatic condition, but this does not imply that these deficiencies were absent prior to the onset of the current declines nor that these deficiencies played a key role in the current decline.

2.3.3. New Zealand

Historical and recent records of Mg deficiency   Mg deficiency in radiata pine (*Pinus radiata*) was first noticed in seedlings in forest nurseries on yellow-brown pumice soils at Whakarewarewa and at Kaingaroa (see Figure 7). In the summer of 1957–58, severe chlorosis and stunted growth in seedlings occurred at both nurseries (Will, 1961). Radiata pine showed the most severe chlorosis. A fertilizer trial was established with various elements and various types of Mg fertilizer. The trial showed conclusively that Mg was required. Mg fertilizers produced green seedlings while other elements were generally ineffective.

In the early 1960s needle tip chlorosis of radiata pine was also noticed in Kaingaroa Forest. Five-year-old trees with varying degrees of yellowing had foliar Mg contents ranging from $0.22\,\text{mg}\,\text{g}^{-1}$ in the most yellow to $0.55\,\text{mg}\,\text{g}^{-1}$ in the least yellow (Will, 1966). It was observed that radiata pine was the most severely affected species. The chlorosis was only found in forests on pumice soils.

New cases of Mg nutrition were recently identified at the southern end of Kaingaroa Forest, but these were found to be, like earlier cases, related to forest

Figure 7. Distribution of areas and locations where Mg deficiency or shortage in radiata pine has been observed in New Zealand

history (see below). The hypothetical coincidence between the temporal develop-ment of Mg deficiency in Europe and New Zealand (Hüttl, 1993) seems fortuitous.

Current spatial distribution of Mg deficiency   Most of the available information on Mg deficiency, and on forest mineral nutrition in general, has been collected by the Soils and Site Productivity Research Field of the Forest Research Institute. Foliar and soil analyses from as far back as 1963 are stored in a computerized database. In the late 1980s, the rapid change and restructuring of the forest industry generated a need to formalize and generalize the knowledge about radiata pine nutrition particu-larly (a total of 28 000 analyses were available for this important species) in the form of an atlas (Hunter *et al.*, 1990).

The distribution of the foliar Mg concentrations in the North Island is shown in Figure 8. The whole area covered by the recent Taupo rhyolitic volcanic ash has marginal concentrations of Mg. The probability that any one group of trees may be deficient is medium. By contrast, the Nelson Forests and Coastal Canterbury, which are in the South Island, are generally satisfactory for Mg. However the West Coast of the South Island has marginal concentrations as do some of the higher inland

44

Figure 8. Distribution of the foliar Mg concentrations in the North Island, New Zealand. The map is based on 4807 analyses. The critical concentrations of Mg used for assigning nutrient ranking were: deficiency <0.7 mg g^{-1}; marginal nutrition 0.71–1.0 mg g^{-1}; satisfactory nutrition >0.10 mg g^{-1} (Hunter *et al.*, 1986)

areas. The low Mg concentrations on the West Coast of the South Island in an area that receives over 2500 mm of rainfall and is in close proximity to a sea over which strong onshore winds blow, are surprising and contradicts experience from Europe where the input of sea spray seems to counteract a tendency to Mg deficiency. The special factors in New Zealand may be the very high rainfall and attendant leaching of soils derived from peat accumulation over glacial fill, and outwash, which nullifies the sea spray effect. This map was drawn using analyses averaged over differing numbers of years, which may introduce some bias, but the interannual variability of Mg values was lower than that reported in some earlier European studies.

Although the symptomatology of Mg deficiency is well known, especially in radiata pine (Hunter, 1992, 1993; Hunter *et al.*, 1986), the degree of discoloration has never been correlated directly to the foliar Mg content, and there was no rating of foliar discoloration parallel to the nutritional survey. Radiata pine has lower Mg concentrations than many other exotic species when grown on the same soil type and is hence more prone to show Mg deficiency (Hunter, 1992). Although faster tree growth and hence higher demand might be a factor in some comparisons, radiata pine took up less Mg than faster-growing eucalyptus in one comparison, hence demonstrating that a genetically induced deficiency in uptake might be present. Beets and Jokela (1994) have, in addition, demonstrated considerable genetic variation within radiata pine populations to the phenomenon known as upper–mid-crown yellowing (UMCY) which is strongly associated with Mg uptake. However, it is recognized that yellow-tipped older needles will only occur for a brief time in spring when foliar Mg concentrations fall below $1.1\,mg\,g^{-1}$. The percentage of the needle yellowed will increase as foliar Mg concentration falls, at a concentration around $0.6\,mg\,g^{-1}$, the needle will have three zones on it: a green base, a yellow middle and a necrotic brown tip (Hunter et al., 1986). These changes would not however be the primary visual indication of the severity of Mg deficiency because commensurate with the colour change is a progressive reduction in needle retention, visual vigour and tree growth which together form the more striking impact of severe deficiency.

New Zealand's exotic forest was fairly intensively monitored by a corps of forest biology observers, concentrating on observing insect and fungal risks to the forest although noting obvious nutritional deficiencies in passing. There is no equivalent of the systematic sampling for forest condition instituted in Europe. However from the mid-1980s onwards there was growing interest in UMCY which was reported from a wide range of sites in New Zealand. Trees affected by UMCY show crown thinning in the upper one third of the crown, typically just below the leading shoot. Side shoots in the affected zone carry few yellowed or bronzed needles and do not extend as fast as normal. Sometimes, in severe cases, side branches die back. The coincidence of the similarity between this phenomenon and that described as typical of 'Type 1 spruce yellowing' (Roberts et al., 1989) was noticed early as was the fact that this phenomenon appeared to be more prevalent on soils with a low Mg status. However its link with Mg deficiency has only been very recently confirmed (Beets and Jokela, 1994). These authors found a strong negative correlation between foliar Mg concentration and degree of UMCY.

Apart from the well-documented tree age effect (Mg requirements are very different according to the stage of development and so is the severity of Mg deficiency) (Hunter, 1992), there was no particular reason to suspect a (negative) long-term evolution of Mg nutrition. This is why no systematic record of going back to the same trial or the same plot for 20 or 30 years is available. In principal, it would be possible to derive average Mg concentrations year by year from the aforementioned data base. However, these averages would be made up of different small numbers of observations from different parts of the forest each year and difficult to interpret.

2.4. Significance of Mg deficiency to forest health

Europe According to the forest health inventories, the yellowing of spruce has never affected more than 10–25% of trees, depending on the regions (Figure 4), while the proportion of 'moderately' and 'severely' discolored trees (>25% and

>60% foliar biomass discolored, respectively) has always remained very small (<5% and <1% of the trees surveyed, respectively). Therefore, some authors have stated that Norway spruce yellowing is not as severe as generally believed, even in reputedly 'severely' affected areas (Rehfuess, 1991).

One may argue that foliage discoloration is only the 'emerged part of the iceberg', and that the more or less generalized Mg subdeficiency found in a number of areas is a threat to forest health. In many reports, Mg deficiency is considered a key factor in the crown thinning observed over large areas in Europe. However, crown thinning and foliage yellowing were found to coincide only partially in all areas where this has been investigated (Harz: Hartmann et al., 1986; Stock, 1988, 1990; Black Forest: Liu and Hüttl, 1991; Vosges: Landmann et al., 1995b; The Netherlands: Henriks et al., 1994).

The best documented area may be the western part of the Harz mountains, where defoliation and foliage yellowing were separately mapped by means of aerial photography (Stock, 1990). The distribution of Mg deficiency coincided well with the base-poorest bedrocks occupying mainly the central plateau at about 500 m and in part of the northern Harz, while stands with heavy needle loss were mainly concentrated on the ridge of the Acker-Bruchberg which exceeds the elevated plains by up to 350 m in altitude, and on the mountain tops of northwestern edge of the range (Stock, 1988, 1990). Thus Mg deficiency, which was presented by the media as an extreme example of 'neuartige Waldschäden', is not involved in the dieback of Norway spruce in the Acker-Bruchberg area. Although soil-mediated effects have been hypothesized, storms and repeated droughts obviously played a key role in this case (Wagner, 1989).

As a matter of fact, acute Mg deficiency has been reported in the several areas (Harz, Fichtelgebirge, Vosges, Ardennes) to lead to the breakdown of some spruce stands. It has not been fully elucidated whether these cases represent extreme cases of Mg deficiencies or reflect interactions with other factors. Although specific studies which carefully analyzed yellowing and defoliation with respect to site and stand conditions are few (see, however, the case of silver fir in the Vosges mountains; Landmann, 1993), it seems reasonable to conclude that severely yellowing and defoliated stands break down not simply because they grow in extreme soil conditions, but more likely because they share both unfavorable soil (chemical, physical or biotic) characteristics and stand history (exacerbated competition for water and nutrients at some stage of their development) (see Becker et al., 1989; Gerecke, 1989; Spiecker, 1990/91, 1991).

Finally, the fact that only acute deficiencies, which are extremely rare, result in growth reductions (Chapter 5) makes it possible to understand why Mg deficiency has not hampered the long-term growth increase of the productivity of Norway spruce over the past decades documented in regional growth analyses carried out in the same areas (Kenk, 1993; Becker et al., 1995).

North America Red spruce decline has attracted greatest interest (Eagar and Adams, 1992) but Mg deficiency does not seem to play a major role in this decline.

Rather, decreased Ca availability because of increased levels of aluminium is thought to be a key factor. Nutritional disturbances, and even less so Mg deficiency, are hardly mentioned in the numerous cases of species-related diebacks recorded (Walker and Auclair, 1989; Walker *et al.*, 1990; Millers *et al.*, 1988). Except in a few localized areas, there does not seem to be any real evidence that Mg deficiency has ever been a major cause of decline. When Mg deficiency does occur it is usually relatively short-lived, implying that other stress factors (drought, root damage by frost or defoliation) strongly influence Mg nutrition.

New Zealand Several cases of serious dieback of Nothofagus and Metrosideros species (Hosking and Hutcheson, 1985; Hosking, 1993; Stewart, 1989) have been reported, but these diebacks are not related to nutritional disturbance. A case of high mortality (at least 30% of the trees) of *Pinus radiata* was reported by Hunter *et al.* (1986) in the unfertilized plots of a fertilization trial in New Zealand. This situation coincided with a severe deficit in magnesium, but also with an attack by the needle cast fungus *Dothistroma pini*. In fact, the infection seemed to be much more severe on unfertilized deficient trees than on fertilized ones. However, this apparent interaction between Mg deficiency and the fungus, may simply reflect the fact that infection is strongly related to tree height with more severe infection occurring in the still, humid, lower layers of the forest, which was the layer where all the unfertilized trees grew by virtue of their slow growth. Overall, the deaths were a long-lasting process, with severely affected trees producing less and less new foliage.

Magnesium deficiency clearly reduces tree growth. The unfertilized trees in the trial described by Hunter *et al.* (1986) grew very little at all and after five years the fertilized trees were over twice as tall and had five times more volume per hectare. Pruning marginally deficient trees can reduce growth by up to 20% relative to an adequately fertilized treatment. It is not yet clear if the kind of deficiency associated with UMCY actually results in slower tree growth.

In summary, Mg deficiency should not be considered an efficient 'tree-killer', which confirms earlier experience of forest ecologists and nutritionists. However, occasionally adverse climatic conditions, specific stand dynamics features, occurrence of pathogens, and other nutritional disturbances may cause mortality by interaction with Mg deficiency.

2.5. Epidemiological approach to the causes of Mg deficiency

2.5.1. Role of soil parent material

Europe Large forest areas in Europe are located on base-poor parent materials. Following the geological history, one can distinguish: (i) granites of the Scandinavian precambrian shield, (ii) base-poor granites, schists and sandstones of the hercynian massifs (mostly granites and other crystalline rocks such as gneiss in the northern French Massif Central, the southern part of the Vosges massif and Black Forest, the Bohemian massif; mostly schistous material in the Harz, Eifel,

Hunsrück, Ardennes and a large part of the Massif Armoricain (western France); complex set of base-poor materials in the Erzgebirge and the Fichtelgebirge), (iii) sedimentary material, especially sandstone of the northern Vosges and Black Forest and the Odenwald, and (iv) quaternary glacial material in northeastern Germany, part of Poland, The Netherlands, and southwestern France (Landes region).

As to the large-scale distribution of Mg deficiency (see Figure 1), all areas affected by Mg deficiency coincide with base-cation-poor parent material. The reverse does not hold true: there are large areas in northern Scandinavia and western and southern Europe which are not or only marginally affected. This is partly explained by the fact that parent materials classified as base poor may actually differ largely in chemical composition. For example, the Scandinavian granites are nutrient-richer overall than the partly extremely base-poor granites found in, for example, part of the Black Forest or the Vosges massif. The other possible reasons for this apparent discrepancy are discussed further.

At the regional scale, i.e. within the main Mg deficiency areas, many field surveys and remote sensing studies confirmed that Mg deficiency symptoms are found on the nutrient-poor bedrocks (Black Forest: Zöttl and Mies, 1983; Vosges: Bonneau, 1985; Harz: Stock, 1988, 1990). Again, it must be kept in mind that all Mg deficiency areas are characterized by variable bedrocks, in terms of Mg contents. Locally, apparent discrepancies have been found, but specific geological features, such as solifluxion in the Black Forest (Zöttl, 1985) and in the Vosges (Bonneau and Fichter, 1991 in Landmann et al., 1995a), or mineralogical peculiarities, as in the Bavarian Forest (Hoffmann-Schiele, 1991), were found to account for this.

North America   The main forest areas of north-eastern North America consist of the Canadian Shield in the north and the Appalachian mountain range that runs from Labrador, to the east of the Canadian Shield and down the east coast of Canada and the USA. The Canadian Shield is composed of Precambrian rocks dominated by intrusive igneous rocks, but also containing metamorphic rocks. The chemical composition of these rocks varies from very acidic granites and syenites to marble ($CaCO_3$).

The areas where Mg deficiency has appeared are concentrated on coarse-textured tills derived from acidic to intermediate igneous rocks containing low concentrations of Mg (and in some cases K). The Appalachian mountains represent a very complex geological setting involving more than one sequence of orogeny. The range includes igneous, metamorphic and sedimentary rocks that vary widely in chemical composition and mineralogy. No systematic study of the relationship between rock type, soil parent material and tree nutrition has been carried out. In the Lower Laurentians, where Bernier and Brazeau (1988) identified areas with severe symptoms of Mg deficiency, soil parent material is naturally low in Mg. In other areas of the Lower Laurentians, north of Montreal, the low Mg levels are also directly related to soils formed from acidic igneous rocks containing relatively abundant Ca but low concentrations of Mg.

New Zealand New Zealand lies at a point on the Earth's surface where two tectonic plates meet and one rides over the other. As a result, New Zealand's geological history consists of periods of rapid mountain chain uplife and periods of strong volcanicity. Simply, New Zealand consists of a mountainous backbone of 'greywacke' – a very hard sandstone – and a string of volcanoes. The volcanic depositions vary from extremely acidic to strongly basic, depending on age and location. Mg-bearing minerals are abundant in the outcrops of basic igneous rock that occur in New Zealand but are less common in acidic rocks and in many sedimentary rocks. Weathering and leaching modify the native state. The most strongly weathered soils of northern New Zealand appear to have reached the state where depletion of an originally abundant Mg status has occurred. Conversely, when leaching is active and weathering slow, as in some of the higher altitude soils, available Mg can be depleted. This seems to be the situation in the higher parts of the pumice plateau (New Zealand Soil Bureau, 1968). Thus the areas most prone to Mg deficiency lie on recent, relatively unweathered but highly leached rhyolitic tephra.

2.5.2. Role of acidic deposition

Europe Acid deposition is widely accepted as an important causal factor of Mg deficiency. In order to test this hypothesis on an epidemiological basis, various atmospheric deposition criteria may be used. A cumulative acid load since pre-industrialized times, such as calculated by Ulrich (1989) for northern Germany, would be of great interest, but is not available at the European scale. In fact, not only the deposition rates of acidity, but also of nitrogen (which may enhance Mg deficiency) and magnesium (which may compensate Mg leaching) must be considered. Finally, as base-poor parent material is a prerequisite of Mg deficiency, a parameter combining atmospheric deposition and site susceptibility would be most relevant. The work recently developed in order to establish the so-called 'critical loads' (deposition rates below which no deleterious effects are observed) provides interesting information in this respect.

When comparing, at the European scale, the distribution of Mg deficiency and the areas of exceedance of N and acid critical loads (De Vries *et al.*, 1994), it appears that the areas affected by Mg deficiency roughly coincide with areas where critical loads for acidity and nitrogen are largely (most of Germany and The Netherlands) or moderately (parts of southern United Kingdom and northern France) exceeded. The areas where acidity critical loads are exceeded coincide with base-poor parent material (except in central Europe where loads are very high). Conversely, other areas characterized by base-poor soils, especially in western and northern Europe, are below the critical loads and free of Mg deficiency.

An interesting feature is the absence of Mg deficiency in areas where critical acidic and N loads are exceeded, especially in the coastal areas of southern Scandinavia, The Netherlands, western France, northwestern Spain, as well as most of England and Denmark. Large marine Mg inputs in these areas (usually

>7 kg ha^{-1} y^{-1}, Roberts *et al.*, 1989) probably prevent the development of Mg deficiency in these areas. By contrast, Mg deposition is naturally very low (<2 kg ha^{-1}) in central Europe, and further decreased as a result of the reduction of anthropogenic particle emissions (Schenck, 1990; Hedin *et al.*, 1994). This may have enhanced acid deposition effects over the past two decades.

Overall, the areas which exceed critical N loads roughly coincide with those of critical acid loads. As the uncertainty of these calculations is relatively large (de Vries *et al.*, 1994), it is difficult to analyze separately these two parameters in relation to Mg deficiency. Temporal development of N deposition and supply may give additional insight. Most historical comparisons and time series show a distinct increase in N foliar concentrations, and many authors have stressed the importance of the N/Mg ratio (see also Chapter 1). Evidence supporting this is available, for example, for Scots pine in northeastern Germany (Hippeli and Branse, 1992), Norway spruce in the Thuringian Forest (Nebe, 1991), the Black Forest (Hüttl, 1990) and Lower Saxony (Reemtsma, 1986), and silver fir in the Vosges (Landmann *et al.*, 1995a).

The year-to-year variations of N and Mg supply may also provide interesting information. For Scots pine stands monitored yearly over 25 years in northeastern Germany, Hippeli and Bramse (1992) found a significant linear negative relationship between foliar N and Mg contents in 6 out of the 7 stands surveyed, the exception being a stand on a nutrient-richer soil. On average, when the N content of current-year needles increases from 1.2 to 1.8 mg g^{-1}, the Mg levels decrease from 1.2–1.4 to 0.7–0.9 mg g^{-1}.

However, there are exceptions to the general trend. The lack of long-term decrease in Mg nutrition in pine stands heavily loaded with N deposits in Bavaria (Sauter, 1991) is noteworthy. Conversely, Mg deficiency only recently developed in areas where N supply is still insufficient. A sharp decline in Mg nutrition was found in the Czech Republic (Kaiserwald) mostly in N-deficient spruce stands. The exceedance of N critical load (in this area) has only resulted in a marginal improvement of the N status up to now (Materna, 1989). Striking is an apparent decline in Mg nutrition in spruce stands on the Plateau de Millevaches (western Massif Central), where probably neither the critical N load nor the critical acid load have ever been reached, and where N supply is relatively poor (Landmann *et al.*, 1995a). The latter observation, however, is based on few data and needs to be confirmed.

In summary, it appears that Mg deficiency is most common in areas where critical acid and/or N loads are exceeded. This suggests that acid deposition (in its broadest sense) at least partly explains the distribution of Mg deficiency. However, one must be aware of two limitations of this analysis. Firstly, the information on the nutritional status of forests from remote areas is scarce overall. Secondly, other important causes of Mg deficiency, such as past land use and forest management (Section 2.5.3), are poorly known, not taken into account in the critical load approach (soil data are taken from FAO Soil Map), and not reflected in the maps shown here.

North America The region with the most severe decline in Quebec receives more wet acid SO_4 deposition (>13 kg ha^{-1} y^{-1} S-SO_4, 1981 CANSAP data in Rennie, 1987) than any other part of eastern Canada. The highest NO_3 deposition (4.5–6.5 kg ha^{-1} y^{-1} N-NO_3) also occurs in the more severely damaged areas of southern Ontario and Quebec. Although the contribution of dry deposited acidic materials is difficult to assess, it probably accounts for an additional 10–100% of the values quoted for wet deposition. Overall, these loads are relatively small compared with those known in large areas of Europe.

Fortuitously or not, the Mg-deficient Norway spruce stands investigated by Ke and Skelly (1990/91) are in an area that receives some of the highest levels of acidic deposition in North America (see Wisniewski and Kinsman, 1988). However, the lack of data on the real extension of Mg deficiency in North America makes it difficult to draw a definite conclusion as to a possible spatial coincidence between acid precipitation and Mg deficiency. Although it is accepted that acid deposition most likely caused Mg depletion of some soils, there is insufficient knowledge to conclude that Mg deficiency in these areas is predominantly caused by acid precipitation. Many authors also stress the possible role of historical factors (Section 2.5.3) and climatic factors (Section 2.5.4). In any case, excessive N supply is almost unknown in the areas affected by Mg deficiency. In fact, N is the most common nutrient deficiency in unmanaged forests of eastern North America.

New Zealand The studies in New Zealand are interesting because they confirm that Mg deficiency can occur without acid deposition, in fact wherever Mg supply is below Mg demand. This may occur naturally because the soil is very low in Mg and the nutrient demand of fast-growing exotic species is rare but convincing examples of natural Mg deficiency in native species have not been found. In New Zealand, Mg deficiency is primarily induced by management practises in the widest sense (see Section 2.5.3).

2.5.3. Past land use and forestry practises

Europe Recent studies of nutrient cycling in stands affected by Mg deficiency have demonstrated that current losses of Mg by leaching are generally greater than losses attributable to tree uptake and harvesting (Johnson *et al.*, 1991; Dambrine *et al.*, 1995). It may therefore be concluded that acidic deposition, by promoting Mg leaching, is (at least currently) the main driving force of this problem.

The influence of past land use and forest history should not be overlooked. In fact, although acidic deposition is *currently* responsible for Mg depletion, Mg deficiency probably would not have developed in a number of sites without the historical (during pre-industrial and early industrial times) impoverishment of the soils by various nutrient-demanding practises: intensive harvesting for charcoal-burning, glass and porcelain industries, litter raking, grazing in forest or on pastures later converted into forests, and replacement of poorly growing hardwood species or mixed stands by pure spruce stands.

It is worth noting that all the main areas of yellowing of spruce occur in forests which were greatly transformed by man. Nutrient-demanding Norway spruce has been widely introduced beyond its natural range (see for example the detailed description for the Fichtelgebirge by Reif (1989)). As already noted (Section 2.3.1), early observations of Mg deficiency often coincided with spruce planted on former agricultural land, especially in mountainous regions. By contrast, Van den Burg (1990) observed that Corsican pine stands in The Netherlands are better supplied in sites previously cultivated, presumably because of the use of fertilizers. The plant genotype may also play a role. For example, Rehfuess (1991) reported that native, autochthoneous spruce populations in the subalpine spruce belt of the Bavarian Forest are much less affected than foreign spruce provenances.

The comparison between the remaining native stands (beech or beech–fir forests in many cases) and the neighboring spruce stands, indicates that whatever the relative weights between the causal factors, Mg deficiency would be much less important – possibly only marginal – in most (if not all) Mg deficiency areas if the natural vegetation had been maintained.

North America   The Appalachians have been extensively cut or burned and most forests in these mountains are the result of natural regrowth after fire or logging. The least accessible parts of the Appalachians, mostly at high elevation, may never have been cut. Forest management is virtually non-existent except for a few plantations on the most productive lands at lower elevations. However, these are commonly old agricultural land that was abandoned because of low productivity, stoniness, distance from markets and poor climatic conditions.

The Laurentians were heavily logged in the 1800s and then wide areas were burned during a disastrous year of drought leading to forest fires in 1923. As a result, large areas of the forests in southern Quebec are now reaching maturity. This, in itself, may have caused a transfer of available nutrients from the soil to the biomass leading to relative scarcity in the soil. There is debate as to whether the sugar maple forest is stable and several researchers believe that it will be replaced with a new forest comprising less maple, more beech (*Fagus grandifolia*) and more hemlock (*Tsuga canadensis*) that are less demanding in terms of mineral nutrition.

The Norway spruce plantations showing Mg deficiency in southwestern Quebec, and studied by Janicki and Jones (1990/91), were established on abandoned farm land and so the cause of the deficiency may be more related to past land use than other causes.

The causes of the most acute cases of Mg deficiency affecting Norway spruce in the northeastern United States have not been firmly established. Ke and Skelly (1990/91) pointed out, that beside other causes, previous land use may have had an influence since the plantations were (as with many of the Canadian ones) established on abandoned farm land that may have been exhausted by poor management.

New Zealand   As the following description will highlight, New Zealand's history of Mg deficiency in forests is closely linked to the history of afforestation (Hunter,

1992 and references therein). Most Mg deficiency was found to occur in Southern Kaingaroa. The high-pumice plateau of Kaingaroa was originally planted with exotic conifers in the 1920s. At that time, the plateau was vegetated largely with heathland and grassland. The lower, warmer parts of the plateau, in the north, were easily established in radiata pine but the southern part of the forest was too high and too frost prone for establishment of radiata pine with the techniques prevalent at the time, so in the first rotation it was planted in ponderosa pine (*Pinus ponderosa*), Corsican pine (*Pinus nigra* var. *maritime*) and lodgepole pine (*Pinus contorta* var. *latifolia*). Fortuitously, the initial species siting distinction happened to correspond broadly with a soil difference: the higher colder parts of the plateau being on Mg-poor deep flow-tephra soils and the lower warmer parts being on slightly richer airfall ash soils. There is no record of any nutritional research on these other species. However there was equally no record of any patches of strongly stunted growth that might indicate deficiency. They grew much more slowly than radiata pine and hence placed less of a demand on the site. The trees planted on the higher plateau were progressively clear felled during the 1960s and 1970s and research carried out at that time, especially improvements in site cultivation and weed control, made it possible to establish radiata pine (which is more productive than the other species) on the cut-over sites. Radiata pine proved to be susceptible to Mg deficiency. The appearance of severe Mg deficiency in radiata pine in New Zealand was clearly a consequence of planting radiata pine, for the first time, on sites that were particularly low in Mg.

Historical features also explain recent observations of Mg deficiency. Patches of severe and persistent Mg deficiency were noted in 1980 in the new crops of radiata pine established on the flow-tephra soils at the southern end of Kaingaroa forest. These areas of severe deficiency occurred on patches of water-sorted tephra and on areas where management had disturbed or displaced the topsoil (Hunter *et al.*, 1986). The affected trees were very yellow in color at all times of the year; they were very small for their age and were malformed.

The first studies in the 1960s showed that in forests on pumice soils where chlorosis of radiata pine was common, the worst affected stands were those which had recently been pruned and thinned. There was a sharp contrast between stands of the same age that had received contrasting treatment, thinned and pruned stands being much more yellow than untreated stands. Will (1966) thought that these observations could be explained by patterns of nutrient demand and supply. Recent observations confirmed that pruning markedly worsens deficiency, while thinning appears to be less damaging.

2.5.4. Role of climatic stresses

Most reports on Mg deficiency mention climatic stress. Although forest monitoring is now well developed in Europe and North America, detailed climatological analyses in relation to temporal variations in Mg deficiency are still scarce. However, some recent analyses allow some insight into this question.

Europe The rapid increase in spruce (and fir) discoloration during the early/mid-1980s in Europe is a striking feature. The view that climate probably is the 'synchronizing' factor for the yellowing of spruce is widely accepted (Prinz and Krause, 1989; Roberts *et al.* 1989; Schulze and Freer-Smith, 1991).

Most authors have stressed the role of droughts in the development of Mg deficiency in central Europe, and concluded that soils were depleted to the point where the soil-driven Mg supply to trees during drought years (summers of 1975/76 and 1981–84) became inadequate. The improvement of Mg nutrition following the wetter summers 1986–88 and its deterioration during the dry period 1989–1992 (see Figure 4), seems to confirm this interpretation. In Norway, Solberg (1993) also found a close correlation between rainfall deficit and the development of spruce yellowing during the 1988–92 period. In northeastern Germany, Hippeli and Branse (1992) found that rainfall showed only a slight insignificant negative trend over the 25 years study period, but a strong, linear correlation between the amount of rainfall during the vegetation season and the foliar Mg concentrations. When this amount increased from 100 mm to 450 mm, the Mg contents increased, depending on the sites, from $0.6-0.8 \, mg \, g^{-1}$ to $1.2-1.6 \, mg \, g^{-1}$. These observations are in agreement with experimental evidence that drought may exacerbate Mg deficiency (Dambrine *et al.*, 1993).

Against the exclusive role of drought is the fact that the peak of 1983–86 was often more pronounced than that of 1989–91, despite the fact that drought stress was more acute during the second period in many places. In fact, the 1983–85 period was not characterized by droughts exceptional in severity (as were for example the years of 1921, 1947, 1976, and more recently 1991) but because they occurred late in the summer after very wet springs. This needs, however, further evaluation. The involvement of climatologically contrasted years would fit with the observation made by several authors that the peak of yellowing coincided with high N foliar contents (Bavarian Forest: Rehfuess, 1989; Black Forest, Hüttl, 1990) indicating enhanced mineralization of organic N and nitrification, which is not typical for dry years. However, a similar trend was not observed in mature fir stands monitored from 1985 to 1991 in the Vosges mountains, although discoloration followed the same time course as in spruce (Landmann *et al.*, 1995a).

North America The cause of the onset of nutrient deficiencies in Quebec is unknown. However periods of extreme climatic conditions (late frosts, low snow cover, droughts) may have contributed to the development of the decline syndrome.

An unseasonal February thaw in 1981 followed by deep frost penetration of the soils in March devastated many apple orchards in the province and initiated unusual stress on sugar and silver maple (*Acer saccharinum*), causing lesions to roots, branches and trunks (Benoit *et al.*, 1982). The winters of 1981 and 1982 were unusual in that there was very little snow cover in the maple forest. Lachance (1985) has suggested that the resulting deep frost penetration may have caused serious damage to the rooting systems of sugar maple. Late frost was also reported for areas between Montreal and Sherbrooke as well as in the Lower Laurentians (north of the

St Lawrence River) between Montreal and Quebec City during the spring of 1983 (Hendershot and Jones, 1989). Lachance *et al.* (1984) note that high temperatures and dryness during July and August of 1983 resulted in leaves drying out, turning brown and falling prematurely, although this damage was mainly restricted to high-elevation sites on thin soils in the Lower Laurentians. There seems to be little relevant information concerning the Appalachians.

The role of these climatic anomalies remains speculative, and there have been few attempts to use this information in relation to the time development of Mg deficiency symptoms. Ke and Skelly (1990/91), in an attempt to explain Mg deficiency affecting Norway spruce in the mid-elevation mountains in north-eastern USA, mentioned that beside acid deposition and past land use, recent periods of extreme drought was one among several factors. They speculated that Mg supply was only temporarily limiting, because of these periods of droughts, but did not provide a more detailed analysis.

New Zealand   Most afforested sites in New Zealand have well-distributed rainfall contributing more than 1000 mm per year. However, at certain phases of the southern oscillation New Zealand is afflicted by strong drying winds and erratic and often lowered rainfall. It was noticeable that there was a greater frequency of mild yellowing in Kaingana in the early 1980s during such a period. These mild symptoms disappeared during wetter years in the mid-1980s. However, the severe symptoms described in Hunter *et al.* (1986) were unaffected by moisture and remained constant and severe from year to year.

2.6.   Summary and conclusions

The current state of the spatial and temporal patterns of Mg deficiency in Europe, North America and New Zealand may be summarized as follows:

–   No unambiguous case of Mg deficiency in natural forests has so far been reported. Although knowledge of the nutritional status of natural forests is generally poor, it seems that Mg deficiency is very rare or even absent in natural forests. Whether the few earlier cases of Mg deficiency of beech and fir stands on sand-stone in Europe may be considered as 'primary Mg deficiencies' is uncertain, as these forests were already altered to some extent by human activity.

–   Whatever the primary causes, Mg deficiency will only show up where the parent material and derived soils are naturally poor in this element.

–   Most of the earlier observations of Mg deficiency are clearly linked to former land use. In the Appalachians and in the middle-elevation mountains in Europe, Mg deficiency often coincides with former agricultural use. The role of forestry in 'promoting' Mg deficiency is most clearly demonstrated in New Zealand where first observations coincide with the introduction of nutrient-demanding species. In Europe, the introduction by humans of Norway spruce on large areas with base-poor soils formerly covered by poorly growing beech or beech–fir stands, and historical nutrient-demanding forestry practises (e.g. litter raking, glass working, etc.) presumably also played an important role (see Chapters 3 and 6). Detailed documentation is not available, as acid rain research has rather neglected this issue, but field observations of remaining native stands suggest that Mg deficiency would currently not be a major issue if the natural vegetation had been maintained.

– The distribution of Mg deficiency in Europe is consistent with the hypothesis that acid deposition played a role in the recent emergence of Mg deficiency in a number of mountainous areas (namely Harz, Fichtelgebirge, Bavarian Forest, Black Forest, Vosges, Ardennes). Severe Mg depletion of forest soils – which seems unlikely to be due to nutrient uptake alone, according to process-oriented and modeling studies – has been documented in several areas where Mg deficiency is currently common. Many of the severe cases of Mg-related yellowing appear as nutritional imbalances between Mg and other growth-stimulating nutrients, especially nitrogen, rather than typical cases of Mg deficiency. The historical increase in foliar N levels, documented in many places in central Europe, is generally attributed to atmospheric N deposition. The analysis of spatial patterns is of little help to investigate the respective roles of acid and N deposition as critical loads for acidity and nitrogen are exceeded in more or less the same areas. In North America, there is also a correlation between the areas where Mg deficiency was recently observed, in the Appalachians (USA and Quebec) and the Laurentians (Quebec). Whether this is fortuitous or not is uncertain as spatial and temporal patterns of Mg deficiency are poorly known. Atmospheric deposition is likely to play a less prominent role compared with Europe: the deposition loads of acidity and nitrogen are generally much lower than the European areas affected by Mg deficiency, there is no clear evidence for a severe recent Mg depletion of forest soils, and N and/or P deficiencies are almost the rule in Mg-deficient trees. Acidic and nitrogen deposition is minimal in New Zealand forested areas and is likely to be of no significance.

– Monitoring of forest health has confirmed that climate is the synchronizing factor for these short-term dynamics of Mg deficiency. In areas only episodically affected by Mg deficiency, such as in southern Scandinavia, climate may be considered as the 'cause' of Mg deficiency although it may be argued that climate only reveals a problem caused by other factors. Drought was found to be the key climatic feature, triggering increase in Mg deficiency for example in the early 1990s in southern Norway and Germany. The unusual severity of yellowing in some European areas during the mid-1980s may be due to a more specific set of climatic conditions but relevant data for this period are insufficient.

– Magnesium deficiency has recently been associated with forest decline, especially in Europe, where the so-called 'Type 1 spruce decline' (yellowing of spruce on acidic substrates, Roberts *et al.*, 1989) was the most intensively studied cause of decline during the 1980s. However, Mg deficiency is not involved in the most severe cases of forest decline. This holds true for Europe (e.g. Norway spruce in the Ore Mountains and at Upper Ridge of the Harz; silver fir decline in Bavaria, Black Forest and Vosges; oak dieback in lowland areas), North America (Red spruce decline in the Appalachians) and New Zealand (*Nothofagus* and *Metrosideros* species).

– So far, Mg deficiency has been found to lead to a severe loss of vitality and a breakdown of conifers stands only in a few European areas during a distinct episode during the 1980s. In the above mentioned areas where Mg deficiency is most widespread, the proportion of trees affected to some degree (>10% foliage) never exceeded 15 –20% at its highest, while the proportion of moderately and severely affected trees (respectively >25% and >60% discolored foliage) remained in the range of 1–5% and less than 1%, respectively. Higher figures were only reported for the Harz.

– One can distinguish the forests stands affected by permanent symptoms from those only affected by temporary symptoms, which seem to represent the largest proportion during 'yellowing episodes'. While the vitality of permanently affected stands is clearly depressed (poor growth), the temporarily affected ones may show only a transitory growth depression expressing an imbalance between the climatic demand for nutrients and the Mg supply. Their growth before and after the episode is generally good, or even better than average. Permanently affected stands may show a stable, although lower level of vitality, or decline further as a result of an interaction with another factor, for example a pathogen.

Increased concern about forest health and exchanges between scientists have recently given rise to the idea of an apparent synchronicity in the temporal development of Mg deficiency in Europe, North America and New Zealand. As we have reported above, there is good evidence of long-term deterioration of Mg nutrition only in Central Europe. In New Zealand, the apparent expansion of Mg deficiency clearly follows forestry management activity. The North American situation is the most difficult to assess, given the relative scarcity of data for these very large forested areas and the occurrence of the different stress factors at intermediate levels.

The large-scale crown condition and nutritional (soil and foliage) surveys recently undertaken in Europe should allow a more complete picture to emerge. The consequences of a possible further depletion of forest soils in some areas and the effects of the soil restoration with Ca/Mg fertilizer, already advanced in Germany and increasingly implemented in a few other countries, must be carefully investigated. Well-focused investigations seem warranted in North America to fill in the gaps in knowledge. In New Zealand, nutritional diagnosis and fertilization are common tools in the silviculture of intensively managed stands. The Mg status of natural forests in all three regions would deserve more attention.

58

In the long term, the distribution of Mg deficiency and its significance to forest health is likely to change. Emissions abatements, and natural (North America) or assisted (Europe) changes in species composition should act positively. Global change seems likely to act negatively; it may increase forest productivity (and nutrient demand) due to CO_2 enrichment and may well be characterized by increased frequency and intensity of droughts.

Acknowledgements   We warmly thank J. Block, G. Bütner, J. Eichhorn, F. H. Evers, M. Ferretti, W. Flückiger, N. Ignatova, J. L. Innes, O. Kandler, M. Manninger, B. Nihlgård, L. Rasmussen, J. Roelofs, K. Rykowski, S. Solberg, K. Stefan, C. O. Tamm, B. Ulrich, J. Van den Burg and F. Weissen for guidance through the partly grey and old literature and unpublished information, and V. Badeau and F. Lebourgeois for drawing Figure 1.

2.7. References

Altherr E, Evers FH. 1974. Unerwartetes Düngungserfolge bei Magnesiummangel in einem jungen Buchenbestand auf mittlerem Bundsandstein des Odenwaldes. Allg. Forst. Jagdztg. 145, 121.

Aronsson A. 1985. Trädens växtnäringstillstånd i områden med skogsskador. Skogsfakta Konferens SLU Uppsala 8, 51–54.

Baule H, Fricker C. 1967. Die Düngung von Waldbäumen. BVL, München, 259 p.

Becker M, Bert GD, Bouchon J, Dupouey JL, Picard JF, Ulrich E. 1995. Long-term changes in forest productivity in northeastern France: the dendroecological approach. In: Forest decline and atmospheric deposition effects in the French mountains. Eds. G Landmann, M Bonneau, Berlin, Springer Verlag, pp. 143–156.

Becker M, Landmann G, Lévy G. 1989. Silver fir decline in the Vosges mountains (France): role of climate and silviculture. Water Air Soil Pollut. 48, 77–86.

Becker-Dillingen J. 1937. Die Gelbspitzigkeit der Kiefer eine Magnesia-Mangelerscheinung. Erd. d. Pflanze. 33, 1–7.

Becker-Dillingen J. 1940. Die Magnesiafrage im Waldbau. Forstarchiv. 16, 88–92.

Beets PN, Jokela EJ. 1994. Upper mid crown yellowing in Pinus radiata: some genetic and nutritional aspects associated with its occurrence. N.Z. J. Forestry Sci. 24, 35–50.

Benoit P, Laflamme G, Bonneau G, Picher R. 1982. Insectes et maladies des arbres – Québec 1981. Supplémenet Forêt Conservation. 48, 7.

Bernier B, Brazeau M. 1988. Magnesium deficiency symptoms associated with sugar maple dieback in a Lower Laurentians site in southeastern Quebec. Can. J. For. Res. 18, 1265–1269.

Bernier B, Pare D, Brazeau M. 1989. Natural stresses, nutrient imbalances and forest decline in southeastern Québec. Water Air Soil Pollut. 48, 239–250.

Blanck K, Matzner E, Stock R, Hartmann G. 1988. Der Einfluss kleinstandörtlicher bodenchemischer Underschiede auf die Ausprägung von Vergilbungsymptomen an Fichten im Harz. Forst Holz. 43, 288–292.

Block J, Bopp O, Gatti M, Heidingsfeld N, Zoth R. 1991 Waldschäden, Nähr- und Schadstoffgehalte in Nadeln und Waldböden in Rheinland-Pfalz. Mitteil. Forstl. Versuchsantalt Rheinland-Pfalz Ministerium für Landwirthschaft, Weinbau und Forsten 17/91, 237 p.

Bondietti EA, Momoshima N, Shortle WC, Smith KT. 1990. A historical perspective on changes in divalent cation availability to red spruce in relation to acidic deposition. Can. J. For. Res. 20, 1850–1858.

Bonneau M. 1985. Le 'nouveau dépérissement' des forêts. Symptômes, causes possibles, importance éventuelle de la nature des sols. Sci. Sol. 4, 239–251.

Bonneau M. 1993. Fertilisation sur résineux adultes dans les Vosges: composition foliaire en relation avec la défoliation et le jaunissement. Ann. Sci. For. 50, 159–175.

Bonneau M, Landmann G. 1986. Analyses foliaires. In: Programme DEFORPA. Etat des recherches à la fin de l'année 1986. Vol 1, pp. 187–204. Ministère de l'Environnement, Paris/INRA Nancy.

Bonneau M, Landmann G, Nys C. 1990/91. Fertilization of declining conifer stands in the Vosges and in the French Ardinnes. Water Air Soil Pollut. 54, 577–594.

Bosch C, Pfannkuch E, Baum U, Rehfuess KE. 1983. Über die Erkrankung der Fichte (*Picea abies* Karst.) in den Hochlagen des Bayerischen Waldes. Forstwiss. Centralbl. 102, 167–181.

Brazeau M, Bernier B. 1973. Composition minérale du feuillage du sapin baumier selon les modalités d'échantillonnage et relations avec quelques indices de croissance. Naturaliste Can. 100, 265–275.

Bridgman HA. 1989. Acid rain studies in Australia and New Zealand. Arch. Environ. Contam. Toxicol. 18, 137–146.

Bruck RI, Robarge WP, McDaniel A. 1989. Forest decline in the Boreal montane ecosystems of the Southern Appalanchians mountains. Water Air Soil Pollut. 48, 161–180.

Brüning D. 1961. Über die Wirkung von Planzennährstoffen auf das Wachstum von Kiefern in Jugendstadium. Allg. Forstz. 132, 168–177.

Brüning D. 1966. Vorzeitiger Nadelabwurf in Kieferndickungen als Folge von Nährstoffmangel. Allg. Forstz. 49, 855–856.

Buffoni A, Giulini P, Schenone G. 1990. Bioaccumulo di elementi negli anelli legnosi nelle Alpi Centro Occidentali e stato nutritivo delle piante. Primi risultati. Monti e Boschi. 5, 55–60.

Bussotti F, Gellini R, Ferretti M, Cenni E, Pietrini R, Sbrilli G. 1992. Monitoring in 1989 of Mediterranean tree condition and nutritional status in southern Tuscany, Italy. For. Ecol. Manage. 51, 81–93.

Côté B, Hendershot WH, O'Halloran I. 1993. Response of sugar maple to seven types of fertilization in Southern Quebec: Growth and nutrient status. In: Forest Decline in the Atlantic and Pacific Regions. Eds. RF Huettl, D. Mueller-Dombois. pp. 162–174. Springer, Berlin.

Daamen WP. 1991. Mineralenbehoefte van het Nederlandse bos. Rapport no 24 Maatschap Daamen, Schoonderwored, Miedema en de Klein.

Dambrine E, Carisey N, Pollier B, Granier A. 1993. Effects of drought on the yellowing status and the dynamics of mineral elements in the xylem sap of declining spruce (*Picea abies* L.). Plant Soil. 150, 303–306.

Dambrine E, Bonneau M, Ranger J, Mohamed AD, Nys C, Gras F. 1995. Cycling and budgets of acidity and nutrients in Norway spruce stands in Northeastern France and the Erzgebirge (Czech Republic). In: Forest Decline and Atmospheric Deposition Effects in the French Mountains. Eds. Landmann G, Bonneau M. pp. 233–258. Springer, Berlin.

De Vries W, Reinds GJ, Posch M. 1994. Assessment of critical loads and their exceedance on European forests using a one-layer steady-state model. Water Air Soil Pollut. 72, 357–394.

Eagar C, Adams MB. 1992. Ecology and decline of red spruce in the eastern United States. Ecological Studies 96. Springer, New York, Berlin, Heidelberg. 417 p.

Edwards GS, Edwards NT, Kelly JM, Mays PA. 1991. Ozone, acidic precipitation, and soil Mg effects on growth and nutrition of loblolly pine seedlings. Environ. Exp. Bot. 31 (1), 67–78.

Evers FH. 1972. Die jahrweisen Fluktuationen der Nährelementkonzentrationen in Fichtennadeln und ihre Bedeutung für die Interpretation nadelanalytischer Befunde. Allg. Forst-Jagdztg. 143, 68–74.

Evers FH. 1984. Lässt sich das Baumsterben durch Walddüngung oder Kalkung aufhalten? Forst-Holzwirt 39, 75–80.

Evers FH. 1994. Magnesiummangel, eine verbreite Erscheinung in Waldbeständen – Symptome un analytische Schwellenwerte. Mitt. Ver. Forstl. Standortkunde u. Forstpflanzenzüchtung. 37, 7–16.

Evers FH, Hausser K. 1973. Ertrags- und ernährungskundliche Ergebnisse von drei Kulturdüngungs-versuchen zu Fichte im Bundsandsteingebiet des Nordschwarzwaldes. Mittl. FVA Baden-Württemberg, H. 54, 1–83.

Evers FH, Schöpfer W. 1988a. Darstellung der Ernährungs- und Belastungsverhältnisse der Fichte. Ergebnisse des Belastungsinventur Baden-Württemberg 1983. Allg. Forst-Jagdztg. 149, 146–147.

Evers FH, Schöpfer W. 1988b. Ergebnisse der Belastungsinventur Baden-Württemberg 1983. Darstelung der Ernährungs- und Belastungsverhältnisse bei Fichte und Tanne. Forstliche Versuchs- und Forschungsanstalt Baden-Württemberg, Freiburg i. Bresigau, FRG. 44 p.

Falkengren Grerup U. 1989. Soil acidification and its impact on ground vegetation. Ambio. 18, 179–183.

Ferretti M, Barbolani E, Grossoni P, Gellini R, Pantani F. 1993. Incipient forest decline in the province of Trento (Northern Italy). Preliminary SEM observations and consideration of inorganic components of leaves and roots. Chem. Ecol. 8, 1–10.

Ferretti M, Cenni E, Cozzi A. 1995. Assessment of tree crown transparency and discoloration as performed by surveyors from six European countries. Results of the 7th Intern. EC/ECE Mediterranean Intercalibration Course, Cagliari (Italy), 1994. Regione Autonoma Della Sardegna – Azienda Foreste Demaniali, Gagliari, 35 p + ann.

60

Fiedler HJ, Nebe W, Hoffmann W, Ilgen G. 1984. Die Ernährung der Fichte mit Mengennährelementen in den Hoch- und Kammlagen des Thüringer Waldes. In: 'Mengen-u. Spurelemente', Arbeitstagung 1984 der K.-M. Univ. Leipzig, Bd. 1, 52–61.

Friedland AJ, Hawley GJ, Gregory RA. 1988. Red spruce (*Picea rubens* Sarg.) foliar chemistry in Northern Vermont and New York, USA. Plant Soil. 105, 189–193.

Gagnon JD. 1965. Effect of magnesium and potassium fertilization on a 20-year-old red pine plantation. For. Chron. 41, 290–294.

Gagnon G. 1988. History and current status of forest decline in Québec. Conference on Forest Decline, Toronto (Québec), 1988/08/8–11. pp. 1–14.

Gerecke KL. 1989. 'Tannensterben' und 'Neuartige Waldschäden' – Ein Beitrag aus der Sicht des Waldwachstumskunde. Allg. Forst-Jagdztg. 161, 81–96.

Glatzel G, Kazda M, Grill D, Halbwachs G, Katzensteiner K. 1987. Ernährungsstörungen bei Fichte als Komplexwirkung von Nadelschäden und erhöhter Stickstoffdeposition – ein Wirkungsmechanismus des Waldsterbens? Allg. Forstz. 158, 91–97.

Hantschel R, Kaupenjohann M, Horn R, Zech W. 1988. Acid rain studies in the Fichtelgebirge (NE-Bavaria). In: Air Pollution and Ecosystems. Int. CEC Symp., Grenoble, France, 1987/05/18–22, Ed. P Mathy, Reidel, Dordrecht, pp. 881–886.

Hartmann G, Uebel R, Stock R. 1985. Zur Verbreitung der nadelvergilbung und Fichte im Harz. Forst-Holzwirt. 40, 286–292.

Hartmann G, Saborowski J, Uebel R. Voretzsch A. 1986. Entwicklung und Verteilung von Waldschäden an Fichte im Harz – Ergebnisse und methodische Aspekte der Luftbild-Waldschadenserhebungen 1983 bis 1985. Forst-Holzwirt. 16, 413–420.

Hauhs M. 1985. Wasser- und Stoffhaushalt im Einzugsgebiet der Langen Bramke (Harz). Diss. Univ. Göttingen, Ber. Forschungszentrums Waldökosysteme Univ. Göttingen. 17, 206 p.

Hedin LO, Granat L, Likens GE, et al. 1994. Steep declines in atmospheric base cations in regions of Europe and North America. Nature. 367, 351–354.

Heinsdorf D. 1993. The role of nitrogen in declining Scots pine forests (Pinus sylvestris) in the lowland of East Germany. Environ. Pollut. 69, 21–35.

Hendershot WH, Jones ARC. 1989. The sugar maple decline in Québec: A discussion of possible causes and the use of fertilizers to limit damage. For. Chron. 65, 280–287.

Hendriks CMA, de Vries W, van den Burg J. 1994. Effects of acid deposition on 150 forest stands in the Netherlands. Relationships between forest vitality characteristics and the chemical composition of foliage, humus layer, mineral soil and soil solution. Report 69.2. DLO Winand Staring Centre, Wageningen (The Netherlands). 55p.

Hippeli P. 1967. Der Einfluss wiederholter NPKCaMg-Düngung auf die Arnährung mittelalter Kiefernbestände auf verbreiteten grundwasserfernen Standorten des nordostdeutschen Tieflandes. Arch. Forstwes. 16, 1073–1086.

Hippeli P, Bramse C. 1992. Veränderungen der Nährelementkonzentrationen in den Nadeln mittelalter Kiefernbeständen auf pleistozänen Sandstandorten Bandenburgs in den Jahren 1964–1988. Forstwiss. Centralbl. 111, 44–60.

Hocke R. 1991. Hessiche Walböden heute. Allg. Forstz. 162(2), 58–61.

Hoffmann-Schielle C. 1991. Beziehungen zwischen den Böden und der Intensität der Fichtenerkrankung in den Hochlagen des inneren Bayerischen Waldes. Forstwiss. Centralbl. 110, 228–239.

Hoshizaki D, Rock BN, Wong SKS. 1988. Pigment analysis and spectral assessment of spruce trees undergoing forest decline in the NE United States and Germany. GeoJournal. 17, 173–176.

Hosking GG. 1993. Nothofagus decline in New Zealand: separating causes from symptoms. In: Forest Decline in the Atlantic and Pacific Region. Eds. RF Hüttl, D. Mueller-Dombois. pp. 275–279. Springer-Berlin.

Hosking GP, Hutcheson JA. 1986. Hard beech (Nothofagus truncata) decline on the Mamaku Plateau, North Island, New Zealand. NZ. J. Bot. 24, 236–269.

Houdijk ALFM, Roelofs JGM. 1993. The effects of atmospheric nitrogen deposition and soil chemistry on the nutritional status of Pseudotsuga menziesii, Pinus nigra and Pinus sylvestris. Environ. Pollut. 80, 79–84.

Hunter IR. 1992. The occurrence and correction of magnesium deficiency in radiata pine plantations in New Zealand. Inaugural Dissertation. Albert-Ludwigs-Universität, Freiburg i. Br. 136 p.

Hunter IR. 1993. The role of nutrition in forest decline – A case study of Pinus radiata in New Zealand. In: Forest Decline in the Atlantic and Pacific Region. Eds. RF Hüttl, D. Mueller-Dombois. pp. 293–306. Springer, Berlin.

Hunter IR, Prince JM, Graham JD, Nicholson GM. 1986. Growth and nutrition of *Pinus radiata* on rhyolitic tephra as affected by magnesium fertiliser. N.Z. J. For. Sci. 16(2), 152–165.

Hunter IR, Rodgers BE, Dunningham A, Prince JM, Thorn A. 1990. An atlas of radiata pine nutrition in New Zealand. For. Res. Inst. Bull. No 165, 61 p.

Hüttl RF. 1989. Liming and fertilization as mitigation tools in declining forest ecosystems. Water Air Soil Pollut. 44, 93–118.

Hüttl RF. 1990. Nutrient supply and fertilizer experiments in view of N saturation. Plant Soil. 128, 45–58.

Hüttl RF. 1991. Die Nährelementversorgung geschädigter Wälder in Europa und Nordamerika. Habilitationsschrift. Freib. Bodenkundl. Abhandl., Heft 28. Freiburg i. Br., FRG, 440 p.

Hüttl RF. 1993. Mg deficiency – A 'new' phenomenon in declining forests – Symptoms and effects, causes, recuperation. In: Forest Decline in the Atlantic and Pacific Region. Eds. RF Hüttl, D Mueller-Dombois. Springer, Berlin. pp. 97–114.

Hüttl RF, Zöttl HW. 1985. Ernährungszustand von Tannenbeständen in Süddeutschland – ein historischer Verleich. Allg. Forstz. 38, 1011–1013.

Innes JL. 1995. Influence of air pollution on the foliar nutrition of conifers in Great Britain. Environ. Pollut. 88, 183–192.

Janicki W, Jones ARC. 1990/91. Nutrient response to diagnostic fertilization of Norway spruce Picea Abies (L.) Karst plantations in Western Québec, Canada. Water Air Soil Pollut. 54, 113–118.

Johnson DW, Cresser MS, Nilsson SI, Turner J, Ulrich B, Binkley D, Cole DW. 1991. Soil changes in forest ecosystems: evidence for and probable causes. Proc. R. Soc. Edinburgh. 97B, 81–116.

Jover J, Barneoud C. 1978. Carence magnésienne sur épicea commun. Ann. AFOCEL. 78, 443–466.

Kandler O, Miller W. 1990/91. Dynamics of 'acute yellowing' in spruce connected with Mg deficiency. Water Air Soil Pollut. 54, 21–34.

Kaupenjohann M, Zech W, Hantschel R, Horn R. 1987. Ergebnisse von Düngungsversuchen mit Magnesium an vermutlich immissionsgeschädigten Fichten (Picea abies (L.) Karst.) im Fichtelgebirge. Forstwiss. Centralbl. 106, 78–84.

Ke J, Skelly JM. 1990/91. Foliar symptoms on Norway spruce and relationships to magnesium deficiencies. Water Air Soil Pollut. 54, 75–90.

Kenk G. 1993. Growth in 'declining' forests of Baden-Württemberg (Southwestern Germany). In: Forest Decline in the Atlantic and Pacific Region. Eds. RF Hüttl, D Mueller-Dombois. pp. 202–215. Springer, Berlin.

Kenk G, Unfried P, Evers FH, Hildebrand EE. 1984. Düngung zur Minderung der neuartigen Waldschäden-Auswertungen eines alten Düngungsversuchs zu Fichte im Buntsandstein-Odenwald. Forstwiss. Centralbl. 103, 307–320.

Knabe W, Cousen G (and members of Working Group 'IWE'). 1988. Regional Verteilung einiger Nähr- und Schadstoffgehalte in Fichtelnadeln. Schätzungen anhand von Analysen 3-jähriger Nadeln der bundesweiten 'Immissionsökologischen Waldzustandserfassung 1983'. Schriftenreihe des Bundesministers für Ernährung, Landwirtschaft und Forsten. Reihe A. Heft 360, 65 p.

Köhl M, Innes JL, Kaufmann E. 1994. Reliability of different densities of sample grids used for the monitoring of forest condition in Europe. Environ. Monit. Assess. 29, 201–220.

Kreuzer K. 1975. IUFRO-Exkursionsführer, Jahrestagung 1975 der Arbeitsgemeinschaft Forstdüngung und der Sektion für Ertragskunde im Deutschen Verband forstlicher Forschungsanstalten. Bischofsgrün 30. Sept.–4. Okt. 1975.

Lachance D. 1985. Répartition géographique et intensité du dépérissement de l'érable à sucre dans les érablières au Québec. Phytoprotection. 66, 83–90.

Lachance D, Benoit P, Laflamme G, Bonneau G, Picher R. 1984. Insectes et maladies des arbres – Québec 1983. Supp. For Conserv. 50, 15–17.

Landmann G. 1993. Role of climate, stand dynamics and past management in forest decline: A review of ten years of field ecology in France. In: Forest Decline in the Atlantic and Pacific Region. Eds. RF Hüttl, D Mueller-Dombois. Springer, Berlin. pp. 18–39.

Landmann G. 1995. Forest decline and air pollution effects in the French mountains: a synthesis 1995. In: Forest Decline and Atmospheric Deposition Effects in the French Mountains. Eds. G Landmann, M Bonneau. pp. 407–452. Springer, Berlin.

Landmann G, Bonneau M, Bouhot-Delduc L, et al. 1995a. Crown damage in Norway spruce and silver fir: Relation to nutritional status and soil chemical characteristics in the French mountains. In: Forest Decline and Atmospheric Deposition Effects in the French Mountains. Eds. G Landmann, M. Bonneau. pp. 41–81. Springer-Heidelberg, Berlin, New York.

62

Landmann G, Bert GD, Pierrat JC, Becker M, Bonneau M, Souchier B. 1995b. Crown damage in Norway spruce and silver fir: Relation to stand factors in the French mountains. In: Forest Decline and Atmospheric Deposition Effects in the French Mountains. Eds. G Landmann, M Bonneau. pp. 82–119. Springer, Berlin.

Landolt W, Bucher JB, Kaufmann E. 1984. Waldschäden in der Schweiz – 1983. Interpretation der Sanasilva-Umfrage und der Fichtennadelanalysen aus der Sicht des forstlichen Immissionsschutzes. Schweiz. Z. Forstwes. 135, 271–287.

Leaf AL. 1968. K, Mg, and S deficiencies in forest trees. In: Forest Fertilization: Theory and Practice. Tennessee Valley Authority, National Fertilizer Development Center, Muscle Shoals, Alabama. pp. 88–122.

Le Tacon F, Toutain F. 1973. Variations saisonnières et stationnelles de la teneur en éléments minéraux des feuilles de hêtres (*Fagus sylvatica*) dans l'Est de la France. Ann. Sci. For. 30, 1–29.

Liu JC, Hüttl RF. 1991. Relations between damage symptoms and nutritional status of Norway spruce stands (*Picea abies* Karst.) in southwestern Germany. Fertil. Res. 27, 9–22.

Liu JC, Trüby P. 1989. Bodenanalytisch Diagnose von K- und Mg-Mangel in Fichtenbeständen (*Picea abies* Karst.). Z. Pflanzenernähr Bodenk. 152, 307–311.

Makkonen-Spiecker K. 1995. Schädigungen durch Trockenstress und Magnesiummangel. Allg. Forstz. 5, 263–267.

Materna J. 1989. Mineral nutrition of Norway spruce stands in the western part of Czechoslovakia. Lesnictvi. 35, 975–982.

Matzner E. 1987. Der Stoffumsatz zweier Waldökosysteme im Solling. Habiltationsschrift Univ. Gottingen. 254 p.

McLaughlin SB, Tjoelker MG, Roy WK. 1993. Acid deposition alters red spruce physiology: Laboratory studies support field observations. Can. J. For. Res. 23, 380–386.

Millers I, Shriner DS, Rizzo D. 1988. History of hardwood decline in the eastern United States. United States Department of Agriculture. Forest Service. Northeastern Forest Experiment Station. 75 p.

Miller-Weeks M, Spruce J, Levesque B et al. 1989. Monitoring of red spruce and balsam fir decline in the Northeastern United States: Symptomatology and mortality mapping. In: Air Pollution and Forest Decline. 14. Int. IUFRO Meeting for Specialists in Air Pollution Effects on Forest Ecosystems. Eds. JB Bucher, I Bucher-Wallin. vol. 2, pp. 483–485. Eidgenössische Anstalt für das forstliche Versuchswesen, Birmensdorf, Switzerland.

Möller H. 1904. Karenzerscheinungen bei der Kiefer. Z. f. Forst-u. Jagdw. 36, 745–756.

Montoya R, Sanchez G, Fernandez J. 1992. Analisis de los resultados de evaluacion de danos sobre especies forestales mediterraneas, Segovia 1992. Ministerior de Agricultura Pesca y Alimentacion. ICONA. Madrid. 27 p. +ann.

Nebe W. 1991. Veränderung der Stickstoff-und Magnesiumversorgung immissionsbelasteter älterer Fichtenbestände in ostdeutschen Mittlegebirgen. Forstwiss. Centralbl. 110, 4–12.

Nebe W, Fiedler HJ, Ilgen G, Hoffmann W. 1987. Immissionsbedingte Ernährungstöhrungen der Fichte (Picea abies (L.) Karst) in Mittelgebirgslagen. Flora. 179, 453–462.

New Zealand Soil Bureau. 1968. Soils of New Zealand. New Zealand Soil Bur. Bull. 26(1–3), 488 p.

Nys C. 1989. Fertilisation, dépérissement et production de l'epicéa commun (Picea abies) dans les Ardennes. Rev. For. Fr. 41, 336–347.

Ouimet R, Fortin JM. 1989. Résultat du projet semi-opérationnel de fertilisation des érablières déprissantes dans la région de l'Estrie-Beauce. In: Cahier des Conférences. Atelier sur le Dépérissement dans les Érablières, Saint-Hyacinthe (Québec), 1989/23–24, Centre acériole. pp. 102–107.

Phu TD. 1975. Potassium et magnésium: deux éléments limitant la croissance en hauteur du pin rouge au Québec. Can. J. For. Res. 5, 73–79.

Prinz B, Krause GHM. 1989. State of scientific discussion about the causes of the novel forest decline in the Federal Republic of Germany and surrounding countries. In: Air Pollution and Forest Decline. 14. Int. IUFRO Meeting for Specialists in Air Pollution Effects on Forest Ecosystems. Eds. JB Bucher, I Bucher-Wallin. vol. 1, pp. 27–34. Eidgenössische Anstalt für das forstliche Versuchswesen, Birmensdorf, Switzerland.

Prinz B, Krause GHM, Jung KD. 1985. Untersuchungen der LIS Essen zur Problematik der Waldschäden. In: Waldschäden: theorie und Praxis auf der Suche nach Antworten. Ed. G von Kortzfleisch. Oldenburg, München. pp. 143–194.

Raitio H. 1993. Calcium and magnesium deficiency in young pines and the stand structure on the affected habitats. In: Forest Decline in the Atlantic and Pacific Region. Eds. RF Hüttl, D Mueller-Dombois. pp. 132–143.

Reemtsa JB. 1986. Der Magnesium-Gehalt von Nadeln niedersächsischer Fichtenbestände und seine Beurteilung. Allg. Forst-Jagdztg. 147, 196–200.

Rehfuess KE. 1967. Standort und Ernährungszustand von Tannebeständen (*Abies alba* Mill.) in der südwestdeutschen Schichtstufenlandschaft. Forstw. Centralbl. 86, 321–348.

Rehfuess KE. 1968. Über den Ernährungszustand nordostbayerischer Tannenbestände (Abies alba Mill.) Forstw. Centralbl. 87, 129–150.

Rehfuess KE. 1989. Acidic deposition – extent and impact on forest soils, nutrition, growth and disease phenomena in Central Europe: a review. Water Air Soil Pollut. 48, 1–20.

Rehfuess KE. 1991. Review of forest decline research activities and results in the Federal Republic of Germany. J. Environ. Sci. Health. A26, 415–445.

Reif A. 1989. The vegetation of the Fichtelgebirge: origin, site conditions, and present status. In: Forest Decline and Air Pollution. Eds. ED Schulze, OL Lange, R Oren. pp. 8–22. Ecol. Studies 77. Springer: Berlin.

Rennie PJ. 1987. The significance of air pollution to forest decline in Canada. In: Forest Decline and Reproduction: Regional and Global Consequences. IIASA/SRI/IUFRO Workshop, Krakow (Poland), 1987/03/23–28. Eds. L Kairiukstis, S Nilsson, A Straszak. 321–334. IIASA, Laxenburg, Austria.

Robarge WP, Pye JM, Bruck RI. 1989. Foliar elemental composition of spruce-fir in the southern Blue Ridge province. Plant Soil. 114, 19–34.

Roberts TM, Skeffington RA, Blank LW. 1989. Causes of type 1 spruce decline in Europe. Forestry. 62, 179–222.

Sauter U. 1991. Zeitliche Variationen des Ernährungszustands nordbayerischer Kiefernbestände. Forstwiss. Centralbl. 110, 13–33.

Schenck GO. 1990. Kann der Umweltschutz der Umweltschutz der Umwelthygiene schaden – unorthodoxen Gedanken zu Umwelt und Waldsterben. Naturwiss. Rundsch. 43, 93–100.

Schulte-Bisping H. 1989. Räumliche und saisonale Variabilität des chemischen Bodenzustand in Buchen- und Kiefern-Waldökosystemen mit Schädigungsgradienten. Berichte des Forschunszentrum Waldökosysteme, Göttingen. Reihe A, Bd 48, 118 p +ann.

Schulze ED, Freer-Smith PH. 1991. An evaluation of forest decline based on field observations focused on Norway spruce, *Picea abies*. Proc. R. Soc. Edinburgh. 97B, 155–168.

Shortle WC, Bondietti EA. 1992. Timing, magnitude and impact of acidic deposition on sensitive forest sites. Water Air Soil Pollut. 61, 253–267.

Solberg S. 1993. Guining av bar hos gran, og betydningten av tørke (Spruce yellowing, and the significance of drought). Research paper of Norwegian Forest Research Institute, Department of Forestry, Agricultural Univ. of Norway. Rapport fra Overvåkingsprogram for skogskader, Ås, 9/93, 13 p.

Solberg S, Solheim H, Venn K, Aamlid D. 1992. Slogkader i Norge 1991 (Forest damages in Norway 1991). Research paper of Norwegian Forest Research Institute, Department of Forestry, Agricultural Univ. of Norway. Rapport fra Overvåkingsprogram for skogskader Ås, 21/92, 31 p.

Spiecker H. 1990/91. Growth variation and environmental stresses: long-term observations on permanent research plots in southwestern Germany. Water Air Soil Pollut. 54, 247–256.

Spiecker H. 1991. Zur Dynamik des Wachstums von Tannen und Fichten auf Plenterwald-Versuchsflächen im Schwarzwald. Allg. Forst. 21, 1076–1080.

Stefan K. 1989. Zur Nährstoffversorgung des Österreichischen Waldes. Nährelementgehalte in Nadelproben des Österreichischen Bioindikatonetzes/1983–87/ und von Dauerversuchsflächen/1968–87/. In: First Workshop on Ecological Monitoring in Forestry, Usti n.L., 1989/09/24–29. Forestry and Game Management Research Institute, Jiloviste-Strnady and Forest Management Institute, Brandys n.L. IUFRO WP 2.05.01, 51–89.

Stefan K. 1991 Hinweise zur Ernährungssituation der Fichte in Österreich. In: Zusammenfassende darstellung der Waldzustandsinventur. pp. 225–247. Mitteilungen der Forstlichen Bundesversuchsan-stalt Wien, Österreichischer Agrarverlag, Wien, Nr 166.

Stefan K. 1992. Der Ernährungszustand der Wälder in Österreich. In: Magnesiummangel in Mittel-europäischen Waldökosystemen. Eds. G Glatzel, R Jandl, M Sieghardt, H Hager. pp. 79–87. Forstl. Schrifenreihe Univ. Bodenkultur, Wien, Nr. 5. (Symp., Salzburg, 1991/04/08–09).

Stewart GH. 1989. Ecological considerations of dieback in New Zealand's indigenous forests. N.Z.J. For. Sci. 19, 243–249.

Stock R. 1988. Aspekte der regionalen Verbreitung 'Neuartiger Waldschäden' an Fichte im Harz. Forst und Holz. 43 (12), 283–286.

Stock R. 1990. Die Verbreitung von Waldschaden in Fichtenforsten des Westharzes – Eine geographische Analyse. Göttinger Geographische Abhandlungen. Vorhand des Geographischen Instituts der

64

Universität Göttingen. Heft 89. Verlag Erich Golze Gmbh & Co. KG, Göttingen. 102 p.

Stone EL. 1953. Magnesium deficiencies of some northeastern pines. Soil Sci. Soc. Am. Proc. 17, 297–300.

Tamm CO, Popovic B. 1989. Acidification experiments in pine forests. National Swedish Environmental Protection Board, Solna, Report Nr 3589. 131 p.

Thimonier A. 1994. Changements de la végétation et des sols en forêt tempérée européenne au cours de la période 1970–1990: rôle possible des apports atmosphériques. Thèse d'Université. Univ. Paris-Sud/INRA Centre de Nancy. 102 p + ann.

Ulrich B. 1989. Effects of acidic precipitation on forest ecosystems in Europe. In: Acidic Precipitation. Vol. 2: Biological and Ecological Effects. Eds. DC Adriano, AH Johnson. pp. 189–272. Springer, New York.

UN-ECE/EC. 1994. Convention on Long-Range Transboundary Air Pollution, International Co-operative Programme on Assessment and Monitoring of Air Pollution Forest Condition in Europe. Results of the 1993 Survey. 1994 Report. Geneva, Brussels. 90 p + ann.

Vallée G. 1967. Nouvelle contribution à l'étude du rôle du manganèse dans la régénération de la sapinière vosgienne. Thèse Faculté des Sciences, Univ. Nancy I. 145 p.

Van Bredow B. Buggert A, Eckhoff A et al. 1986. Vergleichende Untersuchungung der Boden-, Wurzel-und Blatt-Mineralstoffgehalte von Bäumen verschiedener Schadstufen in einem immissionsbelasteten Altbuchenbestand. Allg. Forstz. 22, 551–554.

Van den Burg J. 1990. Stickstoff- und Säuredeposition und die Nährstoffversorgung niederländischer Wälder auf pleistozänen Sandböden. Forst und Holz. 45, 597–605.

Van den Burg J. 1991. Een Inventarisatie van de Mineralevoedingstoestand van Opstanden van Zomereik, Grove den en Douglas op de Nederlandse Zandgronden (Zomer/Najaar 1989). DeDorsch-kamp Instituut vvoor bosbouw en groenbeheer Wageningen Raport nr. 644 ISSN 0924-9141, 162 p + ann.

Van Goor CP. 1970. Fertilization of conifer plantations. Irish Forestry. 27, 80.

Van Miegroet H, Johnson DW, Todd DE. 1993. Foliar response of red spruce saplings to fertilization with calcium and magnesium in the Great Smoky Mountains National Park. Can. J. For. Res. 23, 89–95.

Wagner S. 1989. Letale Waldschäden im Gebiet des 'Ackers' (Hochharz). Forst und Holz. 18, 494–498.

Walker SL, Auclair AN. 1989. Forest Declines in Western Canada and the Adjacent United States. Atmospheric Environment Service, Downsview, Ontario, 150 p.

Walker SL, Auclair AN, Martin H. 1990. History of Crown Dieback and Deterioration Symptoms of Hardwoods in Eastern Canada. Part I, II, III.

Weissen F, Maréchal P. 1991. Le sol, un composant stabilisateur de l'écosystème forestier ardennais. Pédologie. 41(1), 69–87.

Weissen F, Letocart M, Van Praag HJ. 1984. Rapport préliminaire sur les effets de la pollution atmosphérique sur les forêts de l'Ardenne. Bull. Soc. R. For. de Belgique. 91, 61–72.

Will GM. 1961. Magnesium deficiency in pine seedlings growing in pumice-soil nurseries. N Z. J. Agric. Res. 4, 151–60.

Will GM. 1966. Magnesium deficiency: the cause of spring needle-tip chlorosis in young pines on pumice soils. N Z. J. For. 11, 88–94.

Wisniewski J, Kinsman JD. 1988. pH and hydrogen ion deposition patterns in precipitation for the Continental United States and Canada. Water Air Soil Pollut. 38, 1–17.

Wyttembach A, Bajo S, Tobler L, Keller T. 1985. Major and trace element concentrations in needles of Picea abies: levels, distribution functions, correlations and environmental influences. Plant Soil. 85, 313–325.

Zech W, Popp E. 1983. Magnesiummangel, einer der Gründe für das Fichten- und Tannensterben in NO-Bayern. Forstwiss. Centralbl. 102, 50–55.

Zech W, Schneider BU, Röhle H. 1990/91. Element composition of leaves and wood of beech (*Fagus sylvatica* L.) on SO_2-polluted sites of the Ne-Bavarian Mountain. Water Air Soil Pollut. (Eds. HW Zöttl, RF Hüttl. Management of nutrition in forests under stress). 54, 97–106.

Zöttl HW. 1985. Waldschäden und Nährelementversorgung. Düsseldorfer Geobotanisches Kolloquium. 2, 31–41.

Zöttl HW, Mies E. 1983. Die Fichtenerkrankung in Hochlagen des Südschwarzwaldes. Allg. Forst-Jagdztg. 154, 110–113.

Zöttl HW, Stahr K, Keilen K. 1977. Spurenelementverteilungen in einer Bodengesellschaft in Bärhalde-Granit (Südschwarzwald) Mitt. Dtsch. Bodenkundl. Ges. 25, 143–148.

Part II
Magnesium in forest ecosystems

3
Biogeochemistry of magnesium in forest ecosystems

K. H. FEGER

3.1. Introduction

Magnesium deficiency has played a prominent role in recent forest decline phenomena. Symptoms of acute Mg deficiency (conifers: tip yellowing of older needles; deciduous species: intercostal chlorosis in leaves) have been observed in many regions of the world (Hüttl, 1991). In order to understand the differences in Mg supply between sites and temporal developments, a thorough analysis of the biogeochemical cycling of this element and major processes is necessary.

Biogeochemical cycling involves a continuing cyclical exchange of chemical elements between the biota and the physical environment within an ecosystem (Figure 1). Therefore, all biogeochemical cycles show strong interrelations and dependencies with site-internal and -external factors. The main site-internal factor is the soil. Thus, soil properties control the physical, chemical and biological conditions in the rooting zone. External factors include climate, atmospheric pollutants, and forest management.

The cycling of the various nutrient and non-nutrient elements in an ecosystem is highly interrelated. Therefore, Mg cycling can not be isolated from the cycling of other elements. In this chapter, the components of Mg cycling in deficient forest ecosystems and silvicultural implications are discussed.

3.2. Components of Mg cycling in forest ecosystems

3.2.1. General patterns

Figure 2 provides an overview over the cation pools and fluxes in a forest ecosystem. For a given element, the nutritional status of a tree stand depends on the requirement and the uptake. Requirement can be defined as the annual elemental increment associated with all tree components (bole, bark, branches, roots) plus the current foliage production. In addition to requirement, uptake includes also the return (litterfall, stemflow, and crown wash=leaching). Rates of internal recycling can be estimated from the difference, i.e. 'requirement minus uptake' (Cole and Rapp, 1981). Numbers for these parameters can be derived from ecosystem analysis measuring atmospheric inputs (bulk precipitation, canopy throughfall), stemflow, litterfall, and an inventory of the tree biomass and its elemental composition. Rates of dry deposition and leaching are normally calculated from the difference between precipitation and throughfall by means of model assumptions (Ulrich, 1983).

R. F. Hüttl & W. Schaaf (eds): Magnesium Deficiency in Forest Ecosystems, 67–99.
© 1997 *Kluwer Academic Publishers. Printed in Great Britain.*

68

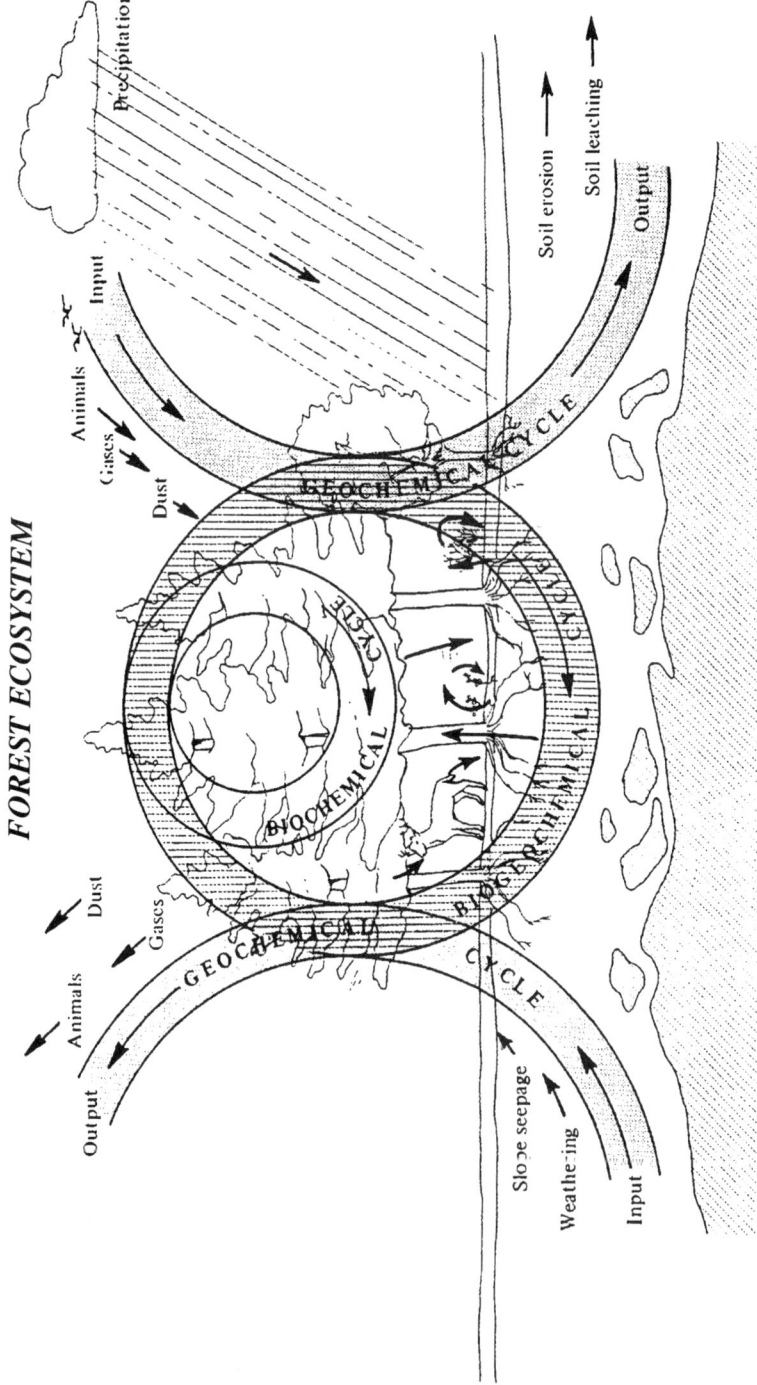

Figure 1. Biogeochemical cycling comprises fluxes of nutrients within a forest ecosystem in contrast to geochemical cycling (nutrient fluxes between ecosystems) and biochemical cycling (within an organism). (Modified from Kimmins, 1987)

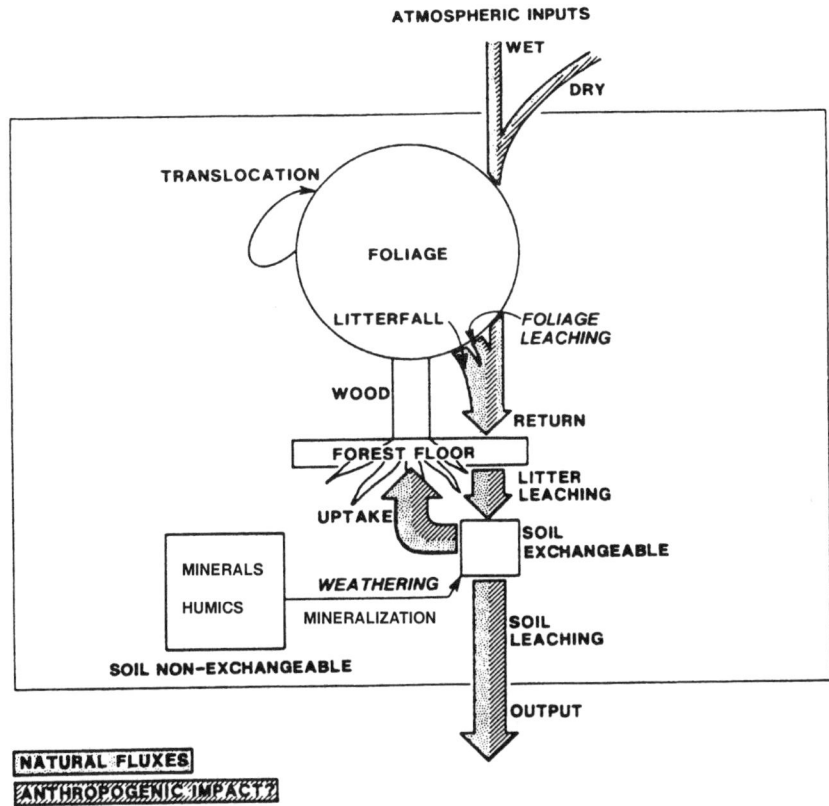

Figure 2. Schematic diagram of base cation cycling in a forest ecosystem. (Modified from Johnson, 1992)

Rates of element uptake, requirement and return for various nutrients including Mg are compared in Table 1. The data set is based on the worldwide analysis of 32 stands within the IBP study (Cole and Rapp, 1981). There are distinct differences in the requirement of the specific nutrients (N > P > Ca > K > Mg = P). Furthermore, the comparison elucidates that the rate of cycling is far more rapid in deciduous than in coniferous species. This is due to the fact that the foliage of deciduous trees is replaced each year. The uptake rates of K, Ca, and Mg in conifers approximate 50% those of deciduous species. For all elements, the return – mainly by litterfall – is a significant flux.

3.2.2. Mg distribution and fluxes in forest ecosystems

Table 2 gives the distribution of Mg pools in the various compartments of several forest ecosystems. As it is typical of forest ecosystems, the total storage in the soil

Table 1. Worldwide comparison of deciduous and coniferous species relative to element uptake, requirement and return for the 32 IBP study sites

Element	Uptake (kg ha^{-1} y^{-1})		Requirement (kg ha^{-1} y^{-1})		Return (kg ha^{-1} y^{-1})	
	Deciduous	Coniferous	Deciduous	Coniferous	Deciduous	Coniferous
N	70	39	94	39	57	30
P	6	5	7	4	4	4
K	48	25	46	22	40	20
Ca	84	35	54	16	67	29
Mg	13	6	10	4	11	4

Data from Cole and Rapp, 1981

is by far the largest amount, constituting more than 90% of total Mg in the eco-system. Vegetation accounts for less than 5% of total pools. The amounts of Mg stored in vegetation and forest floor vary from site to site, but exchangeable pools in the soil vary much more.

In Figure 3 the fluxes of Mg in a typical Mg-deficient system is illustrated for a Norway spruce site in the higher altitudes of the Black Forest (SW Germany). Magnesium deficiency in this stand is reflected by both foliar concentrations and abundant symptoms of needle discoloration (Münch *et al.*, 1990; Feger, 1992a). Typically, there exists a distinct spatial heterogeneity in the intensity of yellowing symptoms within the same stand. Figure 4 presents the results of an elemental inventory of three individual spruce trees. There exists a distinct relation-ship between the intensity of yellowing and the Mg contents of several tree compon-ents. Differences in the Mg level are most apparent for the older needles and the bark.

First observations of acute Mg deficiency in this region were made during the mid-1970s (Zöttl *et al.*, 1977). A maximum of discoloration was reached in the early 1980s (Zöttl and Mies, 1983). Since 1985 a slow but continuing decrease in symp-toms of acute Mg deficiency has been evolving (Münch *et al.*, 1990). Comparing the fluxes and reserves in the ecosystem makes an insufficient supply of Mg under-standable. The amount stored in the above-ground stand biomass clearly exceeds the reserve of exchangeable Mg in the rooting zone which, in Norway spruce mono-cultures, is normally very shallow (see Raspe, this volume). The amount of total Mg in the fine earth is up to 100 times greater. This is an enormous Mg pool, especially if the reserves in the gravel fraction are also considered. However, the reserves of silicate-bound Mg are of minimal significance in the ecosystem Mg cycling because of the limited rates of silicate weathering (see 3.2.7.). The rate of Mg provided through weathering is clearly in the range below the average uptake of the aggrading stand. When the fluxes of Mg recycling, litterfall, and leaching from the canopy are considered, a tightly closed cycling of Mg becomes evident. In fact, a spruce stand suffering from acute Mg deficiency lives from 'hand to mouth' with respect to Mg (e.g. Schluchsee, Figure 4). The Mg output from the watershed at the weir (2.3 kg ha^{-1} y^{-1}) nearly equals the seepage loss from the solum at 80 cm depth. This indicates that the Mg release from the extremely Mg-poor bedrock ('Bärhalde' granite: see Tables 6 and 7) in the deeper infiltration zone is minimal.

Table 2. Magnesium pools and fluxes in several coniferous and deciduous forest ecosystems

	Conifers					Deciduous				
	DF1	NS1-R1	Soll-S	Sch1	ST-1	RA	SB1	TL	Soll-B	CH
Amounts (kg ha⁻¹)										
Foliage total	7	14	6	13	4	6	3	5	3	11
Branches	8	5	20	13	8	11	14	15	8	20
Boles	8	17	31	12	29	40	21	28	57	32
Roots	5	6		13	6	12	20	8	15	89
Understorey vegetation	6	2			8	11	8	1		11
Forest floor	31	53	35	29	57	97	13	49	29	56
Soil-rooting zone, exchangeable	100	28	37	17	62	74	50	56	35	1016
Soil-rooting zone, total	3200	48475	15500	2400	20280	20300	17900	25300	15500	85457
Fluxes (kg ha⁻¹ y⁻¹)										
Bulk precipitation	0.6	0.4	1.6	0.9	0.7	0.6	0.5	0.6	1.7	0.4
Canopy throughfall + stemflow	1.1	1.6	4.7	1.3	3.5	1.8	1.5	1.7	4.0	2.0
Total deposition	0.8	1.2	3.8	1.1	2.3	0.8			2.9	
Canopy leaching	0.3	0.4	0.9	0.2	1.2	0.9	1.0	1.1	1.1	
Seepage output, rooting zone	1.5	1.7	5.8	2.6	4.2	7.3	1.8	5.5	3.1	2.2
Litterfall	1.5	1.4	0.7	0.6	1.4	6.3	1.9	4.0	1.5	10.3
Uptake	1.6	2.8	2.7	3.4	0.7	7.2	4.2	4.3	4.3	12.2
Requirement	1.6	3.5			0.8	8.0	4.7	5.3		16.7
Stand increment	0.1	1.4	1.1	2.4	-0.1	0.6	1.7		1.7	0.9

Site		Tree species	References
DF1	= Thompson Forest, NW USA	Douglas fir	Johnson and Lindberg (1992)
NS1-R1	= Nordmoen/SE Norway	Norway spruce	Johnson and Lindberg (1992)
Soll-S	= Solling/N Germany	Norway spruce	Matzner (1988)
Sch1	= Schluchsee/SW Germany	Norway spruce	Feger (1993)
ST-1	= Great Smoky Mountains SE USA	Red spruce	Johnson and Lindberg (1992)
RA	= Thompson Forest, NW USA	Red alder	Johnson and Lindberg (1992)
SB1	= Great Smoky Mountains SE USA	American beech	Johnson and Lindberg (1992)
TL	= Turkey Lakes, Ontario/Canada	Northern hardwoods	Johnson and Lindberg (1992)
Soll-B	= Solling/N Germany	European beech	Matzner (1988)
CH	= Coweeta, SE USA	Southern hardwoods	Johnson and Lindberg (1992)

Magnesium

Watershed Schluchsee 1 Period 11/1987 - 10/1990
Norway spruce 50 yr-old

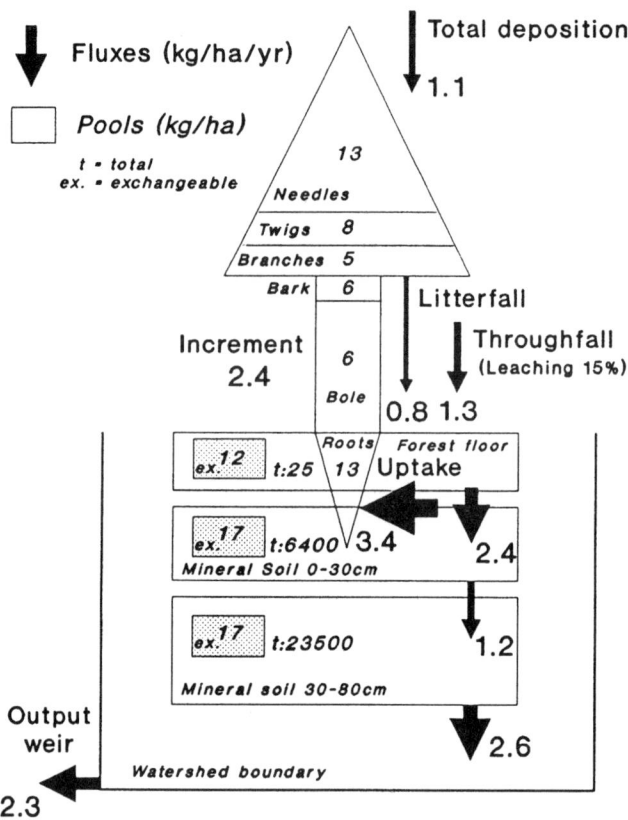

Figure 3. Magnesium cycling in a Mg-deficient Norway spruce stand in the Black Forest (ARINUS-site Schluchsee; for site description, see Feger, 1992a)

In the following sections the principal processes and components of Mg cycling in forest ecosystems are discussed with respect to their implications in Mg deficiency.

3.2.3. Deposition of Mg from atmospheric sources

As for many other elements, input from the atmosphere can be an important component in the cycling of Mg in forest ecosystems, notably at sites where the supply in the soil (via mineral weathering and/or mineralization) is low and the requirement

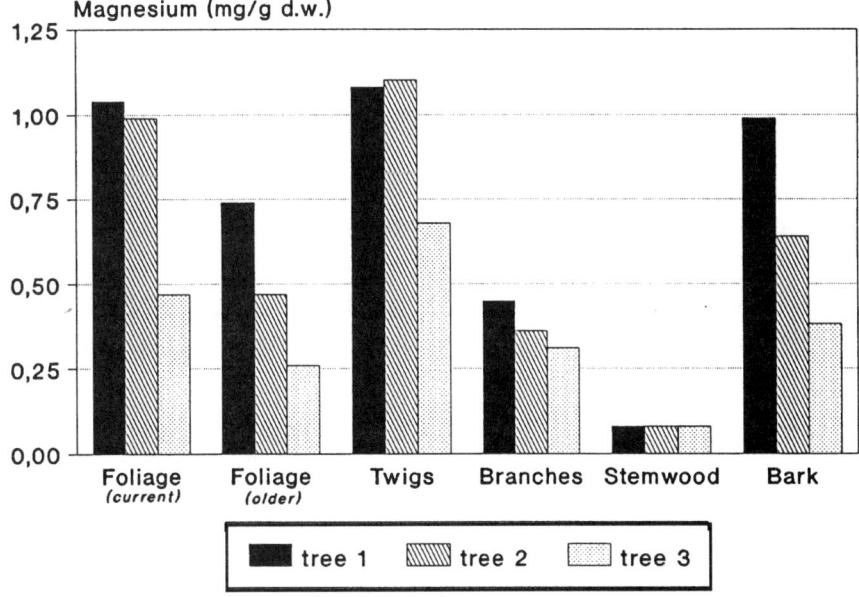

Figure 4. Distribution of Mg in three individual spruce trees growing on the same soil in the Black Forest (Arinus-site Schluchsee). Trees differ considerably in the intensity of yellowing symptoms (tree 1: green, no obvious discoloration; tree 2: slight tip-yellowing of the older needles; tree 3: strong tip-yellowing of the older needles). The threshold value for acute Mg deficiency in Norway spruce is 0.7 mg g^{-1} dw (current foliage). (Data from Münch *et al.*, 1990).

of the aggrading stand biomass is high. Besides the nutritional implications, the rate of cation input from the atmosphere is also important in the acid–base status of the soil. Nevertheless, the accurate estimation of atmospheric deposition of base cations to plant canopies is still a problem. The higher fluxes in canopy throughfall and stemfall of these elements compared with bulk precipitation result from both leaching from plant tissues and scavenging from the atmosphere through the forest surface. Since there is no direct measurement, the rates of total deposition must be calculated. A common method uses 'canopy differences' (canopy throughfall minus bulk precipitation) where Na is normally used as an index element for canopy inter-actions (Ulrich, 1983). Other methods for the estimation of interception deposition are based on the measurement of the size and chemical composition of particles and appropriate deposition velocities (e.g. VDI, 1987; Lindberg *et al.*, 1988).

Unlike N and S, Mg and the other base cations, Ca, K, and Na, have no gaseous phase in the atmosphere and are deposited by rainfall (wet deposition), cloud water deposition, and particulates (dry deposition). Coarse particles (>2 μm diameter) are deposited on canopy surfaces primarily by sedimentation and impaction, while deposition of fine aerosols occurs mainly through turbulent mixing and diffusion through boundary layers of surfaces (Georgii, 1985; VDI, 1987). There are three main sources of Mg in the atmosphere:

1. **Sea-salt.** The marine influence is highest in coastal areas. At such sites, elevated Mg inputs are primarily associated with wet deposition and fine aerosols. In marine aerosols Mg occurs together with Cl^- and SO_4^{2-} as counter-ions.

2. **Mineral dusts** originate from agricultural tillage practices, traffic on unpaved roads, industrial and mining activities (e.g. cement production), and natural wind erosion of arid soils. Hence, the Mg content of aerosols mainly depends on the mineral composition of the land surface in the source area.

3. **Fly ashes** result from point-source emissions from fuel combustion and industrial processes. Furthermore, volcano eruptions and forest fires can release significant amounts of ash particles.

Base cation deposition typically varies on a regional scale. The variability exists with respect to absolute rates, sources, and ratios to other elements, as well as to the relative importance of the different deposition processes. For the IFS sites, Ragsdale *et al.* (1992) reported rates of total Mg deposition ranging between 0.2 and 2.6 kg $ha^{-1} y^{-1}$ with wet and dry deposition being the predominant processes. Cloud water deposition was only relevant at the high elevation sites. The ratio of wet to dry Mg deposition was highly variable (0.1–2.9). The predominant elements in total atmospheric deposition of base cations were Na and Ca which together contributed between 70% and 85%. Potassium and Mg ranked nearly equal in relative contribution among all IFS sites. Figure 5 shows the distribution of total Mg deposition among Norway spruce stands in Germany. Elevated rates in the northern part are due to the marine influence. In the central and eastern parts of the country, deposition of Mg may primarily originate from the emission of alkaline fly ash resulting from lignite burning in power plants and private households.

Atmospheric deposition is, however, not a time-constant site factor. It is a fact that the scavenging efficiency increases when the stands grow older. This applies to the deposition of acids, but also to Mg-containing sea-salt aerosols and neutralizing mineral particles. Furthermore, it is important to note that the reduced emission of particles in vast regions of Europe and North America has led to steep declines in the atmospheric concentrations of base cations over the past two decades (Hedin *et al.*, 1994). Data from 30 rainfall stations distributed throughout Sweden revealed continued linear declines during 1983–91 for both non-marine Ca ($-7\% y^{-1}$; $p < 0.01$) and Mg ($-9\% y^{-1}$; $p < 0.02$). Hedin *et al.* (1994) explain these observed trends primarily by the strong declines in particulate emissions from urban and industrial point sources. This obviously results from technical improvements in air pollution control and fits in with published emission and air quality data (Umweltbundesamt, 1994). Unfortunately, long-term deposition measurements in forests exist only for a few sites. Figure 6 demonstrates the temporal development of Mg in canopy throughfall at four Norway spruce sites in Germany. At the Solling site, which has the longest record starting in 1969, both the Mg and Ca fluxes have been markedly decreasing. The Heidelberg site follows a similar pattern. The

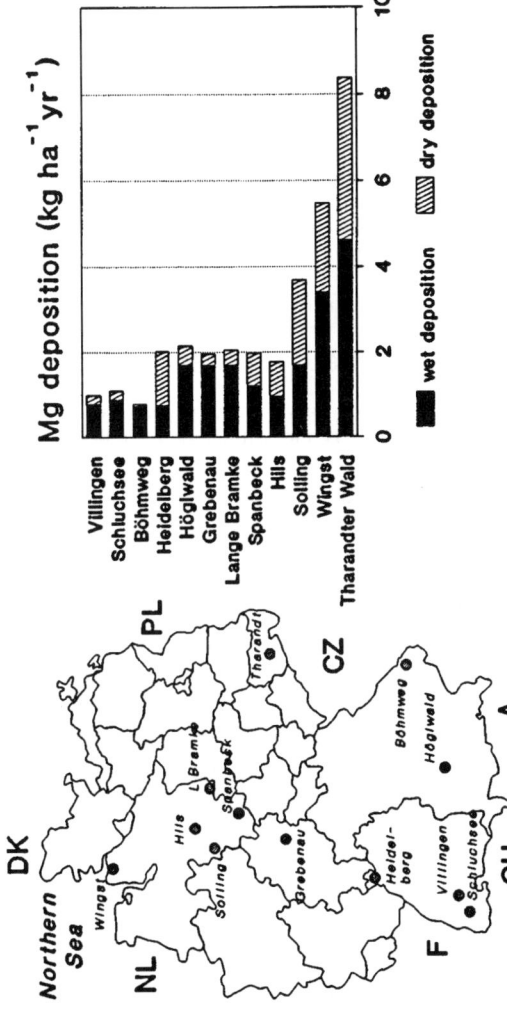

Figure 5. Distribution of wet/dry deposition in total Mg deposition to Norway spruce stands in Germany (calculated from literature bulk precipitation and throughfall data as 'canopy differences' according to Ulrich, 1983). Data source: Göttlein and Kreutzer (1991), Feger (1993), Lux and Börtitz (1990), Hildebrand and Hochstein (1992)

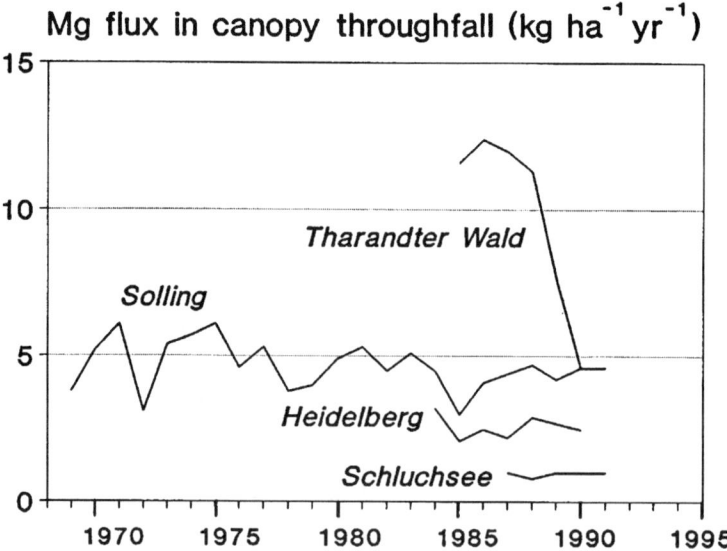

Figure 6. Temporal development of Mg in canopy throughfall of four Norway spruce sites in Germany (data from Manderscheid *et al.*, 1994; Feger, 1993; Lux and Börtitz, 1990; Hildebrand and Hochstein, 1992; for location of sites, see Figure 5)

steepest decline can be seen at the Tharandter Wald site in eastern Germany. This decreasing trend apparently results from the drastic reductions in lignite burning and gradual increase in filter installation which both commenced in the late 1980s. It is interesting to note that the decrease in Mg deposition rates is paralleled by a gradual lowering of the Mg nutritional status at base-poor mountain sites in eastern Germany (Nebe, 1991). In contrast to sites in eastern and northeastern regions of Germany, which are characterized by a moderate-to-high level of S and base cation deposition, remote sites in extended forest regions of SW Germany (e.g. Schluchsee in the Black Forest) have not hitherto displayed a similar decrease in cation deposition.

3.2.4. *Uptake in the rhizosphere*

Tree roots satisfy much of their nutrient requirement by direct uptake from the soil solution. The concentration of nutrients in, and their uptake from, soil water is influenced by a wide variety of physical and chemical soil properties. Four major factors controlling Mg uptake can be separated:

1. Diffusion transfer;
2. Mass transfer;
3. Root growth into unutilized nutrient pools in the soil;
4. Ion antagonism.

Diffusion transfer by which nutrients move from the surrounding soil to the root may limit uptake when structure-dependent differences in available nutrient pools exist and the diffusion velocity of the specific nutrient ion is small. In Norway spruce stands in the Black Forest (SW Germany), Hildebrand (1990) found evidence for a selective depletion of water-soluble K, Ca and Mg on aggregate surfaces. As aggregate surfaces bind soil water with low tension this might be the preferential location of root growth (Kaupenjohann and Hantschel, 1989). This phenomenon, however, can explain only the recent outbreak of K deficiency in well-aggregated loamy soils (Zöttl and Hüttl, 1985) because K^+ translocation to the roots occurs mainly by diffusion which is too slow to satisfy the K requirement of the tree. Even though structure-dependent heterogeneities also exist for plant-available Mg, this effect might be compensated by the relatively high mobility of Mg^{2+} in the soil solution. This may explain the observation that exchangeable Mg in the bulk soil is much better correlated with the foliar Mg concentration, compared with K (Liu and Trüby, 1989).

The mass flow of nutrients to the roots depends on the amount of water in the soil and the rate at which it is moving toward the root. This in turn is influenced by the transpiration rate and soil moisture conditions. In pot experiments with young Norway spruce, Makkonen-Spiecker and Evers (1993) could induce Mg deficiency under drought conditions. They observed a rapid recovery after re-wetting. The role of water supply in the rate of Mg uptake and outbreak of acute deficiency is evident in a number of studies (Kandler et al., 1987; Lange et al., 1987; Feger et al., 1996). Also Will (1985) reported that Mg deficiency of *Pinus radiata* in New Zealand was more widespread and intense during dry summers. A key factor in mass transport of a nutrient is its total quantity in the rooting zone. Thus, the available Mg pool is a function of Mg input from deposition sources as well as soil-internal release from cation exchange, mineral weathering and mineralization, and finally leaching from the rooting zone. The growth and distribution of fine roots, which is also a determinant in nutrient uptake, is strongly interrelated with the level of Mg nutrition of the tree (Gonzalez-Cascon et al., 1990; Raspe, this volume). The uptake rate is not only controlled by the Mg concentration in the soil solution, but also by the concentrations of other ions. With decreasing soil pH, aqueous Al^{n+} activities increasingly impede Mg^{2+} uptake by cation antagonism (Jorns and Hecht-Buchholz, 1985; Marschner, 1989).

There are only a few studies elucidating the role of mycorrhiza in Mg uptake. In contrast to P, Mg uptake appears not to be intensified by mycorrhiza fungi. Microprobe analysis in Norway spruce seedlings showed hardly any difference in Mg uptake between mycorrhizal and non-mycorrhizal spruce seedlings (Kuhn, 1993). Processes of Mg uptake are discussed in detail in Chapters 4 and 10.

3.2.5. Foliar leaching

The suggestion that increased leaching of Mg from the canopy is an important contributing factor to the outbreak of Mg deficiency has been the subject to extensive laboratory and field studies. Leaching in this context means the removal by solution of material that is physically or chemically bound to plant surfaces or interiors. Thus, it can be separated from the wash-off of dry-deposited material that is not bound to the surfaces. For a variety of forest ecosystems in the IFS study, Ragsdale et al. (1992) found a Mg leaching rate between 0.4 and 1.6 kg ha^{-1} y^{-1}. This is also the range for other sites (Table 2). For the IFS sites, leaching of Mg^{2+} was generally less than that of Ca^{2+}.

Both acid deposition and ozone have been suggested as the main factors which increase leaching of Mg, and also of Ca and K, from tree canopies. Other possible effects include:

1. Acute tissue necrosis from severe pollutant exposure, which could cause leakage of cell contents and subsequent leaching from the plant;
2. Effects of solution pH on cuticle and cell-wall permeability; and
3. Indirect effects, such as altering cold tolerance or leaf-surface microbial communities (Lovett and Schaefer, 1992).

For acid deposition, the mechanism is identified as cation exchange between deposited protons and foliar base cations on exchange sites on or in the leaf (Tukey, 1980). By this mechanism a distinct part of rainwater acidity is buffered. However, the plant must compensate for base cation losses resulting from cation exchange processes in the canopy by additional root uptake. This can cause an accelerated acidification of the soil/root interface (Ulrich, 1991) which in turn negatively influences Mg availability. However, experimental studies have not shown significant effects of acidity on foliar leaching until the pH is decreased to 3.3 or below (Wood and Bormann, 1975; Scherbatskoy and Klein, 1983; Schier, 1987). Even then, the quantitative effect on Mg cycling at the ecosystem level is small. Mengel et al. (1987) concluded that leaching of spruce seedlings subjected to a pH 2.75 mist for 6 weeks removed only 1% of the foliar Mg. In open-top chamber experiments with 10–15-year-old spruce trees at the 'Edelmannshof' site in SW Germany, Evers et al. (1993) found less cation leaching when SO$_2$ was excluded from the ambient air. This applies mainly to Mg, but also to Ca, Mn, Zn, and other cations. Interestingly, the differences in cation leaching existed even when levels of SO$_2$ concentrations were very low. Nevertheless, Evers et al. (1993) observed no reduction in foliar Mg concentrations in the trees with higher rates of foliar leaching. The trees were able to compensate for leaching losses by cation uptake in the rhizosphere. In addition, a generous supply of Mg did not result in increased leaching rates of this element from the foliage.

For ozone, the proposed mechanism is damage to foliage cellular membranes. This causes leakage of cell contents that are then leached by subsequent rainfall. Early laboratory experiments by Krause et al. (1983) showed that ozone in combination with acid mist could increase the foliar leaching of Mg. However, subsequent experiments by Brown and Roberts (1988) have revealed that the accelerated leaching was mainly due to the deposition of anhydrous nitric acid, which had been produced in the ozonated air as an impurity. In these studies, acid mist (pH < 4) accelerated leaching losses by 10% from Norway spruce needles without, however, lowering their Mg content. Hence, the additional effect of ozone was small. Similar results on the relative importance of acid mist vs. ozone were obtained by Lovett and Schaefer (1992). In both white pine (Figure 7) and sugar maple, there were clear pH effects whereas variations in ozone concentration did not cause statistically significant differences in Mg leaching. This is consistent with the results from the open-top chamber experiments by Evers *et al.* (1993) which likewise did not yield a correlation between ambient air ozone concentrations and base cation leaching.

3.2.6. *Litterfall, litter decomposition and mineralization*

Losses of nutrients from plants via litterfall can be substantial but they are characterized by great year-to-year variability. Litterfall generally accounts for the majority of the N, P, Ca, and Mg losses from forest vegetation, whereas leaching is the dominant process for K and Mn (Cole and Rapp, 1981; Johnson, 1992).

Of all the pathways of nutrient loss from forest vegetation above-ground litterfall has received the most attention. On the other hand, below-ground litter production (the annual death of large quantities of fine roots) is one of the least studied pathways of nutrient removal from plants, in spite of the fact that in some forests the production of fine root necromass may exceed above-ground litterfall by several times. In Table 3, the average above-ground litterfall and associated nutrient return is given for various forest regions. Litterfall losses are generally greatest on moist, warm, fertile, and other high-productivity sites and least on dry, cold, infertile, and other low-productivity sites (Cole and Rapp, 1981; Kimmins, 1987).

The importance of litter production from fine root mortality varies according to stand age and site. Vogt *et al.* (1983) reported rates of below-ground litter production which were double the above-ground litterfall in a 23-year-old Pacific silver fir stand. In contrast, the below-ground production of litter in a 180-year-old stand of the same species was 4 times lower. Notably, the soil moisture regime is a determining factor for the quantity of below-ground litter production. Keyes and Grier (1981) calculated a rate of fine-root litter production in a dry ridge-top Douglas-fir stand as being four times higher than that of a stand on a moist lower-slope site (5.6 vs. 1.4 $kg\,ha^{-1}\,y^{-1}$ of dry matter).

The rate of nutrient release from litter decomposition is a function of the litter biomass, the type (leaves, branches, bark, etc.) and the nutrient concentrations in the litter, all of which vary from site to site. Often litter is exposed to leaching, fragmentation, and fungal attack while it is still attached to the tree. This litter arrives at the forest floor at a stage of partial decomposition. Release of nutrients from litter decomposition is often the critical link in the biogeochemical cycling in forests. If decomposition is too slow, a significant amount of nutrients is removed from active circulation for a long period of time. Excessive accumulation of litter can lead to deep forest floors which normally are extremely acid, often excessively wet, and thus, microbiologically inactive. In contrast, excessively rapid decomposition of litter can release nutrients faster than vegetation and soil are able to retain them (temporal

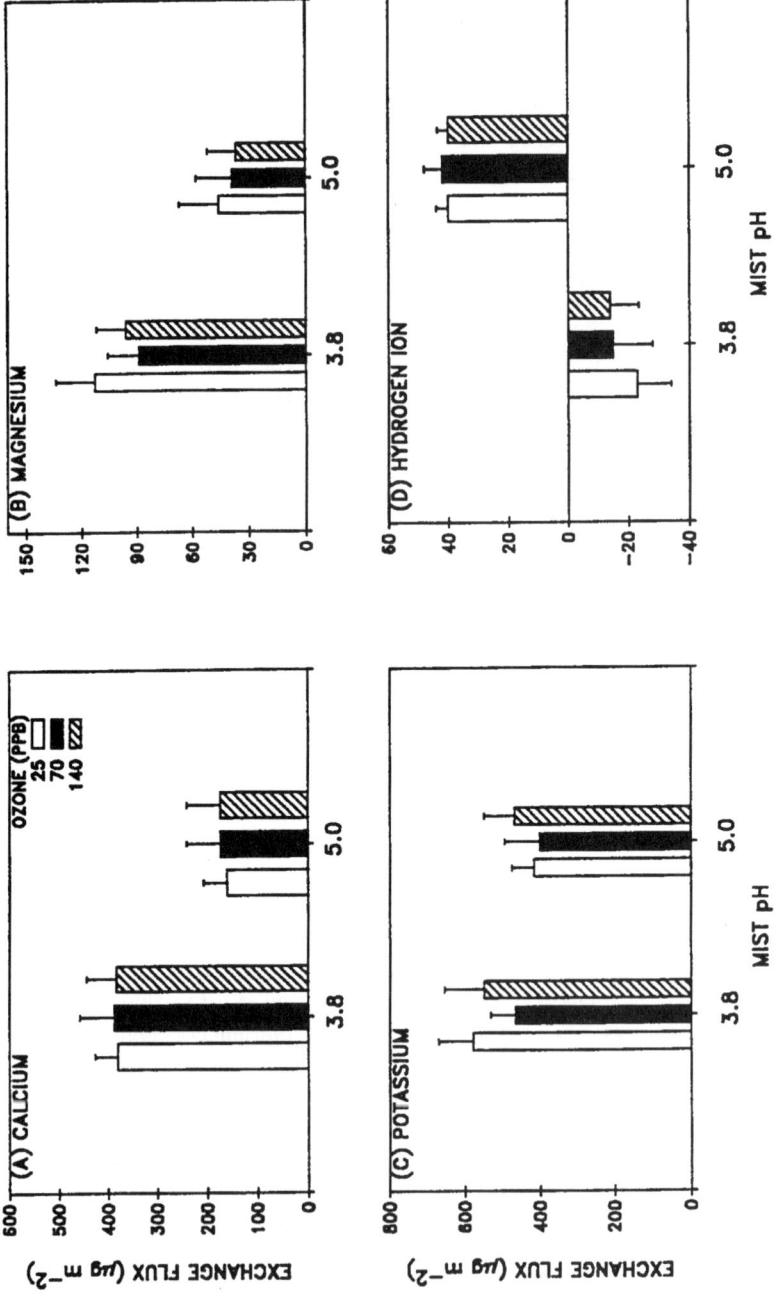

Figure 7. Group means and standard error bars for Mg leaching from foliage resulting from ozone and acid mist treatments on white pine. (Modified from Lovett and Schaefer, 1992)

Table 3. Average litterfall and nutrient return by forest regions as determined at the 32 IBP study sites

Forest region	No. of sites	Litterfall ($kg\,ha^{-1}\,y^{-1}$)	Nutrient return ($kg\,ha^{-1}\,y^{-1}$)				
			N	P	K	Ca	Mg
Boreal coniferous	3	322	2.9	0.7	1.1	3.8	0.3
Boreal deciduous	1	2645	20.2	5.2	9.8	35.5	9.7
Temperate coniferous	13	4377	36.6	4.4	26.1	37.3	5.6
Temperate deciduous	14	5399	61.4	4.0	41.6	67.7	11.0
Mediterranean	1	3842	34.5	4.7	44.0	95.0	9.0
All sites	32	4373	43.3	3.7	27.9	47.3	7.4

Data from Cole and Rapp, 1981

Table 4. Mean residence (turnover) time in years of the forest floor and its mineral elements by forest regions as determined at the 32 IBP sites

Forest region	No. of sites	Organic matter	N	P	K	Ca	Mg
Boreal coniferous	3	353	230	324	94	149	455
Boreal deciduous	1	26	27.1	15.2	10.0	13.8	14.2
Temperate coniferous	13	17	17.9	15.3	2.2	5.9	12.9
Temperate deciduous	14	4.0	5.5	5.8	1.3	3.0	3.4
Mediterranean	1	3.0	3.6	0.9	0.2	3.8	2.2
All sites	32	12.0	34.1	46.0	13.0	21.8	61.4

Values calculated by dividing annual return into total forest floor accumulation; a steady-state condition is assumed. Data from Cole and Rapp, 1981

'discoupling' of nutrient cycling). Over the long term, such an enhancement of nutrient leaching out of the rooting zone will generally lower soil fertility and gradually deplete the pool of Mg available to plants.

The rate of litter decomposition and associated release of nutrients varies enormously. Most estimates of litter decomposition are based on short-term studies of litter confined in litterbags or on a comparison of the forest floor mass and the annual litterfall. As an example of the latter approach, Table 4 gives an overview of forest floor nutrient turnover for the 32 IBP sites. Such calculations are based on the assumption that the forest floor and the rate of return are both in a steady state. Even though this assumption may not be correct for most sites, the calculation still provides a relative index useful for comparisons. The more northerly boreal forests have an exceedingly long mean residence time for both organic matter and all five nutrients. There is no consistent trend in the residence times of the nutrients. Among the base cations, however, Mg has the longest residence time, ranging from 455 years in the boreal coniferous forests to 1 year in tropical forests.

Litter decomposition is accomplished by the combined activity of soil animals of various sizes (mesofauna and microfauna) and soil microorganisms. Intense activity of soil animals is normally a prerequisite for the soil microflora which is responsible for the most of the actual decomposition and mineralization. Hence, any explanation

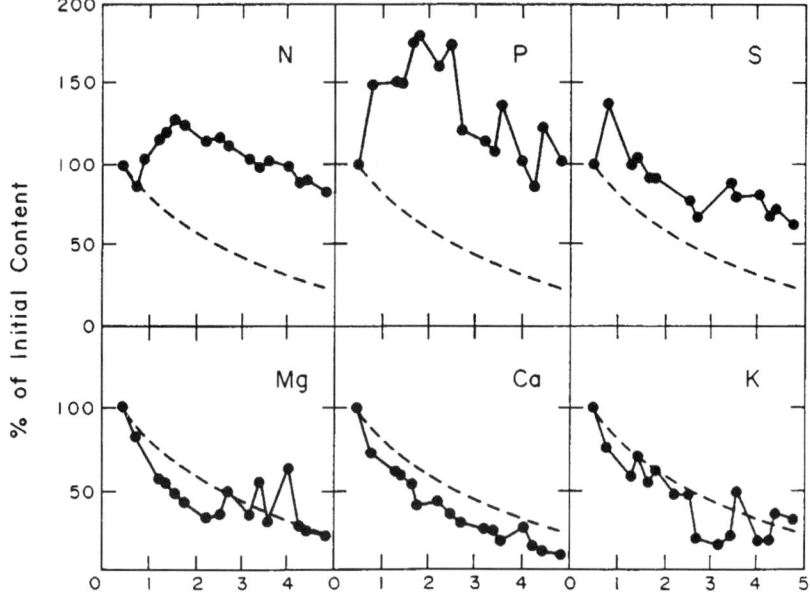

Figure 8. Changes in absolute amounts of Mg and various other plant nutrients in decomposing Scots pine needle litter. The dashed line shows the fitted exponential decay function for weight loss of organic matter. (Modified from Staaf and Berg, 1982)

of the variation in litter decomposition rates must consider the factors that control the abundance, species composition, and activity of the microflora (Millar, 1974; Kimmins, 1987). Such factors are highly interrelated and include:

1. Soil microclimate (moisture, temperature);
2. Chemical composition of the litter depending on the vegetation producing litter (tree species, understory and ground vegetation);
3. Chemical status of the soil and the forest floor (pH, base saturation, quality of humic substances, availability of N and P);
4. Activity of the soil fauna.

Only few experimental data on the dynamics of Mg release from litter decomposition exist. In most instances, relatively rapid losses of K, Ca, Mn, and Mg were observed (Parmentier and Remacle, 1981; Anderson *et al.*, 1983). The same pattern was found in litter bag experiments by Staaf and Berg (1982) who followed losses of nutrients from needle litter of Scots pine (*Pinus sylvestris*). They observed that K, Ca, and Mg were lost more rapidly than the disappearance of the litter mass, whereas N, P, and S were retained during litter decay (Figure 8). Therefore, base cations are not a major constituent of humic substances which are continually produced by the transformation of plant residues and soil microbial biomass. The humus component of soils nevertheless provides a large quantity of negatively charged exchange sites. These cation exchange sites are capable of holding a considerable pool of cationic nutrients, such as K, Ca, and Mg, which are rapidly released from forest litter and/or enter the soil solution via canopy throughfall and stemflow (see 3.2.8.). In the case of Mg and Ca, the organic soil matter can also bind these elements as weak complexes. In contrast to N and P, Mg mineralization rates do not reflect the available amounts of this element in the rooting zone over short intervals. The analysis of exchangeable cations in the soil is therefore a useful tool to quantify the plant-available pools of base cations (Liu and Trüby, 1989).

3.2.7. Mineral weathering

Although external inputs from the atmosphere can provide a distinct portion of the Mg uptake in an aggrading forest, the ultimate source of this inorganic elemental nutrient is supplied through weathering of primary minerals in the soil. Table 2 and Figure 3 demonstrate that the largest amount of Mg in a forest ecosystem is stored within the soil minerals and, thus, is not plant-available. The most common Mg-containing mineral in calcareous soils is dolomite [$CaMg(CO_3)_2$]. In soils derived from siliceous parent materials, there is a number of primary silicates which contain Mg. The most important of them are:

Biotite [$KAl(Mg,Fe)_3Si_3O_{10}(OH)_2$]
Hornblende [$Ca_2Al_2Mg_2Fe_3Si_6O_{22}(OH)_2$]
Augite [$Ca_2(Al,Fe)_4(Mg,Fe)_4Si_6O_{24}$]
Cordierite [$Mg_2Al_3(AlSi_5O_{18})$]
Olivine [$(Mg,Fe)_2SiO_4$]

The given formulas are only approximate since the elemental composition of many minerals is highly variable. In areas where Mg deficiency is a common phenomenon, the bedrock normally contains only minimal portions of these minerals. Table 5 gives the mineral composition of various soil-forming crystalline rocks in the Black Forest. In all rock types, quartz forms the most abundant mineral. The 'Bärhalde'-granite has the lowest quantity of Mg-containing minerals. The variability in mineral composition is reflected in the Mg contents ranging between 3% by weight MgO in metamorphic rocks and 0.2% by weight MgO in quartz-rich plutonites (e.g. 'Bärhalde'-granite: Schreiner and Wimmenauer, 1979).

Weathering is basically a combination of destruction of primary minerals and synthesis of secondary minerals. Rock materials are first broken down physically into smaller pieces and eventually into the individual minerals of which they are composed. Simultaneously, rock fragments and the minerals therein are attacked by chemical forces and are changed to new minerals either by minor alterations or by complete decrease in particle size and by the release of soluble constituents, which are subject to losses in seepage water or recombination into new (secondary) minerals (clay minerals, hydrous oxides of Fe and Al). Thus, the mineral composition of a soil is the result of complex processes during the whole time span of soil formation.

Table 5. Mineral composition of various soil-forming rocks in the Black Forest (Germany) according to Schreiner and Wimmenauer (1979)

	Paragneisses	Metatexites	Anatexites	'Bärhalde'-granite
Quartz	30	29	31	35
Orthoclase	2	4	11	33
Plagioclase	43	20	36	25
Biotite	20	24	15	4
Muskovite	< 1	< 1	< 1	4
Cordierite	4	22	6	< 1
Accessory minerals	1	2	2	< 1

In the temperate humid climate of Europe and North America where soils are much younger than in the tropics, even the extremely acid soils normally contain large quantities of unweathered or just slightly altered primary minerals (April and Newton, 1992; Dultz, 1993). This is due to the fact that the glacial and periglacial processes removed most of the highly weathered soil material and enhanced the physical weathering of the bedrock. As an example, Table 6 presents the mineral composition of an iron–humus–podzol of the typically Mg-deficient Schluchsee site which was glaciated during the last (Würm) glaciation period (Figure 3). In this soil, biotite is the only Mg-containing primary mineral. Even the bleached horizon (Ahe) contains a considerable amount of biotite.

The most important driving force of weathering is the proton activity in the percolation water. Protons are provided from various sources: atmospheric deposition of acids, respiration of soil microorganisms and plant roots (carbonic acid), surplus uptake of cations by roots, mineralization of organic S compounds and oxidation of sulfide materials (sulfuric acid), nitrification (nitric acid), and formation of organic acids during the process of organic matter decomposition. Table 7 shows the strong influence of pH on element release from silicate weathering. In an artificial weathering experiment Stahr *et al.* (1993) percolated unweathered cubes of 'Bärhalde'-granite with solutions of pH 5, 4, and 3 adjusted with sulfuric acid. Solutions were removed continuously to avoid equilibrium conditions. As could be expected, the most acid percolation yielded the highest rates of element release. The release of elements occurred in a distinct sequence $(Si > Na > Ca > Al > K > Mg > Fe > Mn)$ which is due to the fact that the individual minerals have a different ability to resist protolysis. The selectivity of silicate weathering can also be seen from a comparison between the element ratios in the rock material and in the percolation water. Plagioclase dissolves more rapid than orthoclase or biotite. This means that Mg is released from silicate weathering at slower rates as could be expected from the chemical composition of the bedrock. From the mineral inventory and geochemical budget in the Schluchsee podzol (derived from 'Bärhalde'-granite: Tables 5 and 6), Stahr *et al.* (1993) calculated an average loss of base cations of $1.2 \, kmol(c) \, ha^{-1} \, y^{-1}$ under the assumption of 10000 years of soil formation. However, only 2.5% of this consisted of Mg. Similar results were obtained by Hofmann-Schielle (1993) who studied Mg

Table 6. Mineral composition and weathering degree of rock-forming minerals in an iron–humus–podzol in the Black Forest (ARINUS site, Schluchsee)

Horizon	Depth (m)	pH (CaCl$_2$)	Quartz	Orthoclase			Plagioclase			Biotite			Muscovite			Fe-oxides/clay
				f	+w	++w	f	+w	++w	f	+w	++w	f	+w	++w	
Ahe	0–30	3.3	49.2	0.0	24.2	6.1	0.0	3.0	5.0	0.6	3.3	4.0	0.2	2.5	1.0	0.8
Bsh	30–40	3.8	45.9	0.0	22.3	11.0	0.0	1.3	5.6	0.0	2.2	2.5	0.1	3.3	1.4	4.5
Bs	40–60	4.2	43.9	1.3	24.0	10.9	0.0	1.9	4.2	0.0	1.8	3.2	0.3	3.3	2.0	2.5
BvCv	60–80	4.3	45.0	1.0	26.0	10.9	0.0	2.1	3.1	0.0	2.2	3.1	0.5	4.1	1.2	1.0
Cv	80–>100	4.4	43.7	2.3	25.0	7.2	0.0	3.0	4.6	1.1	3.5	2.2	2.1	2.9	1.7	0.7

Mineral numbers are % counts on thin sections; f = fresh mineral; +w = slightly weathered; ++w = strongly weathered. Data from Zarei et al., 1992

Table 7. Element losses during an experimental weathering of 'Bärhalde'-granite using different pH levels of percolation solution

pH	Na	K	Ca	Mg	Mn	Fe	Al	Si	Si/Mg
5	17	5	8	0.1	0.3	0.4	0.2	115	1150
4	35	9	30	7	1.4	2	2	430	61
3	109	21	65	29	3.8	20	195	1570	54
Rock	1130	1076	110	108	7.3	318	7890	53000	490

Material: unweathered rock cubes; 6 months duration; data from Stahr *et al.*, 1994

release in mineral soils derived from various crystalline rocks in the upper Bavarian Forest.

Laboratory weathering experiments using either rock material or mineral soil provide valuable information on the kinetics and relative importance of the weathering of individual mineral constituents. However, weathering experiments cannot give accurate flux rates on the ecosystem level. The same is true for mineral budgets which integrate over the whole period of holocene soil formation (Mazzarino *et al.*, 1983; Dultz, 1993). Furthermore, this concept is based on the ideal assumption of a homogeneous parent material for the formation of the entire profile. As a consequence, weathering rates resulting from profile budgets are of limited use for models of the present cation cycling. On the other hand, element budgets based on the analysis of the present element cycling in an ecosystem, including element uptake by the vegetation, can only provide rates of element release in a defined soil compartment. Hence, the relative contribution of the various individual processes (mineral weathering, cation exchange, mineralization) is hardly possible to determine. From element flux measurements in the Mg-deficient spruce stand at Schluchsee (Figure 3), Feger (1993) calculated an annual Mg release in the mineral soil (0–80 cm) of 2.3 kg ha^{-1} from which a considerable part is assumed to originate from organic matter decay. For forests in the Solling Mountains (N Germany), Matzner (1988) reported an annual rate of Mg release of 3.3 kg ha^{-1} under beech and 4.6 kg ha^{-1} under Norway spruce. The author attributes nearly all Mg released in the mineral soil to silicate weathering. However, this ignores inherent contributions of other processes, such as mineralization, especially that resulting from root decay (see 3.2.6).

3.2.8. Cation exchange and drainage loss

In contrast to mineral weathering processes which release nutrient cations only at limited rates in a long-term perspective, cation exchange is the direct link between soil solution and plant uptake. The cation exchange complex of forest soils is a mixture of constant and variable charge on surfaces both of organic and inorganic soil particles. Many forest soils typically have conditions in which exchange sites are saturated with Al. Exchangeable Al commonly accounts for more than 75% of effective cation exchange capacity. Moreover, the ionic strength of the soil solution

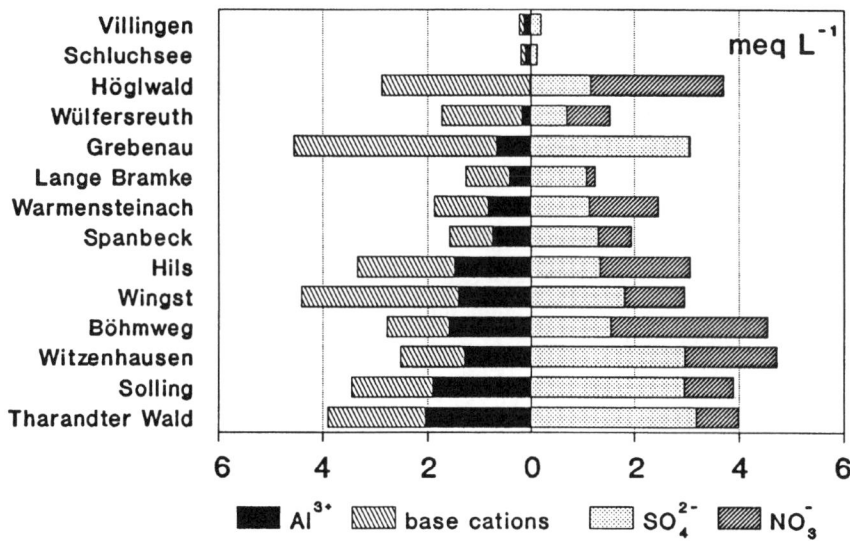

Figure 9. Chemical composition of seepage water beneath the rooting zone of several spruce forest ecosystems in Germany. (For data source and location of sites, see Figure 5)

in forest ecosystems is very low compared with agricultural soils. Cation concentrations are highly controlled by the concentrations of strong acid anions ('mobile' anions SO_4^{2-} and NO_3^- ; Figure 9). Several models of soil solution chemistry suggest that Ca^{2+}, Mg^{2+}, and Al^{3+} are exchanged and leached from acid forest soils as a result of increases in ionic strength (e.g. Cosby *et al.*, 1985; Reuss and Johnson, 1986). According to these models, the exchange and leaching of cations in forest soils appear highly sensitive to small changes in the concentrations of mobile anions which dominate the ionic strength of the soil solution. However, concentrations of SO_4^{2-} and NO_3^- vary markedly as a function of many ecosystem processes (deposition, evapotranspiration, mineralization, weathering etc.). With respect to Mg cycling in the ecosystem, there are two main consequences of increased anion concentrations:

1. Elevated concentrations of inorganic Al species which might be toxic to plant roots and/or inhibit Mg^{2+} uptake antagonistically;
2. Increased seepage losses of Mg^{2+} from the rooting zone.

Richter *et al.* (1992) studied cation exchange reactions for a variety of acid forest soils. In nearly all soils tested they found that increasing solution ionic strength by even small amounts increased exchange of the polyvalent cations Al^{3+}, Ca^{2+}, and Mg^{2+} whereas the exchange of monovalent cations Na^+ and K^+ was much less dependent on ionic strength. For a given ionic strength of soil solution, base saturation is the key factor controlling Al mobilization. With a simulation model, Reuss and Johnson (1986) demonstrated that fairly minor changes in base saturation within the 10% to 20% range can result in large increases in soil solution Al.

It has been well documented in numerous studies that SO_4^{2-} deposition has increased electrolyte concentrations in forest soils in the northern hemisphere of the order of 0.1 to $0.5\,mmol(c)\,l^{-1}$ (Reuss and Johnson, 1986; Göttlein and Kreutzer, 1991; Johnson and Lindberg, 1992). In some regions of central Europe with an extreme level of SO_2 emission, the electrolyte concentration of the soil solution is even higher and is strongly controlled by SO_4^{2-} (Figure 9). In addition, NO_3^- can significantly contribute to the ionic strength of the soil solution. In contrast to SO_4^{2-}, for which soil-internal sources normally play only a minor role, NO_3^- concentrations in soil solutions are mediated by a large number of biotic processes. Plant uptake and microbial N transformations, including mineralization/immobilization, nitrifi-cation, N_2-fixation, and denitrification, are key processes defining the level of NO_3^- leaching (Vitousek et al., 1982; van Miegroet and Cole, 1984). The complex N turnover in forest soils is strongly influenced by seasonal and year-to-year variations of climatic conditions. As a consequence, there normally exists no clear relationship between atmospheric N input and NO_3^- leaching (Hauhs et al., 1989). An exception might represent sites subject to excess N deposition mainly in the form of NH_4^+ (e.g. The Netherlands: van Breemen et al., 1987). In regions characterized by low-to-moderate levels of atmospheric N input the ability to retain N in the ecosystem often depends greatly on the site management history. For forests in central Europe it has been demonstrated that former land-use practices coupled with a high removal of biomass (litter raking, fuelwood coppicing, etc.) have led to a distinct impoverishment of the N reserves in the soils (Glatzel, 1991; Tamm, 1991; Katzensteiner and Glatzel, this volume). In such cases, all mineral N originating from the atmospheric deposition and/or litter decomposition is retained in the biomass either of the forest vegetation or soil microorganisms, resulting in hardly any NO_3^- leaching (Feger, 1992b). On the other hand, forest management can also increase NO_3^- leaching rates. The silvicultural system (e.g. cutting and thinning regime) and the choice of tree species are important factors in this respect. After conversion of deep-rooting species (beech, fir) into shallow-rooting stands (spruce), a spatial and temporal dis-coupling of nutrient uptake and mineralization evolves. Kreutzer (1989) reported excess nitrification from spruce plantations in Bavaria, where spruce was in the third rotation after beech or mixed deciduous stands. The same pattern of excess pro-duction of mobile anions in humic horizons of the mineral soil was found in the Black Forest after a former change in tree species (Feger, 1992b). Notably at sites where atmospheric deposition is relatively low, the soil-internal production of mobile anions is a key factor controlling the leaching both of base cations and inorganic Al.

In the discussion about the widespread Mg deficiency which is more or less restricted to Mg-poor soils, it has hitherto not been clarified whether this insufficient supply of Mg has already existed for a long period or whether it is a fairly recent phenomenon. For the specific areas in Germany with acute Mg deficiency (e.g. Black Forest, Bavarian Forest, Fichtel Mountains) there are no preceding analyses of exchangeable cations, nor can a significant change be verified (Rehfuess, 1988). On the other hand, many surveys of forest soils in central and northern Europe have alluded to a significant decrease in pH over the last 1–6 decades (e.g. Butzke, 1981; Lochman 1981; Wittmann and Fetzer, 1982; Klimo and Kulhavy, 1984; Hallbäcken and Tamm, 1986). An accepted explanation for these pH reductions is the impact of acid deposition. Changes were greatest in the top soils and did not depend on site parameters, such as stand age and tree species. However, a further decrease in pH was not found in soil types which were originally highly acidic (e.g. podzols, dystric cambisols) (Wittmann and Fetzer, 1982; Fiedler and Hofmann, 1985; Grimm and Rehfuess, 1986).

It is difficult, however, to evaluate pH values or reported decreases in pH < 4.5 with respect to potential consequences for the exchange properties of forest soils. The reason is that, at the same pH, base saturation and the processes of cation mobilization vary considerably. Even though there is a lack of observational data on the recent development of the exchange properties of forest soils, it is plausible to assume that acid deposition has been accelerating the export of Ca^{2+} and Mg^{2+} in those forest soils which have been subjected to high deposition rates and where the replenishment of the exchangeable pool by mineral weathering is low (Rehfuess, 1988). Significant increases in base cation leaching have been demonstrated in soil manipulation experiments using artificial acid rain (Abrahamsen, 1984, 1993; Kreutzer et al., 1991). On the other hand, there are many experimental data supporting the fact that soils which have always been strongly acid, even under the impact of high acid deposition rates, do not further acidify and continue losing base cations. Acid buffering in such soils is mainly accomplished by dissolution of Al and Fe compounds whereas the pool of exchangeable base cations remains at a very low but constant level. In a soil inventory in Mg-deficient spruce stands in the Odenwald Mountains near Heidelberg (relatively high level of acid deposition) and in the NE Black Forest (Dornstetten: low to moderate level of acid deposition), Hildebrand (1986) compared samples from upper mineral soils taken in 1968/69 and 1983/84, respectively. Soils were dystric cambisols derived from a base-poor quartz sandstone (base saturation < 5%). Hildebrand (1986) observed no changes in the exchangeable pools of Ca^{2+}, Mg^{2+}, and K^+, whereas Al^{3+} saturation decreased and H^+ increased (Figure 10). Obviously, if the rate of proton input exceeds the rate of proton consumption by buffering processes, protons begin to dominate soil solution chemistry and are consequently adsorbed to exchange sites. Under such soil chemical conditions, the relative selectivity of mineral soils for Ca^{2+} and Mg^{2+} increases as a function of proton saturation. These findings from repeated soil inventories are supported by element flux measurements. There is clear evidence for a net retention of base cations in forest ecosystems with extremely acid soils (Solling beech: Matzner, 1988; Villingen spruce: Feger, 1993). Retention is highest for Ca^{2+} and K^+ whereas the budget for Mg^{2+} is mostly balanced or slightly negative.

3.2.9. Losses by forest biomass harvesting

A major avenue of loss of nutrients from managed forest ecosystems is through removal of the harvested crop. The annual accumulation of nutrients is dependent

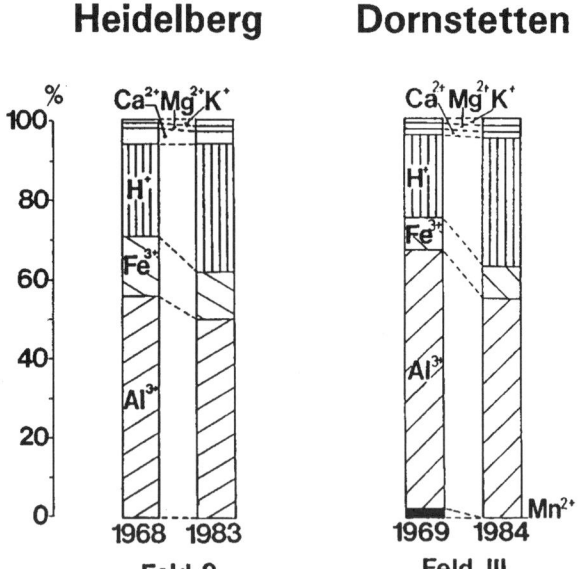

Heidelberg **Dornstetten**

Figure 10. Results of repeated soil inventories (exchangeable cations: NH_4Cl extraction) in Mg-deficient spruce stands in the Odenwald Mountains near Heidelberg (relatively high level of acid deposition) and in the NE Black Forest (Dornstetten: low to moderate level of acid deposition). Both soils are sandy-loamy dystric cambisols. (Modified from Hildebrand, 1986)

on the growth rate and the age of the stand. Generally, the accumulation of nutrients in the aggrading biomass is highest in the first third of the rotation (Kimmins, 1987). A relatively small amount of nutrients is stored in the stemwood and bark. More than 70% returns to the nutrient pool in the soil via foliage, branches, fruits, and roots. Thus, the extent of the nutrient export from the site depends on the duration of the rotation period and the intensity of biomass utilization. According to the principle of sustained yield, forestry in central Europe is characterized by long rotations and conventional stem harvesting. However, the trend toward mechanization of logging operations has locally also led to intensified forms of forest biomass harvesting (whole-tree: removal of all above-ground tree biomass from the site; complete-tree: whole-tree + tree stumps and coarse root systems). Feger (1993) calculated for the Mg-problem Schluchsee site (see Figure 3) an average Mg export rate (based on a 130-year rotation period) of $0.2\,kg\,ha^{-1}\,y^{-1}$ (stemwood without bark), $0.4\,kg\,ha^{-1}\,y^{-1}$ (stemwood + bark), $1.5\,kg\,ha^{-1}\,y^{-1}$ (whole-tree), and $1.8\,kg\,ha^{-1}\,y^{-1}$ (complete-tree), respectively. Similar figures were reported by Kreutzer (1979) and Nebe and Herrmann (1987). Such calculations reveal that an intensified utilization of biomass on susceptible sites over the long term would inevitably lead to Mg deficiency. Moreover, the numbers elucidate the deleterious effect of intense biomass export practices in former centuries (e.g. litter removal) for present site quality (Glatzel, 1991; Feger, 1993; Katzensteiner and Glatzel, this volume).

3.3. Synthesis: an ecosystem-based explanation of acute Mg deficiency in central Europe

Based on the discussion of the major components of biogeochemical cycling of Mg in forests, a complex model to explain acute Mg deficiency can be established. The diagram in Figure 11 originates from Roberts *et al.* (1989) who thoroughly reviewed the causes and relationships of this type of forest decline in mountainous sites in central Europe. With this model, the sudden and more or less synchronous outbreak can be elucidated.

Hence, an acid soil derived from a Mg-poor parent material is an inevitable precondition for acute Mg deficiency. At many sites in central Europe, the intense utilization of forests in former times has considerably added to an unfavorable soil chemical status. Slow Mg release by mineral weathering and simultaneously low atmospheric deposition of Mg results in a low availability of Mg. Furthermore, the depletion of the exchangeable pool has been accelerated by planting monocultures of Norway spruce which is known as a very shallow-rooting species.

More recently, the atmospheric input of protons and associated mobile anions have enlarged soil acidification and Mg leaching from the rooting zone. However, there are distinct differences in the regional pattern as well as in the temporal development of acid deposition. This also applies to atmospheric Mg which mainly originates from mineral dusts and has markedly decreased between 1975 and 1990. Increased soil acidity could reduce Mg uptake by lowering the Mg^{2+}/Al^{3+} ratio in soil solution. Numerous field studies have confirmed that direct Al toxicity is unlikely at most sites (Rehfuess, 1988).

In spruce stands showing symptoms of acute Mg deficiency, a pronounced tree-to-tree and stand-to-stand variation in the intensity of tip-yellowing on the same soil has been observed (Figure 4; Rehfuess, 1987; 1988). This may be explained by genetic differences within these stands. Autochthonous populations suffer considerably less than obviously foreign spruce provenances which were introduced, presumably from lower elevations, when vast clear-cut areas on former pasture land were afforested in the 19th and 20th century (Rehfuess, 1987; Löchelt *et al.*, 1993).

The sudden outbreak which occurred more or less synchronously over a wide area can be attributed to strongly unfavorable climatic conditions starting in the mid-1970s and continuing until the mid-1980s. Some of those summers (notably 1976, 1983) exhibit the extremely unusual pattern of a drought in a very early phase of the growing season (Feger *et al.*, 1996). In dry years, there is a reduced Mg release by litter decomposition, less Mg uptake from the mineral soil, poorer growth of fine roots, and, potentially, more Mg leaching and Al mobilization due to a climate-induced nitrification pulse. In addition, there might also be some impact of other types of climatic stress (Löchelt *et al.*, 1993; Rehfuess, 1995). In the late 1970s there were a couple of extreme frost shock events which resulted in a marked loss of younger needles. These needles may represent a crucial tree-internal Mg pool with respect to the supply of new shoots. Ozone may be involved in reducing root development in dry periods but the effect on foliar Mg leaching appears to be small. In

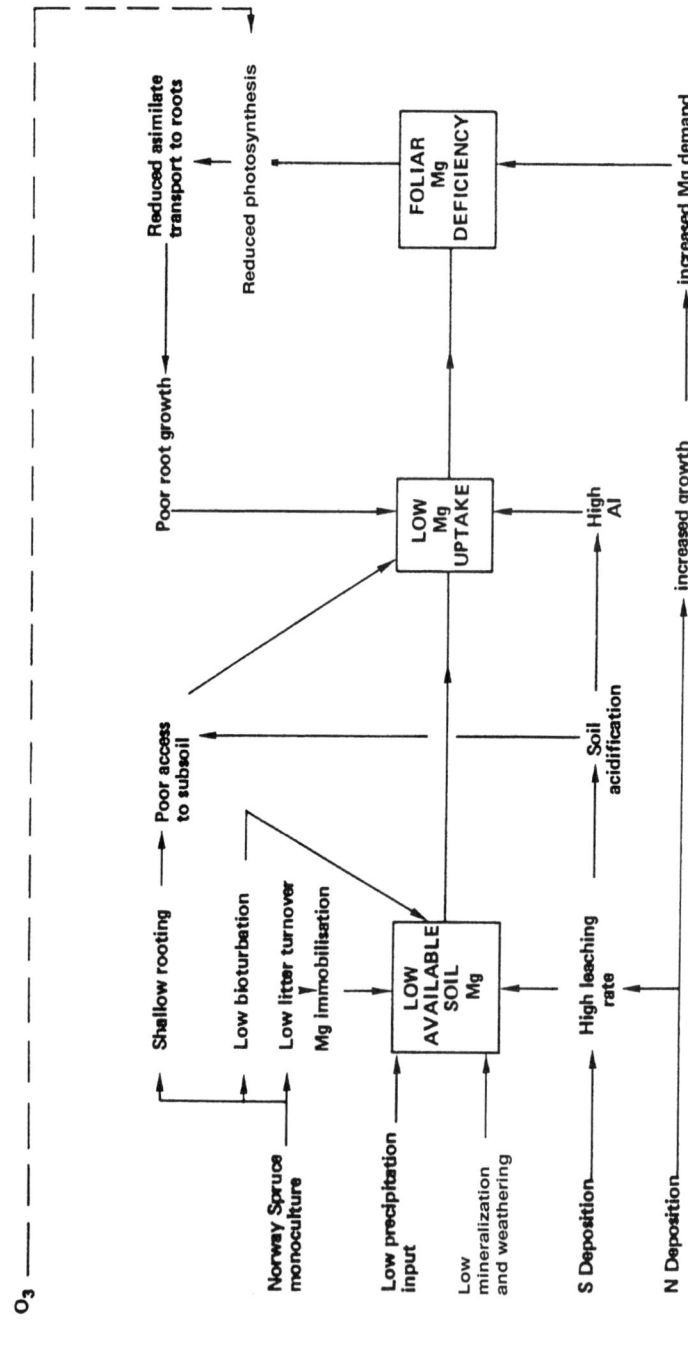

Figure 11. Causes and relationships in forest decline resulting from acute Mg deficiency. (Modified from Roberts *et al.* 1989)

years with a more favorable soil moisture regime the supply of Mg from litter decomposition and mineral weathering occurs at higher rates. As a consequence, symptoms are gradually reduced. This could explain the reversibility of this type of forest decline in the western parts of Germany since 1985 (Kandler *et al.*, 1987) whereas the gradual increase in symptoms in mountainous regions of eastern Germany (Nebe, 1991) may be the result of a sharp decrease in Mg deposition from fly ash sources. The importance of atmospheric Mg input is evident from the fact that, in countries with a maritime climate (e.g. Great Britain, Norway), symptoms of Mg deficiency are relatively rare (Roberts *et al.*, 1989).

The relative contribution of acid deposition in the evolution of acute Mg deficiency is not easy to determine. Roberts *et al.* (1989) concluded that Mg deficiency would probably have occurred in dry summers irrespective of acid deposition. But there is no doubt that S deposition, and more recently an increased N deposition, has accelerated this development. Another factor may be the improved increment which has been reported for vast forest areas in central Europe (Kenk *et al.*, 1991). Elevated rates of N deposition and intensified forest management (e.g. stronger thinnings) may be important factors in this respect. It should be considered that a better growth is also associated with a higher Mg requirement.

3.4. Conclusions and implications for forest management

If forest soils continue to acidify and the levels of Mg supply decrease still further, symptoms of acute Mg deficiency will occur more widely and more frequently. The complex nature of Mg deficiency necessitates a versatile strategy for nutritional management (Seitschek, 1990; Evers and Hüttl, 1990). All measures of mitigation and/or restabilization must aim at a better Mg supply in the soil, improved soil chemical conditions including soil solution, and an increased integration of subsoil horizons into the nutrient cycle. With respect to atmospheric deposition not only protons and associated mobile anions but also the variable input of Mg from atmospheric sources must be considered. Another important aspect for the prognosis of Mg supply is the ongoing progress in the availability and nutrition of N which is both due to elevated rates of N deposition and a changed forest management (improved thinning, hydromelioration etc.).

In the short term, acute Mg deficiency can efficiently be mitigated by amending various forms of Mg. This recommendation is confirmed by the fact that the application of additional Mg leads to a rapid disappearance of the specific symptoms and that stands which had received Mg prior to the evolution of acute deficiency are more vigorous and resistant against environmental impacts than the unfertilized stands (Evers and Hüttl, 1990). Hence, specific fertilization with Mg can be used as a preventative as well as a therapeutic tool. Furthermore, the amendment of Mg is unlike former amelioration practices, not aiming at increasing productivity but at improving the stability of the nutrient cycling in the ecosystem and the vitality of the trees. Nevertheless, forest fertilization measures bear site-specific ecological risks and have definite limitations. As a consequence, all operations must be designed on a site-specific basis in order to avoid negative side-effects.

By the application of readily soluble Mg salts ($MgCl_2$, $MgSO_4$), a relatively quick response of the Mg nutrition of the trees can be expected (Hüttl, 1991). Furthermore, a considerable amount of the fertilized Mg is retained in the rooting zone by cation adsorption processes (Feger, 1992a). Due to the quick migration and adsorption of Mg^{2+} ions in the mineral soil, the growth of fine root biomass in the mineral soil is enhanced (Raspe, 1992). On the other hand, the introduction of a neutral salt into the soil solution leads to cation exchange processes which are reflected in an initial decrease in soil solution pH and a mobilization of inorganic Al species. But this 'acidification pulse' is only a temporary phenomenon and does not negatively affect fine roots and mycorrhiza (Haug and Feger, 1990). A greater risk is represented by the fact that a distinct part of exchanged Al and acidity may be transported into the ground water and stream water, notably if neutralization capacity in deeper compartments of the mineral soil and the bedrock aquifer is limited. Temporary acidification of the drainage water is also likely to occur in watersheds where the water follows shallow lateral pathways (Feger, 1992a; Prietzel and Feger, 1996).

Liming of forest soils has become a common countermeasure to acid rain effects. In contrast to former ameliorative liming practices, the recent liming activities aim to buffer protons originating from deposition already at the soil surface (Gussone, 1987; Evers and Hüttl, 1990). In recent years, dolomitic materials have increasingly been applied in order to simultaneously improve the nutritional level of Mg in the forest ecosystem. The assessment of sites which had been limed in forestry practice during the 1960s and 1970s revealed that the application of Mg-containing limestone materials has a beneficial and sustained effect on tree vitality and nutrition (Aldinger, 1987). However this study demonstrated too, that due to the slow dissolution of carbonate materials liming effects in terms of soil melioration are normally restricted to the forest floor and the uppermost mineral soil, at least for a couple of years after the application. The application of slowly dissolving carbonates to the forest floor results in sharp chemical gradients stimulating microbial activity in the zone of distinctly higher pH values. Accordingly, the transformation of organic matter is enhanced which is often indicated in the higher concentrations and changed functional properties of dissolved organic matter in the soil solution (Göttlein et al., 1991). This may increase the mobility of heavy metals (Schierl and Kreutzer, 1991) which in turn are transported in higher rates from the O-horizon into the mineral soil. Another common observation is a drastic shift in N transformations in the top soil after liming. Notably nitrification is strongly enhanced whereas the rates of net N mineralization are not necessarily higher (Lang, 1986; Wölfelschneider, 1994). Accordingly, the risk of NO_3^- leaching and combined effects on the chemistry of other elements will increase. However, there is no clear pattern of elevated NO_3^- concentrations after liming operations. But it appears that liming induces excess nitrification and subsequent leaching of NO_3^- at sites with high N availability and/or elevated N deposition rates (Sauter and Meiwes, 1990). The stimulation of microbial nutrient release from the O-horizon has also appeared to affect the distribution of fine roots which tend to be more shallow after soil liming (Murach and

Schünemann, 1985; Schaaf and Zech, 1991; Raspe *et al.*, 1994). The susceptibility of limed stands toward drought stress and wind cast may therefore be increased, at least during the years subsequent to application.

Generally, the strategy of using readily soluble Mg-salts vs. carbonate forms should not merely be based on the soil chemical status (pH, base saturation) but should also integrate biotic aspects including microbial turnover and fine root distribution, as well as potential risks for drainage water quality. To avoid triggering of an excess nitrification and NO_3^- leaching, the application of Mg-carbonates should be restricted to N-limited sites with unfavourable soil biological conditions (inactive humus forms: raw-humus type) (Rehfuess, 1995). In this context, it should be noted that the number of N-limited sites will likely be reduced due to the recent levels of N deposition rates.

In the long term, silviculture can substantially improve Mg nutrition by minimizing the rate of soil-internal formation of protons and mobile anions. This will decrease both losses of Mg with the seepage water and elevated concentrations of inorganic Al species. Monocultural forestry which has mainly been focused on Norway spruce as a main species has proved to markedly enhance the rate of natural soil acidification (Feger, 1993). Besides the effect that, in evergreen conifer stands, the scavenging of acid components from the atmosphere ('dry' deposition) is much higher than that in deciduous stands, the main reason is a spatial and temporal discoupling of biotic accumulation and mineralization of elements. Such discoupling processes favor the internal production of mobile anions, notably of NO_3^-. Excess mineralization and nitrification can be minimized by avoiding forest floor accumulation and 'drastic' forest management operations (e.g. clear-cutting, insufficient thinning). A tighter and more efficient element cycling can be achieved by establishing vertically structured stands with a high percentage of deep-rooting species In this respect, deciduous species that produce easily decomposable litter should be favored. Such stands do not undergo a period of forest floor accumulation and nutrient release from litter mineralization is more continuous. The integration of subsoil horizons into the nutrient cycling increases Mg uptake from mineral sources and will therefore minimize drainage losses due to temporal and/or spatial discoupling processes. Furthermore, such stands are less susceptible to unfavorable climatic conditions such as summer drought.

In addition, it might be beneficial to support the silvicultural transition from conifer plantations to structured mixed stands which have a definitely higher nutrient requirement with amendments of critical nutrients. In Germany, forest management has recently launched related programs on a large scale (e.g. Seitschek, 1990). In the long-term perspective, the management of a balanced and harmonized tree nutrition will therefore be a key factor, notably at sites with a continuously elevated N availability.

Finally, the harvest of forest biomass should be restricted to the stemwood component. The removal of additional tree components does not lead to a self-sustained system with respect to Mg supply and necessitates a permanent amendment of this and other nutrients.

96

3.5. References

Abrahamsen G. 1984. Effects of acidic deposition on forest soil and vegetation. Phil. Trans. R. Soc. Lond. B. 305, 369–382.

Abrahamsen G. 1993. Long-term acidification experiments in forest stands. In: Experimental Manipulations of Biota and Biogeochemical Cycling in Ecosystems. Eds. L Rasmussen, T Brydges, P Mathy. pp. 192–194. Commission of the European Communities Ecosystems Research Report 4.

Aldinger E. 1987. Elementgehalte im Boden und Nadeln verschieden stark geschädigter Fichten-Tannen-Bestände auf Praxiskalkungsflächen im Buntsandstein. Freiburger Bodenkundl. Abh. 19, 266.

April R, Newton R. 1992. Mineralogy and weathering. In: Atmospheric Deposition and Forest Nutrient Cycling. Eds. DW Johnson, SE Lindberg. pp. 378–425. Springer-Verlag.

Anderson JM, Inneson P, Huish SA. 1983. Nitrogen and cation mobilization by soil fauna feeding on leaf litter and soil organic matter from deciduous woodlands. Soil Biol. Biochem. 15, 463–467.

Brown KA, Roberts TM. 1988. Effects of ozone on foliar leaching in Norway spruce (*Picea abies* (L.) Karst.): confounding factors due to NO_x production during ozone generation. Environ. Pollut. 55, 55–73.

Butzke H. 1981. Versauern unsere Wälder? Forst. Holzwirt. 36, 542–548.

Cole DW, Rapp M. 1981. Elemental cycling in forest ecosystems. In: Dynamic properties of forest ecosystems. Ed. DE Reichle. pp. 341–410. Cambridge University Press, Cambridge.

Cosby BJ, Hornberger GM, Galloway JN, Wright RF. 1985. Modeling the effects of acid deposition: assessment of a lumped-parameter model of soil water and streamwater chemistry. Water Resour. Res. 21, 51–63.

Dultz S. 1993. Verwitterungsbilanzen an sauren Waldböden aus Geschiebesand. PhD Thesis, Faculty of Geosciences, University of Hannover (Germany), 96 p. + 35 tables.

Evers FH, Hüttl RF. 1990. A new fertilization strategy in declining forests. Water Air Soil Pollut. 54, 495–508.

Evers FH, Hoyer V, Kottke I, Quian M, Schnepf M, Seufert G. 1993. Schadgasausschluß-Experiment bei Fichten am Edelmannshof – Grundlegende Messungen und Stoffbilanz. KfK/PEF-Berichte. 104, 127–153.

Feger KH. 1992a. Bilanzierung von Stofflüssen in Mg-gedüngten Fichtenökosystemen im Schwarzwald. In: Magnesiummangel in mitteleuropäischen Waldökosysemen. Eds. G Glatzel, R Jandl, M Sieghardt, H Hager. pp. 88–101. Forstliche Schriftenreihe Universität für Bodenkultur Wien 5.

Feger KH. 1992b. Nitrogen cycling in two Norway spruce (*Picea abies*) ecosystems and effects of a $(NH_4)_2SO_4$ addition. Water Air Soil Pollut. 61, 295–307.

Feger KH. 1993. Bedeutung von ökosysteminternen Umsätzen und Nutzungseingriffen für den Stoffhaushalt von Waldlandschaften. Freiburger Bodenkundl. Abh. 31, 237 p.

Feger KH, Raspe S, Zimmermann L. 1996. Boden- und ernährungskundliche Untersuchungen zum Schadtyp "montane Nadelvergilbung" bei Fichte. Verh. Gesellsch. Ökologie 26, 53–60.

Fiedler HJ, Hofmann H. 1985. Ältere und neuere Messungen zur Bodenazidität in Fichtenbeständen des Erzgebirges. Berichte Jahrestagung Agrarwiss. Ges. der DDR, 20. Nov. 1985 in Dresden, 64–97.

Georgii HW. 1985. Chemische Reaktion von Gasen und Aerosolen in Regen- und Wolkentropfen und ihre feuchte und nasse Deposition. In: Atmosphärische Spurenstoffe und ihr physiaklisch-chemisches Verhalten. Eds. KH Becker, J Löbel. pp. 129–152. Springer-Verlag, Berlin.

Glatzel G. 1991;. The impact of historic land use and modern forestry on nutrient relations of Central European forest ecosystems. Fertil. Res. 27, 1–8.

Gonzalez-Cascon MR, Alcubilla M, Rehfuess KE. 1990. Wirkungen von Magnesium- und Calcium-Sulfate und -Carbonat auf Sproß- und Wurzelentwicklung junger Weißtannen (*Abies alba* Mill.) im Topfversuch mit sauren Böden. Allg. Forst. u. Jagdz. 161, 21–28.

Göttlein A, Kreutzer K. 1991. Der Standort Höglwald im Vergleich zu anderen ökologischen Fallstudien. In: Ökosystemforschung Höglwald. Eds. K Kreutzer, A. Göttlein. pp. 22–29. Forstw. Forschungen 39, Munich (Germany).

Göttlein A, Kreutzer K, Schierl R. 1991. Beiträge zur Charakterisierung organischer Stoffe in wäßrigen Bodenextrakten unter dem Einfluß von saurer Beregnung und Kalkung. In: Ökosystemforschung Höglwald. Eds. K Kreutzer, A Göttlein. pp. 212–221. Forstw. Forschungen. 39, Munich (Germany).

Grimm R, Rehfuess KE. 1986. Kurzfristige Veränderungen von Bodenreaktion und Kationen-austauschereigenschaften in einem Meliorationsversuch zu Kiefer auf Podsol-Pseudogley in der Oberpfalz. Allg. Forst. u. Jagdztg. 147, 205–213.

Gussone HA. 1987. Kompensationskalkungen und die Anwendung von Düngemitteln im Walde. Forst. Holzwirt. 42, 158–163.

Hallbäcken L, Tamm CO. 1986. Changes in soil acidity from 1927 to 1982–1984 in a forest area of South-West Sweden. Scand. J. For. Res. 1, 219–239.

Haug I, Feger KH. 1990. Effects of fertilization with $MgSO_4$ and $(NH_4)_2SO_4$ on soil solution chemistry, mycorrhiza and nutrient content of fine roots in a Norway spruce stand. Water, Air, and Soil Pollut. 54, 453–467.

Hauhs M, Rost-Siebert K, Raben G, Paces T, Vigerust B. 1989. 5. Summary of European data. In: The Role of Nitrogen in the Acidification of Soils and Surface Waters. Eds. JL Malanchuk, J Nilsson. pp. 1–37. Nordic Council of Ministers, Miljörapport 1989/10, Copenhagen.

Hedin LO, Granat L, Likens GE et al. 1994. Steep declines in atmospheric base cations in regions of Europe and North America. Nature. 367, 351–354.

Hildebrand EE. 1986. Zustand und Entwicklung der Austauschereigenschaften von Mineralböden aus Standorten mit erkrankten Waldbeständen. Forstw. Cbl. 105, 60–76.

Hildebrand EE. 1990. Die Bedeutung der Bodenstruktur für die Waldernährung, dargestellt am Beispiel des Kaliums. Forstw. Cbl. 109, 2–12.

Hildebrand EE, Hochstein E. 1992. Stand und Entwicklung der Stoffeinträge in Waldbestände von Baden-Württemberg. Allg. Forst.u. Jagd-Ztg. 163, 21–26.

Hofmann-Schielle C. 1993. Silikatbestand und Verwitterungssimulation in den Hochlagen des Inneren Bayerischen Waldes. Z. Pflanzenernähr. Bodenk. 156, 333–339.

Hüttl RF. 1991. Die Nährelementversorgung geschädigter Wälder in Europa und Nordamerika. Freiburger Bodenkundl. Abh. 28, 440 p.

Johnson DW. 1992. Base cation distribution and cycling. In: Atmospheric Deposition and Forest Nutrient Cycling. Eds. DW Johnson, SE Lindberg. pp. 275–340. Springer-Verlag, Berlin.

Jorns A, Hecht-Buchholz C. 1985. Aluminiuminduzierter Magnesium- und Calciummangel im Laborversuch bei Fichtensämlingen. Allg. Forstz. 40, 1248–1252.

Kandler O, Miller W, Ostner R. 1987. Dynamik der akuten Vergilbung bei Fichte. Allg. Forstz. 42, 715–723.

Kaupenjohann M, Hantschel R. 1989. Nährstoffreisetzung aus homogenen und in situ Bodenproben: Bedeutung für die Waldernährung und Gewässerserauerung. Kali-Briefe (Büntehof). 19, 557–572.

Kenk G, Spiecker H, Diener G. 1991. Referenzdaten zum Waldwachstum. KfK/PEF-Berichte 82, Kernforschungszentrum Karlsruhe (Germany), 59 p.

Keyes MR, Grier CC. 1981. Above- and below-ground net production in 40-yr-old Douglas-fir stands on low and high productivity sites. Can. J. For. Res. 11, 599–605.

Kimmins JP. 1987. Forest Ecology. Macmillan, 531 p.

Klimo E, Kulhavy J. 1984. Acidification of forest soils in the region of the Moravskoslezské Beskydy Mountains. In: Proceedings of Symposium on Air Pollution and Stability of Coniferous Forest Ecosystems Ostravica (Czechoslovakia), pp. 93–98.

Krause GHM, Jung KD, Prinz B. 1983. Neuere Untersuchungen zur Aufklärung immissionsbedingter Waldschäden. VDI-Berichte 500, 257–266.

Kreutzer K. 1979. Ökologische Fragen zur Vollbaumernte. Forstw. Cbl. 98, 298–308.

Kreutzer K. 1989. Änderungen im Stickstoffhaushalt der Wälder durch anthropogen verursachten Auswirkungen auf die Qualität des Sickerwassers. DVWK-Mitteilungen 17, 121–132.

Kreutzer K, Göttlein A, Pröbstle P. 1991. Auswirkungen von saurer Beregnung auf den Bodenchemismus in einem Fichtenaltbestand (Picea abies [L.] Karst.) In: Ökosystemforschung Höglwald. Eds. K Kreutzer, A. Göttlein. pp. 174–186. Forstw. Forschungen 39, Munich (Germany).

Kuhn AJ. 1993. Mikrosonden-Analysen zur Ionenaufnahme in Fichten (Picea abies [L.] Karst.). Berichte des Forschungszentrums Jülich 2744, KFA Jülich (Germany), 204 p.

Lang E. 1986. Heterotrophe und autotrophe Nitrifikation untersucht an Bodenproben von drei Buchenstandorten. Göttinger Bodenkundl. Ber. 89, 199 p.

Lange OL, Zellner H, Gebel J, Schramel P, Kostner B, Czygan FC. 1987. Photosynthetic capacity, chloroplast pigments, and mineral content of the previous year's spruce needles with and without the new flush: analysis of the forest decline phenomenon of needle bleaching. Oecologia 73, 351–357.

Lindberg SE, Lovett GM, Schaefer DA, Bredemeier M. 1988. Coarse aerosol deposition velocities and surface-to-canopy scaling factors from forest canopy throughfall. J. Aerosol Sci. 19, 1187–1190.

Liu JC, Trüby P. 1989. Bodenanalytische Diagnose von K- und Mg-Mangel in Fichtenbeständen (Picea abies) Karst.). Z. Pflanzenern. Bodenk. 152, 307–311.

Lochman V. 1981. Changes in soil ecological conditions with special respect to soil chemical status in forests under the impact of industrial air pollution (in Czech). Lesnictvi 27, 699–714.

Löchelt S, Moosmayer HU, Rehfuess KE. 1993. Bestimmung der genetischen Konstitution von Fichten

98

(*Picea abies* [L.] Karst.) mit unterschiedlich ausgeprägten Vergilbungsgraden aus mehreren Waldgebieten. Forstw. Cbl. 113, 236–244.

Lovett GM, Schaefer DA. 1992. Canopy interactions of Ca^{2+}, Mg^{2+}, and K^+. In: Atmospheric Deposition and Forest Nutrient Cycling. Eds. DW Johnson, SE Lindberg. pp. 253–275. Springer-Verlag, Berlin.

Lux H, Börtitz S. 1990. Das ökologische Meßfeld der Sektion Forstwirtschaft der TU Dresden: XI. Untersuchungen zur atmosphärischen Deposition (1985 bis 1989). Wiss. Z. Techn. Univers. Dresden 39, 149–153.

Makkonen-Spiecker K, Evers FH. 1993. Untersuchungen zur Reaktionsweise junger Klonfichten (*Picea abies* [L.] Karst.) auf Trockenstreß und Magnesiummangel. KfK/PEF-Berichte 114, Kernforschungszentrum Karlsruhe (Germany), 81 p.

Manderscheid B, Matzner E, Meiwes KJ, Xu Y. 1995. Long-term development of element budgets in a Norway spruce (*Picea abies* [L.] Karst.) forest of the German Solling site. Water Air Soil Pollut. 79, 3–18.

Marschner H. 1989. Effect of soil acidification on root growth, nutrient and water uptake. In: Internationaler Kongreß Waldschadenforschung: Wissenstand und Perspektiven, Friedrichshafen am Bodensee, 2.–6. Oktober 1989, Proceedings Vol. I, Ed. B Ulrich. pp. 381–404.

Matzner E. 1988. Der Stoffumsatz zweier Waldökosysteme im Solling. Berichte des Forschungszentrums Waldökosysteme/Waldsterben Reihe A 40, Göttingen (Germany). 217 p.

Mazzarino MJ, Heinrichs H, Fölster H. 1983. Holocene versus accelerated actual proton consumption in German forest soils. In: Effects of Accumulation of Air Pollutants in Forest Ecosystems. Eds. B Ulrich, J Pankrath. pp. 113–123. D. Reidel Publishers, Dordrecht.

Mengel K, Lutz HJ, Breininger MT. 1987. Auswaschung von Nährstoffen durch sauren Nebel aus jungen intakten Fichten (*Picea abies*). Z. Pflanzenernähr. Bodenk. 150, 61–68.

Millar CS. 1974. Decomposition of coniferous leaf litter. In: Biology of Plant Litter Decomposition. Eds. CH Dickinson, GJF Pugh. Vol. I, pp. 105–128. Academic Press, London and New York.

Münch D, Feger KH, Zöttl HW. 1990. Nadelvergilbung, Elementverteilung und Wachstum von Fichten eines Hochlagenstandortes im Südschwarzwald. Allg. Forst. u. Jagdz. 161, 210–217.

Murach D, Schünemann E. 1985. Die Reaktion der Feinwurzeln von Fichten auf Kalkungsmaßnahmen. Allg. Forstz. 40, 1151–1154.

Nebe W. 1991. Veränderung der Stickstoff- und Magnesiumversorgung immissionsbelasteter älterer Fichtenbestände in ostdeutschen Mittelgebirgen. Forstw. Cbl. 119, 4–12.

Nebe W, Herrmann UJ. 1987. Das ökologische Meßfeld der Sektion Forstwirtschaft der TU Dresden: VI. Zur Verteilung der Nährelemente in der oberirdischen Dendromasse eines 100jährigen Fichtenbaumholzes. Wiss. Z. Techn. Univers. Dresden. 36, 235–241.

Parmentier G, Remacle J. 1981. Production de litière et dynamisme du retour au sol des éléments minéraux par l'intermédiaire des feuilles de hêtre et des aiguilles d'épicea en Haute Ardenne. Rev. Ecol. Biol. Sol. 18, 159–177.

Prietzel J, Feger KH. 1996. Dynamik von Aluminium und ökotoxischen Al-Bindungsformen in kleinen Fließgewässern nach Forstdüngung mit sulfatischen Magnesiumsalzen. Vom Wasser 87, 387–408.

Ragsdale HL, Lindberg SE, Lovett GM, Schaefer DA. 1992. Atmospheric deposition and throughfall fluxes of base cations. Eds. DW Johnson, SE Lindberg. pp. 235–253. Springer-Verlag, Berlin.

Raspe S. 1992. Biomasse und Mineralstoffgehalte der Wurzeln von Fichtenbeständen (*Picea abies* Karst.) des Schwarzwaldes und Veränderungen nach Düngung. Freiburger Bodenkundl. Abh. 29, 197 p.

Raspe S, Feger KH, Zöttl HW. 1994. Projekt ARINUS: VIII. Feinwurzelverteilung und -ernährung nach experimenteller Düngung. KfK/PEF-Berichte 117, 13–27.

Rehfuess KE. 1987. Perceptions on forest diseases in Central Europe. Forestry 60, 1–11.

Rehfuess KE. 1988. Übersicht über die bodenkundliche Forschung im Zusammenhang mit den neuartigen Waldschäden. KfK/PEF-Berichte 35(1), 1–26.

Rehfuess KE. 1995. Gefährdung der Wälder in Mitteleuropa durch Luftschadstoffe und Möglichkeiten der Revitalisierung durch Düngung. Ber. d. Reinh.-Tüxen-Ges. (Hannover) 7, 141–156.

Reuss JO, Johnson DW. 1986. Acid deposition and the acidification of soils and waters. Ecol. Stud. 59, 119 p. Springer-Verlag, Berlin.

Richter DD, Johnson DW, Dai KH. 1992. Cation exchange reactions in acid forest soils: Effects of atmospheric pollutant deposition. In: Atmospheric Deposition and Forest Nutrient Cycling. Eds. DW Johnson, SE Lindberg. pp. 341–358. Springer-Verlag, Berlin.

Roberts TM, Skeffington RA, Blank LW. 1989. Causes of type 1 spruce decline in Europe. Forestry 62, 179–222.

Sauter U, Meiwes KJ. 1990. Auswirkungen der Kalkung auf den Stoffaustrag aus Waldökosystemen mit dem Sickerwasser. Forst. Holzwirt 45, 605–610.

Schaaf W, Zech W. 1991. Einfluß unterschiedlicher Löslichkeit von Düngern. Allg. Forstz. 46, 766–768.

Scherbatskoy T, Klein RM. 1983. Response of spruce and birch foliage to leaching by acidic mists. J. Environ. Qual. 12, 189–193.

Schier GA. 1987. Throughfall chemistry in a red maple provenance plantation sprayed with 'acid rain'. Can. J. For. Res. 17, 660–665.

Schierl R, Kreutzer K. 1991. Einfluß von saurer Beregnung und Kalkung auf die Schwermetalldynamik im Högwaldexperiment. In: Ökosystemforschung Högwald. Eds. K Kreutzer, A. Göttlein. pp. 204–211. Forstw. Forschungen. 39, Munich (Germany).

Schreiner A, Wimmenauer W. 1979. Geologisch-tektonischer Aufbau und Gesteinsverbreitung. Mitt. Dtsch. Bodenkundl. Ges. 28, 23–39.

Seitschek O. 1990. Waldbauliche Maßnahmen in immissionsbelasteten Wäldern. In: Neuartige Waldschäden – Erkenntnisse und Folgerungen. Ed. J Jositz. pp. 183–202. Berichte und Studien der Hanns-Seidel-Stiftung e.V. München 56.

Staaf H, Berg B. 1982. Accumulation and release of plant nutrients in decomposing Scots pine needle litter. Long-term decomposition in a Scots pine forest. II. Can. J. Bot. 60, 1561–1568.

Stahr K, Feger KH, Zarei M, Papenfuß KH. 1983. Estimation of weathering rates in small catchments on Bärhalde granite (Black Forest, SW Germany). In: Clays controlling the Environment. Eds.: CJ Churchman, Fitzpatrick RW, Eggleton RA. Proc. 10th Inter. Clays Conference, Adelaide, Australia, July 18–23, 1993, 494–498.

Tamm CO. 1991. Nitrogen in terrestrial ecosystems. Questions of productivity, vegetational changes, and ecosystem stability. Ecol. Stud. 81, 115 p. Springer-Verlag, Berlin.

Tukey JB Jr. 1980. Some effects of rain and mists on plants, with implications for acid deposition. In: Effects of Acid Precipitation on Terrestrial Ecosystems. Eds. TC Hutchinson, M Havas. pp. 141–149. Plenum, New York.

Ulrich B. 1983. Interaction of forest canopies with atmospheric constituents: SO_2, alkali and earthalkali cations and chloride. In: Effects of Accumulation of Air Pollutants in Forest Ecosystems. Eds. B Ulrich, J Pankrath. pp. 1–29. D. Reidel Publishing Company.

Ulrich B. 1991. An ecosystem approach to soil acidification. In: Soil Acidity. Eds. B Ulrich, ME Sumner. pp. 28–79. Springer-Verlag, Berlin.

Umweltbundesamt (Ed.) 1994. Daten zur Umwelt 1992/93. Erich Schmidt Verlag, Berlin, 688 p.

van Breemen N, Mulder J, van Grinsveen JJM. 1987. Impact of acid deposition to woodland soils in the Netherlands. II. Nitrogen transformations. Soil Sci. Soc. Am. J. 51, 1634–1640.

van Miegroet H, Cole DW. 1984. The impact of nitrification on soil acidification and cation leaching in a Red Alder ecosystem. J. Environ. Qual. 13, 586–590.

VDI Kommission Reinhaltung der Luft (Ed.) 1987. Acid precipitation: Formation and impact to terrestrial ecosystems. VDI-Kommission, Düsseldorf (Germany), 281 p.

Vitousek PM, Gosz JR, Grier CC, Melillo JM, Reiners WA. 1982. A comparative analysis of potential nitrification and nitrate mobility in forest ecosystems. Ecol. Monogr. 52, 155–177.

Vogt KA, Grier CC, Meier CE, Keyes MR. 1983. Organic matter and nutrient dynamics in forest floors of young and mature *Abies amabilis* stands in western Washington, as affected by fine-root input. Ecol. Monogr. 53, 139–157.

Will G. 1985. Nutrient deficiency and fertilizer use in New Zealand exotic forests. New Zealand Forest Service FRI Bull. 97.

Wittmann O, Fetzer KD. 1982. Aktuelle Bodenversauerung in Bayern. Materialien Bayer. Staatsmin. f. Landesentwicklung u. Umweltfragen 20.

Wölfelschneider A. 1994. Einflußgrößen der Stickstoff- und Schwefel-Mineralisierung auf unterschiedlich behandelten Fichtenstandorten im Südschwarzwald. Freiburger Bodenkundl. Abh. 34, 191 p.

Wood T, Bormann GH. 1975. Increases in foliar leaching caused by acidification of an artificial mist. Ambio 4, 169–171.

Zarei M, Stahr K, Papenfuß KH. 1992. Veränderungen der gesteinsbildenden Minerale und des Elementbestandes von Böden der Standorte Schluchsee und Villingen. KfK/PEF-Berichte 94, 235–250.

Zöttl HW, Hüttl R. 1985. Schadsymptome und Ernährungszustand von Fichtenbeständen im süddeutschen Alpenvorland. Allg. Forstz. 40, 197–199.

Zöttl HW, Mies E. 1983. Die Fichtenerkrankung in den Hochlagen des Südschwarzwaldes. Allg. Forst- u. Jagdz. 154, 110–114.

Zöttl HW, Stahr K, Keilen K. 1977. Bodenentwicklung und Standortseigenschaften im Gebiet des Bärhaldegranites (südlicher Hochschwarzwald). Allg. Forst. u. Jagdz. 148, 185–196.

4
Tree physiology

S. SLOVIK

Tree physiology is defined as 'physiology of woody plants'. It focuses on physiological questions of fruit production and grafting in orchards, of forest botany and of growing ornamental trees and shrubs. This part of plant science recently became most important in the context of 'forest decline' research since the early 1980s in central Europe. This is reflected by the recent foundation of new periodicals like *Trees* and *Tree Physiology*. Still, tree physiology is not a novel science. Theophrastos Eresios (372–287BC), the most famous disciple and successor of Aristotle at the peripatetic school of Athens and the most important botanist of antiquity and the Middle Ages, studied wild trees, the progagation of trees, timber wood, undershrubs, resin production, frost hardiness of trees, oil production, juices of plants, etc.

Since there are plant families (e.g. *Fabaceae*, *Rosaceae*) which cover both herbaceous and woody plants, fundamental physiological differences may hardly be expected between both types of cormus habitus, especially at the level of cell physiology. Knowledge of woody plants is sparse compared with herbaceous plants, which are most important in human nutrition (cereals, vegetables etc). Herbaceous plants can be grown and handled much more easily for physiological experiments. The lack of knowledge in tree physiology is still most evident for forest trees. Table 1 shows that recent research on forest trees is focused on only a few tree species. Most of the knowledge which has accumulated is for Norway spruce (*Picea abies* (L.) Karst.), which is the economically most important woody plant species in the northern hemisphere (Schmidt-Vogt, 1986).

This chapter summarizes important data and perspectives of the role of magnesium in plant physiology. The aspects of tree physiology focus on Norway spruce and on the ecophysiological fate of magnesium deficiency in the context of forest decline research.

4.1. Cell physiology and metabolism of magnesium

4.1.1. General survey

According to Aikawa (1981), magnesium was the key cation in prebiotic and early biotic evolution of life. The tendency of Mg^{2+} cations to form organic complexes (e.g. with porphyrins) and chelates (especially with phosphorus-containing metabolites and dicarboxylic acids) enabled Mg^{2+} to strengthen enzyme–substrate interactions and thereby to catalyze the formation and reduplication of organic

R. F. Hüttl & W. Schaaf (eds): Magnesium Deficiency in Forest Ecosystems, 101–214.

Table 1. Percentage of communications dealing with different tree species

Tree genus	Percentage of papers
Picea abies	54.4
Fagus sylvatica	17.6
Pinus sylvestris	11.2
Quercus spp.	4.0
Other tree genera	12.8

Data are based on the Proceedings of the 'International Kongress Waldschäden: Wissensstand und Perspektiven', Friedrischshafen 1989

matter. Still, all procaryotic and eucaryotic organisms demand Mg^{2+} as an essential key nutrient, which is directly involved in most known metabolic and regulatory pathways of cells as a cofactor. The role of Mg^{2+} is well known as the central cation of all the chlorophylls on earth, including procaryotic bacteriochlorophylls. Thus, chlorophyl synthesis and photosynthesis of autotrophic organisms strongly depend on the availability of magnesium. No other cation can substitute for Mg^{2+} in chlorophyls. Since the demand for Mg^{2+} to form (i) apoplasmic Mg^{2+}-pectinate in primary cell walls, (ii) Mg^{2+}-phytate (e.g. in caryopses of cereals), and (iii) rarely, vacuolar Mg^{2+}-oxalate is relatively small, more than half of the total Mg^{2+} content in leaves is usually available for the catalytic needs of plant metabolism. The water-soluble magnesium content of Norway spruce needles is a minimum of 70% of the total needle content (Hoeß, 1986). Most of this soluble nonstructural Mg^{2+} is reversibly bound to proteins (partly as an allosteric activator), or is chelated with numerous phosphorus-containing cytoplasmic compounds or divalent organic acids. In liver cells, less than 10% of the total Mg^{2+} appears to be 'free' (Veloso *et al.*, 1973). The equilibrium ion activity ($\gamma \leq 0.1$) of soluble 'free' Mg^{2+} cations in the cytoplasm, which results from association and dissociation reactions of Mg^{2+} with numerous cytosolic compounds, is for all tested eucaryotic cells about 1.0–1.3 mmol/L (Thaler, 1991). Most important for metabolism is the high 'affinity' of Mg^{2+} to all phosphorus-containing metabolites, e.g. to ATP. 'Wherever there is ATP, there is an obligatory need for magnesium' (Aikawa, 1981). All ATP-dependent phosphorylation reactions in any metabolic pathway consume the $Mg-ATP$ chelate as an obligatory substrate. The phosphorylation potential of ATP is dependent on the Mg^{2+} activity. Günther (1981) summarized 300 enzyme reactions (and pathways) which essentially demand Mg^{2+} (see also Aikawa, 1981; Bergmann, 1988; Kirby and Mengel, 1976; Kiss, 1981; MacKintosh and MacKintosh, 1993; Marschner, 1986; Mengel, 1984; Michal 1978; Welte and Werner, 1959):

(i)   Energy metabolism (photophosphorylation, Calvin cycle, glycolysis, oxidative decarboxylation of pyruvate, respiratory phosphorylation, citrate cycle): Mg^{2+} is a cofactor of ATP (adenosine triphosphate), UTP (uridine triphosphate: polycondensation of sugars) and CTP (cytidine triphosphate: phospholipid synthesis);

(ii)   Assimilation metabolism (gluconeogenesis, polycondensation of sugars to sucrose, starch etc. (ADP- and UDP-glucose derivatives), fatty acid synthesis (acetyl-CoA:CO_2-ligase));

(iii) Amino acid synthesis (glutamine: glutamine synthase (ammonium assimilation); asparagine: asparagine synthetase; methionine: N^5-methyltetrahydrofolic acid–homocysteine–methyltransferase; threonine: homoserine kinase; histidine: ribosephosphate pyrophosphokinase; valine, leucine and isoleucine: 2-hydroxy-3-ketoacid lyase, 2-hydroxy-3-ketoacid reductoisomerase, dihydroxyacid dehydratase; tryptophan: anthranilate synthase; degradation of storage proteins: amino-peptidases;

(iv) Sulfur metabolism: glutathione synthesis (L-γ-glutamylcysteine synthase, glutathione synthase);

(v) Chloroplast pigments and isoprenoids (steroid, carotenoid, xanthophyll and phytol synthesis: mevalonate kinase, phosphomevalonate kinase, isopentenyl diphospho-Δ-isomerase, dimethylallyl transferase, geranyl transferase, farnesyl transferase, squalene synthase);

(vi) Membrane transport (H^+-pumping ATPases and pyrophosphatases, membrane 'stabilization' via Mg^{2+}-complex formation with phospholipids, synthesis of phospholipids, association of extrinsic membrane proteins with the membrane surface etc.);

(vii) Synthesis of purine nucleotides (six Mg^{2+}-dependent enzymes of the inosine acid synthesis pathway), or pyrimidine nucleotides (orotate phosphoribosyl transferase: this enzyme is synthesized only in the S-phase of the mitosis cycle; Mg^{2+} links both subunits of this enzyme) and synthesis of desoxyribosyl nucleotides (DNA-monomers: ribonucleotide reductase, thymidylate synthase);

(viii) Transcription and RNA synthesis (RNA nucleotidyl transferase), DNA replication (DNA nucleotidyl transferase), RNA turn over (polynucleotide phosphorylase);

(ix) Translation and protein synthesis (aminoacyl-t-RNA-synthesis, association of ribosome subunits, association of ribosomes at the rough endoplasmic reticulum rER, dissociation of synthesized proteins from ribosomes), posttranslational protein modifications and glycoprotein synthesis (uridine diphosphate glucosamine pyrophosphorylase);

(x) Regulation of metabolism via protein kinases and protein phosphatases;

(xi) Others: pollen germination, phytohormone (auxin) synthesis (shikimi acid pathway: anthranilate synthase), synthesis of enzyme cofactors: tetrahydrofolic acid: dihydrofolate synthase; NAD^+ and $NADP^+$ synthesis: chinolinate phosphoribosyl transferase, NAD^+ synthetase, NAD^+ kinase.

The enzyme cofactor activities of magnesium can be substituted in part of the mentioned metabolic pathways by manganese Mn^{2+} ions (see Bergmann, 1988; Marschner, 1986; Mengel, 1984; Welte and Werner, 1959). Mn^{2+} cations are usually not as affine and catalytically effective as Mg^{2+} cations. Additionally, the plant cell content of Mn^{2+} is lower than Mg^{2+} usually by one or two orders of magnitude. Partially, Mg^{2+} can be substituted in some reactions also by Zn^{2+}, Co^{2+} and Ca^{2+}, but all these cations are suboptimal compared with Mg^{2+}. Manganese is more effective than Mg^{2+} only in catalyzing some reactions of the citrate cycle (isocitrate dehydrogenase, α-ketoglutarate dehydrogenase). Nevertheless, Mn^{2+} can at least partially substitute for Mg^{2+} in plants growing on Mg^{2+}-depleted acidic soils which can supply Mn^{2+} (see Kaupenjohann, this volume). Quast (1981) found that Mg^{2+} deficiency symptoms of the apple cultivar Cox Orange occurred only if the leaf contents of Mg^{2+} and Mn^{2+} were both suboptimal. Fankhausen et al. (1976) reported that combined leaf fertilization with Mg^{2+} and Mn^{2+} synergistically avoided chlorotic and necrotic leaf spots and leaf loss in Cox Orange.

Phosphorus uptake from the soil and the physiologically effective metabolic use of phosphorus both depend on the Mg^{2+} supply of plants. It is possible to enhance the relative phosphorus supply via Mg^{2+} fertilization (Marschner, 1986; Mengel, 1984). Since many in-vivo catalytic reactions of Mg^{2+} are related to the cytoplasmic presence of rate-limiting $Mg-P$ complexes, it is possible to increase the cytosolic activity of these $Mg-P$ complexes by increasing either the cytoplasmic Mg^{2+} content or the cytoplasmic P content. If phosphorus is available, some Mg^{2+}-deficient plants do indeed increase the phosphorus content of their vegetative organs (Bergmann, 1988), apparently partly in order to compensate for lack of Mg^{2+}. Accordingly, Mg^{2+}-deficiency symptoms of plants occasionally resemble phosphorus deficiency symptoms if phosphorus supply is not luxurious. Anthocyane synthesis (rubinosis of leaves) and delay in the formation of reproductive organs are symptoms which can be induced in some plants by either Mg^{2+} deficiency or P deficiency.

4.1.2. Photosynthesis

Photosynthesis is defined here as assimilation and photoreduction of CO_2. Spruce trees use 40% of the annual CO_2 fixation of needles for growth demands (formation of dry matter); 60% of the assimilates are essentially consumed in mitochondrial respiration (Hager, 1975). Photosynthesis is a highly regulated process, which can be roughly divided into: (i) 'biophysical light reactions' at the chloroplast thylakoids, and (ii) 'biochemical dark reactions' in the Calvin cycle of the chloroplast stroma. Both essentially depend on sufficient supply of magnesium. Intact chloroplasts contain $0.4-1.0\,\mu mol\,Mg^{2+}\,(mg\,chlorophyl)^{-1}$ (Gimmler et al., 1974; Portis and Heldt, 1976), i.e. roughly 50% or more of the total Mg^{2+} of leaves is located in chloroplasts. Only about $10-20\%$ of the total Mg^{2+} content of leaves is bound to chlorophyl molecules. This fraction increases to about 30% in leaves of Mg^{2+}-deficient plants (Michael, 1941) which indicates that there is massive Mg^{2+}

depletion of the cytoplasm before finally also the chlorophyl content of the leaves is reduced.

There are four enzymes of the Calvin cycle, which demand stromal Mg^{2+} cations (Buchanan, 1980; Jensen and Bassham, 1967; Kelley et al., 1976; Lilley et al., 1974; Portis et al., 1977; Walker, 1976). The involvement of Mg^{2+} in regulating the Calvin cycle is not clear. Other existing regulatory factors of chloroplast enzymes (thioredoxin-dependent thiol-disulfide interchange, phosphorylation of thylakoid membrane proteins etc.) are not discussed here:

(i)   The CO_2-fixing ribulose-1,5-diphosphate carboxylase is the main enzyme protein of leaves (up to 50% of the total leaf protein). This enzyme is activated by high stromal pH values (dark: 7.5; light: 8.0), and by increasing stromal Mg^{2+} concentrations (dark: 1–3 mmol/L; light: 2–5 mmol/L; Portis, 1981). Under these conditions, the K_m value for CO_2 (usually 10–20 μmol/L) decreases. Hence, the affinity of ribulose-1,5-diphosphate carboxylase to dissolved stromal CO_2 rises. Additionally, V_{max} of this enzyme increases. Both effects lead to a potential enhancement of stromal CO_2 fixation. Since there are also measurements which show that the stromal Mg^{2+} increase is less than 10% compared with the stromal dark situation (Krause, 1977; Portis and Heldt, 1976), there is still doubt about whether Mg^{2+} is indeed involved in the regulation of the Calvin cycle (Buchanan, 1980).

(ii)   The second substrate of the ribulose-1,5-diphosphate carboxylase besides CO_2 is ribulose-1,5-diphosphate. Dependent on the stromal Mg^{2+}-ATP content, it is synthesized from D-ribulose-5-phosphate by the stromal enzyme phosphoribulo-kinase. After activation of photosynthesis in the light, phosphate is supplied from the cytosol and imported into the chloroplast stroma via the phosphate translocator. This is an intrinsic membrane protein of the inner chloroplast envelope, which also exports stromal assimilates (triose-phosphates) in exchange. Both increased phosphate P_i and magnesium supply, promote the photon quantum $h\nu$-dependent formation of Mg-ATP from ADP, P_i and Mg^{2+}:

$$ADP + P_i + Mg^{2+} + nh\nu \rightarrow Mg^{2+}\text{-ATP}$$

(iii) Two further enzymes of the Calvin cycle, fructose-1,6-diphosphatase and the analogous enzyme sedoheptulose-1,7-diphosphatase, are activated in the presence of increasing stromal Mg^{2+} concentrations (See Kelley et al., 1976; Walker, 1976; Woodrow et al., 1984).

After onset of light, there is, according to Mitchell's theory, light-dependent net transport of stromal protons across the thylakoids into the intermembrane space. Thus, stromal pH increases, and the pH in the intermembrane space decreases. With rising acidity in the intermembrane space the strength by which Mg^{2+} is bound to thylakoidal protein surfaces decreases. Charge compensation of stromal proton loss after the onset of illumination first mainly proceeds via efflux of Mg^{2+} cations from the intermembrane space into the stroma. This $2H^+/Mg^{2+}$ exchange rate was estimated to be about $20-100 \, nmol \, Mg^{2+} (mg \, chlorophyl)^{-1}$ (Barber *et al.*, 1974; Ben-Hayyim, 1978; Dilley and Vernon, 1965; Hind *et al.*, 1974). It is still an open question whether there is light-dependent Mg^{2+} transport across the chloroplast envelope in vivo from the cytosol into the stroma. In vitro, the Mg^{2+} permeability of the chloroplast envelope is too small to significantly contribute to stromal Mg^{2+} supply (Gimmler *et al.*, 1974; Pfluger, 1973; Portis and Heldt, 1976), but electro-physiological measurements with Mg^{2+}-sensitive microelectrodes reveal that the Mg^{2+} activity in the cytosol decreases in the light in vivo (Thaler, 1991). After depletion of the $2H^+/Mg^{2+}$ ion exchange capacity of thylakoids (e.g. in continuous light), proton transport into the intermembrane space of thylakoids is compensated more and more by accompanying anions instead (e.g. chloride). In contrast to mitochondrial cristae, the proton motive force generated across thylakoid membranes predominantly consists of the transmembrane ΔpH component. If this component of proton motive force is sufficiently high, protons recirculate back into the stroma passing the thylakoidal ATP–synthetase complex. This thylakoid enzyme consists of an intrinsic (F_o) and an extrinsic protein part (CF_1). Both parts of the ATP–synthetase complex are linked by electrostatic forces of Mg^{2+} ions. Part of the Mg^{2+} released from the thylakoidal intermembrane space after onset of light remains associated with the stromal face of the thylakoids (Ben-Hayyim, 1978). Of course, the ATP–synthetase additionally needs soluble stromal Mg^{2+} cations for $Mg^{2+}-ATP$ chelate synthesis, which is consumed in the Calvin cycle and resynthesized at thylakoids in the light.

4.1.3. Assimilation of sulfate

There is still a lack of information about the Mg^{2+} dependency of sulfate reduction. It is only known that both enzymes of glutathione biosynthesis, L-glutamylcysteine synthase and glutathione synthase, are strongly Mg^{2+} dependent (Michal, 1978). Glutathione GSH (Bergmann and Rennenberg, 1993; Rennenberg, 1984; Rennenberg and Brunold, 1994) is the main storage and transport compound of reduced sulfur in phloem and xylem sap of herbaceous and woody plants. As a shoot-to-root messenger, it is involved in the regulation of sulfate uptake from soil. Glutathione GSH is involved in oxygen radical (superoxide anions O_2^-) detoxi-fication in photosynthesis. GSH synthesis and turn-over is one important key reaction of ozone detoxification (De Kok and Stulen, 1993) and of reductive SO_2 detoxification (net GSH synthesis and GSH export from SO_2 burdened leaves via the phloem sap). More information about the suspected interdependency between Mg^{2+} deficiency and the magnitude of tolerance mechanisms of trees against the air pollutants, SO_2 and O_3 is urgently needed.

4.1.4. Enzyme modulation

There is evidence that Mg^{2+}-dependent enzyme protein phosphorylation controls numerous physiological processes in eukaryotic cells. Some of the following have already been confirmed also with plant cells (MacKintosh and MacKintosh, 1993 and personal communication): regulation of electron transport in thylakoids, regulation of mitosis and the cell cycle, regulation of fatty acid synthesis, formation of the nucleus membrane, involvement in amino acid and nucleotide metabolism, organization of microtubuli and the cytoskeleton, protein synthesis, RNA synthesis, phytoalexin formation, and synthesis of mevalonate (precursor of all isoprenoids). There are numerous different MgATP-dependent protein kinases in eukaryotic cells, which all belong to the same gene family, and there are four major families of protein phosphatases (PP1, PP2A, PP2B, PP2C) in eukaryotic cells, which differ in several biochemical features (MacKintosh and MacKintosh, 1993). All but PP2B, which is Ca dependent, have been identified also in plants. PP2C is strongly Mg^{2+} dependent. PP1 and PP2A are not dependent on metal cations (MacKintosh and Cohen, 1989; MacKintosh *et al.*, 1991). At least seven plant enzymes have been reported to be regulated by protein phosphorylation. Most interesting for tree physiologists and forest decline researchers are the following cytosolic leaf enzymes (MacKintosh and Cohen, 1989; MacKintosh *et al.*, 1991; MacKintosh and MacKintosh, 1993): nitrate reductase, sucrose phosphate synthase and phosphoenolpyruvate carboxylase. All these cytosolic enzymes are phosphorylated by MgATP-dependent protein kinases, and reversibly dephosphorylated by the soluble Mg-independent protein phosphatase PP2A. PP2A is also involved in all tested eukaryotic cells regulating the mitotic cell cycle. Since there is no detectable activity of PP2A in chloroplasts and in mitochondria, PP2A seems to be exclusively active in the cytosol.

Nitrate reductase   The nitrate reductase (NR) of leaves is a cytoplasmic NADH-dependent enzyme, which reduces nitrate to nitrite. Nitrite is then converted to ammonia by reduced ferredoxin-dependent nitrite reductase (NiR) in the chloroplast. The expression of NR in leaves is highly regulated by both light and external nitrate at the level of gene transcription and possibly protein turnover (Campbell, 1989). The catalytic activity of both enzymes is not directly dependent on Mg^{2+} in vitro. NiR is activated by Mn^{2+}. In trees, nitrate and nitrite reduction usually takes place mainly in root cells. Still, nitrate and nitrite reduction in leaves of trees is possible. Both enzymes are active in detoxifying air-borne nitrate and nitrite, which are produced after hydration of nitrous air pollutants taken up via leaf stomata:

$$2NO_2 + H_2O \rightarrow HNO_2 + HNO_3$$

Nitrate is phytotoxic (Wellburn, 1990), but NO_2-fumigated Norway spruce needles readily reduce air-borne NO_3^- and NO_2^- anions. There is no accumulation of nitrate in needles of spruce trees growing on well-fertilized soils in phytotron-like growth chambers (Kaiser *et al.*, 1993). Only fumigation with 500 ppb NO_2 (11 weeks), which is 30–60 times higher than annual means of NO_2 pollution in native spruce forests (see Figure 6), caused an accumulation of nitrate in the shoots of spruce seedlings (Tischner *et al.*, 1988) despite enhanced metabolic activities of the nitrate-assimilating enzymes. There is recent evidence that modulation of nitrate reductase activity in leaves requires the presence of Mg^{2+}. NR is reversibly activated upon illumination of leaves due to hydrolytic dephosphorylation of this enzyme by the protein phosphatase PP2A (MacKintosh and MacKintosh, 1993):

$$NR_{inactive} - P_n \rightarrow NR_{active} + nP_i$$

The cytosolic protein phosphatase PP2A is not Mg^{2+} dependent. There is inactivation of the NR in the dark by phosphorylation of the NR enzyme by a MgATP-dependent protein kinase (Kaiser et al., 1992; MacKintosh and MacKintosh, 1993). The NR kinase phosphorylates the NR at defined serine and threonine residues of the NR polypeptide:

$$NR_{active} + nMgATP^- \rightarrow NR_{active} - P_n + nADP^{3-} + nMg^{2+}$$

Under conditions of either Mg^{2+} depletion or inhibition of the protein phosphatase PP2A, the NR remains more active in the dark compared with the control (MacKintosh, personal communication). Since NO_2^- reduction in the chloroplasts is light dependent (but not the cytosolic NO_3^- reduction), there was indeed accumulation of phytotoxic nitrite in the dark under conditions of Mg^{2+} deficiency in vitro (MacKintosh, personal communication). Since there is an increase in the $N_{org} : Mg^{2+}$ ratio in spruce tissues in the Fichtelgebirge (Schulze et al., 1989), Mg^{2+} deficiency seems to disturb the regulation of nitrate assimilation in Norway spruce trees.

The product of the NiR is ammonium. The rapid and effective assimilation of ammonium cations by the glutamine synthase is also Mg^{2+} dependent:

$$glutamate^- + NH_4^+ + MgATP^- \rightarrow glutamine + ADP^{3-} + Mg^{2+} + P_i$$

Besides MgATP, free Mg^{2+} is necessary to stabilize protein association of the intermediate phosphoglutamyl anhydride. This fact may become important at Mg^{2+}-deficient sites with high NH_3 pollution or NH_4^+ deposition.

Sucrose phosphate synthase The key enzyme of sucrose synthesis in illuminated leaves is sucrose phosphate synthase (SPS), which is regulated largely in parallel with nitrate reductase (MacKintosh and MacKintosh, 1993). SPS in spinach leaves is converted from a phosphate-inhibited less-active phosphorylated form into a phosphate-insensitive more-active dephosphorylated form in the light. Presumably SPS and NR share not only the same protein phosphatase (PP2A), but also the same protein kinase (MacKintosh and MacKintosh, 1993). This synchronous regulation is apparently an important factor in keeping molar $C : N_{org}$ ratios in the phloem sap, and finally of the whole plant cormus, largely constant. A necessary supposition is sufficient Mg^{2+} supply. In fact, Mg^{2+}-deficient plants tend to accumulate sugars in their vacuoles (see below).

Phosphoenolpyruvate carboxylase The cytosolic enzyme phosphoenolpyruvate carboxylase (PEPcase) was shown to be dephosphorylated by PP2A, converting it from a malate-insensitive active into a malate-sensitive less-active form (MacKintosh and MacKintosh, 1993). The PEPcase is: (i) a key enzyme photosynthesis in CAM plants, (ii) in either plant the PEPcase is the cytosolic key enzyme of anaplerosis of the mitrochondrial citrate cycle (net supply of C-skeletons for net synthesis of amino acids, organic acids etc.), and (iii) involved in cellular pH-stat mechanisms. Under conditions of Mg^{2+} deficiency, the PEPcase is tentatively expected to be more and more dephosphorylated because there is a lack of MgATP for protein phosphorylations, and the malate-sensitive less-active state of the PEPcase should dominate in the cytosol. Thus, the synthesis rate of organic acids is expected to be small from these deductions. In fact, Mg^{2+} deficiency symptoms are usually accompanied by reduced contents of organic acids in leaf dry matter (Bergmann, 1988). A low level of PEPcase activity in Mg^{2+}-deficient plants also reduces the supply of C-skeletons for amino acids. Reduction of growth can be one consequence. The PEPcase is an important key enzyme in forest decline research.

4.1.5. pH-stat mechanisms

All enzyme reactions in vivo essentially depend on the physicochemical properties of their compartments, e.g. ionic strength, water potential, redox potential and the H_3O^+ potential (pH). Biochemical reactions, which keep compartmental pH values within (e.g. circadian) amplitudes constant, are so-called pH-stat mechanisms (Raven, 1985). The most important pH-stat mechanisms available to plants exposed to chronic pollution of potentially acidic air pollutants are (see also Slovik et al., 1992a,b):

(i)   Production of OH^- anions via assimilation of nitrate,

(ii)   Production of OH^- via decarboxylation of organic salts or amino acids,

(iii) Extrusion of H_3O^+ into the soil in exchange for cations, and

(iv)   Sequestration of H_3O^+ into leaf vacuoles together with air-borne anions (e.g. SO_4^{2-}).

All these pH-stat mechanisms require the presence of Mg^{2+}. The content of organic anions available for decarboxylation reactions is reduced in Mg^{2+}-deficient plants. Mechanisms (iii) and (iv) are endergonic transmembrane H_3O^+ transport processes, which are catalyzed by intrinsic membrane proteins: H^+-ATPases or H^+-pyrophosphatases (H-PPase). These enzymes consume MgATP or $MgPP_i$ as substrates. Transmembrane pH gradients are thoroughly regulated in vivo. Since all important pH-stat mechanisms depend on Mg^{2+}, it is expected that Mg^{2+}-deficient plants are less capable of neutralizing air-borne proton burden. This is shown in Figure 1 for hydroponically grown Mg^{2+}-deficient spinach leaves. Leaves are fed with pyranine, a fluorescent dye which records cytosolic pH values. After fumigation with even 5000 ppb SO_2 for 5 min, control plants are able to re-establish the original cytosolic pH value within 10 min after transient cytosolic acidification despite this extreme SO_2-burden, which does not occur in the field. Mg-deficient spinach leaves suffer from rapid cytosolic acidification. Resaturation of the original cytosolic pH was not

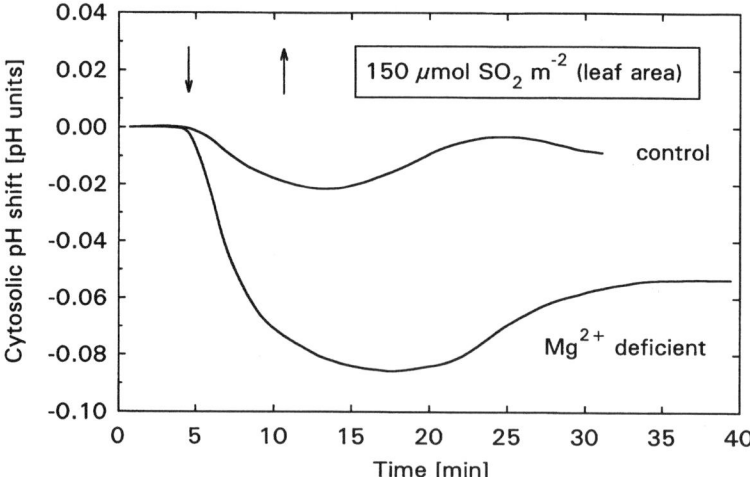

Figure 1. Cytosolic pH shift of Mg^{2+}-deficient and well-fertilized control leaves of spinach after fumigation with SO$_2$ for 5 min (arrows). Measurements were performed in the light (160 W m^{-2} red light, 500 ppm CO$_2$). The SO$_2$ concentration (about 5000 ppb) was chosen to achieve the same SO$_2$ dose of 150 μmol m^{-2} leaf area for both Mg^{2+}-deficient and control leaves. (After Yin, 1990)

observed. The mean rate of cytosolic pH-stat reactions in Mg-deficient leaves was only 46% of the control (Yin, 1990).

4.1.6. Magnesium deficiency symptoms

At many sites in Central Europe, Mg^{2+} deficiency symptoms are associated with forest decline symptoms (Hunger, 1964; Mies and Zöttl, 1985; Zech and Popp 1983). Most visual and latent Mg^{2+} deficiency symptoms can be interpreted on the basis of cell physiological Mg^{2+} demands (see below).

Leaf pigment synthesis Chloroplast pigment synthesis of both chlorophyl and carotenoids strongly depends on sufficient Mg^{2+} supply (Michael, 1941; Mothes and Baudisch, 1958). Figure 2 shows the dependency of the chlorophyl content on the Mg^{2+} content of Norway spruce needles in the Fichtelgebirge (NE Bavaria, Germany). Below 12 mmol Mg^{2+} (kg dw)$^{-1}$ needle chlorosis gradually develops. Deduced from indirect evidence, this observation was explained mainly as a consequence of reduced chloroplast pigment synthesis (Köstner, 1989; Köstner *et al.*, 1990; cf. Lange *et al.*, 1989a,b). Compared with green needles, the deficit in chlorophyl and carotenoid content of chlorotic needles was not due mainly to a degradation of pigments, but rather to reduced pigment resynthesis after depression of the pigment content in early spring. The ratios of the different chloroplast pigments (chl a : chl b : total carotenoids) remained surprisingly stable in Mg^{2+}-deficient chlorotic

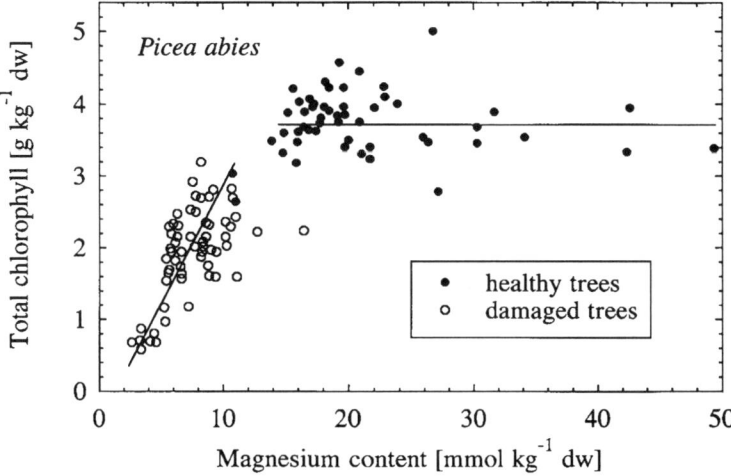

Figure 2. Total chlorophyl (a + b) content of 1 – 4-year-old needles from Mg^{2+}-deficient (damaged) and healthy Norway spruce trees in the Fichtelgebirge (NE Bavaria, Germany) measured in fall 1985, 1986 and 1987 in relation to the Mg^{2+} content (5 healthy and 5 deficient trees; dw = dry weight basis). After Köstner (1989); see Lange *et al.* (1989a,b)

needles in the Fichtelgebirge (15 ppb SO$_2$). Rabe and Kreeb (1980) found that after chronic SO$_2$ pollution, the chlorophyl a and chlorophyl b contents were reduced in a manner which keeps the chl a : chl b ratio constant. Only carotene (the biochemical precursor of the carotenoids) is more often depressed than are the epoxy-carotenoids or luteine (Köstner *et al.*, 1990). A typical feature of Mg^{2+} deficiency of *Picea abies*, *Pinus sylvestris* and *Pseudotsuga spp.* is loss of chloroplast pigments, which starts at the needle tips (German: 'Goldspitzigkeit'; Bergmann, 1988). This is most obvious in the sun crown of spruce trees. In the typical case, yellowing is confined to the older needle age classes, while the youngest growth remains green. In the Fichtelgebirge, the pigment content was stable in the winter season. Chlorosis of older needles usually starts in late spring, when the new growth is in first flush. In the summer, slow recovery of the pigment content starts in the Mg^{2+}-deficient older needle age classes, but there is no consistent increase in chlorophyl content from year to year as there is for needles which do not suffer from Mg^{2+} deficiency (Köstner *et al.*, 1990, cf. Lange *et al.*, 1989b). Additionally, spruce needle bleaching ('acute yellowing'; Kandler *et al.*, 1987) is observed mainly on sunlight-exposed upper needle surfaces (Baule, 1984; Buchner and Isermann, 1984; Buchner 1985; Fiedler *et al.*, 1984; Hüttl, 1987; Isermann, 1985; Lichtenthaler and Buschmann, 1984). Thus, visual Mg^{2+} deficiency symptoms are enhanced by illumination (Bergmann, 1988; Lange *et al.*, 1989a,b; Wild, 1988; Winter *et al.*, 1989). Also drought stress in warm and dry years (e.g. 1982 and 1983 in central Europe) increased the occurrence of needle chlorosis of spruce (Hüttl, 1987). In contrast to similar K$^+$-deficiency symptoms of needles, the very tip cells: (i) do not decline to a brown color as early, and (ii) show a sharp gradient between green and yellow mesophyll cells. The 'novel type forest decline', which developed after the end of the 1970s, is often characterized by mineral-deficiency symptoms, which can be roughly simplified for gymnosperms as follows: Mg^{2+} deficiency is characterized by yellow needle tips (German: 'Goldspitzigkeit'), K^+ deficiency by red tips (German 'Rotspitzigkeit'), and Zn^{2+} deficiency by stork's nest symptoms (German 'Storchennestbildung'). Visible needle symptoms have been associated with magnesium deficiencies on silicate soils, potassium deficiencies on limestone, and manganese deficiencies on dolomite (Abrahamson, 1980; Anonymous, 1986; Hauhs and Wright, 1986; cf. Schulze et al., 1989). Mg^{2+}-deficient deciduous trees exhibit leaf chlorosis and necrosis symptoms of the intercostal areoles formed by veins, but the edges and veins still remain green for some time (Bergmann, 1988).

Photosynthesis Since photosynthesis strongly depends on sufficient Mg^{2+} supply, reduction of CO_2 assimilation is often observed in Mg^{2+}-deficient plants (Keller, 1972; Lange et al., 1986; Lyr et al., 1967; Mengel, 1984). Beyschlag et al. (1987) compared needle photosynthesis rates of Mg^{2+}-deficient and Mg^{2+}-fertilized Norway spruce trees in the Fichtelgebirge. They found the CO_2 assimilation rate to be 20–30% higher in Mg^{2+}-fertilized spruce ($\approx 41\,mmol\,Mg^{2+}\,kg^{-1}\,dw$ of one-year-old needles) than that of needles of unfertilized control trees ($\approx 11\,mmol\,Mg^{2+}\,kg^{-1}\,dw$) at equal light intensities. Reduction of photosynthesis was not the consequence of reduced stomatal conductance of Mg^{2+}-deficient needles. The intercellular CO_2 concentration necessary to obtain similar CO_2 assimilation rates was higher in Mg^{2+}-deficient needles, i.e. the Calvin cycle was, other things being equal, not fully active compared with Mg^{2+}-fertilized spruce trees (Beyschlag et al., 1987). Consequently, Mg^{2+}-deficient needles showed reduced water use efficiency (WUE), which is defined as the net CO_2 photosynthesis rate per H_2O transpiration rate of leaves. Since growth of Norway spruce is more sensitive to water supply than that of other tree species (Schmidt-Vogt, 1986), reduced WUE results in reduced annual carbon gain if water supply is growth limiting for spruce trees growing on Mg^{2+}-depleted soils. There are independent data from other plant species which indicate that Mg^{2+} deficiency may cause wilting symptoms resembling K^+ deficiency (Bergmann, 1988). But there were only small and hardly significant differences in photosynthesis rates between bleached and healthy needles of unfertilized spruce trees in the field growing in the same stand in the Fichtelgebirge (Oren and Zimmermann, 1989). Apparently, reduction of light-absorbing chloroplast pigments is more sensitive to Mg^{2+} deficiency than photosynthesis rates at ambient CO_2 concentrations in the air. Since recovery of bleached needles of formerly 'damaged' trees is possible after Mg^{2+} fertilization (Beyschlag et al., 1987; Kaupenjohann et al., 1985), there is reversible reduction of the number of light-harvesting protein complexes in chloroplast thylakoids of Mg^{2+}-deficient plants rather than real pigment or thylakoid 'damage'. In fact, the turnover of photosystem II is impaired mainly in the spring season in mineral-deficient and air-pollution-burdened needles of Norway spruce (Godde, 1991, 1992; Godde and Buchhold, 1992; Lütz et al., 1992). This reduction of light-absorbing chlorophyls presumably adapts the photon absorption of needles to the decreased capacity of the Calvin cycle in Mg^{2+}-deficient spruce needles. Steady-state measurements revealed a lower maximum photosynthetic capacity of Mg^{2+}-deficient spruce needles. Maximum photosynthesis rates at saturating light and saturating CO_2 concentrations were lower in Mg^{2+}-deficient needles compared with fertilized or healthy looking control needles (Lange et al., 1987, 1989a,b; Beyschlag et al., 1987). Also the carboxylation efficiency was lower than in control needles. The carboxylation efficiency is determined from the plot of 'photosynthesis rate vs. intercellular CO_2 concentration' at saturating light intensity as the initial slope of the saturation curve [μmol CO_2 assimilation m^{-2} leaf area s^{-1} μbar^{-1} CO_2]; the saturation plateau is the photosynthetic capacity. The reduction of photosynthetic capacity by 50% in Mg^{2+}-deficient needles and the reduction of the carboxylation efficiency again indicate that the Calvin cycle is not fully active in illuminated Mg^{2+}-deficient spruce needles (Lange, 1987, 1989a,b; Beyschlag, 1987).

Phloem loading and growth Since H-ATPase dependent-endergone phloem loading processes in the leaf veins are MgATP dependent, Mg^{2+}-deficient leaves of many plant species indeed exhibit a reduced ability to export assimilates (sucrose, amino acids etc.) via the phloem sap to heterotrophic plant organs. Consequently, accumulation of starch in chloroplasts or increasing sugar concentrations in vacuoles can be measured in leaves of many Mg^{2+}-deficient plant species. Herbaceous leaves may become very turgescent and even brittle in some cases (Bergmann, 1988). Since, also, the polycondensation of sugars and amino acids is Mg^{2+} dependent, the fraction of compounds with low molecular weights increases in Mg^{2+}-deficient leaves. On the other hand, the content of starch, sugar, fat and proteins in sink organs (fruits, sperms) of Mg^{2+}-deficient plants is reduced. Apples for instance, remain sour, contain less sugar and cannot be readily stored. Like in phosphorus-deficient herbaceous plants, there is also delay of the generative phase in plant's ontology (Bergmann, 1988). Concerning trees, the electron microscope pictures of Parameswaran et al. (1985) show that there is morphological damage in the Strassburger cells immediately surrounding the phloem cells of damaged spruce needles. Thus, it seems possible that assimilates are restricted in spruce needles, but there is no accumulation of at least starch in needles of Mg^{2+}-deficient spruce trees growing in the Fichtelgebirge (Oren and Zimmermann, 1989). The reduction of needle photosynthesis is much too small to explain the observed depression in stem wood production rate per leaf area index (Oren et al., 1988). Fiedler and Höhne (1965) reported that apical growth of Norway spruce is reduced at limiting Mg^{2+} supply. There are reports that Mg^{2+}-deficient dicotyl plants have reduced growth rates of side roots (Bergmann, 1988). Schneider et al. (1989) summarized that the level of non-structural carbohydrates in roots was quite similar at both the declining and the healthy site in the Fichtelgebirge. The seasonal course of carbohydrate concentrations was opposite to what would be expected if assimilate starvation had occurred. Also there were no significant differences in total fine root or large root biomass between the Mg^{2+}-deficient site and the healthy control in the Fichtelgebirge. Thus, the observed accumulation of roots in the organic soil surface layer and the limitation of root growth into the mineral soil horizons at the declining site must have been induced by other factors than assimilate starvation. Schulze *et al.* (1989) report that the production rates of needles and twigs in Norway spruce crowns of Mg^{2+}-deficient trees and healthy control trees were similar in the Fichtelgebirge, but the declining stand produced less trunk wood than the healthy stand at a similar leaf area index. Hence, Mg^{2+} deficiency reduced the mitotic cambium activity. Bergmann (1988) reported that Mg^{2+}-deficient apple trees show bark symptoms which are similar to frost damage. Thus, it appears that the production of stemwood in the Fichtelgebirge is governed by the limited Mg^{2+} supply (Meyer *et al.*, 1988). As a tentative result, there seems to be no dominant inhibition of phloem loading in Mg^{2+}-deficient spruce, but rather a direct impact of Mg^{2+} deficiency on the mitotic activity of the trunk cambium (see section 4.7). This assumption is supported by the fact that Mg^{2+} is essentially involved in numerous processes of the cell cycle (protein synthesis, RNA turnover, DNA synthesis). We urgently need more information on phloem loading, source–sink relations and mitosis activity in Mg^{2+}-deficient trees.

Needle senescence In the long term, individual Mg^{2+}-deficient leaf cells will decline. Necrotic spots and areoles which occur mainly in older needle age classes will become larger (Lyr *et al.*, 1967). Finally, entire preferentially older leaves decline early, and become brown and dry in all woody plant genuses investigated so far (*Malus, Pyrus, Prunus, Citrus, Thea, Laurus*, and numerous deciduous and evergreen forest trees; Clijsters, 1971; Bergmann, 1988). Concerning deciduous trees, there is early leaf senescence in late summer ('Neuartige Herbstfärbung'; 'new type autumn coloring'), perhaps because of early Mg^{2+} retranslocation into the twigs and branches. Mg^{2+} deficiency symptoms of deciduous trees apparently seem to be focused in the fall (Baule, 1978). The early leaf fall of deciduous trees and shrubs has been interpreted as the consequence of anthropogene air pollutants (Bergmann, 1988 and section 4.7).

4.2. Transport of magnesium

4.2.1. Stomatal conductance

The transport of Mg^{2+} cations in the xylem sap (see below) depends on the transpiration rate, which is a function of stomatal conductance (and of other factors, see below). Additionally, stomatal conductance governs the leaf uptake of potentially phytotoxic trace gases (SO_2, NO_2, O_3), which partly interfere with cellular Mg^{2+} supply and demand (section 4.7). The stomatal water conductance g_{H_2O} ($mmol\,m^{-2}\,s^{-1}$) of leaves or needles directly or indirectly depends on numerous exogenous ecophysiological factors (air temperature T_{air} and humidity RH_{air}, apparent leaf water vapour pressure difference ΔW between ambient air and the leaf intercellular space, light intensity $h\upsilon$ and its spectral composition, day length, leaf and soil water potential $\Delta\Psi_{leaf}$, $\Delta\Psi_{soil}$, soil temperature T_{soil} etc.), and endogenous metabolic factors (intercellular CO_2 concentration $[CO_2]_i$, blue light receptor (cryptochrome system) of guard cells, abscisic acid (ABA) redistribution within leaves, ABA signals from the root system, supply with metal cations $[Cat^+]$ etc.), which nonlinearly regulate the stomatal aperture in a complex manner (Hartung and Slovik, 1991; Hedrich and Schröder, 1989; Keller and Häsler, 1986, 1987; Mansfield, 1976; Slovik et al., 1992c; Slovik and Hartung, 1992a,b,c):

$$g_{H_2O} = g_{H_2O}(T_{air}, h\upsilon, \Delta W, \Delta\Psi_{leaf}, \Delta\Psi_{soil}, T_{soil}, [CO_2]_i, [Cat^+] \ldots)$$

Important aims of this complex regulation are mainly the control of stomatal water loss relative to net CO_2 assimilation ('water use efficiency'), and control of the transport of dissolved compounds in the xylem water stream (ion uptake, 'root–shoot communication' etc.). No significant differences in stomatal leaf conductance of Norway spruce could be observed in the Fichtelgebirge between: (i) chlorotic Mg^{2+}-deficient needles, (ii) green healthy needles, which are sufficiently supplied with Mg^{2+}, and (iii) needles of Mg^{2+}-fertilized spruce trees (Beyschlag et al., 1987; Lange et al., 1989a; Pfeiffer, 1987; Wedler, 1986). Stomatal performance and maximum needle conductance were comparable. The range of needle conductance was similar to that reported for other sites with lower air pollution than in the Fichtelgebirge with ca. 15 ppb SO_2 (annual mean; see Weikert, 1986; Weikert et al., 1989). Stomatal responses to environmental factors were also not impaired. Needles of all relevant age classes and degrees of damage with different Mg^{2+} contents responded well to changed conditions of air humidity or photon flux density. There is no stomatal missfunctioning of Mg^{2+}-deficient spruce trees in the Fichtelgebirge. Keller (1972) found that the transpiration rate of spruce growing on Mg^{2+}-deficient soils was only slightly reduced.

Körner et al. (1995) studied the complex dependencies of the stomatal water conductance g_{H_2O} of Picea abies and Pinus sylvestris in the field and deduced statistical response functions, which allow the calculation of g_{H_2O} for any day of the year if fundamental meteorological data (light intensity, air temperature, air humidity, day length, day time etc.) are available in the field. In the following, boundary layer resistances are not discussed separately, but respected as the 'effective, apparent' mean stomatal conductance in the field. Application of these findings to native Norway spruce canopies allowed the calculation of the annual kinetics of stomatal aperture at six sites in Germany at which continuously measured complete meteorological data sets are available (Slovik et al., 1995). Typical daily means of the stomatal water conductance in the course of a year of 'mean' needles of Norway spruce canopies are shown in Figure 3. A large fraction of the annual pattern of stomatal aperture can be explained simply by the varying daylength and the determined maximum stomatal conductance which can be achieved in different seasons (for details see Körner et al., 1995). Table 2 summarizes 'typical' relative stomatal aperture data of different tree species as compared with Norway spruce, which is set to 1.0 relative units. Data are calculated from leaf transpiration rates of

116

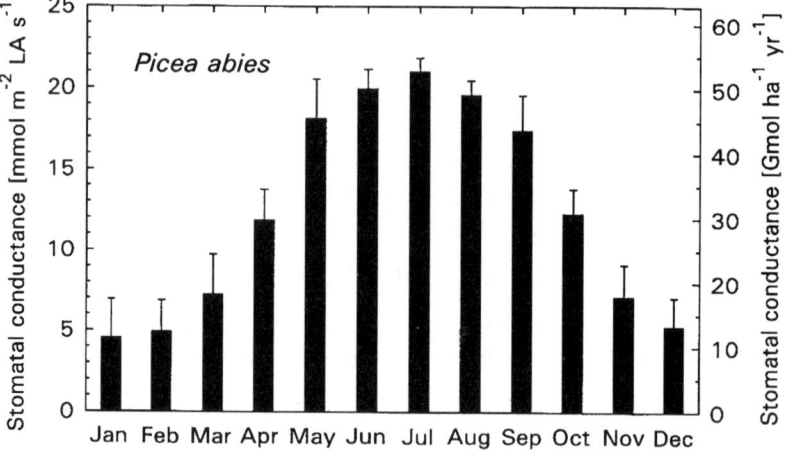

Figure 3. Monthly means of the stomatal water conductance of Norway spruce canopies given either on a leaf area (LA) basis of 'mean' canopy needles (left ordinate) or estimated on a hectare basis (right ordinate). Data are means of the whole day including the night. Error bars indicate the standard deviations of data from different spruce stands and years ($n = 23$). Data are based on meteorological field data and empirically found statistical response functions (Körner *et al.*, 1995) of stomatal behavior after changing environmental conditions (calculated after Slovik *et al.*, 1995)

Table 2. Stomatal water conductances of different tree species relative to Norway spruce

Tree species	Relative stomatal water conductance (*Picea abies* = 1.0 relative unit)	
	'Leaf basis'	'Canopy basis'
Abies grandis	0.70* ±0.10	n.d.
Pseudotsuga menziesii	0.93	1.26
Abies procera	0.98* ±0.04	n.d.
Picea abies	1.00±0.13	1.00±0.29
Abies alba	1.06* ±0.13	n.d.
Picea sitchensis	1.07* ±0.23	n.d.
Pinus strobus	1.15* ±0.38	n.d.
Thuja plicata	1.22* ±0.11	n.d.
Pinus sylvestris	1.50±0.20	1.03±0.44
Tsuga heterophylla	1.72* ±0.26	n.d.
Larix kaempferi	2.03* ±0.45	n.d.
Fagus sylvatica	2.71±1.01	0.93±0.10
Larix decidua	2.93±0.71	1.69±0.46
Quercus robur	4.86±0.81	1.10
Betula pendula	6.25±0.76	1.24±0.22

'Leaf basis' data derive from the daily sum of transpiration per leaf matter from tree species growing in the same atmosphere [g(H$_2$O) g(leaf fresh weight)$^{-1}$]; Dry weight based on [g(H$_2$O) g(leaf dry weight)$^{-1}$] are marked with asterisks. 'Canopy basis' data ('big leaf model') have been determined from water balance data [mm H$_2$O ha^{-1} y^{-1}] in the field. Data after Eidmann (1943), Eidmann and Schwenke (1967), Kirwald (1950), Ladefoged (1963), Pisek and Cartellieri (1939, 1941), Pisek and Tranquillini (1951), Polster (1954), Schubert (1939), Sonn (1960); see also Polster (1950), and Schmidt-Vogt (1986); n.d. = not determined in the cited literature. Standard deviations were calculated based on available relative data from different authors different tree species growing in the same air. Relative conductance data are given on a leaf biomass basis in Table 2 ($mmol\,H_2O\,kg^{-1}(fw)\,s^{-1}$) and on the basis of the total canopy transpiration ('big leaf approach'). Deciduous trees have considerably higher stomatal conductances than evergreen gymnosperms but the differences become smaller if whole canopies are regarded.

Since stomatal aperture is an inevitable necessity of photosynthesis, synchronous stomatal uptake of ambient anthropogene trace gases cannot be avoided. Stomatal conductances of different atmospheric (trace) compounds vary considerably. Table 3 summarizes the relative diffusion coefficients (H_2O is set to 1.000 relative units) of potentially acidic air pollutants, primary and secondary photo-oxidants and of main compounds of exhaust fumes of vehicles. The Maxwell theory defines that the kinetic energy (translation plus rotation) of any gaseous molecule species is constant at a given gas temperature. The translational velocity of molecules is proportional to its diffusion coefficient. For a given stomata anatomy (i.e. plant species), the diffusion coefficient is proportional to the stomatal conductance of that compound if the degrees of freedom for rotational energy are the same for water and the gas molecule species regarded. This is approximately true for many small molecules. The estimated relative stomatal conductance, Rel_X [$mol\,X\,per\,mol\,H_2O$] of a compound X is found by:

$$Rel_X = \sqrt{M_r(H_2O)/M_r(X)},$$

where M_r is the molecular mass [$g\,mol^{-1}$] of either water $M_r(H_2O)$ or of the gaseous compound X regarded. Table 3 also partially compares calculated data with diffusion coefficients from the literature. The maximum deviation between both approaches is only 16% for nitrogen monoxide (NO). Thus, data given in Table 3 are sufficiently accurate for most physiological purposes. Thus, the stomatal conductance g_X [$mmol\,X\,m^{-2}s^{-1}$] of a gaseous compound X can be calculated if the stomatal water conductance g_{H_2O} is known:

$$g_X = Rel_X \cdot g_{H_2O}$$

The flux J_X ($mmol\,m^{-2}s^{-1}$) of a trace gas X from the atmospheric source into the intercellular sink within the leaf or needle mesophyll is given by the flux equation:

$$J_X = g_X \cdot \Delta[X],$$

where $\Delta[X]$, given in the dimension of a mole fraction [$mol\,mol^{-1}$] or partial pressure [$Pa\,Pa^{-1}$] in the air, is the concentration or activity difference between the atmosphere and the intercellular air. Usually, the concentration in the air can be readily measured, but the intercellular concentration of trace gases can be determined only using fumigation experiments, which revealed that the intercellular gas concentration largely depends on the water solubility of the gaseous compound X (see Table 4). Gases which either readily dissolve in the apoplasmic cell wall water within the substomatal cavity (HCl, NH_3, HF, SO_2), or which are extremely reactive (e.g. ozone O_3) have intercellular concentrations close to zero, i.e. the 'mesophyll resistance' of these gases is small and the compensation point of these trace gases

Table 3. Relative stomatal conductance of different gaseous compounds in the atmosphere relative to water (= 1.000 relative unit)

Compound (symbols)	Molecular mass ($g\,mol^{-1}$)	Stomatal conductance G_x (relative units; $H_2O = 1.00$)
H_2O	18.02	1.000
H_2	2.02	2.987
H_2O_2	34.01	0.728
HO_2^-	33.01	0.739
OH^-	17.01	1.029
O_3	48.00	0.613 (0.612)
O_2	32.00	0.750 (0.808)
CO_2	44.01	0.640 (0.642)
CO	28.01	0.802 (0.810)
N_2	28.01	0.802 (0.803)
NH_3	17.03	1.029 (0.937)
NO	30.01	0.775 (0.667)
NO_2	46.01	0.626 (0.577)
N_2O_3	76.01	0.487
N_2O_4	92.01	0.443
N_2O	44.01	0.640
HNO_3	63.01	0.535
SO_2	64.06	0.530 (0.509)
SO_3	80.06	0.474
H_2S	34.08	0.727 (0.693)
COS	60.07	0.548
CS_2	76.14	0.486
HCl	36.46	0.703 (0.666)
Cl_2	70.91	0.504
$HClO$	52.46	0.586
HF	20.01	0.949
ethane, C_2H_4	28.05	0.802
acetylene, C_2H_2	26.04	0.832
propene, C_3H_6	42.08	0.654
butene, C_4H_8	56.10	0.567
PAN	121.05	0.386
PPN	135.08	0.365
PBN	149.10	0.348
CH_3CCl_3	133.41	0.368

The estimated relative stomatal conductance of a compound x $(G_x) = \sqrt{M_r(H_2O)/M_r(x)}$, where M_r is the molecular mass [$g\,mol^{-1}$]. Numbers in brackets are calculated from literature data of gas diffusion coefficients (see Andrussow, 1969; Nobel, 1983; Winner and Mooney, 1980). The maximum deviation between both approaches is 16% (for NO). PAN = peroxyacetyl nitrate, PPN = peroxypropionyl nitrate, and PBN = peroxybutyryl nitrate; all are photochemical smog compounds

is close to zero. Thus, just the known potentially most phytotoxic trace gases, SO_2, HF and O_3, readily produce 'full' stomatal gradients $\Delta[X]$ and ceteris paribus maximum stomatal uptake rates. Trace gases, which are neither extremely reactive, nor exhibit outstanding high water solubility, are in fact not extremely phytotoxic (CO, NO, NO_2) and their intercellular concentration is not close to zero. The 'typical' intercellular NO_2 concentration (= compensation point) in Norway spruce intercellulars equals approximately 3.2 ppb NO_2 (field data from the Schwarzwald, Germany, after Baumann and Baumbach, Stuttgart-Hohenheim, personal communication; H. Rennenberg, Freiburg, personal communication, found similar

Table 4. Water solubility of atmospheric (trace) gases at 20°C

Compound (symbols)	Henry coefficient (mol m^{-3} hPa^{-1}) K_H (20°C)	Ostwald coefficient (mol m^{-3}/mol m^{-3}) α (20°C)
N$_2$	0.692	0.0169
CO [1]	1.008	0.0246
[2]	1.035	0.0252
O$_2$ [1]	1.477	0.0360
[2]	1.366	0.0333
NO	2.072	0.0505
NO$_2$	9.847	0.240
O$_3$	11.04	0.269
CO$_2$ [1]	39.55	0.964
[2]	38.67	0.943
H$_2$S [1]	100.3	2.444
[2]	114.5	2.792
SO$_2$	1437	35.02
HF	14670	357.6
NH$_3$	72030	1756
HCl	> 100000	> 10000

Data marked with [1] are calculated using the standard heat $\Delta H°$ [kJ mol^{-1}] and the entropy $\Delta S°$ [J K^{-1} mol^{-1}] (Weast, 1989) on the basis of the solubilization reaction GAS$_{air} \leftrightharpoons$ GAS$_{water}$. The Henry coefficient K_H obeys the equation ln $K_H = -\Delta H°/RT + \Delta S°/R$, where R is the gas constant ($= 8.3144$ J K^{-1} mol^{-1}), and T is the absolute temperature in Kelvin. Data marked with [2] are table data from the literature. Concerning all other gases, calculated data and literature data match closely. The Ostwald coefficient α, which is a special case of the Nernst partition coefficient of matter between two phases (water : air), and the Henry coefficient K_H are interrelated by the equation $\alpha = RTK_H$, where $R = 83.144 \cdot 10^{-6}$ m^3 hPa K^{-1} mol^{-1} is the gas constant (given in rearranged dimensions), and T is the absolute temperature in Kelvin. Data basis: Bennet and Hill (1975), D'Ans and Lax (1949), Farquhar *et al.* (1980), Heath (1975), Hocking and Hocking (1977), Kruis and May (1962), Nobel (1983), Weast (1989), Wilhelm *et al.* (1977)

results in unpublished NO$_2$ fumigation experiments). Since the water solubility of NO$_2$ and NO are comparable (see Table 4), a similar compensation point of roughly 3.2 ppb NO may be estimated also for NO. Thus, there is net uptake of ambient NO$_2$ (and presumably NO) only above about 3.2 ppb. Below 3.2 ppb, there is observed NO$_2$ (NO) emission by spruce needles.

The most important trace gases in forest decline research are SO$_2$, O$_3$ and NO$_2$. Regarding the measured irregular daily and annual patterns of SO$_2$ pollution and meteorological data in native Norway spruce forests, Slovik *et al.*, (1995) deduced the annual stomatal SO$_2$ dose D$_{SO_2}$ of mean needles in six native Norway spruce canopies (Figure 4). Each data point is the result of the stomatal SO$_2$ flux summation of a whole year with 48 iteration steps (= measured data sets) per day. The SO$_2$ uptake by 'mean' needles of native spruce canopies per 'mean' day of the vegetation period equals ca. 0.255 μmol SO$_2$ per m^2 (total) needle surface and per ppb SO$_2$ (annual mean in the ambient air). The apparent intercept of D$_{SO_2} = 0.647$ μmol SO$_2$ m^{-2} (needle surface) day^{-1} (vegetation period VP) ppb^{-1} (annual mean of SO$_2$) at zero ppb SO$_2$ in the ambient air is just the consequence of a negative background correlation (not shown here) between the annual mean of SO$_2$ pollution and the length of the vegetation period: In years with warm winters (long vegetation period), there is reduced emission of SO$_2$ from fuels and vice versa. Of course, D$_{SO_2}$

120

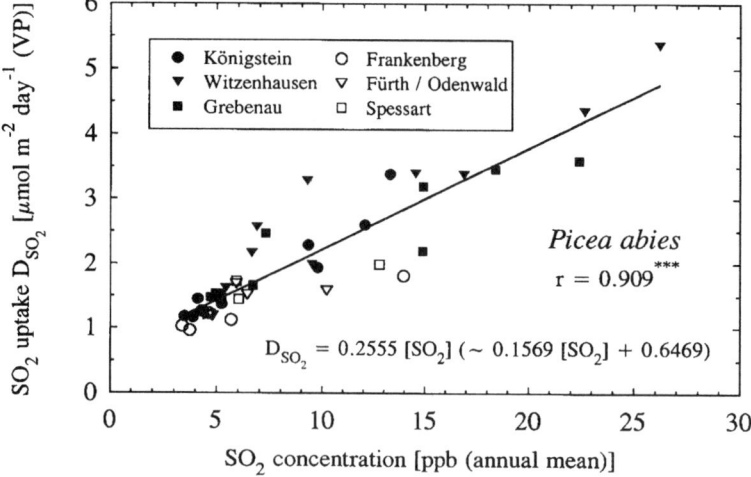

Figure 4. Integrated annual stomatal SO_2 uptake of six native Norway spruce canopies (Hessen, Germany) in different years between 1984 and 1992 per m^2 (total) needle surface and recalculated per day of the vegetation period VP of the corresponding years. VP is defined here after Wiersma (1963) as the number of days per year with a mean air temperature of at least 6.0°C. This mean daily (VP) uptake is plotted versus the annual mean of the SO_2 pollution. The given regression formula $D_{SO_2} = f([SO_2])$ allows the estimation of the annual stomatal SO_2 uptake of native Norway spruce canopies in the field if just the length of the vegetation period VP and the annual mean of SO_2 pollution [ppb = nPa Pa^{-1}] are known. Below ca. 3 ppb SO_2 D_{SO_2} must be extrapolated to zero (after Slovik *et al.*, 1995, supplemented here also with data determined between 1990 and 1992)

at zero ppb SO_2 must be zero, and below 3 ppb SO_2 the regression function must be extrapolated to D_{SO_2} = zero in Figure 4. The annual uptake of ozone by spruce needles, which is calculated using mutatis mutandis the same approach after Slovik *et al.* (1995), is shown in Figure 5. The ozone uptake by 'mean' needles of native spruce canopies per 'mean' day of the vegetation period equals 0.495 μmol O_3 per m^2 (total) needle surface and per ppb O_3 (annual mean in the ambient air). As expected, the intercept is close to zero here. In Figure 6, the dose D_{NO_x} per day VP of NO_x (NO_2+NO) as depending on the pollution of the dominant nitrogen oxide NO_2 (annual mean) is shown. Given data points already respect the synchronous needle uptake (respective emission) of both NO_2 and NO, employing a separate compensation point of 3.2 ppb for each NO_x species (see above). There are often pollution situations in the field, where high NO_2 above 3.2 ppb results in net NO_2 uptake, but synchronous NO is below 3.2 ppb, which results in net NO release from spruce needles. Since in Figure 6 the total net NO_x dose per day of the vegetation period VP is plotted versus NO_2, the abscissa intercept shifted from 3.2 ppb (approximate compensation point of both NO_2 and NO) to ca. 7 ppb. There is net uptake of NO_x-nitrogen only above ca. 7 ppb NO_2. The slope of 0.477 μmol NO_x uptake per day VP per m^2 needle surface and per ppb NO_2 (annual mean) above ca. 7 ppb is the detailed result of the complex patterns of stomatal conductance, NO_2 pollution and

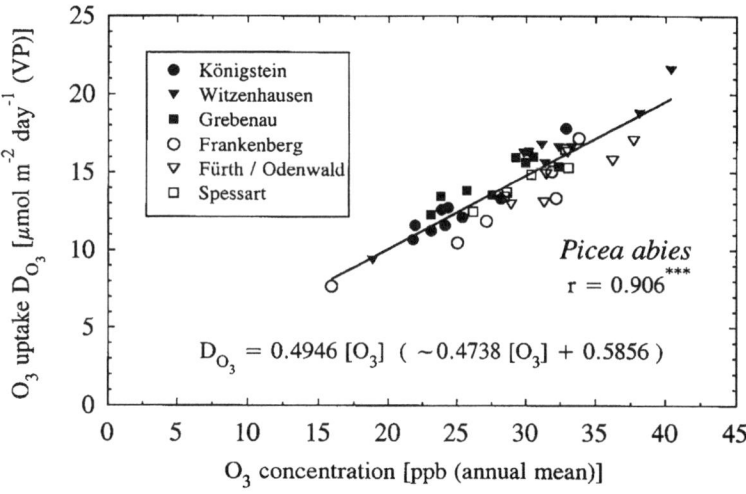

Figure 5. Mean stomatal O_3 uptake of six native Norway spruce canopies between 1984 and 1992 as determined by the method of Slovik *et al.*, 1995. The O_3 compensation point is close to zero (see Laisk *et al.*, 1989). The inserted calculus allows an estimation of the annual stomatal O_3 uptake by native spruce canopies if just VP [days y^{-1}] and the annual mean of the ozone pollution [nPa Pa^{-1}=ppb] are known. For more details, see legend of Figure 4 (Slovik, unpublished)

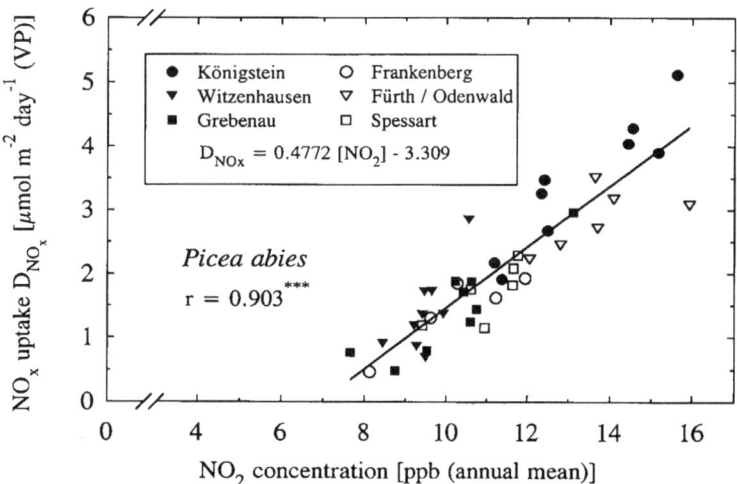

Figure 6. Mean stomatal NO_x (NO_2 plus NO) uptake of six native Norway spruce canopies in Hessen (Germany) between 1984 and 1992 as determined by the approach after Slovik *et al.*, 1995. The NO_2 compensation point is ca. 3.2 ppb NO_2 (after field data from Baumann and Baumbach, University of Stuttgart and after Rennenberg, unpublished personal communication). A similar compensation point is assumed for NO (see Table 4). The inserted calculus allows an estimation of the total annual stomatal NO_x uptake by native spruce canopies if just VP [days y^{-1}] and the annual mean of the NO_2 (!) pollution [nPa Pa^{-1}=ppb] are known. For more details, see text y and legend of Figure 4 (Slovik, unpublished)

concomitant NO pollution, which all behave largely chaotically in the field. Still, integration of NO_x fluxes over whole years largely compensated the chaotic patterns resulting in a highly significant and tight correlation with $r = 0.903$. This is true also for SO_2 and O_3. On the basis of Figures 4, 5 and 6, it is possible to estimate realistic data of the stomatal trace gas net uptake into mean needles of native spruce canopies if only the length of the vegetation period VP and the annual mean of the pollution of SO_2, NO_2 and O_3 are known. The annual trace gas uptake rates J_{SO_2}, J_{NO_x} and J_{O_3} [$\mu mol\, m^{-2}$ (needle surface) $year^{-1}$] equal:

$$J_{SO_2} = D_{SO_2} \cdot VP$$
$$J_{NO_2} = D_{NO_2} \cdot VP$$
$$J_{O_3} = D_{O_2} \cdot VP,$$

where VP is the number of days per year with a mean air temperature above 6°C (see Wiersma, 1963), and D_{SO_2}, D_{NO_2} and D_{O_3} are calculated using the inserted formula in Figures 4, 5 and 6. The reliability of the given formula has been independently confirmed on the basis of measured sulfate accumulation rates in needles at different sites with different annual means of SO_2 pollution (Slovik et al., 1995). At the Kahleberg in the Erzgebirge, we measured sulfate accumulation rates in spruce needles of $12.4 \pm 4.8\, mmol\, SO_4^{2-}\, kg^{-1}(dw)\, year^{-1}$ (ca. 34 ppb SO_2 annual mean, 905 m asl, 50,76°C northern latitude, $VP \approx 166$ days $year^{-1}$). This measured sulfate accumulation rate strongly matches the calculated value of $14.3\, mmol\, SO_2$ $kg^{-1}(dw)\, year^{-1}$; specific needle surface at the Kahleberg: $14.4\, m^2 kg^{-1}(dw)$. In the Erzgebirge, Norway spruce needles apparently detoxify SO_2 almost exclusively ($100 \cdot 12.4/14.3 = 87\%$) via oxidation to sulfate, which is sequestered into vacuoles and neutralized by metal cations (K^+, less important: Mg^{2+}; see Dittrich et al., 1991a,b; Kaiser et al., 1993; Lanzl et al., 1989). Knowledge of the annual stomatal dose of potentially acidic air pollutants, respective plant nutrients (SO_2, NO_2), is necessary to estimate the interference of air pollutants with the annual Mg^{2+} supply and demand (see section 4.7).

There is no simple pattern of the response of stomatal water conductance to anthropogen trace gases. Norway spruce cuttings subjected to 25 ppb SO_2 for almost half a year reduced their stomatal transpiration in the light, and stomata reacted sluggishly to changed light conditions (Keller and Häsler, 1986). Since SO_2-dependent sulfur tends to accumulate in or close to guard cells in fumigation experiments with labeled $^{35}SO_2$ (Weigl and Ziegler, 1962), accumulation of SO_2-dependent sulfate in guard cell vacuoles possibly may interfere with the osmotic regulation of stomatal aperture. Fumigation of spruce with 47 ppb ozone for 45 h per week for all the vegetation period showed similar dysfunctions. Stomatal water conductance was higher in the night than that of the control and stomatal responses tended to become insensitive to light–dark changes (Keller and Häsler, 1987). In the Fichtelgebirge, where similar O_3 and SO_2 concentrations can be measured in the field, no such observations could be made (see above).

4.2.2. Canopy transpiration

The transpiration of tree canopies strongly depends on complex interdependencies and gradients of meteorological factors within native canopies (light intensity, relative humidity, air temperature, wind velocity, boundary layer resistances etc.), and on ecophysiological factors and their gradients (stomatal aperture, canopy structure etc.). The transpiration rate J_{H_2O} ($\mu mol\,H_2O\,m^{-2}\,s^{-1}$) of leaves is the product of the 'apparent' stomatal water conductance g_{H_2O} ($mmol\,H_2O\,m^{-2}\,s^{-1}$) and the needle water vapour mol fraction difference ΔW [$mPa\,Pa^{-1}$] between the substomatal cavity and the ambient air:

$$J_{H_2O} = g_{H_2O} \cdot \Delta W$$

ΔW depends on the: (i) air temperature(T)-dependent water vapour mol fraction $N_{sat}(T)$ at saturation (i.e. at 100% relative humidity), (ii) air temperature (T_{air}) and (iii) needle temperature (T_{leaf}), (iv) relative humidity RH_{air} of the ambient air, and (v) RH_{leaf} of the substomatal cavity which depends on the leaf's water potential $\Delta\Psi_{leaf}$ (e.g. Nobel, 1983):

$$W_{air} = RH_{air} \cdot N_{sat}(T_{air})$$
$$W_{leaf} = RH_{leaf} \cdot N_{sat}(T_{leaf})$$
$$\Delta W = W_{leaf} - W_{air}$$

Since (i) the relative humidity of leaf intercellulars, RH_{leaf}, is usually close to 1.0 relative units, and (ii) the leaf temperature, T_{leaf}, does not differ too much from the air temperature, T_{air}, if time ranges of weeks up to years are regarded, it is sufficiently accurate to calculate ΔW on the basis of the following approximation:

$$\Delta W = N_{sat}(T_{air}) \cdot (1 - RH_{air})$$

This equation allows the estimation of the stomatal water vapour mol fraction difference ΔW between leaf intercellulars and the ambient air on the mere basis of easily measurable meteorological air properties within forest canopies. Together with stomatal conductance data (Körner et al., 1995), daily and annual kinetics of transpiration rates of exposed spruce needles can be calculated (Slovik et al., 1995). Since also the ecophysiological regulation of stomatal aperture depends on meteorological signals (Körner et al., 1995), transpiration rates of leaves depend, in a 'synergistic' or 'autocorrelative' manner, strongly on weather conditions. In Figure 7, 'typical' transpiration rates of exposed spruce needles J_{H_2O} are already recalculated for 'mean' needles of whole Norway spruce canopies, which transpire only ca. 32% of water per unit leaf surface area (LA) relative to exposed leaves at the top of the canopy (100%) (for details, see Slovik et al., 1995). The very small transpiration rates in winter are the synergistic consequence of (i) long night lengths, (ii) small water gradients ΔW in the 'wet' and 'cold' season, and (iii) small daily stomatal aperture in the winter (and respectively vice versa for the summer season). Figure 8 shows the annual stomatal transpiration rate of Norway spruce canopies

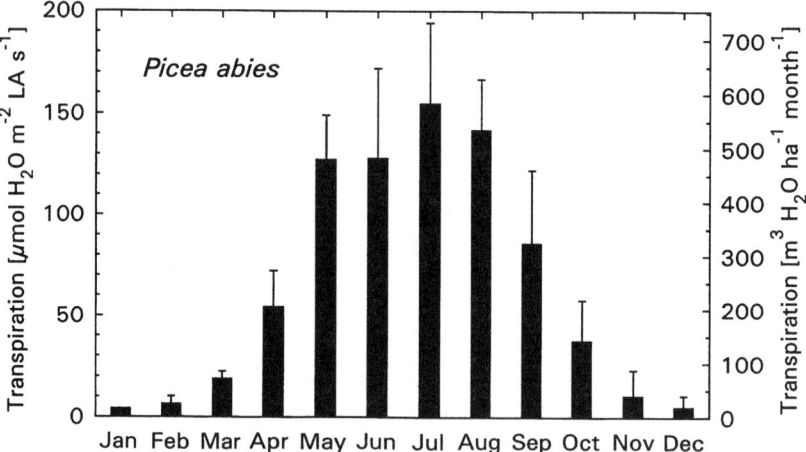

Figure 7. Typical 'mean' stomatal transpiration rates of 'mean' Norway spruce needles in native canopies as given on a leaf area LA basis (total needle surface) [μmol $H_2O\,m^{-2}\,s^{-1}$] or given in [$m^3\,H_2O\,ha^{-1}\,month^{-1}$]. Data are based on six German sites in Hessen and several years per site (see Slovik *et al.*, 1995). Error bars represent standard deviations ($n=23$ sites·years)

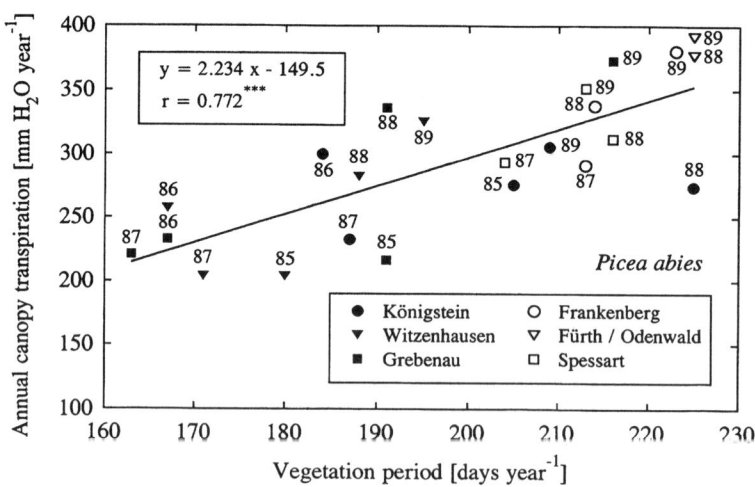

Figure 8. Dependency of the stomatal Norway spruce canopy transpiration on the length of the vegetation period, VP. Symbols indicate different sites in central Germany (Hessen). Adjacent numbers indicate the year. The linear regression is significant at the 99.9% level. Data after Slovik *et al.* (1995)

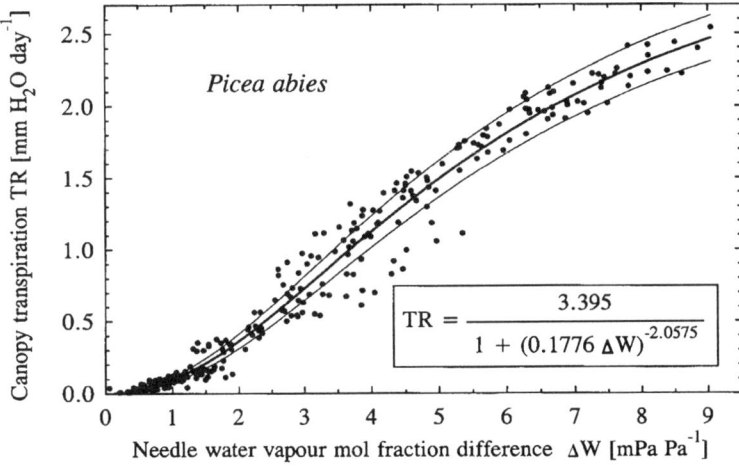

$$TR = \frac{3.395}{1 + (0.1776\,\Delta W)^{-2.0575}}$$

Figure 9. Canopy transpiration TR of Norway spruce in the field in relation to the needle water vapour mol fraction difference ΔW between the substomatal cavity and the ambient air. The given statistical response function TR(ΔW) allows an estimation of the canopy transpiration TR if just air temperature and air humidity are continuously measured for at least a few days, for calculating a 'mean' of ΔW. TR(ΔW) was fitted to data points after Slovik *et al.* (1995) by the method of the least squares using a Marquardt–Levenberg algorithm. The thin lines cover the standard deviation area of the fitting function TR(ΔW). This area was calculated from the standard errors of the fitting parameters on the basis of the Gauss law of error progression

(TR) in different years and at different sites as depending on the length of the vegetation period (VP) (same data basis as in Figure 7 after Slovik *et al.*, 1995). Sites or years with long vegetation periods transpire more water than stands with short vegetation periods. Stomatal canopy transpiration rates vary between 200 and 400 mm H_2O per year. Benecke (1978) found the mean annual evapotranspiration rate of 90-year-old Norway spruce stands in the Solling (Lower Saxony, Germany) to be 335 mm H_2O per year (data basis from 1968 to 1975). Since about 10% of the evapotranspiration rate is direct water evaporation from the forest soils, the independently measured transpiration data of Benecke fit very well with the data presented in Figure 8 which were obtained by a completely different approach (Slovik *et al.*, 1995). Detailed field data of spruce canopy transpiration rates are also available from Werk *et al.* (1988).

There is a strong independence of numerous meteorological field data at least if time scales of weeks or months are regarded. The relative air humidity decreases just by increasing the air temperature. The light intensity correlates with the air temperature. Consequently the relative humidity correlates with the light intensity. The light intensity itself and its spectral composition correlate with the season etc. Thus, it is not surprising that there is a rather tight but nonlinear correlation between the apparent needle water vapour pressure difference $\Delta W = f(T_{air}, RH_{air})$ and the canopy transpiration TR. This is shown for Norway spruce in Figure 9 (data after Slovik *et al.*, 1995). TR$=f(\Delta W)$ conjointly respects the (i) weather-dependent stomatal regulation and (ii) stomatal 'water gradient' ΔW. The given dependency is sufficiently tight only if sufficiently long time intervals (at least a few days) are regarded. Time intervals with small mean monthly ΔW values are either cold winter periods or those with high precipitation (high air humidity, wet canopies). Consequently, there are small stomatal (!) transpiration rates under these conditions. Evaporation of interception water in the spruce canopy does not confuse the given function because it influences only the abscissa ΔW in Figure 9, which is based on measured data. With increasing ΔW, the stomatal canopy transpiration TR increases. At very high values of ΔW, stomatal closure reactions become dominant and transpiration rates saturate in a logistic manner. Presumably, the absolute canopy transpiration will be reduced at prolonged times of extremely high ΔW values after passing a maximum beyond about $\Delta W > 10\,\mathrm{mPa\,Pa^{-1}}$, but this is outside the range of available data. The thin lines in Figure 9 cover the standard deviation area of the function, $\mathrm{TR} = f(\Delta W)$. The statistical parameters of the inserted formula in Figure 9 have no ecophysiological meaning. The given logistic function allows an estimation of the mean actual canopy transpiration rate of native Norway spruce canopies if just continuously measured data of air temperature and air humidity are available for time intervals of at least a few days. If direct transpiration measurements in the field are not available, then the presented logistic function is sufficiently accurate to estimate xylem water flow rates in native Norway spruce stands if time scales of weeks, months or years are in scope. Mainly in spring and fall, with 'mean' ΔW values between 2.5 and $5.0\,\mathrm{mPa\,Pa^{-1}}$, there are larger residuals, but they are evenly distrib-uted around the fitting function.

On the basis of cautiously combining data of Figure 9 (*Picea abies*) and additional information on stomatal canopy ('big leaf') conductances of other tree species relative to *Picea abies* (see Table 2), it is possible to tentatively and roughly estimate stomatal canopy transpiration rates TR of other important forest tree species (*Pinus sylvestris*, *Fagus sylvatica*, *Quercus robur*, *Larix decidua*, *Pseudostuga menziesii* and *Betula pendula*) if air humidity and air temperature data are available in these forests. Detailed individual dependencies for these species, as given for *Picea abies* in Figure 9, are not available to my knowledge.

4.2.3. Xylem sap transport

Xylem sap transport in wood is a mass transport of dissolved compounds in moving tracheidal water of sap wood. The actual transport rate J (e.g, $J_{Mg^{2+}}$) is defined by the product of the flow rate of the xylem sap, which approximately equals the transpiration rate TR (roughly 5% of the xylem water is usually re-exported by leaves via the phloem sap) times the actual xylem sap concentration of a dissolved compound, e.g. of Mg^{2+} cations $[Mg^{2+}]_{xyl}$:

$$J_{Mg^{2+}} = TR \cdot [Mg^{2+}]_{xyl}$$

Both the stomatal canopy transpiration rate TR and the concentration $[Mg^{2+}]_{xyl}$, depend (and interdepend) in a complex manner on meteorological, pedological and

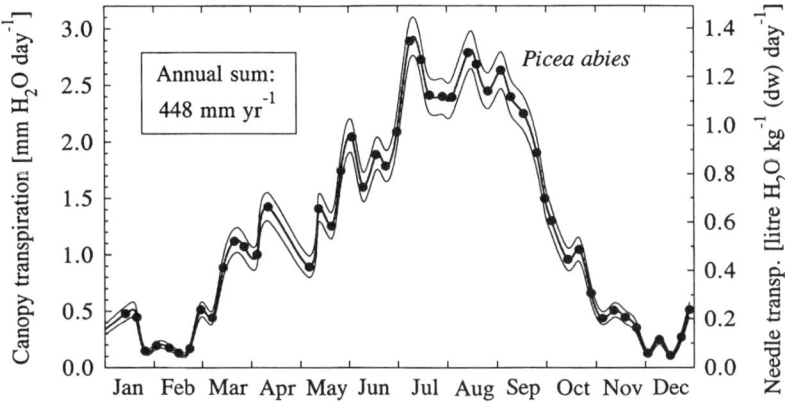

Figure 10. Stomatal canopy transpiration (at Würzburg, 1991) as estimated using continuously measured air temperature and air humidity data (Slovik, unpublished), from which ΔW was calculated. The obtained annual kinetics of ΔW were employed to calculate the annual canopy transpiration kinetics using the calculus given in Figure 9. The thin lines represent uncertainties of data (statistical error progression after Gauss)

ecophysiological conditions. Figure 10 shows the annual kinetics of the stomatal transpiration TR of spruce trees in the field (Würzburg in 1991) as estimated from continuously measured $\Delta W = f(T_{air}, RH_{air})$ data using the sigmoid response function supplied in Figure 9. Transpiration rates are closely interrelated to weather conditions. Synchronously, we also measured the annual kinetics of the xylem sap composition of twigs of seven visually healthy Norway spruce trees (Würzburg in 1991) using the pressure bomb method as described by Osonubi *et al.* (1988); see Kindermann *et al.*, 1992 and Slovik *et al.*, 1992a). The kinetics of the $[Mg^{2+}]_{xyl}$ concentration, which was measured using ICP spectrophotometry, are shown in Figure 11. The kinetics of Figure 10 (TR) and Figure 11 ($[Mg^{2+}]_{xyl}$) are largely different. Comparing both graphs in detail, there is a rough tendency that Mg^{2+} concentration is high if transpiration is small, e.g. in the winter season, and vice versa. But this is not very clear since there is a synchronous dependency of Mg^{2+} uptake by roots on the soil temperature, precipitation etc., and there is recirculation of Mg^{2+} from the phloem sap into the xylem, consumption of Mg^{2+} for growth etc. Additionally, there are time delay phenomena because of the long transport distances in large trees and since ion exchange processes occur along the tree's trunk (see section 4.3). In Figure 12, the Mg^{2+} transport kinetics and the annual integral is obtained by multiplying the corresponding data in Figure 10 and Figure 11. The total annual xylem transport of Mg^{2+} in Norway spruce trees in Würzburg in 1991 was $870 \pm 130 \, mol \, Mg^{2+} \, ha^{-1}$. Compared with the annual Mg^{2+} demand of Norway spruce crowns in Würzburg, which is about $182 \, mol \, Mg^{2+} \, ha^{-1}$ (see section 4.4), the annual xylem flux rate is 4.8 times higher. In Würzburg, about 21% of the xylem sap import into the crown is consumed for growth and 79% of the Mg^{2+} transport rate in the xylem sap is Mg^{2+} recycling via phloem export from the canopy or Mg^{2+}

128

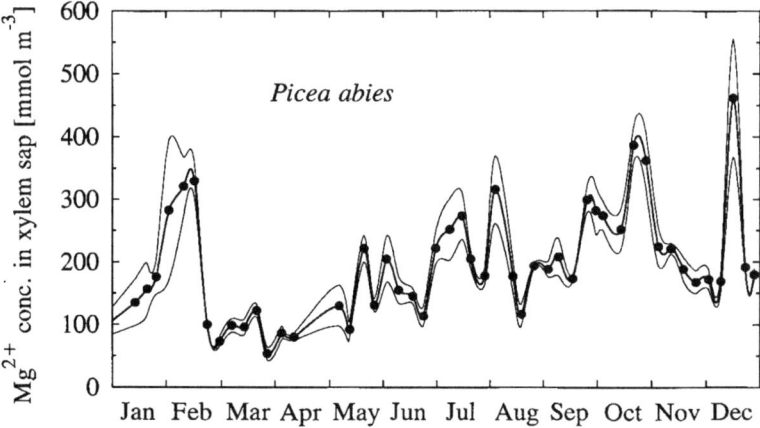

Figure 11. Annual kinetics of the Mg²⁺ content in the xylem sap of Norway spruce trees (Würzburg, 1991). Xylem sap was extracted with a pressure bomb from spruce twigs. Analysis of cations was performed by ICP spectrophotometry. The standard deviation is twice as high ($n = 4$ measurements per point) as the given thin lines indicate. They represent the standard errors of the mean Mg²⁺ concentrations, since these standard errors! must be taken for the Gauss law of error progression (see Figure 12; Kindermann and Slovik, unpublished)

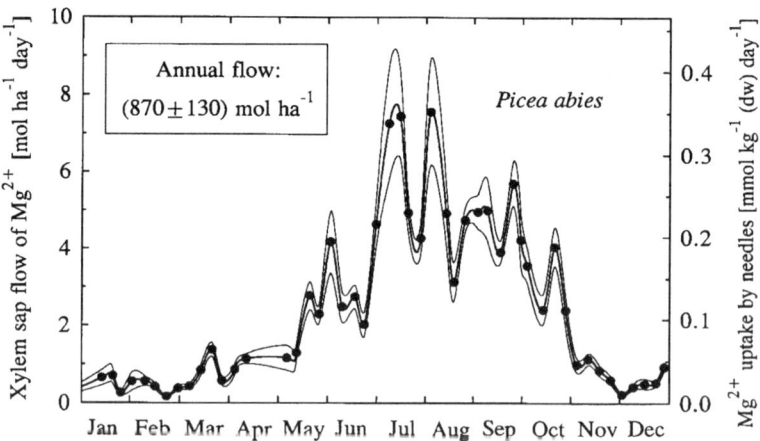

Figure 12. Xylem sap flow rate of Mg²⁺ in Norway spruce (Würzburg, 1991) as estimated from the product of the corresponding Mg²⁺ concentration and water flux data (see Figures 10 and 11). The thin lines represent standard errors (Gauss law of error progression) of the Mg²⁺ flux rate kinetics. The total annual Mg²⁺ flux is the area below the given curve. The integral was 870 ± 130 mol Mg²⁺ ha⁻¹ y⁻¹ in 1991

Table 5. Estimate of the total mass of magnesium transported via the xylem sap during the growth period, of the Mg demand for growth and canopy leaching within the growth period, and of the total magnesium content in the nonstructural xylem water pool of the stems of healthy and Mg-deficient Norway spruce trees in the Fichtelgebirge

Magnesium balance	Healthy trees	Declining trees
Xylem sap content $(mol\,Mg^{2+}\,m^{-3})$	0.25–0.44	0.17–0.35
Magnesium transport $(mmol\,Mg^{2+}\,y^{-1})$	62–108	50–102
Annual Mg^{2+} demand of spruce crowns $(mmol\,Mg^{2+}\,y^{-1})$	16	12
Mg^{2+} pool in stem and branch water $(mmol\,Mg^{2+}\,tree^{-1})$	4–7	2–4

After Schulze *et al.*, 1989

leaching (see section 4.6). In the Fichtelgebirge, Schulze *et al.* (1989) compared individual Mg^{2+}-deficient and healthy spruce trees (see Table 5). They found that about 16–19% of the xylem sap transport of Mg^{2+} is consumed to supply the crown growth and leaching demands. This is similar to the Würzburg situation. Schulze *et al.* (1989) found no significant difference in the capability of Mg^{2+}-deficient and healthy spruce individuals to recycle Mg^{2+} via the phloem sap. It is necessary to mention here that recycling fractions of Mg^{2+} depend on the leaf age, on growth rates and on the plant species. This is well known for herbaceous plants. Jeschke *et al.* (1985, 1987) and Jeschke and Pate (1991) found a relative xylem–phloem Mg^{2+} recycling rate of $21 \pm 2\%$ in shoots of *Lupinus albus*, and $61 \pm 35\%$ in 'mean' leaves of *Rhizinus communis*. Retranslocation was almost 100% in old leaves in *Rhizinus*.

There are numerous valuable communications which present data of the xylem sap composition of several tree species from different sites and gained by diverse methods (e.g. Glavac *et al.*, 1989, 1990; Kazda and Weilgony, 1988; Osonubi *et al.*, 1988) but complete annual concentration kinetics and synchronously measured transpiration rates are rare in order to obtain Mg^{2+} flux balances in detail (but see Dambrine *et al.*, 1992, 1995).

4.2.4. Phloem sap transport

Mg^{2+} cations are readily mobile in the xylem sap and in the phloem sap of plants, but it was concluded that the acropetal Mg^{2+} transport in the xylem sap is facilitated over the basipetal Mg^{2+} transport in sieve cells and tubes of the phloem (Bergmann, 1988). These commonly accepted statements are based on herbaceous plants. Concerning woody plants there is only indirect evidence that Mg^{2+} is readily mobile in phloem cells, since Mg^{2+} deficiency symptoms occur first in the oldest leaves of many trees and there is much more annual xylem transport of Mg^{2+} than necessary to supply the annual demand of above ground organs. Only a small part of the flux difference supplies a moderate Mg^{2+} accumulation in spruce needles from Würzburg

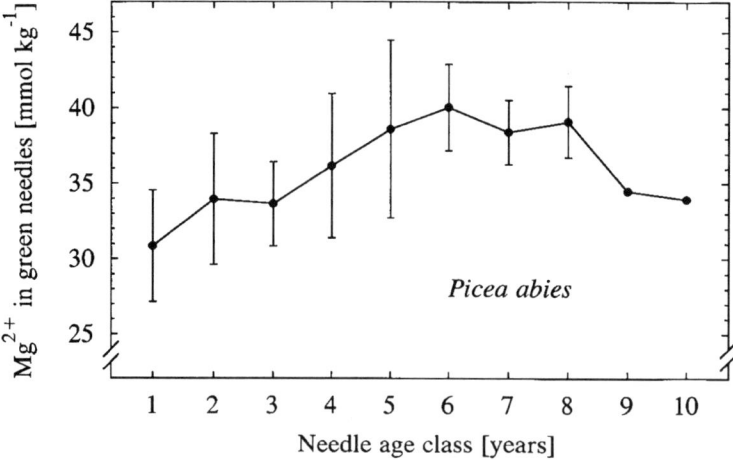

Figure 13. Content of Mg^{2+} in different needle age classes of Norway spruce needles from Würzburg in 1992 (Kindermann and Slovik, unpublished)

(see Figure 13). Most Mg^{2+} must have been re-exported. Young plant organs (developing leaves, sperms and fruits) are potent sinks for Mg^{2+}. Orchard trees with high production rates of fruit tend to develop Mg^{2+} deficiency symptoms more often than vegetative trees. It is worth mentioning that 'oak decline' (*Quercus robur*) in central Europe is accompanied by high production rates of acorns and early leaf senescence. There are indeed hints that the 'brush disease' (German: 'Pinsel-krankheit') of oaks, which is characterized by early loss of old leaves while young leaves at the very top remain alive, is a phenomenon attributed to Mg^{2+} deficiency (Bergmann, 1988). We urgently need fundamental tree physiological information on source sink relations of Mg^{2+}, and on the role of Mg^{2+} mobility in the phloem sap.

At the moment it is possible to estimate Mg^{2+} transport rates in the phloem sap of trees only indirectly and with maximum caution. If the total Mg^{2+} content of 37.3 ± 6.6 mmol kg^{-1} (dw) in Norway spruce bark from Würzburg were symplasmic, a cytoplasmic Mg^{2+} content of about 37 mol m^{-3} would result at ca. 50% relative water content of bark. Only the physicochemically 'free' Mg^{2+} can be transported via mass flow of the phloem sap. Most of the Mg^{2+} remains bound to histological structures and surfaces. Since the cytoplasmic activity of Mg^{2+} is at maximum 10% of the total cytoplasmic Mg^{2+} content (Veloso *et al.*, 1973), the cytoplasmic Mg^{2+} activity in bark cells would be less than 3.7 mol m^{-3}. Norway spruce bark of about 40-year-old trees in the Guttenberger Forst (close to Würzburg, moderate cutting, yield class I–II; Schörry, forest officer, personal communication) consists of 67 ± 13% of living phloem tissue (bast, *n* = 24, Slovik, unpublished). Usually, 20% of the phloem consists of sieve elements (Ziegler, 1982). On the basis of the mentioned (i) age of the stand, (ii) yield class and (iii) moderate cutting regime, it is possible to estimate that the total trunk area is about $34\,m^2\,ha^{-1}$ (see Schober, 1987). About 10% of the trunk volume is bark (Mette and Korell, 1989). Thus, the area per ha of active sieve cells in spruce trunks can be estimated to be $34 \cdot 0.1 \cdot 0.67 \cdot 0.2 = 0.45\,m^2\,ha^{-1}$. The mean longitudinal flow velocity of the phloem sap is roughly $0.1\,m\,ha^{-1}$ in gymnosperms (Ziegler, 1982). Employing the mean length of the vegetation period in Würzburg of 210 days per year, there are 210 days times 24 hours per day = 5040 hours of annual phloem sap transport. Thus, the estimated annual Mg^{2+} transport rate via the phloem sap is $3.7\,mol\,m^{-3} \cdot 0.1\,m\,h^{-1} \cdot 0.45\,m^2\,ha^{-1} \cdot 5040\,h\,y^{-1} = 840\,mol\,Mg^{2+}\,ha^{-1}\,y^{-1}$. From data mentioned above, a Mg^{2+} recycling rate via the phloem sap = 79% of $870\,mol\,ha^{-1}\,y^{-1} = 687\,mol\,ha^{-1}\,y^{-1}$ can be estimated. Eschrich et al. (1988) showed that Mg^{2+}, K^+ and phosphorus were readily retrieved from beech leaves prior to shedding, and deposited mainly in cortex and pith tissues of the stem. After all, there is no doubt that there is a sufficiently high phloem mobility of Mg^{2+} in forest trees. Dambrine et al. (1995) produced data for Norway spruce which indicate that there is much more recycling of Mg^{2+} within the spruce crown than between roots and shoots. Xylem sap (and perhaps phloem sap) concentrations are higher in spruce crowns than in the trunk (Dambrine et al., 1995). Böttcher (1987) found that the electrical conductance of spruce bark, which is a measure of the unspecific overall ion mobility in the bark tissue, is reduced by ca. 10% in damaged spruce trees relative to healthy trees. More and detailed information on the phloem sap of trees is urgently needed.

4.3. Mobilization of magnesium

There are two main ecophysiological sources of Mg^{2+}: (i) uptake from soil and Mg^{2+} transport in the xylem sap (see also Chapters 6, 8 and 10), and (ii) remobilization of Mg^{2+} from different tree organs in the spring season (Norway spruce) or from senescent leaves in the autumn (deciduous trees like beech: see Eschrich et al. 1988). Also the extent of ion exchange between sap wood and the xylem sap is discussed as a means of Mg^{2+} mobilization and Mg^{2+} buffering.

4.3.1. Magnesium uptake by roots

In contrast to most other cationic plant nutrients, uptake of Mg^{2+} from the soil by fine roots and via mycorrhiza is extremely sensitive to cation competition (Bergmann, 1988; Marschner, 1986; Mengel, 1984; for details see Chapter 8 by Kaupenjohann, this volume). Typically, uptake of Mg^{2+} by Norway spruce trees is reduced by soil podsolization, which is facilitated e.g. on sandy and readily leaching acidic substrates and by acid precipitation. Physiological adaptation strategies of plants to acidic soils are summarized by Marschner (1991). Schulze et al. (1989) found in the Fichtelgebirge that Mg^{2+} in the soil solution and the Mg^{2+} content in spruce needles showed only weak correlations, but there was a positive correlation between Mg^{2+} content in needles and the number of ectomycorrhizal root tips per leaf area (Meyer *et al.*, 1988). Also too high soil pH values reduce the Mg^{2+} availability (Baule and Fricker, 1967; Lyr *et al.*, 1967). Mg^{2+} uptake is also reduced in cold spring seasons with intensive precipitation.

Competition with potassium Intensive potassium fertilization produces two inversely acting effects on the Mg^{2+} supply of plants. (i) K^+ cations substitute Mg^{2+} from cation-exchanging clay surfaces in the soil (Németh and Grimme, 1974; Hahlin, 1973), and (ii) there is competition of Mg^{2+} uptake with K^+ (Baumeister and Ernst, 1978; Bussler, 1979). The net effect of potassium nutrition depends on the specific field situation. In typical cases, oversupply with potassium reduces the Mg^{2+} uptake from the soil. Hence, Buchner (1985) stressed the importance of balanced K:Mg ratios in soils and in plant fertilizers. Very sensitive to imbalanced K^+ and Mg^{2+} fertilization are fruit trees (*Malus* spp., *Citrus* spp.). The apple cultivar 'Golden Delicious' is known to sensitively respond with Mg^{2+}-deficiency symptoms and early leaf loss if potassium fertilization is imbalanced (Mantinger, 1974; Schumacher, 1976), especially if the root space is restricted in compact orchards. Concerning Norway spruce, Schneider *et al.* (1989) found differences neither in the K^+ and Mg^{2+} contents in fine roots, nor in the mycorrhizal infection between healthy and declining trees on Mg^{2+}-depleted soils in the Fichtelgebirge. Hence, observed Mg^{2+} deficiency in the Fichtelgebirge seems not to be a consequence of K^+ and Mg^{2+} competition.

Competition with calcium Usually, the Mg^{2+} concentration in soil water is roughly one order of magnitude smaller than the Ca^{2+} concentration. This fraction drops, especially on alkaline Mg^{2+}-poor rendzina soils. Possible reduction of Mg^{2+} uptake by strongly increased Ca^{2+} concentrations in the soil solution must be considered if acidic forest soils are to be limed (see Schaaf (Chapter 11) and Kaupenjohann (Chapter 8), this volume). An optimum Ca:Mg ratio in plants must be achieved if Ca^{2+}-dependent Mg^{2+} deficiency is to be avoided (Wichmann, 1976).

Competition with aluminium The inhibition of Mg^{2+} uptake by different Al ion species, Al^{3+}, $Al\cdot(OH)^{2+}$ etc., is a well-known phenomenon which has been repeatedly found with herbaceous plants and tree seedlings (e.g. Godbold 1991; Godbold *et al.*, 1988; Grimme, 1981, 1982, 1983, 1984; Hecht-Buchholz *et al.*, 1987; Jentschke *et al.*, 1991; Jorns and Hecht-Buchholz, 1985; Kuhn *et al.*, 1995; Marschner, 1992; Schimansky, 1991; cf. Zöttl, 1983). Competitive aluminium cations are potentially phytotoxic (Foy *et al.*, 1978; Parker *et al.*, 1989). Paulus and Bresinsky (1989) studied in vitro the competition of Mg^{2+} uptake by hyphae of the fungus *Suillus variegatus*, which is an important mycorrhizal partner of *Pinus sylvestris* roots in the field, in the presence of 0.37 mmol/L Al in the culture medium. The Mg^{2+} content of the still-growing mycelium was reduced to 18.5% relative to the Al-free control. In the field, sufficiently high Al concentrations in the soil solution occur on acidic soils which may result from 'acid rain' and 'acid mist' deposition. Especially in Mg^{2+}-depleted acidic soils, the ratio of soluble Al:Mg ions may become sufficiently high to significantly inhibit Mg^{2+} uptake and to contribute to Mg^{2+} deficiency (for details, see Chapter 6). The situation is confused in acidic soils by synchronous solubilization of Mn^{2+} which also inhibits Mg^{2+} uptake, but an increased uptake of Mn^{2+}, on the other hand, substitutes Mg^{2+} in many cell physiological processes (see section 4.1). Fertilization of Mg^{2+}-depleted acidic soils with $MgSO_4$ was not a sufficient means of Mg^{2+} supply if there was no synchronous liming (Jung and Dressel, 1977). Schneider et al. (1989) found no differences in the Al contents of fine roots of either damaged or healthy Norway spruce trees growing in the Fichtelgebirge on Mg^{2+}-depleted soils. In Würzburg, we found a total annual uptake rate of $45.5 \, mol \, Al \, ha^{-1} y^{-1}$ and $352 \, mol \, Mg \, ha^{-1} y^{-1}$ of healthy Norway spruce trees (yield class I, after Wiedemann 1939) growing on shell lime soils (Slovik, unpublished). Thus, there is considerable Al uptake also from alkaline soils, which does not cause aluminium toxicity symptoms. The potential role of Al in Mg^{2+} nutrition and in forest decline is still a matter of dispute.

Interactions with nitrogen supply   Ammonium NH_4^+ cations are potent competitors of Mg^{2+} uptake by roots. This has been repeatedly shown for numerous plant species (Baumeister and Ernst, 1978; Palaniyandi and Smith, 1978, 1979) and was confirmed for barley, spinach and *Fagus sylvatica* by W.M. Kaiser and his coworkers (personal communication). Boxman (1988) found that ammonium uptake increases the efflux mainly of Mg^{2+} from the roots. In the Fichtelgebirge, there are NH_4^+ concentrations in the soil solution up to $0.8 \, mol \, m^{-3}$. Jung and Dressel (1977) detected increased Mg^{2+} leaching rates from soils by $15-25\%$ in the presence of NH_4^+ cations, which release Mg^{2+} from clay surfaces in exchange. Since nitrification of ammonium to nitrate by soil bacteria is inhibited mainly in acidic soils, there are synergetic effects of soil podsolization on the Mg^{2+} supply of forest trees which can hardly be quantified in the field. Mg^{2+} deficiency symptoms and early leaf loss was induced by imbalanced nitrogen fertilization also with apple trees (cultivar 'Golden Delicious'; Schumacher, 1976; Mantinger, 1974). It is necessary to pay more attention to the role of NH_4^+ competition in Mg^{2+} nutrition of forest trees.

4.3.2.   Retranslocation from other organs

In the spring, the Mg^{2+} supply for young growth in the spruce crown partly originates from retranslocation via the phloem sap from older leaves (Baule and Fricker, 1967; Heinsdorf, 1963; Kozlowski, 1971; Lyr et al., 1967). This is the case even in Mg^{2+}-deficient plants, but the Mg^{2+} content of current-flush spruce needles (and fine roots) remains smaller than the Mg^{2+} contents in healthy control trees growing in the Fichtelgebirge (Schulze et al., 1989). In Norway spruce, Mg^{2+} mobilization is focused mainly on one-year-old needles in Mg^{2+}-deficient trees. These needles export Mg^{2+} independent of their absolute Mg^{2+} content and of their health status, even if their own magnesium content drops below the level where yellowing begins (Oren et al., 1988; cf. Ingestad, 1959). Pruning of terminal buds of spruce twigs removes the most dominant Mg^{2+} sinks and thus indeed avoids development of

Mg^{2+}-deficiency symptoms in older needle age classes (Lange *et al.*, 1987, 1989a,b; Lippert, 1988; Oren *et al.*, 1988). As a consequence of pruning, the previous year's needles retained minerals that were lost from the needles of intact branches and one-year-old needles may now even become sinks for Mg^{2+} (Ca^{2+}, Mn^{2+}) instead of Mg^{2+} sources. Consequently, the chlorophyl and carotenoid content of the Mg^{2+}-deficient previous year's needles increased. The photosynthesis capacity and, to a small extent, the CO_2 assimilation rate was raised. At sites without Mg^{2+} deficiency (e.g. Würzburg) there is usually a more or less steady increase in the Mg^{2+} content in ageing needles (see Figure 13). Rauterberg and Miraftabi (1970) found that the ratio of the Mg^{2+} contents of old and young leaves is a potent means to early identification of latent Mg^{2+} deficiency.

Bussler (1973) observed that the Mg^{2+} retranslocation via the phloem sap is less potent than the retranslocation of organic nitrogen compounds N_{org} in numerous plants, i.e. the relative retardation of N_{org} in source tissues is smaller. As a consequence of the annual leaf and needle turnover, this may lead to an increasing 'dilution' of Mg^{2+}, i.e. to a reduction of the $Mg^{2+}:N_{org}$ ratio in needles after numerous years if there is either Mg^{2+} shortfall or N oversupply (Schulze, 1989). In contrast to mainly cytosolic protein N_{org}, apoplasmic and vacuolar Mg^{2+} in needles is not as readily available as N_{org} for retranslocation (Bergmann, 1988). The reason is that some of the Mg^{2+} anions are either covalently bound to structural cell wall compounds (e.g. apoplasmic Mg-pectate), or the charge compensating anions of Mg^{2+} are not sufficiently phloem mobile (e.g. vacuolar SO_4^{2-} in Norway spruce: see Slovik *et al.*, 1992a,b; Kaiser *et al.*, 1993). There are observations with herbaceous plants that transient Mg^{2+} deficiency symptoms can be induced if growth rates and supply rates with nitrogen are higher than the Mg^{2+} retranslocation rates from older tissues into the rapidly growing young tissues, i.e. there are transient kinetic problems of transport 'harmony'. Transient Mg^{2+} deficiency symptoms may indeed occur with fast-growing cereals but chlorotic leaves will later become green again and there is no reduction in the final grain yield in the field (Bergmann, 1988).

In the spring, there is a second main source of Mg^{2+} in Norway spruce besides older needles. Figure 14 shows the annual kinetics of the total Mg^{2+} content of the sap wood of spruce twigs (Würzburg in 1991, Kindermann and Slovik, unpublished). Starting in March, the Mg^{2+} content in wood is reduced by about $4\,mmol\,kg^{-1}$ dw (almost 50%) within two months. In May, slow recovery of the Mg^{2+} content in sap wood starts. The Mg^{2+} content in bark of the same Norway spruce trees and twigs showed no pronounced annual kinetics but there was a synchronous increase in the Mg^{2+} content in bark by about $6\,mmol\,kg^{-1}$ in March to April, which was the steepest concentration change of the whole annual kinetics (Figure 15). Since the dry matter ratio of bark to wood in the analyzed twigs was approximately 4 : 5, there seems to be a transient retranslocation of Mg^{2+} from twig wood into the twig cortex. Apparently, there was no significant net contribution of the bark itself in supplying the young growth of spruce buds in May. As already shown, e.g. in the Fichtelgebirge, we confirm with healthy trees that there is only a small contribution from older needle age classes in supplying the young flush with Mg^{2+} cations

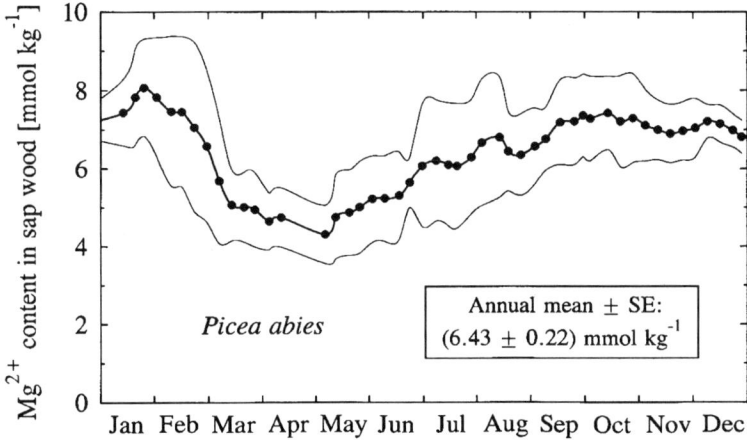

Figure 14. Annual kinetics of the Mg^{2+} content in twig sap wood of seven different Norway spruce trees (Würzburg, 1991), which have been analyzed repeatedly (every seven weeks four twigs of the same tree). The running averaging (7 points per running mean) compensated for possible genetic differences of the trees. Thin lines represent the standard error of the obtained 'function'. The annual mean (±SD) of the Mg^{2+} content in sap wood was 6.43 ± 1.54 mmol Mg^{2+} kg^{-1} (dw); after Kindermann and Slovik, unpublished

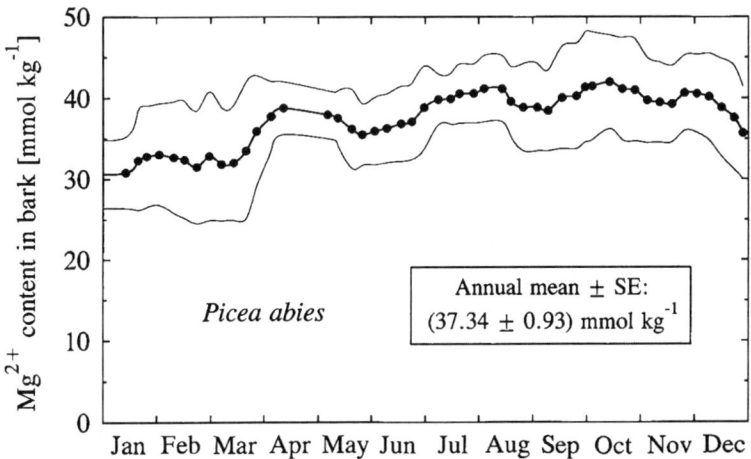

Figure 15. Annual kinetics (1991) of the Mg^{2+} content in twig bark (phloem + rhytidoma) of Norway spruce growing in Würzburg. For details, see legend of Figure 14; after Kindermann and Slovik, unpublished

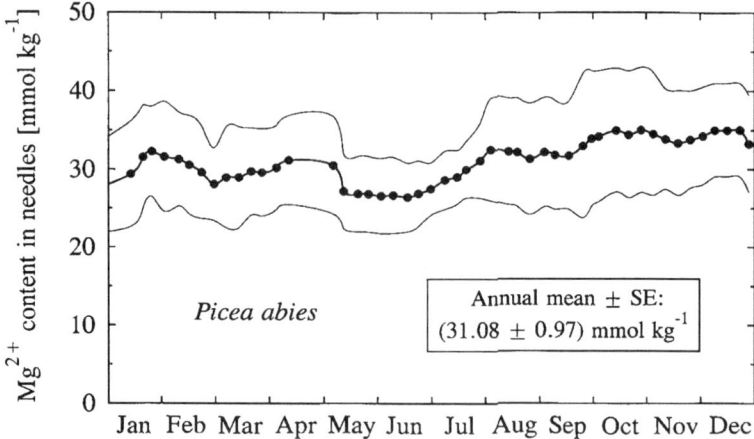

Figure 16. Annual kinetics of the mean Mg^{2+} content of 'mean' Norway spruce needles (Würzburg, 1991). For details, see legend of Figure 14; after Kindermann and Slovik, unpublished

(Figure 16). The Mg^{2+} content of older needle age classes dropped in May only from approximately 30 to 26 mmol Mg^{2+} (kg dw)$^{-1}$, which corresponded to the most pronounced change rate of the Mg^{2+} content in older spruce needles in 1991. On the basis of the mean relative composition of mean Norway spruce branches (needles : wood : cortex = 33% : 49% : 18%; see Table 6 and section 4.5), tentatively the relative contribution of different tissues in supplying the young flush can be estimated if, for simplicity's sake, it is assumed that in the first winter season young needles would achieve approximately the Mg^{2+} content of older needles age classes. Transient mobilization of Mg^{2+} in sap wood is 0.49 kg wood (kg branch)$^{-1}$ times 4 mmol Mg^{2+} (kg wood)$^{-1} \approx 2.0$ mmol Mg^{2+} (kg branch)$^{-1}$ per year, and the transient Mg^{2+} mobilization from 'older' needle age classes is roughly maximum 0.33 kg needles (kg branch)$^{-1}$ times 4 mmol Mg^{2+} (kg needles)$^{-1} \approx 1.3$ mmol Mg^{2+} (kg branch)$^{-1}$ per year. The estimated sum of both is 3.3 mmol Mg^{2+} (kg branch)$^{-1}$ for healthy twigs. The mean needle age in Würzburg is 6.1 ± 0.6 y for *Picea abies* (Slovik, unpublished), 5.7 ± 0.5 y for *Picea pungens* and 2.3 ± 0.5 y for *Pinus sylvestris* (Hüve and Slovik, unpublished). Since the mean spruce needle age in Würzburg equals NA – 6.1 y, the net Mg^{2+} demand for just replacing the annual needle litter production (keeping the volume of the spruce crown constant) would be 0.33/6.1 times 31 mmol Mg^{2+} (kg needles)$^{-1} \approx 1.7$ mmol Mg^{2+} (kg branch)$^{-1}$ per year. Thus, the annual Mg^{2+} retranslocation from sap wood and from older needles is sufficient to transiently supply the annual Mg^{2+} demand of two times the actual needle turn-over rate. Apparently, Mg^{2+} uptake from the soil is not yet dominant in directly supplying the Mg^{2+} demand of the new growth in the spring. The soil supply mainly just resubstitutes the transient Mg^{2+} loss of branch wood and of older needle age classes

Table 6. Mg^{2+} contents in xylem sap, wood and cortex (phloem + rhytidoma) of twigs, and in green leaves and leaf litter of 17 different tree species growing in the botanical garden of Würzburg

Tree species	Xylem sap (μmol/L)	Sap wood (mmol kg^{-1})	Cortex (mmol kg^{-1})	Green leaves (mmol kg^{-1})	Leaf litter (mmol kg^{-1})
Abies alba	172±35	6±1	41±10	70±5	69±5
Acer pseudoplatanus	157±55	15±3	53±12	141±22	115±39
Alnus glutinosa	264±192	13±6	33±4	130±38	132±22
Betula pendula	209±41	13±2	27±1	101±14	91±11
Carpinus betulus	240±106	11±4	51±8	76±8	66±13
Fagus sylvatica	146±71	43±36	32±11	111±12	75±16
Fraxinus excelsior	295±131	24±6	51±17	172±103	154±48
Larix decidua	254±21	12±2	39±2	120±20	120±27
*Picea abies**	194±89	6.4±1.6	37.3±6.6	31.1±6.8	37.6±5.1
Picea pungens	183±3	8±2	44±14	24±7	24±7
Pinus sylvestris	202±75	17±8	44±9	54±8	56±19
Populus nigra	209±21	17±8	51±4	86±3	85±21
Quercus robur	152±30	22±11	54±13	120±47	115±27
Salix alba	121±46	13±7	56±10	129±14	145±33
Sorbus aucuparia	216±73	15±4	75±1	183±42	152±35
Tilia cordata	168±66	36±4	50±8	147±28	162±16
Ulmus glabra	248±159	13±2	55±8	144±6	152±7

Harvest in fall 1991; means ± SD, $n = 3$ for all species but Norway spruce ($n = 200$) (see Table 10)
*Means from entire annual kinetics. The Mg content in root wood of spruce is 15.2 ± 1.4 mmol kg^{-1}; in the root cortex 43.9 ± 9.4 mmol kg^{-1}

(mainly age class 1) in the summer (Figure 14). These deductions from the Würzburg situation fully agree with field observations made in the Fichtelgebirge (see e.g. Lange *et al.*, 1989a,b). Thus, retranslocation of Mg^{2+} in the spring is not a consequence of Mg^{2+} deficiency but it is a common phenomenon of Norway spruce.

A rough estimation is possible concerning the possible time range until mobilization of Mg^{2+} from trunk sap wood is depleted if the annual Mg^{2+} retranslocation in branches and from older needles were fully compensated only from steady and linear reduction of the Mg^{2+} content in the trunk at stands with completely Mg^{2+}-depleted soils. The total stock of sap wood can be estimated on the basis of data given in Schober (1987), page 62, e.g. for the Guttenberger Forst in Würzburg (35-year-old spruce trees, yield class I, moderate cutting): the stock of trunk wood ('Derbholz') is about 240 m^3 fresh volume per hectare. Since the specific density is about 360 kg dry wood (m^3 fresh volume)$^{-1}$ (see Figure 33 in section 4.5), the stock of wood dry matter is ca. 86 tonnes per ha in the Guttenberger Forst. The mean lifetime of sap wood is 10 – 37 years in spruce trunks (Bernhart, 1965; Bertog, 1895; Hartig, 1892; Langner, 1932). A 'typical' sap wood age of spruce is roughly 25 years. Thus, about 8% of the wood stock is dead heart wood if the thickness of the annual circles in the wood is approximately equal for all years. Hence, 0.92 times 86 tonnes ≈ 79 tonnes is the amount of trunk sap wood per ha. The Mg^{2+} content of sap wood in Würzburg is ca. 31 mmol kg^{-1}. Therefore, the total stock of Mg^{2+} equals 31 mol Mg^{2+} tonnes^{-1} times 79 tonnes ha^{-1} ≈ 2450 mol Mg^{2+} ha^{-1} in trunk sap wood of 35-year-old Norway spruce trees in Würzburg. From Figure 14, we learn that about 50% of the Mg^{2+} content can be mobilized, i.e. 1225 mol Mg^{2+} ha^{-1}. The total annual

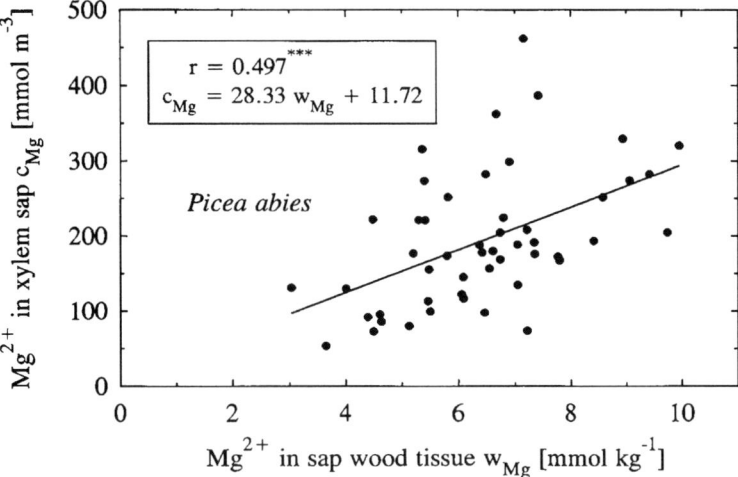

Figure 17. Correlation of the Mg^{2+} concentration in the xylem sap of Norway spruce with the Mg^{2+} content in twig wood from which the xylem sap was extracted (Würzburg in 1991); after Kindermann and Slovik, unpublished

Mg^{2+} demand for needle growth of these 35-year-old trees equals about $143 \, mol \, Mg^{2+} ha^{-1} y^{-1}$ (see Figure 38). Thus, the ratio 1225/143 reveals that 8–9 years would pass before Mg^{2+}-deficiency symptoms started to develop if, over all that time, there was no Mg^{2+} uptake at all from the soil. This is an extreme assumption. In the field, the situation is of course much more complicated. Nevertheless, the given figures impressively stress the high potential importance of the Mg^{2+} buffer in the trunk sap wood to compensate transient Mg^{2+} deficiency. On the other hand, it is important to supply Mg^{2+} as early as possible on Mg^{2+}-depleted soils in order to allow trees to fill their trunk buffers. This can be time-consuming and may delay recovery from Mg^{2+} deficiency in the field after Mg^{2+} fertilization.

4.3.3. Ion exchange between xylem sap and sap wood

The evident potential importance of sap wood (besides one-year-old needles) as a main source of Mg^{2+} in the spring leads to the question of whether available Mg^{2+} mainly originates from living xylem parenchyma cells from ion exchange processes, or directly from the xylem sap. Figure 17 shows a significant correlation ($p > 9.99\%$) between the Mg^{2+} concentration in the xylem sap of Norway spruce and the total Mg^{2+} content in Norway spruce sap wood dry matter of the same twigs from which the xylem sap has been obtained. There is significant Mg^{2+} exchange between the xylem sap and the sap wood. Since the correlation coefficient $r = 0.497$ is small ($r^2 = 0.247$), roughly only $100 \cdot r^2 = 25\%$ of the Mg^{2+} concentration in the xylem sap can be explained by Mg^{2+} exchange processes. The slope in Figure 17 approximates the native partition coefficient or 'equilibrium', constant $K = [Mg^{2+}]_{sap}/[Mg^{2+}]_{wood}$ $\approx 28\,(\mu mol/L)/(mmol\,kg^{-1})^{-1}$ between xylem sap and sap wood, which is a moderate Mg^{2+} buffer that interferes with measured xylem sap concentrations. The smallest annual Mg^{2+} concentrations in the xylem sap (see Figure 11) are measured from March to April. This coincides with the sharp reduction of the Mg^{2+} content in spruce sap wood (Figure 14) when Mg^{2+} is apparently transported into the bark (phloem) and from there together with stored assimilates via the phloem sap into the young needle flush. The ion exchange properties of the wood of *Abies alba* have been analyzed by Korte (1985).

4.4. Magnesium content of tree tissues

The total Mg^{2+} content in different plant species strongly depends on the plant organ, soil conditions, tissue age, climatic conditions, etc. High Mg^{2+} contents and 'sink capacities' of Mg^{2+} are usually found in young vegetative and generative organs (flowers, fruits, seeds), which are supplied with assimilates and minerals mainly via the phloem sap. Nuts of different tree species contain $60-130\,mmol\,Mg^{2+}\,kg^{-1}\,dw$ in the eatable cotyledons (Bergmann, 1988). Thus, fruit production strongly competes with other tissues for phloem-mobile Mg^{2+} mainly in years with excessively high fruit production (= alternance).

4.4.1. Contents in leaves, wood and bark of different tree species

Table 6 alphabetically summarizes the Mg^{2+} concentrations (xylem sap) and tissue contents (wood, bark, green leaves and fresh leaf litter) in autumn 1991 of 17 different almost evenly aged European tree species growing in the botanical garden of our institute in Würzburg (Kindermann and Slovik, unpublished). The Mg^{2+} content in twig wood ranges from 6 (*Abies alba*) to 43 (*Fagus sylvatica*) $mmol\,Mg^{2+}\,kg^{-1}\,dw$. Bark contents, ranging from 27 (*Betula pendula*) to 75 (*Sorbus aucuparia*) $mmol\,Mg^{2+}\,kg^{-1}\,dw$ are not as heterogeneous. Leaf contents vary between 24 (*Picea pungens*) and 183 (*Sorbus aucuparia*) $mmol\,Mg^{2+}\,kg^{-1}\,dw$. Since all tree species were growing on the same soil and in the same climate in Würzburg, observed differences of the Mg^{2+} content may be attributed mainly to species differences in Mg^{2+} acquisition from the soil and in Mg^{2+} partitioning. There seems to be no significant Mg^{2+} retranslocation before leaf abscission in green leaves harvested in September are compared with senescent leaves shed in October or November. There is no dominant Mg^{2+} retranslocation in the late stages of leaf senescence or it is too moderate to be readily detectable in the field. Table 6 summarizes the mean Mg^{2+} contents of 'old' but still green leaves before leaf senescence, with respect to the oldest still green needle age classes of evergreen gymnosperms in September 1991. Since the dry matter per needle considerably increases in the first year, young spruce needles may contain more Mg^{2+} on a leaf dry weight basis than older needle age classes (e.g. Höhne, 1964a,b,c). In contrast to Figure 13, Mg^{2+} contents of needles (*Picea*, *Pinus*, *Abies*, *Pseudotsuga*) usually

Table 7. Mg²⁺ minimum contents in young leaves of different tree species which are assumed to be sufficient to support optimal growth

Tree species	Green young leaves (mmol kg⁻¹)
Abies alba	60–170
Acer spp.	60–120
Betula spp.	60–120
Fagus spp.	60–120
Fraxinus spp.	80–170
Larix decidua	50–120
Picea abies	40–100
Pinus radiata	40–50
Pinus sylvestris	40–80
Populus spp.	80–120
Pseudotsuga taxifolia	40–100
Quercus spp.	60–120
Taxus baccata	40–100
Tilia spp.	60–120

From Bergmann, 1988, S. 381

decrease with increasing leaf age in the field on poor soils (Reemtsma, 1964) but the opposite situation can also be observed (see Leyton, 1948 and Figure 13), especially in well-supplied stands. Usually, there are hardly any proven influences of the location of twigs within the trees on the Mg²⁺ content (Höhne, 1964a,b,c; Leyton, 1948; Wehrmann, 1959b) but there are also data which indicate the highest Mg²⁺ concentrations can be found close to the top of the canopy where xylem sap transpiration is focused within canopies (Fiedler and Höhne, 1965; Pfanz [Würzburg], personal communication).

4.4.2. *Necessary minimum contents in leaves*

Literature data of 'necessary' minimum contents of Mg²⁺ in thoroughly defined young leaves of 'healthy' trees are extremely heterogeneous (see e.g. Table 7). We mention here only a few gymnosperms and will focus on *Picea abies*.

Abies alba   Usually, the Mg²⁺ content in young needles varies between 45 and 65 mmol kg⁻¹ (Lyr *et al.*, 1967), i.e. Mg²⁺ deficiency is expected to occur at almost any site if Table 7 is universal.

Larix spp.   Themlitz (1958) reports 'typical' Mg²⁺ contents in young needles of 50–120 mmol kg⁻¹, which match the interval given in Table 7.

Pinus sylvestris   'Typical' Mg²⁺ contents in young needles are 40–100 mmol kg⁻¹. This interval matches the values given in Table 7. Mg²⁺ deficiency is assumed to become evident below 20–40 mmol Mg²⁺ kg⁻¹; the optimum supply is

50–70 mmol $Mg^{2+}kg^{-1}$ (Baule and Fricker, 1967; Heinsdorf, 1963; Ingestad, 1962; Wehrmann 1959a).

Picea abies   There is a broad variety of Mg^{2+} contents in young spruce needles ranging from 8 (Mg^{2+} deficient) to 100 (Mg^{2+} luxurious) mmol $Mg^{2+}kg^{-1}$ dry matter (e.g. Bosch *et al.*, 1983; Hunger, 1964; Nebe, 1963, 1967; Strebel, 1961; Zech and Popp, 1983; Zöttl and Mies, 1983). Fiedler *et al.* (1984) analyzed in one-year-old needles from the tops of spruce trees 30–60 mmol $Mg^{2+}kg^{-1}$ in the high altitudes of the Erzgebirge, and 20–30 mmol $Mg^{2+}kg^{-1}$ in the Thüringer Wald (800–980 m above sea level (asl)). The optimum supply ranges between 60–90 mmol $Mg^{2+}kg^{-1}$ in one-year-old needles (Fiedler *et al.*, 1984) or between 40 and 70 mmol $Mg^{2+}kg^{-1}$ according to Ingestad (1958, 1959, 1962). It is worth mentioning that the postulated optimum supply with Mg^{2+} is higher in the SO_2-polluted areas of the former GDR (Fiedler *et al.*, 1984) compared with data from Sweden and West Germany with much lower chronic SO_2 pollution. The minimum Mg^{2+} supply (tolerance limit) is even more heterogeneous: most data range between 8 and 30 mmol $Mg^{2+}kg^{-1}$, (Ingestad, 1958, 1959, 1962; Wehrmann, 1963). According to Themlitz (1958), the Mg^{2+} content should be at least 25–45 mmol $Mg^{2+}kg^{-1}$ in spruce needles; below 16–20 mmol $Mg^{2+}kg^{-1}$, deficiency symptoms develop. Very concrete data are published by Zech (1968) and Zech and Popp (1983) who found a limit of 14 mmol $Mg^{2+}kg^{-1}$ in the Fichtelgebirge, which was confirmed by Lange *et al.* (1989a; see Figure 2).

The broad variation of the minimum Mg^{2+} content in young spruce needles (8–30 mmol $Mg^{2+}kg^{-1}$ dw) which is necessary if needle chlorosis symptoms are to be avoided apparently depends on numerous factors. First, there is the complex interdependence of Mg^{2+} supply and phosphorus supply on the one hand, and the possibility to substitute for Mg^{2+} with Mn^{2+} on the other (see section 4.1). Mg^{2+} content limits are low (14 mmol kg^{-1}) at sites where, besides Mg^{2+} depletion, there is synchronous restricted phosphorus supply and limited Mn^{2+} supply. This is the situation in the Fichtelgebirge (Lange *et al.*, 1989a,b; Schulze *et al.*, 1989). Second, the subcellular availability of Mg^{2+} for cytosolic and chloroplast demands is another confusing factor. Vacuolar Mg^{2+} cations do not catalyze or regulate metabolic reactions, nor can they be incorporated into chlorophyl porphyrins, but they can be analyzed in crude needle homogenates. In Figure 18, the sulfate-dependencies of Mg^{2+} content in spruce needles is shown (at Würzburg, 1991). Roughly 40% (slope ≈ 0.41 mol mol^{-1}) of the sulfate content in Norway spruce needles correlates with Mg^{2+} in healthy needles analyzed in the oldest available needle age classes. At 30 mmol $SO_4^{2-}kg^{-1}$ dw, there is 45% more Mg^{2+} in spruce needles compared with the intercept with the ordinate ($100 \cdot 0.415 \cdot 30/27.4 = 45\%$, see equation in Figure 18). This intercept approximates the sulfate-independent Mg^{2+} content in spruce needles, and is therefore a better parameter of the nutritional status than the total Mg^{2+} content in needles, especially in SO_2-polluted areas. Materna (1962) summarizes that there is accumulation of K^+, Mg^{2+} and Ca^{2+} in spruce needles of the Czech republic compared with healthy sites. Consequently, it is necessary to analyze also

142

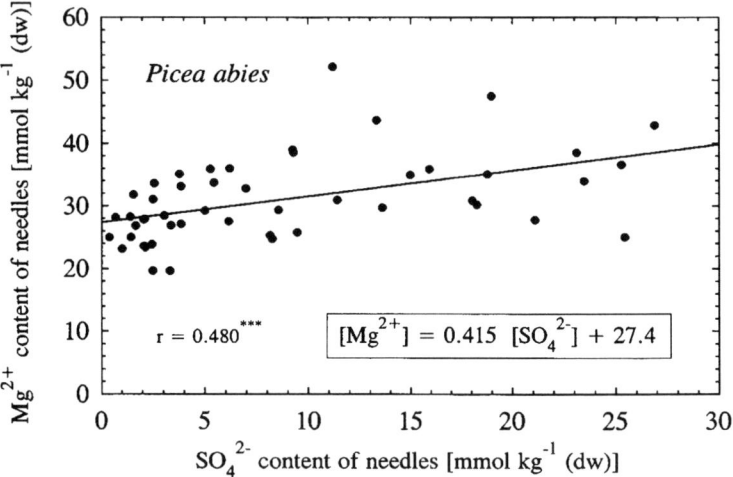

Figure 18. Correlation of the total Mg^{2+} content in Norway spruce needles (Würzburg in 1991) with the water-soluble sulfate content of these needles. Each point is the mean of needles from 4 different twigs per tree. Points are from seven spruce trees and from all seasons. There is an apparent SO_4^{2-} dependency of the Mg^{2+} content by $0.415\,mol\,Mg^{2+}\,mol^{-1}\,SO_4^{2-}$ (Kindermann and Slovik, unpublished)

anions that are sequestered into vacuoles and subsequently neutralized by cations (e.g. SO_4^{2-}) if leaf analysis data are to represent the real nutritional status of plants. Regarding the Würzburg situation, the physiologically available Mg^{2+} content in spruce needles is about $27\,mmol\,kg^{-1}\,dw$ (intercept in Figure 18).

4.5. Growth rates and magnesium demand of different tree tissues

4.5.1. General scope

Knowledge of the individual growth rates of different tree tissues and organs is a fundamental prerequisite of determining source–sink relationships within individual trees and the flux rates and flux balances of Mg^{2+} cycling in native forest ecosystems. On the basis of: (i) the ion content $[Mg^{2+}]_i$ of a tree organ or tissue $[mol\,Mg^{2+}\,kg^{-1}]$, and (ii) the actual growth rate R_i of this organ or tissue $[kg\,ha^{-1}\,y^{-1}]$, the partial annual Mg^{2+} 'demand' D_i $[mol\,Mg^{2+}\,ha^{-1}\,y^{-1}]$ of a defined tissue or organ can be calculated from the product:

$$D_i = R_i \cdot [Mg^{2+}]_i$$

'Demand' is defined here not as the physiologically necessary absolute minimum demand but as the demand that is necessary to explain the observed ion content at a given growth rate at concrete sites. Different tree species contain different amounts of $[Mg^{2+}]_i$ in different tree organs (see Table 6 for the Würzburg situation). Much more complicated to obtain are the individual tissue growth rates R_i for an actual site or stand. In the field, growth rates depend on practically all pedological,

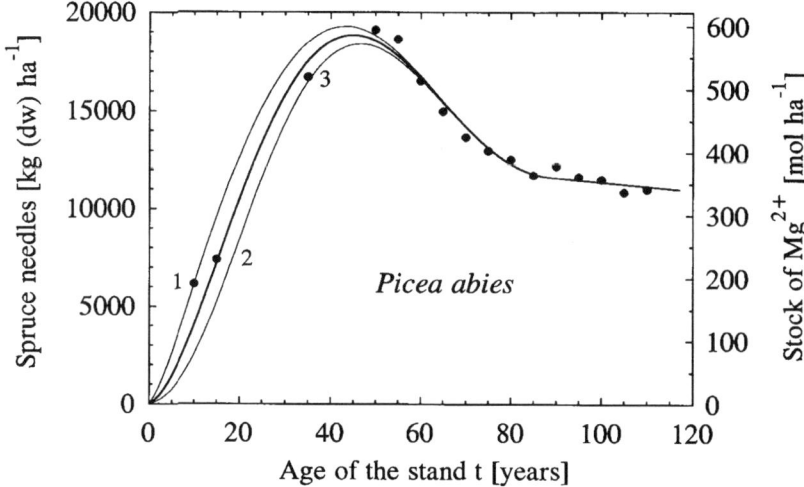

Figure 19. Field data from Vanselow (1951) allow an estimation of the 'typical' stand age dependency of the needle stock in native Norway spruce canopies. Point 1 is data from a Danish Christmas tree (Slovik, unpublished); point 2 is after Møller (1946); and point 3 is from Burger (1939a). Data from Von Droste zu Hülshoff (1969, 1970) indicate that the stock of needles in stands older than 60 years can remain almost as high as in ca. 40-year-old stands (if there is weak cutting of trunks). The right ordinate is an estimate of the Mg^{2+} content in needles per hectare if they contain $31 \, mmol \, Mg^{2+} kg^{-1}$ (dw) as observed in Würzburg (see Figure 15)

meteorological, phytopathological, ontological and genetic influence factors (Kozlowski, 1971; Linder and Flower-Ellis, 1992; Mitscherlich, 1975; Schmidt-Vogt, 1986; Thomasius, 1964). These complex interdependencies can be investigated in some detail only with enormous investment of time and money for particular sites or projects, and even then part of the growth parameters are determined only indirectly, or they must be thoroughly estimated in the field, e.g. root growth. Usually, the annual nutrient demand is estimated simply by summation of the measured demands for trunk growth, litter production and canopy leaching. Nevertheless, there is an urgent demand to know the annual production rates R_i of leaves, leaf litter, twig wood, twig bark, twig litter, trunk wood, trunk heart wood, trunk sap wood, trunk bark, bark litter, root wood, root bark and fine roots in more detail and as exactly as possible. For spruce trees, the following sections will deduce and graphically demonstrate 'typical' growth rates of all mentioned spruce organs and the corresponding partial annual Mg^{2+} demands.

4.5.2. *Stock of spruce needles*

Figure 19 shows the stock of dry matter of spruce needles per hectare depending on the age, t, of a 'typical' spruce stand. The right ordinate represents the Mg^{2+} content of spruce needles per hectare for $[Mg^{2+}]_{needles} = 31.1 \, mmol \, kg^{-1} \, dw$ (Würzburg location). There is a maximum of almost 20 tonnes of needle dry matter per hectare.

In ageing stands, the needle mass per hectare decreases due to cutting trees (thinning). The needle expansion growth rate of the surviving trees only partially compensates the cutting-dependent loss of needle matter per hectare.

The stomatal uptake of air pollutants depends on the total needle surface per hectare. The specific surface S_m of spruce needles [m^2 (total needle surface) $kg^{-1} dw$ (needles)] depends on the light exposure within the canopy, and on the yield class of the stand ('growth conditions'). Mean S_m values usually range between 11 and 13 $m^2 kg^{-1}$ (dw) in healthy stands. Schöpfer (1961) divided the spruce crown into three equal-sized zones, the illuminated 'sun crown', the intermediate 'transient crown' and the 'shallow crown'. S_m is smallest in the 'sun crown', ($S_m = 10.2 m^2 kg^{-1}$), where needles are thick, and highest in the 'shadow crown', where needles are flat (Schmidt-Vogt, 1949). Kerner et al. (1977) present a variable model to compute the spruce needle surface in detail. Burger (1953) found that needles of suppressed trees within the shaded canopy have about 1.18 times higher specific needles surfaces S_m than tall dominant spruce trees. The specific surface S_m tends also to become higher at sites with smaller wood yield, i.e. higher yield classes (Schmidt-Vogt, 1949). Summarizing the data given by Schmidt-Vogt (1949, 1986), the following 'typical' S_m values of 'mean' needles within the crown can be estimated for healthy trees (YC = wood yield class I–V, after Wiedemann; see Schober, 1987):

$$S_m = 0.49 \cdot YC + 10.6$$

In Figure 20, data from Figure 19 are recalculated for yield classes I, II and III in order to obtain the total needle surface, S(t) [m^2 (total) m^{-2} (soil)] and the leaf area index, LAI(t) [m^2 (needle projection area) m^{-2} (soil)] as a function of the age t of the stand. From the total needle surface S(t), the leaf area index LAI can be obtained since the spruce needle surface ratio (m^2 total needle surface) : (m^2 needle projection area) is 2.6 $m^2 m^{-2}$ (Oren et al., 1986) or 2.7 $m^2 m^{-2}$ (Ch. Körner, Basel, personal communication). Thus, LAI(t) S(t)/2.65 typically ranges from about 8 $m^2 m^{-2}$ in 40-year-old canopies to only 5 $m^2 m^{-2}$ in 100-year-old canopies. The maximum total needle surface S is approximately 22 $mm^2 m^{-2}$. A very similar maximum value of S = 21 $m^2 m^{-2}$ saturates close to the trunk of a healthy 10-year-old Norway spruce 'Christmas tree' (Slovik, unpublished; see Figure 21). Typical LAI values for individual Norway spruce trees in the field are 8–11 $m^2 m^{-2}$ (Körner, personal communication). The value LAI = 12 $m^2 m^{-2}$ (Mitscherlich, 1970) is an extreme. The deviation between these LAI values below individual spruce trees and field data from below entire spruce canopies (Figure 20) mainly depends on the magnitude of illuminated gaps between individual trees (See Perterer and Körner, 1990 for details).

4.5.3 Litter production (needles, twigs, bark)

Needle litter There is an annual needle growth–needle litter production balance in native Norway spruce canopies. This balance is negative only at sites with progressive forest decline. Usually, it is positive, i.e. there is net growth of needles per

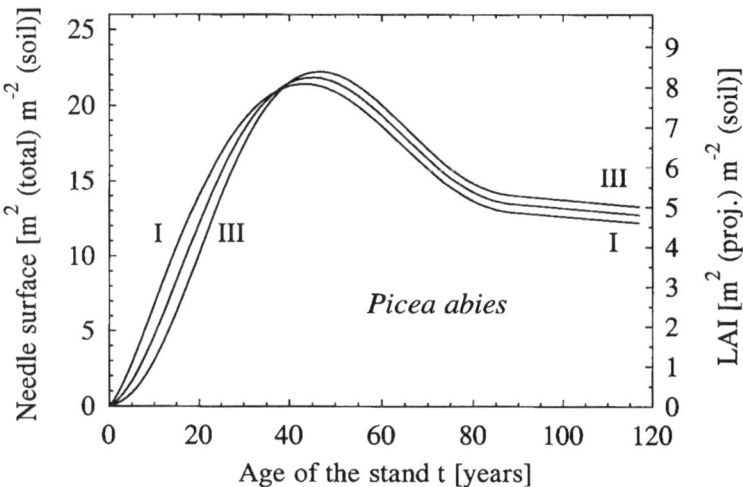

Figure 20. Total spruce needle surface area in relation to the age of the stand calculated from Figure 19 for three trunk wood yield classes after Wiedemann (see Schober, 1987; indicated by roman numbers) if the following specific needle surface data (mean of the canopy) of healthy trees after Schmidt-Vogt (1986) are employed: $I = 11.1\,m^2\,kg^{-1}$ (dw); $II = 11.6\,m^2\,kg^{-1}$ (dw); $III = 12.1\,m^2\,kg^{-1}$ (dw). The right ordinate represents the corresponding leaf area index (LAI = m^2 leaf projection area per m^2 soil area). There are $2.6-2.7\,m^2$ total needle surface area per m^2 needle projection area (Oren *et al.*, 1986; Ch. Körner, Basel, personal communication)

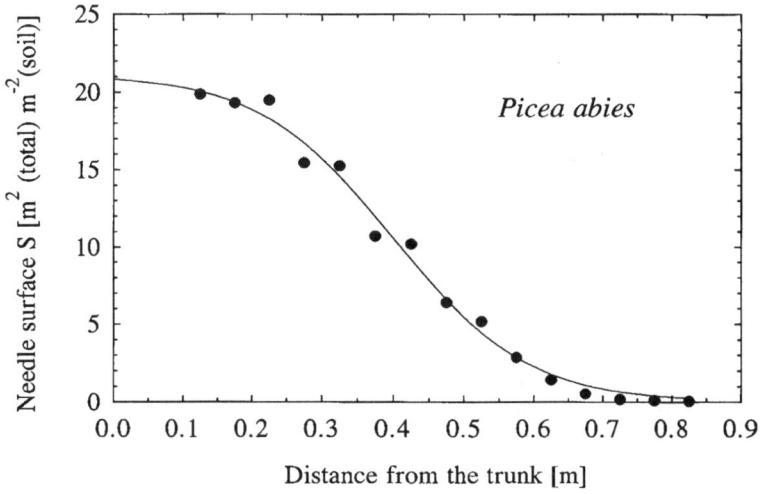

Figure 21. Total needle surface area per ground area of a Danish Christmas tree in relation to the distance from the trunk. The needle surface area is maximum at $20-21\,m^2$ (needles) m^2 (soil), which corresponds to a LAI of $7.5\,m^2\,m^{-2}$. This tree was the same as indicated by point 1 in Figure 19 (Slovik, unpublished)

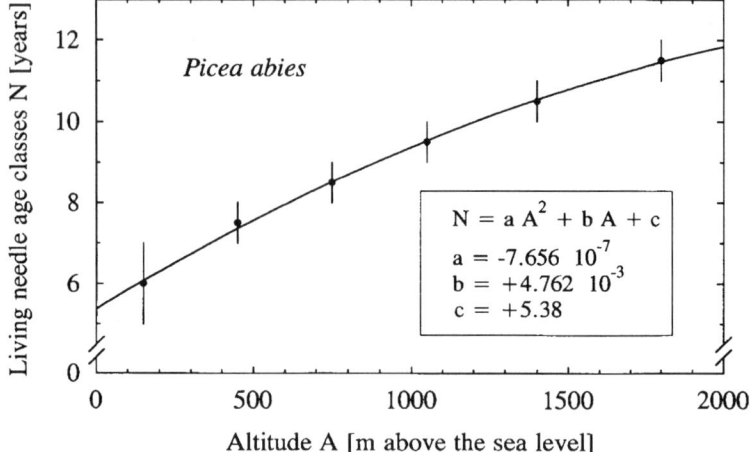

Figure 22. The number, N, of living needle age classes in healthy Norway spruce stands in Switzerland in relation to the altitude above sea level (Burger, 1927). The second-order polynome equation given in the insert is an estimate of the expected needle age, NA, of healthy stands if only the altitude is known

tree, which overcompensates the need to replace needles that are lost via needle litter formation. Thus, there is net expansion growth of healthy spruce crowns. The necessity to cut trees in spruce forests is the consequence and it is the most important reason for positive needle net growth in ageing stands. The total annual needle growth rate (new flush) is the sum of needle net growth (expansion growth) and needle turn-over growth (substitution growth). The needle turn-over growth rate equals the annual needle litter production rate. The needle turn-over rate R_{NL} (index NL=needle litter, [kg dw ha^{-1} y^{-1}]) is the quotient of the stock of needles N [kg dw ha^{-1}] (Figure 19) divided by the mean needle age NA [years]:

$$R_{NL} = N/NA$$

The mean needle age NA of healthy spruce trees, which is approximated by the number of 'living needle age classes N', depends on the altitude A above the sea level (see Figure 22, data after Burger 1927). According to these findings, the mean needle age in Würzburg (190 m asl) is expected to be NA $\approx -7.656 \times 10^{-7} \cdot 190^2 + 4.762 \times 10^{-3} \cdot 190 + 5.38 = 6.3$ y. As already mentioned, the measured NA equals 6.1 ±0.6 y in Würzburg, which is close to expectations. Reduction of the mean needle age in the field can be used to quantify spruce vitality and the state of crown health (Eichhorn and Ackerbauer, 1987; cf. Arndt *et al.*, 1987). Further factors besides the altitude, which influence the mean needle age, are: (i) the location within the canopy, (ii) the twig order within the branch (Burger, 1927), (iii) the provenience of the trees (genetics), (iv) the sociological status (Zederbauer, 1916), (v) the meteorological site conditions (Münch, 1928; Wachter, 1985; Zederbauer, 1916), (vi) edaphic and (vii) hydrological factors (Wiedemann, 1923). These uncertainties

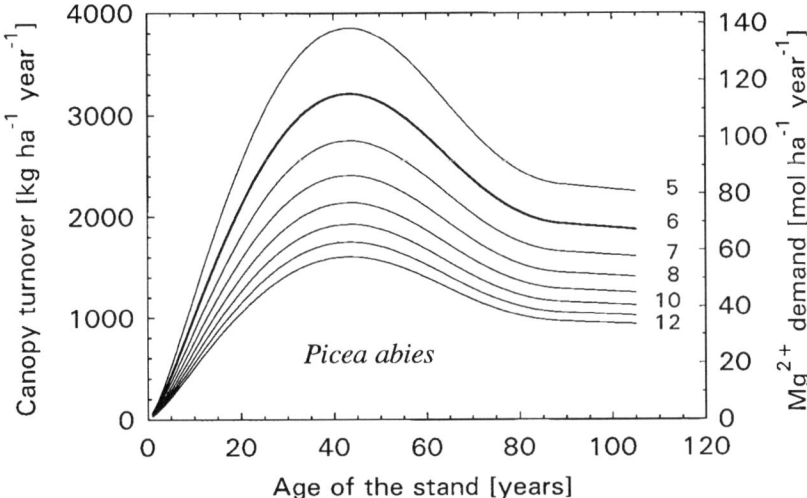

Figure 23. Annual needle turnover growth rates of spruce in relation to the stand age, t, and the mean needle age, NA = 5–12 years. The bold line approximates the Würzburg situation (NA = 6.1 years). The right ordinate corresponds to the Mg^{2+} turnover via needle litter formation if the mean Mg^{2+} content in spruce needle litter from Würzburg is employed. This Figure is calculated from Figure 19 (see text)

are partially respected by the bars in Figure 22, which represent the interval of observations of healthy spruce trees, but not error bars.

Figure 23 shows the needle turn-over growth rates R_{NL} for NA = 5 y to NA = 12 y. Stands at high altitudes above the sea level considerably reduce their annual needle litter production rate R_{NL}, which is plotted on the left ordinate and given on a needle dry matter basis. The necessary Mg^{2+} net demand for needle turn-over equals the remaining ('lost') Mg^{2+} content in needle litter. About 74% of needle litter consists of brown needles and 26% are green needles which mainly fall due to mechanical rupture by the wind (Gruber, 1990). The Mg^{2+} content of brown needles equals 37.6 mmol Mg^{2+} kg⁻¹ in Würzburg. Green needles contain 31.1 mmol Mg^{2+} kg⁻¹ (see Table 6). Thus, the mean $[Mg^{2+}]_{NL}$ content of native needle litter in the field (in Würzburg) equals approximately:

$$[Mg^{2+}]_{NL} = 0.26 \cdot 31.1 + 0.74 \cdot 37.6 = 35.9 \, mmol \, kg^{-1} dw$$

The partial annual Mg^{2+} demand D_{NL} of needle turn-over growth is plotted in Figure 23 on the right ordinate, which corresponds to Mg^{2+} contents in Würzburg:

$$D_{NL} = R_{NL} \cdot [Mg^{2+}]_{NL}$$

Non-needle litter   Dietrich (1963) analyzed the composition of Norway spruce litter in detail and found that 68–86% of the total litter mass consists of needles. This was confirmed by Gruber (1990) who found a mean fraction of 84% needles. The remaining fraction of 16% consists of twigs, bark, parts of cones, detritus from

Table 8. Means±SD of the relative composition of three classes of branches (German 'Faserholz', 'Sägeholz' and 'Hackschnitzel') of Norway spruce *Picea abies* (L.) Karst.

Organ or tissue	Relative composition of harvested branches (% (dw/dw))	Stock ratio (tissue/needles) (kg kg⁻¹)	Growth ratio (tissue/needles) (kg y⁻¹ (kg y⁻¹)⁻¹)
Needles	32.9±9.7	—	—
Wood	49.5±5.6	1.505	0.333
Cortex, bark	17.6±11.5	0.535	0.118

Recalculated after Mette and Korell, 1989. The stock ratio of wood and bark, given on a needle matter basis, is calculated from the first column. The third column summarizes the relative growth rates of wood and bark per needle growth rate. The ratio $0.118/0.333 = 0.35$ kg (dw bark) kg⁻¹ (dw wood) was used to estimate the cambium mitosis activity of twigs and branches within the spruce canopy, assuming that (small) loss of bark litter was absent

the canopy etc. Dietrich (1963) found that the non-leaf litter (10–26% of total litter) consists of about 75% dead twigs and 25% the 'remainder' (bark, detritus, etc.). On the basis of these data, the following 'typical' relative litter composition of Norway spruce canopies can be estimated: 22% = green needles, 62% = brown needles, 12% = dead twigs and 4% the 'remainder'. Consequently, the annual production rate of twig litter R_{TL} is:

$$R_{TL} = R_{NL} \cdot 12/84$$

i.e. ca. 14% of the needle litter production rate R_{NL} is twig litter production. The production rate of 'remaining' litter R_{RL} is:

$$R_{RL} = R_{NL} \cdot 4/84$$

The production rate of needle litter R_{NL} is given in Figure 23. The mean relative tissue (wood, bark) composition of branches and twigs within spruce canopies is summarized in Table 8. About 49.5±5.6% of twigs and branches is wood, and 17.6±11.5% is bark. Hence, $49.5/(49.5 + 17.6) = 0.74$ relative units is wood in twig litter. Thus, the production rate of wood litter R_{WL} is:

$$R_{WL} = 0.74 \cdot R_{TL}.$$

The relative fraction of bark in twig litter equals $17.6/(49.5 + 17.6) = 0.26$ relative units (see Table 8). Additionally, the production rate of the 'remaining' litter R_{RL} may be attributed mainly to 'bark' (=definition for simplicity's sake). The production rate of bark litter R_{BL} is estimated to be:

$$R_{BL} = 0.26 \ R_{TL} + R_{RL}.$$

The production rates of wood litter and bark litter are approximately proportional to the needle litter production rate.

4.5.4. *Net growth of needles (crown expansion)*

Net growth of needles consists of two components: (i) the change of the stock of needles per hectare, which can be either positive (in young stands) or negative (in

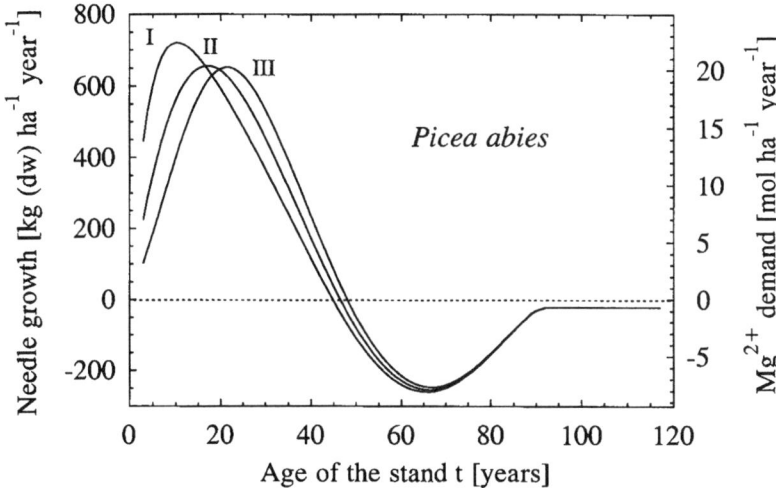

Figure 24. Crown expansion growth rates of Norway spruce needles calculated from the first derivation of the curves given in Figure 19. Roman numbers estimate the yield class, after Wiedemann (see Schober, 1987). There is apparent negative needle growth in stands beyond ca. 40 years old (on a hectare basis). This just indicates that needle loss via cutting of trees in the stand has become higher than the net crown expansion growth rate of the remaining trees per hectare. The differentials presented here are the balances between gain of needles by crown expansion growth and loss of needles by cutting trees. The cutting-enhanced crown expansion growth component of the surviving trees is shown in Figure 25. The right ordinate represents the partial annual Mg^{2+} demand of this crown expansion growth fraction (see Table 6)

ageing or declining stands), and (ii) needle net growth into canopy gaps by surviving trees after cutting of the stand in order to (partially) replace the lost needles of the cut trees. Both needle growth components must be regarded separately. The sum of both is the total annual (crown net) expansion growth rate of needles. The first component of needle net growth equals the first derivation $\partial N/\partial t$ [kg ha^{-1} y^{-1}] of the curves N(t) given in Figure 19. Results are shown in Figure 24. Needle growth rates are high and positive only in young stands. Stands older than 40 years reduce their stock of needles. This is no net needle loss per tree, but a net needle loss per hectare by cutting. This loss is higher than can be compensated by the positive needle net growth of the surviving trees per hectare. This cutting-induced additional needle growth of surviving trees only partly compensates the loss of needle matter per hectare. This second component of needle net growth is 'substitution growth' R_{NS} which just re-establishes the really observed stock (Figure 19) of needles despite cutting:

$$R_{NS} = N(t) \cdot \partial n/n(t)\partial t$$

N(t) is the stock of needles [kg ha^{-1}] shown in Figure 19; $\partial n/n(t)\partial t$ is the change rate of trees per hectare $\partial n/\partial t(t)$ and per surviving trees n(t). This term equals the stand-age t-dependent additional soil space per tree [m^2 (afterwards) m^{-2} (before)]. This mean relative additional space per surviving tree (after cutting) is used to extend the crown into these cutting-dependent gaps just re-establishing the given stock of needles N(t), which net decreases. Depending on the yield class of trunk wood, and on the cutting regime employed ('moderate', 'successive' or 'strong'), Wiedemann (1938, 1939; cf. Wiedemann and Schober, 1957; Schober, 1987) defined cutting and growth scenarios, to which forest officers and forest administrators in Central Europe usually refer. These tables supply the necessary data of trees per hectare n(t) (column 2, p. 62–75 in Schober, 1987) in order to calculate the first derivation $\partial n/\partial t(t)$, which is the mean annual cutting rate [trees ha^{-1} y^{-1}]. Thus, the 'tree sub-stitution' needle growth rate R_{NS} can be obtained on the basis of Figure 19 and on the tables of Wiedemann, which do not supply needle data (the consistency of both data sets is successfully tested below). Results are shown in Figure 25 for all available cutting regimes and for the individual yield classes I, II and III. The needle growth component R_{NS} depends on the cutting regime (one graph per regime), on the yield class (one curve per graph) and on the age of the stand (abscissa). Data below 20 years are not supplied by the yield tables of Wiedemann (cf. Schober, 1987) since there is no cutting yet. Cutting massively starts with ca. 20-year-old stands. Later, the mean annual cutting rate (cut trees per hectare) is reduced. In ageing spruce canopies, expansion growth of the crowns is governed by available illuminated gaps within the stand. Of course, there is no regular cutting in the field, neither every year nor e.g. exactly every five years etc. The yield tables of Wiedemann (cf. Schober, 1987) supply 'smooth' data of a 'mean' continuous reduc-tion of trees per ha. These tables 'distribute' irregular distinct cutting events in the manner of 'running averages'. This is the reason why also Figure 25 presents steady and smooth functions.

The total annual needle (crown) expansion growth rate R_{NE} is the sum of the data given in Figure 24 and 25:

$$R_{NE} = \partial N/\partial t + N(t) \cdot \partial n/n(t)\partial t$$

The total annual needle flush R_{NT} is the sum of needle turn-over growth R_{NL} (see Figure 23) and the total needle expansion growth rate R_{NE}. This sum R_{NT} is shown in Figure 26 for a mean needle age of NA = 6.1 years (Würzburg). All curves show the expected discontinuity at 20 years, which represents the onset of cutting. As desired by forest management, the employed high cutting rates between 20 and 40 years extrapolate almost linearly the juvenile phase of high needle production rates, which is one important criterion of tree vitality. Needle expansion growth is the most important means of repairing crown damage by frost, drought, wind, patho gens etc. If needle expansion growth is restricted by Mg^{2+} deficiency, air pollutants etc., the 'overall tree vitality' must decrease. The total needle flush growth rate (and 'vitality') decreases in ageing stands. The relative fraction of needle expansion growth R_{NE} per total needle flush growth ($= R_{NE} + R_{NL}$) is visualized in Figure 27. In young stands needle net growth is dominant. Above 60 years, needle net growth is only 5–10% of the total needle flush. Details of the curves for above-70-year-old stands are not significant in the field. The age-dependent percentage of needle net

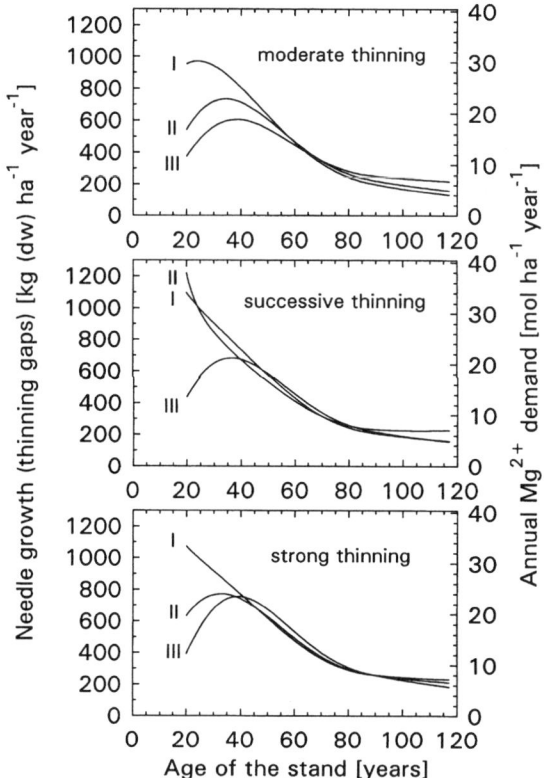

Figure 25. Cutting-induced crown expansion rate of surviving spruce trees (left ordinate) which grow twigs with needles into the gaps formed by cutting the trees in stands older than 20 years (start of cutting). Three cutting scenarios are shown after Wiedemann (see Schober, 1987, pp. 62–75, column 2): moderate, successive and strong cutting. Successive cutting was defined as "first strong, later only moderate". Roman numbers indicate the trunk wood yield class, after Wiedemann. The observed stock of needles (see Figure 19) is kept constant despite cutting losses of needles, which are partially compensated for by enhanced canopy net expansion growth until just the observed needle stock of Figure 19 is re-established. The right ordinate is an estimate of the corresponding annual Mg^{2+} demand of this cutting-dependent expansion growth into canopy gaps (see Table 6 for Mg^{2+} contents in Würzburg). The calculus of these curves is explained in the text

growth $R_{NE}/(R_{NE}+R_{NL})$ can be employed to tentatively quantify the actual 'tree vitality'.

4.5.5. Growth of branches

Needle growth is not independent of branch growth since it is accompanied by growth of twigs which carry the new flush and by synchronous secondary dilatation growth of already existing twigs and branches within the canopy. Hence, branch wood and branch bark production are interrelated with the total needle growth R_{NT}.

152

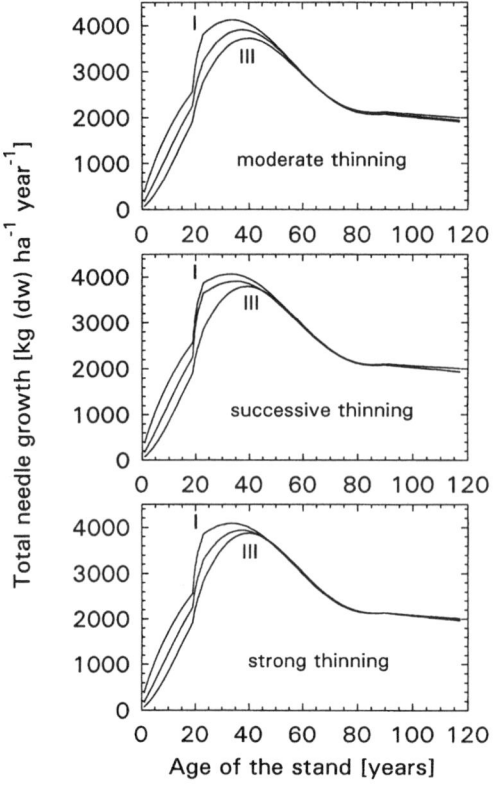

Figure 26. The total annual needle flush growth rate is the sum of the needle growth that just replaces needle litter losses (see Figure 23) plus both components of crown expansion growth of the remaining trees after each cutting event (Figures 24 and 25)

In Table 8 (see above), the relative 'typical' composition of spruce branches is summarized (see Mette and Korell, 1989). On this basis, the ratio of wood and bark mass per needle mass can be estimated as 49.5% (wood)/32.9% (needles) = 1.505 for wood, and 17.6% (bark)/32.9% (needles) = 0.535 for bark. These relative values are ratios of the stock of tissues and organs within the canopy. But we need ratios of production rates in order to estimate the annual twig wood and twig bark production rate if the total annual needle production rate R_{NT} is given (see below). These rates can be obtained from the stock ratios in Table 8 if the mean age of the needles on the one hand, and of branch wood and bark on the other hand are known. The mean needle age has been repeatedly mentioned to be 6.1 years in Würzburg. The mean wood and bark tissue age within the canopy equals the mean age of branches until they decline and become branch litter. This mean branch age can be determined by counting the number of age classes of living branches inserted in the trunk. This is easy to do with spruce trees, but more complicated for other tree species, which

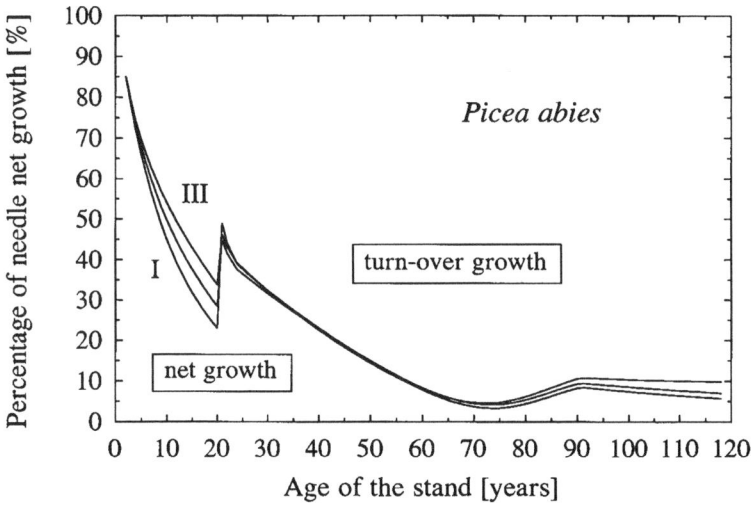

Figure 27. Percentage of needle net (=crown expansion) growth (sum of the data given in Figures 24 and 25) per total annual flush growth given in Figure 26. The areas below the curves indicate net growth, above canopy turn-over growth. With increasing stand age, the canopy turnover growth becomes dominant and net crown expansion growth is approximately only ca. 10% of the total annual flush. Integrated over the whole time span of 120 years, 19% of the total needle flush is needle expansion growth, 81% is needle turnover. Young stands are vital with dominant expansion growth. After onset of cutting trunks in 20-year-old stands, there is a transient peak of desired 'canopy revitalization'. Roman numbers indicate the corresponding yield class, after Wiedemann (see legends to Figures 24 and 25 for details)

grow in a more complex manner. In the Guttenberger Forst (Würzburg), the mean age of branches (i.e. of branch wood and bark) equals 27.4 ± 3.6 y ($n=50$ trees, determined in March 1993, Slovik, unpublished). The ratio of the mean needle age to the mean branch age equals $6.06/27.4 = 0.221$ y (needles) y^{-1} (branch). Mean needles become only 0.221 times as old as mean branch wood and bark. Consequently, 0.221 times the stock ratios in Table 8 equal the ratios of wood and bark growth rates relative to the needle growth rate. Thus, the mean annual branch wood production rate R_{BW} is $0.221 \cdot 1.505 = 0.333$ times the total annual needle production rate R_{NT}, and the mean annual branch bark production rate R_{BB} is $0.221 \cdot 0.535 = 0.118$ times R_{NT}. These corrections are necessary, since branches accumulate wood and bark dry matter until they decline. The complication of bark, twig and branch litter production must not be regarded here, since it is respected separately (see below). Thus, branch wood and branch bark production rates are approximately proportional to the total annual needle growth rates R_{NT}, which are given in Figure 26. The ratio of the branch bark growth rate R_{BB} per branch wood growth rate R_{BW} equals $0.118/0.333 = 0.354$ kg (bark) kg^{-1} (wood). This ratio estimates the mean mitotic activity of the branch cambium if branch bark litter production within the canopy did not exist. In old canopies, the branch litter production rate must approximate the tissue age ratio 0.221 (needle : branch) times the needle litter production rate. If the needle litter production rate is one unit, then the total litter production is 1.221 units. Hence, the percentage of the non-needle litter per total litter production is $0.221/1.221 = 18\%$. Dietrich (1963) found that the non-leaf litter accounts for $10-26\%$ of the total litter. This is exactly the median of the observed percentage interval.

4.5.6. Properties of trunk growth

Concerning trunk production rates of different tree species, there are yield tables, which quantify annual increments as depending on the stand age, trunk yield class and the applied cutting regime (e.g. Schober, 1987). As already mentioned, the commonly used tables of Wiedemann (see Schober, 1987, pp. 62–75) are employed for Norway spruce here. Besides the number of trees $n(t)$ per hectare (see below), these tables also supply consistent time-dependent trunk volume data given in m^3 (fresh volume) ha^{-1}. Since the number of 'surviving' trees per hectare steadily drops by cutting in ageing stands, these hectare-based table data are recalculated on a tree individuum basis in order to get the growth rates per tree of the surviving trees. Results are given in Figure 28, which shows the trunk volume per spruce tree V_T [m^3 (living trunk volume) $tree^{-1}$] as a function of the age of the spruce stand. The most productive stands (yield class I) are growing on the most favourable sites (soil, climate etc.). The first derivation of the curves shown in Figure 28 $\partial V_T/\partial t = V'_T(t)$ [m^3 (living trunk volume) $tree^{-1} y^{-1}$] is the annual volume growth increment per 'surviving' tree as a function of the age t of the stand.

In the field, the annual trunk growth rate is usually determined on the basis of (i) trunk length increments and (ii) trunk diameter increments. The tables of Wiedemann (see Schober, 1987) relate this information to consistent trunk volume growth data. The magnitude of trunk volume growth defines the 'yield class' (I, II, III), which estimates the 'economic profit' of a stand. The observed yield class in the field may be expressed by interpolation if appropriate (e.g. Guttenberger Forst, Würzburg: I to II). One important characteristic of the Wiedemann tables is that the stem is defined as a 'trunk' (German 'Derbholz') only if the trunk diameter is at least 7.0 cm. This definition has economic reasons. It is based on the utilization of harvested wood. As a consequence, trunk wood apparently starts to grow instantaneously in these tables only in about 20-year-old stands when the trunk diameter just had exceeded this artificial 7.0 cm limit. For a trunk diameter smaller than 7.0 cm, trunk growth is out of the scope of the Wiedemann tables. This economic definition is to be mentioned here in order to avoid confusion. It is extrapolated later in this communication in order to obtain trunk growth rates of stands younger than 20 years.

On the basis of the trunk growth rates $\partial V_T/\partial t = V'_T(t)$ [$m^3 tree^{-1} y^{-1}$] derived from $V_T(t)$ in Figure 28, the individual annual growth rates of trunk wood and trunk bark must be separated. The total trunk volume $V_T(t)$ is the sum of the trunk bark volume $V_{TB}(t)$ and the trunk wood volume $V_{TW}(t)$. As a crude estimation, 10% of the trunk volume is bark volume, and 90% of the trunk volume is wood. The trunk diameter

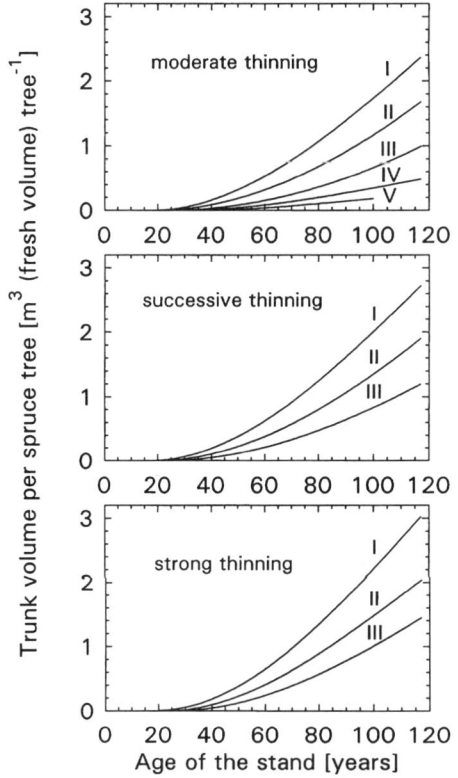

Figure 28. Mean stock of the spruce trunk volume per tree in relation to the age of the stand, the cutting regime and the trunk wood yield class, after Wiedemann (calculated after Schober, 1987; see text for details)

(including bark) is about 1.05–1.06 times thicker than the diameter of the wood inside (=c-factor; Liese and Dujesiefken, 1986; Mette and Korell, 1989; Schmidt-Vogt, 1986; Von Guttenberg, 1915). A more detailed view is given in Figure 29, which shows the relative bark volume $V_{bark}(d)$ of spruce trunks as depending on the trunk diameter d. The minimum percentage that can be observed in the field is about 6% (Liese and Dujesiefken, 1986). Typical twig bark volume percentages are 26% (=17.6%/[49.5% + 17.6%]; see Table 8), 30% (Hakkila, 1970) or exceptionally only 13–17% (Blossfeld *et al.*, 1963). This latter data approximate the maximum value for trunks (see Liese and Dujesiefken, 1986). With increasing age (or trunk diameter), the absolute bark thickness increases but the bark percentage decreases. A similar tendency is observed within individual trunks. The more juvenile upper parts of spruce trunks possess a higher bark percentage (12.8%) than close to the soil (8.5%; Altherr *et al.*, 1978). All these findings are tentatively respected in Figure 25, which estimates the situation of 'mean' trunks in the field. The bark percentage calculated after Liese and Dujesiefken (1986) is about 11.6% for d=10cm trunk

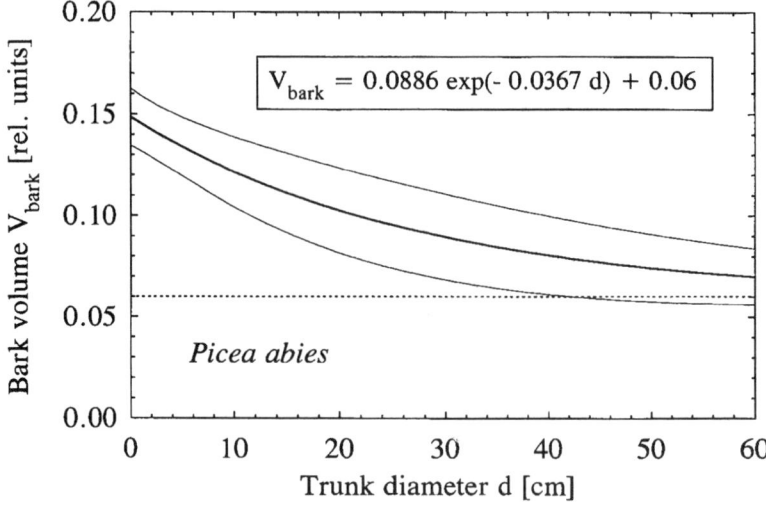

Figure 29. The relative bark percentage V_{bark} (relative units) of native Norway spruce trunks depends on the trunk diameter, d. The exponential formula, $V_{bark} = f(d)$, given in the insert is a tentative attempt to summarize numerous and heterogenous quantitative and semiquantitative literature data (see text). Its aim is to approximate the 'typical, mean' field situation. The lines indicate the error of the mentioned formula (Gauss law of error progression)

diameter, and about 7.8% for d = 50 cm trunk diameter. Thin trunks have higher bark percentages, which come close to the branch situation (see Eh, 1961; Flury, 1987; Hoffmann, 1958; Nagel, 1968; Östlin, 1963; Von Guttenberg, 1915).

Again, trunk diameter data d(t) [cm] are supplied by the Wiedemann tables (see Schober, 1987, pp. 62–75, table column 8). Consequently, the diameter dependency of the relative trunk bark volume fraction $V_{bark}(d)$ on the trunk diameter d(t) can be recalculated by employing the exponential formula given in the insert of Figure 29 in order to get the corresponding time-dependency curves $V_{bark}(t)$. Results of this recalculation by the following formula (see Figure 29):

$$V_{bark}(t) = 0.0886 \cdot \exp(-0.0367 \cdot d(t)) + 0.06$$

are shown in Figure 30. There are high bark percentages in slow-growing stands (high yield class, i.e. small trunk yield) and vice versa (see Figure 30). This fact has been repeatedly shown in the field (Blossfeld *et al.*, 1963; Koltzenburg, 1981; Korell, 1972; Liese and Parameswaran, 1971; Poller, 1968; Von Guttenberg, 1915; Zieger, 1960; cf. Schmidt-Vogt, 1986). Interestingly, these findings were already 'mathematically predefined' in the applied formula and the Wiedemann table data. Thus, these are neither 'new', nor 'independent' field observations. Hence, they need – and must – not be regarded separately in quantitative trunk growth physiology.

On the basis of Figure 28 and Figure 30, the annual volume growth rates of trunk wood and trunk bark can now be separated. The stock of trunk bark volume $V_{TB}(t)$

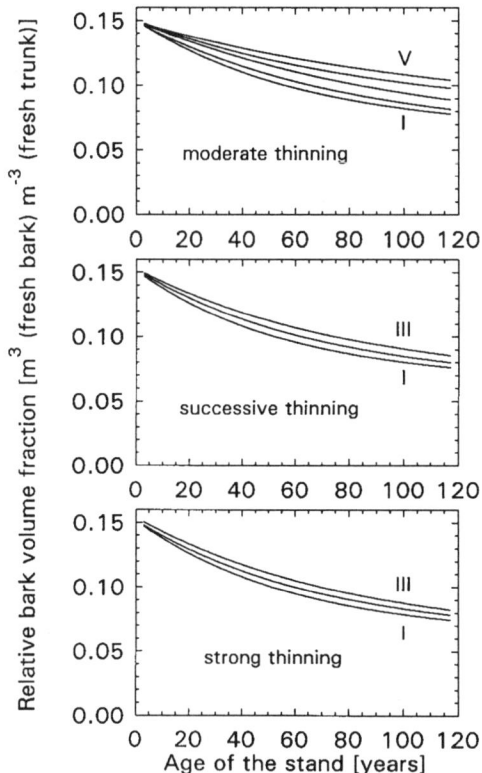

Figure 30. The relative fraction of spruce bark volume of trunks in relation to the age of the stand, the applied cutting regime and the yield class. Data are calculated on the basis of the first derivation of Figure 28, on data given in Figure 29, and on the consistent trunk diameter increment tables of Wiedemann (see Schober, 1987; and text for calculus)

$[m^3 (bark) tree^{-1}]$ equals the relative fraction of trunk bark per total trunk $V_{bark}(t)$ $[m^3 (bark) m^{-3} (trunk)]$ (see Figure 30) times the total stock of trunk volume $V_T(t)$ $[m^3 (trunk) tree^{-1}]$ (see Figure 28):

$$V_{TB}(t) = V_{bark}(t) \cdot V_T(t)$$

Since $V_{TB}(t)$ is the product of the two fractions $V_{bark}(t)$ and $V_T(t)$ that both depend on the time t, the bark volume growth rate $\partial V_{TB}/\partial t(t)$ is the differential of $V_{bark}(t)$, which equals:

$$\partial V_{TB}/\partial t(t) = V_{bark}(t) \cdot \partial V_T/\partial t(t) + V_T(t) \cdot \partial V_{bark}/\partial t(t)$$

This formula is the 'product law' of differentials. The bark volume growth rate $\partial V_{TB}/\partial t$ is the sum of the products (i) of the relative bark fraction (Figure 30) times the first derivation of the trunk volume per tree (differential of Figure 28), and (ii) of the trunk volume per tree (Figure 28) times the first derivation of the relative bark fraction (differential of Figure 30). The corresponding similar equations for calculating the volume growth rates of trunk wood $V_{TW}(t)$ are:

$$V_{TW}(t) = (1 - V_{bark}(t)) \cdot V_T(t)$$

and $\partial V_{TW}/\partial t(t) = (1 - V_{bark}(t)) \cdot \partial V_T/\partial t(t) + V_T(t) \cdot \partial(1 - V_{bark})/\partial t(t)$.

The sum of the volume growth rates of trunk bark and trunk wood equals the first derivation of the curves shown in Figure 28. Before these volume growth rates are recalculated into dry matter growth rates (see below), which are not supplied by the Wiedemann tables, the consistency of trunk volume growth data with needle data (Figure 19) must be examined by calculating the specific trunk growth rate μ_N [litre (fresh trunk) kg^{-1} (needle dw) y^{-1}], which is the annual trunk growth rate (see Figure 28) per photosynthetically active stock of needle dry matter present within the spruce canopy:

$$\mu_N = \partial V_T/\partial t(t) \cdot n(t) \cdot N(t) \cdot 1000$$

where $\partial V_T/\partial t(t)$ is the volume growth rate of the trunk [m^3 tree^{-1} y^{-1}], n(t) is the number of spruce trees per hectare [ha^{-1}], N(t) is the stock of needles [kg (dw) ha^{-1}] (see Figure 19), and there are 1000 L wood per m^3 wood. Figure 31 shows that there is roughly an annual production rate of 1.0 ± 0.4 L trunk volume per kg^{-1} living needles (dry weight); see also Meyer (1956). The absolute values in Figure 31 are almost identical to the detailed data published by Vanselow (1951). He found that the specific trunk growth in the field is highest in yield class I, and smallest in yield class III. The better the growth conditions of a spruce stand (German: 'Bonität'), the more trunk wood per needle surface or per needle matter (dry or fresh) is annually produced (Burger, 1953; Møller, 1946; Schmidt-Vogt, 1949, 1953; Vanselow, 1951). Thus, μ_N is a measure of the 'growth efficiency' of the annual needle net photosynthesis, which depends on the length of the growth period (Burger, 1937). Von Droste zu Hülshoff (1969, 1970) determined 1.01 L kg^{-1} y^{-1} in a ca. 78-year-old canopy of yield class I. Burger (1953) investigated 50-year-old stands and found 1.0 L kg^{-1} y^{-1} for yield class III and 1.25 L kg^{-1} y^{-1} for yield class I. Schmidt-Vogt (1953) found 1.0 L kg^{-1} y^{-1} in a 50-year-old stand of yield class I. The trunk productivity per needle dry matter depends on the age of the stand. Burger (1952) determined 1.01 L kg^{-1} y^{-1} in stands between 40 and 80 years old and 0.67 L kg^{-1} y^{-1} in stands older than 80 years. There are certain deviations between different literature data on the one hand and between the literature data and results in Figure 31 on the other. These deviations are mainly the consequence of field observations which show that the stock of canopy needles can be reduced by 20–25% before the trunk volume growth rate starts to respond (Kramer, 1976).

The productivity of 1 kg needles (dw) drops with increasing altitude above the sea level. Burger (1937, 1939b, 1952, 1953) found 0.93 L kg^{-1} y^{-1} up to 900 m asl, 0.79 L kg^{-1} y^{-1} between 900 and 1500 m asl and only 0.50 L kg^{-1} y^{-1} above 1500 m asl in Switzerland. This has been confirmed by Schmidt-Vogt (1953) and by Von Droste zu Hülshoff (1970). The main factor which defines the specific trunk growth rate is the length of the vegetation period (Burger, 1937). Figure 32 summarizes the

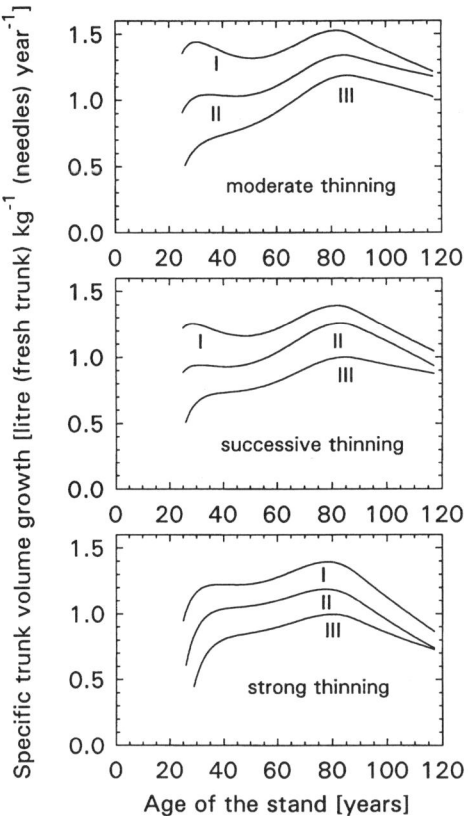

Figure 31. Specific trunk volume growth rates per stock of photosynthesizing needles in the canopy in relation to the age of the stand, the cutting regime and the trunk wood yield class. Data are calculated from the first derivation of Figure 28 and the stock of needles given in Figure 19

'needle efficiency' E_N [relative units] as a function of the site altitude above sea level (asl) (calculated after Burger, 1937). The value of μ_N is set to 1.0 relative units at the control site at 550 m asl, to which all other given μ_N data relatively refer. Close to the altitude limit of spruce in the alps, the specific trunk growth rate μ_N is only ca. $E_N \approx 50\%$ relative to control trees.

Spruce decline, which can be the consequence of Mg^{2+} deficiency at various sites, is characterized – as well as by needle chlorosis – by early needle senescence. A consequence of increased needle loss is reduction of the leaf area index LAI (reduction of spruce crowns). The specific trunk growth rate μ_N and the needle efficiency E_N are expected to be affected in a complex manner by these canopy processes. First, loss of needles allows increasing illumination and photosynthesis of remaining needles within the canopy. This effect largely compensates for the loss of needles in terms of assimilate production and 'needle efficiency' E_N. Kramer (1976) found that a reduction of the spruce crown by 20% did not reduce the trunk

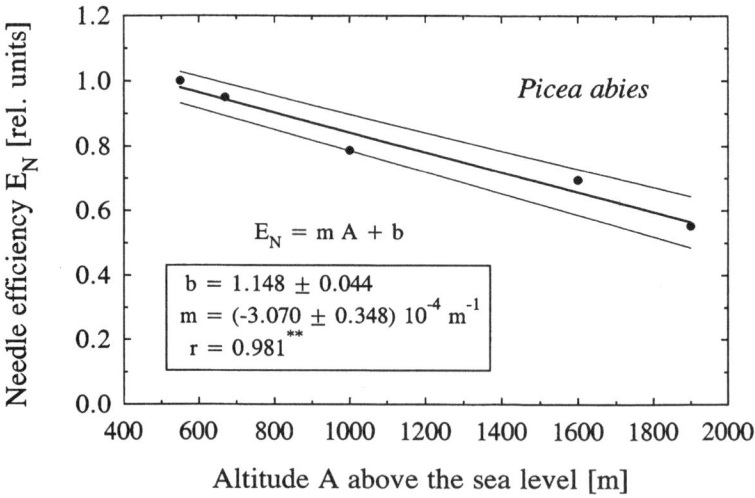

Figure 32. Trunk growth efficiency of a given stock of needles in relation to the site altitude above the sea level. The inserted formula estimates that, at high altitudes, one unit of needle dry matter produces only 50% of trunk wood matter relative to the reference site (1.00 relative units) at 540 m asl (data from Burger, 1937, 1939b, 1952, 1953)

diameter growth. This critical crown reduction of 20% before trunk growth is depressed was confirmed by Moosmayer (1984) for *Picea abies* and *Abies alba*. Numerous authors confirmed that about 25% of spruce branches can be removed (starting from the bottom of the trunk) before trunks respond with volume growth reduction (Burger, 1951; Hartig, 1896; Heil, 1970; Kramer, 1976; Ladefoged, 1946; Meyer, 1959, 1968; Mitscherlich and Von Gadow, 1968; Olischläger, 1969; Vuokila, 1968; Wilhelmi, 1968). Hence, there is a 'luxurious' stock of 20–25% 'ineffective' needles in healthy spruce canopies. Below 75% of needles relative to 'healthy' spruce trees, there is an increasing shortfall of assimilates. Pechmann (1958) found that loss of needles is lethal to spruce trees only if the loss of needles permanently exceeds 60–70%. Below a remaining needle stock of ca. 40% the annual net photosynthesis becomes insufficient to supply the annual assimilation demand for mitochondrial CO_2 respiration and for essential e.g. fine root (turn-over) growth rates etc. Hence, this 40% limit of remaining needles in spruce crowns is in fact an appropriate borderline to characterize 'severe damage' (onset of damage class 3) in forest decline evaluation. Together with data in Figure 19, these findings can be employed to tentatively quantify effects of needle loss on trunk growth and tree survival.

4.5.7. Growth of trunk wood and bark

By 1885, Hartig had already found that volume growth rates of trunks do not proportionally represent dry matter production, especially if different tree species

Table 9. 'Typical' absolute and relative specific densities of trunk wood, trunk bark, root wood and root bark of Norway spruce

| Spruce tissue | Index | Specific tissue density (ρ) | |
		kg m^{-3} (fresh) (fresh)	Relative (fresh)
Trunk			
Wood	TW	378	1.00
Bark	TB	446	1.18
Root			
Wood	RW	472	1.25
Bark	RB	336	0.89

Absolute values vary with an amplitude of ca. ±30% in the field. Relative data are obtained by setting the trunk wood density ρ_{TW} to 1.0 relative units. The volume loss of dry tissues is ca. 12% for wood and ca. 30% for bark relative to fresh matter. Data are from Anonymous (1939), Liese and Dujesiefken (1986), Mette and Korell (1989), Scholze and Slovik (unpublished), and from Volz (1974)

are compared. The specific density (German 'Raumdichte') of spruce trunk wood ρ_{TW} [kg (dw) m^{-3} (fresh wood)] depends on (i) the thickness of the annual trunk cycle, and (ii) the relative volume shrinkage of dried fresh wood. This relative shrinkage amounts to ca. 12%. The remaining dry wood volume is 0.88 m^3 (dry wood) m^{-3} (fresh wood); see Anonymous (1939); Liese and Dujesiefken (1986). The specific density ρ_{TW} varies between 264 and 563 kg (dw) m^{-3} (fresh wood); a typical mean value for Norway spruce trunks is 378 kg (dw) m^{-3} (fresh wood) (Anonymous, 1939; see Table 9). The specific density of branch wood is higher than of trunk wood (Wandt, 1937). The specific sap wood density of dry trunk wood depends on the thickness of the annually grown sap wood circles Δd [cm] (Trendelenburg, 1934, 1936). With increasing thickness Δd, the relative fraction of early wood increases. About 80% of the variation in specific density of spruce trunk wood can be explained by the relative composition of trunk wood by early wood (low specific density) and late wood (high specific density). The data of Trendelenburg (1936) are recalculated on a fresh wood volume basis employing the above-mentioned factor 0.88 m^3 (dry wood) m^{-3} (fresh wood). Results are shown in Figure 33, which also supplies the statistical response function $\rho_{TW} = f(\Delta d)$

$$\rho_{TW} = 191.5 \cdot \exp(-5.532 \cdot \Delta d) + 307.0$$

which allows the calculation of the specific trunk wood density ρ_{TW} [kg (dw) m^3 (fresh volume)] if the thickness of the actual annual wood cycle Δd [cm] is known. $\Delta d(t)$, which is time dependent, can be determined from the derivation of the trunk diameter $d(t)$, which is given in column 8 of the Wiedemann tables (see Schober, 1987) in combination with the actual volume growth ratio of trunk wood per total trunk volume growth (see below):

$$\Delta d(t) = \frac{\partial V_{TW}/\partial t(t) \cdot \partial d/\partial t(t)}{2 \cdot \partial V_T/\partial t(t)}$$

162

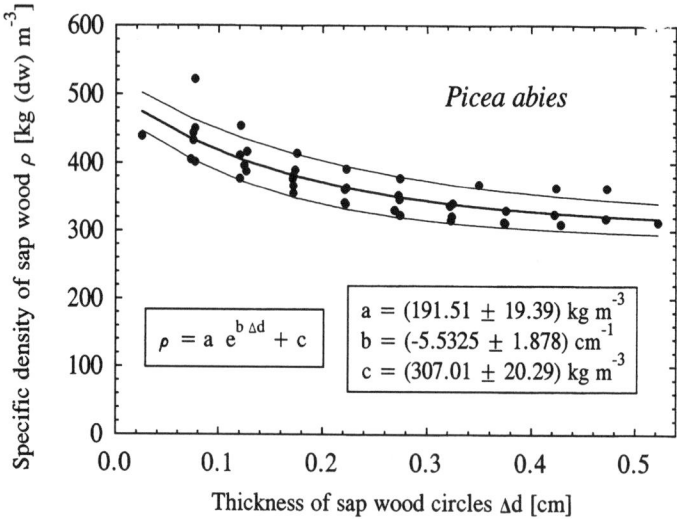

Figure 33. Specific density ρ (kg wood dry matter per m^3 fresh volume) of native sap wood in relation to the growth increment of sap wood circles (from Trendelenburg, 1934, 1936). The given exponential formula estimates ρ if the trunk diameter growth rate, after Wiedemann (see Schober, 1987), is known

The trunk wood volume growth rate $\partial V_{TW}/\partial t(t)$ is given below. The annual trunk diameter increment $\partial d/\partial t(t)$ is the first derivation of $d(t)$ (Wiedemann table, column 8, after Schober, 1987). The total trunk volume growth rate $\partial V_T/\partial t(t)$ is the first derivation of the data given in Figure 28. The factor 2 respects that only half of the trunk wood diameter increment contributes to the actual thickness $\Delta d(t)$ of the growing trunk wood circle. Of course, in the field, there are varying values of $\Delta d(t)$, which do not ideally match the Wiedemann table data. Additionally, there may be complete depletion of annual wood circles in years with the most unfavourable growth conditions (Athari and Kramer, 1983a,b; Schweingruber, 1980). Nevertheless, based on the consistent trunk diameter and trunk volume tables of Wiedemann, the specific trunk wood density $\rho_{TW}(t)$ is accessible as a mean value. It is time dependent, since $\partial d(t)$ depends on the age t of the spruce stand. Additionally, ρ_{TW} decreases with increasing altitude. It varies between 350 and 400 kg(dw)m^{-3} (fresh wood) at sea level and between 325 and 350 kg(dw)m^{-3} fresh wood) in the high mountains (Bernhart, 1967; Hildebrandt, 1954; Leinert, 1962; Trendelenburg, 1936). Their published 'dry volume' data can be recalculated into 'fresh volume' data. The relative maximum difference of ρ_{TW} between different altitudes asl is roughly about 10% (A=altitude asl [m]; $0.90 < \rho_{rel} < 1$, given in relative units, corrects ρ_{TW} if demanded):

$$\rho_{rel} = 1 - 5 \cdot 10^{-5} \cdot A$$

Employing the mentioned growth scenarios and the altitude-dependent wood density functions, the annual trunk wood growth rate R_{TW} [kg(dw)ha^{-1}y^{-1}] now

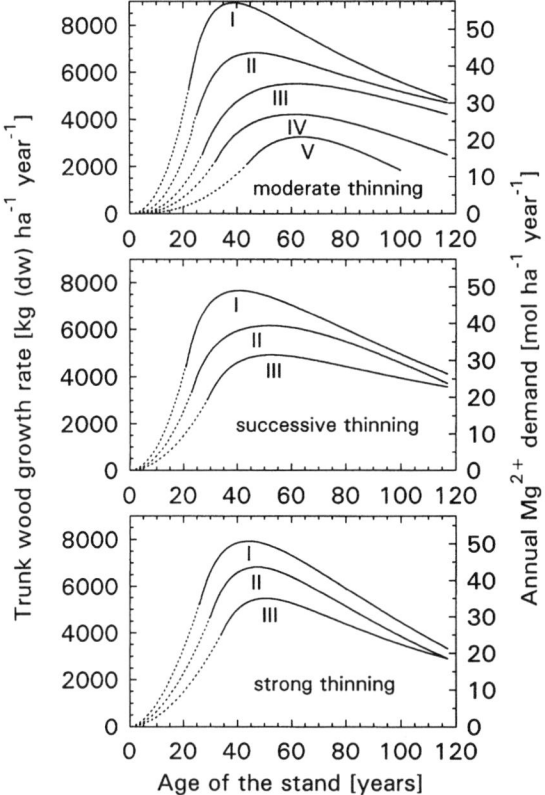

Figure 34. Calculated trunk wood growth rates in relation to the stand age, the cutting regime and the yield class, after Wiedemann (left ordinate), and the corresponding annual Mg^{2+} demand (see Table 6 for the Würzburg situation; right ordinate). Dotted lines are nonlinear extrapolations to obtain data which, for economical reasons, are not supplied by the Wiedemann tables (see text)

becomes available:

$$R_{TW} = \rho_{TW}(t) \cdot n(t) \cdot \partial V_{TW}/\partial t(t)/1000$$

The actual tree number per hectare $n(t)$ $[ha^{-1}]$ is given in column 2 of the Wiedemann tables, the production rate of trunk wood $\partial V_{TW}/\partial t(t)$ $[L(fresh wood) tree^{-1} y^{-1}]$ is given below, and the factor 1000 just converts litres into m^3. Figure 34 shows the dependency of the annual trunk wood production rate R_{TW} on the employed cutting regime and on the yield class (roman numbers). Data for yield classes IV and V are available only for moderate cutting starting at 30 years (class IV) and 40 years (class V).

The dotted lines are extrapolations to zero, since (i) trunk wood data below 20 years are not supplied by the Wiedemann tables, and (ii) table data between 20 and ca. 30 years only take account of those trees per hectare with a trunk diameter of at least 7.0 cm. Thus, a consistent extrapolation of trunk growth rates to zero must be found. The dotted nonlinear extrapolation in Figure 34 is based on the mathematical substitution of the trunk wood volume (or matter) by a cone with a cyclic trunk area at the bottom and the height being the length of the trunk. In the field, trunk altitude growth and trunk diameter growth vary between different growth scenarios. It can be shown (based on cone geometry) that under these variant conditions the annual trunk volume increment $\partial V/\partial t$ generally obeys a function of the type:

$$\partial V/\partial t = a \cdot t^b$$

where a and b are constants, and t is the age of the trunk (or stand). This potence function type takes into consideration that at $t = 0$ years there is still no trunk and no trunk growth. At the point where these extrapolation functions (dotted curves in Figure 34) match the recalculated Wiedemann table data (solid curves), the function value and the slope of the dotted curves must be identical to the 'last' function value and slope of the solid curves. This postulate is sufficient to determine the constants a and b of the dotted extrapolation curve:

$$b = \frac{x \cdot \partial R_{TW}/\delta x(x)}{R_{TW}(x)}$$

$$a = R_{TW}(x)/x^b$$

where x [years] is defined as the smallest solid plotted stand age point between the solid and the dotted curves, $\partial R_{TW}/\delta x(x)$ is the last available (solid) slope at the interface abscissa x, and $R_{TW}(x)$ is the last available (solid) R_{TW} value at the interface abscissa x. Three of the visualized (dotted) extrapolations (yield class I) could be successfully tested on the basis of the morphometrically determined trunk wood dry matter growth rate of a 10-year-old Norway spruce Christmas tree (grown in Denmark, yield class I assumed), which was 0.139 kg (dw) trunk wood tree^{-1} y^{-1} in the year of harvest. Using the number of trees per hectare after the tables of Wiedemann before onset of cutting (5917 trees ha^{-1}), the annual trunk wood production rate per hectare yields 822 kg (dw) ha^{-1} y^{-1} (Slovik, unpublished; the tree is the same as in Figure 21). This is very close to the values of yield class I presented in Figure 34 at a stand age $t = 10$ years. The mean value of b for all 11 shown (dotted) extrapolation curves is $b = 2.15 \pm 0.32$. Theoretically, the exponent b equals 2.00 if trunk diameter growth and trunk altitude growth were proportional at any time of the extrapolation interval.

'Typical' absolute and relative specific density values of trunk bark, root wood and root bark are summarized in Table 9, but deviations in the field can be considerable. The annual trunk bark matter growth rate R_{TB} [kg (dw) ha^{-1} y^{-1}] equals (see Figure 35):

$$R_{TB} = 1.18 \cdot \rho_{TW}(t) \cdot n(t) \cdot \partial V_{TB}/\partial t(t)/1000$$

The actual tree number per hectare n(t) [ha^{-1}] is given in column 2 of the Wiedemann tables; the production rate of trunk bark volume $\partial V_{TB}/\partial t(t)$

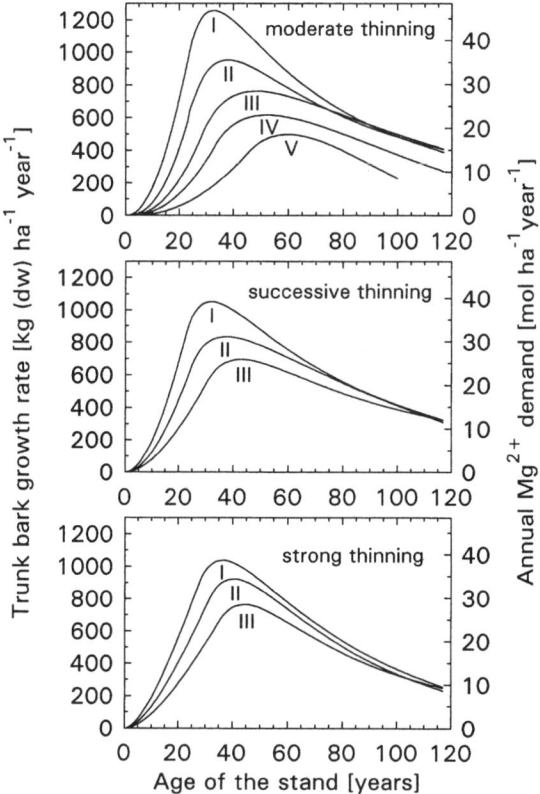

Figure 35. Trunk bark growth rates in relation to the stand age, the cutting regime and the yield class, after Wiedemann (left ordinate), and the corresponding annual Mg^{2+} demand (see Table 6 for the Würzburg situation; right ordinate); see legend to Figure 34

[L (fresh wood) $tree^{-1} y^{-1}$] is given above; the factor 1000 just converts litres into m^3, and the factor 1.18 reflects that the specific trunk bark density is 1.18 times higher than the density of trunk wood (fresh volume basis). Figure 35 shows the annual trunk bark dry matter growth rate R_{TB} correctly indicating the increasing bark fraction of young trunks (see Figure 29 and 30) within the interval of extrapolation below 20–30 years (see Figure 34).

4.5.8. *Growth of root tissues*

Growth rates of root tissues and fine roots are extremely difficult to determine in the field. According to Mette and Korell (1989), the dry matter ratio between spruce roots (including the stump) and spruce trunks is about 100:25 for harvested 100-year-old Norway spruce trees. Within the UNESCO project, 'Man and Biosphere',

Vyskot *et al.* (1981) dissected 45 complete Norway spruce trees (25–69 years old) into their separate parts. They found that the total root dry matter was $18.6 \pm 4.1\%$ of the trunk dry matter (see also Schmidt-Vogt, 1986). Assuming that the harvested stock of roots has developed proportional to the trunk matter growth rate all the time, the total root dry matter growth rate $R_{RT} = R_{RW} + R_{RB}$ (root wood plus root bark) can be estimated:

$$R_{RT} = 0.186 \cdot (R_{TW} + R_{TB})$$

R_{TW} and R_{TB} are the actual trunk wood and trunk bark matter production rates. The real root dry matter growth rate in the field must be somewhat higher because some roots may have been overlooked in the field and – more important – the percentage of once-grown but already decomposed root matter is not known. Additionally, young trees produce more root matter relative to their trunk production rate than do old trees (Schmidt-Vogt, 1986). An exact quantification of these findings is difficult. For simplicity's sake, the branch matter production rate $(R_{BW} + R_{BB})$ may be proportional to this unknown additional root matter production component. Since branch wood is proportional to the needle production rate, the following equation tentatively takes account of (i) higher root production rates of young trees that have high needle growth rates, and (ii) growth of root matter which will be decomposed before harvest:

$$R_{RT} = 0.186 \cdot (R_{TW} + R_{TB} + R_{BW} + R_{BB})$$

This is a reasonable estimate of the total Norway spruce root dry matter growth rate. With increasing root age, the relative bark content of Norway spruce roots drops from 72% (m/m) in 1-year-old roots to 23% (m/m) in 7-year-old roots. The relative bark formation rate is 1.23 times higher in Norway spruce roots relative to Norway spruce branches (Scholze and Slovik, unpublished). The dry weight ratio of bark growth to wood growth of spruce branches equals 26%:74%. Thus, 1.23 times 26% = 32% (m/m) of the total annual root dry matter production is root bark growth R_{RB}

$$R_{RB} = 0.32 \cdot R_{RT}$$

and the remaining 68% (m/m) of the total annual root dry matter production equals root wood growth R_{RW}:

$$R_{RW} = 0.68 \cdot R_{RT}$$

Figure 36 shows the annual root wood dry matter growth rates R_{RW} [kg ha^{-1} y^{-1}]. From data in Figure 36, the root bark production rate can be calculated as 0.32/0.68 = 0.47 times the root wood production rate. A tentative estimate of the annual fine root growth rate is based on the postulate that crown (needle) net growth must be accompanied by a proportional fine root growth in healthy stands, since additional needles need an additional and proportional supply of water and nutrients. This is a reasonable physiological necessity. Assuming that the fine root growth R_{FR} : needle expansion growth R_{NE} ratio equals 0.186, which is the root-to-

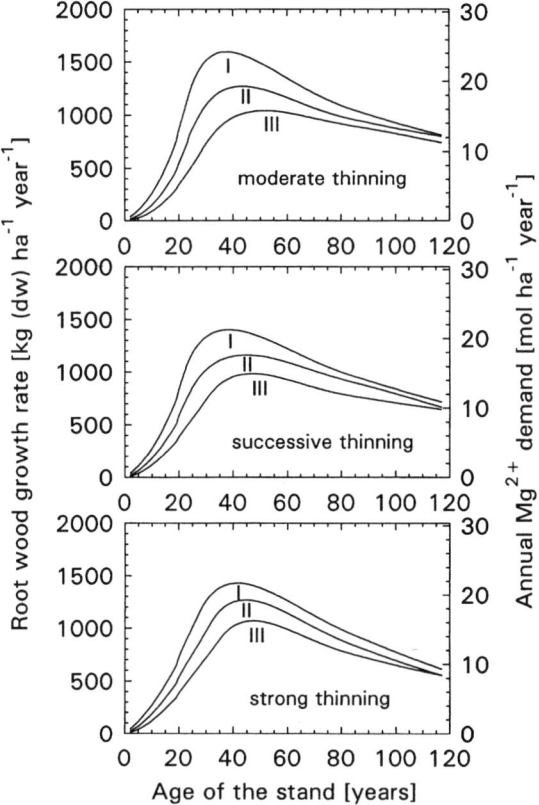

Figure 36. Root wood growth rates in relation to the stand age, the cutting regime and the yield class, after Wiedemann (left ordinate), and the corresponding annual Mg^{2+} demand (see Table 6 for the Würzburg situation; right ordinate); see text for calculus

shoot growth ratio, then the fine root growth rate R_{FR} [kg ha^{-1} y^{-1}] becomes somewhat smaller than the annual root bark growth rate:

$$R_{FR} = 0.186 \cdot R_{NE}$$

4.5.9. The annual magnesium demand of spruce trees

The total annual Mg^{2+} growth demand of Norway spruce trees can be determined if tissue analysis data and tissue growth rates are available. The leaching demand of Mg^{2+} from spruce canopies, which is an additional Mg^{2+} demand that is not available for growth, is discussed in section 4.6. The total annual Mg^{2+} growth demand of whole spruce trees [given in mol ha^{-1} y^{-1}] is the sum of all partial Mg^{2+} demands of all different tissues and trees organs (needle litter, wood litter, bark litter, net growth of needles, branch wood, branch bark, trunk wood, trunk bark, root wood, root bark and fine roots). The different tissue-dependent Mg^{2+} demands must be calculated

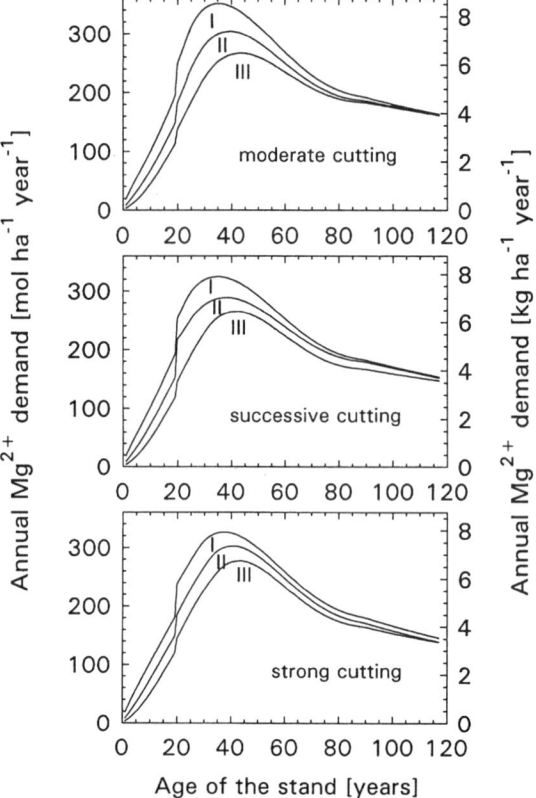

Figure 37. The annual Mg²⁺ demand of Norway spruce stands in relation to the age of the stand, the cutting regime and the yield class, after Wiedemann (left ordinate), as calculated on the basis of (i) the Mg²⁺ content in different spruce organs and tissues in Würzburg (Table 6) and (ii) the specific spruce tissue growth rates derived in Figures 23–36 (see text for calculus)

(see Table 6 for Würzburg) before summation of the partial Mg^{2+} demands. The annual Mg^{2+} demand is important information for tree nutrition. Results are shown in Figure 37 using tissue analysis data from Würzburg (Table 6). The presented curves are valid for any 'typical' spruce stand with a mean needle age of NA = 6.1 years if tissue analysis data reflect approximately the Würzburg situation. For all other conditions, Figure 37 must be calculated anew. The maximum annual Mg^{2+} demand is associated with moderate cutting, but the difference from heavy cutting is only up to $0.5\,kg\,Mg^{2+}\,ha^{-1}\,y^{-1}$. Stands with high wood production rates need more Mg^{2+} than stands of lower trunk yield (higher yield class). The maximum difference between the yield classes I, II and III is about $2.0\,kg\,Mg^{2+}\,ha^{-1}\,y^{-1}$. The maximum annual Mg^{2+} supply from the soil is needed in 20–40-year-old stands. In ageing stands (>60 years), the remaining annual Mg^{2+} demand is only half the corresponding maximum demand, i.e. the possible soil-dependent Mg^{2+} supply rate per hectare is now not fully utilized. Thus, Mg^{2+} fertilization usually is most important in spruce stands which are younger than ca. 60 years, or at SO_2-burdened sites (see sections 4.1. and 4.7.).

4.6. Ecophysiological flux balance of magnesium

On the basis of the presented approach of integrating yield table and literature data for the example of 'typical' Norway spruce stands, detailed and, in most cases, sufficiently accurate ecophysiological flux balance data and nutrient cycling rates of any nutrient element of interest (e.g. Mg^{2+}) can be deduced if chemical tissue analysis data of green needles, needle litter, trunk wood, trunk bark, root wood and wood bark are available. All mentioned spruce tissue and organ growth rates can be estimated for Norway spruce stands in central Europe if the following fundamental information is supplied for actual sites: (i) the mean needle age NA of spruce trees, (ii) the age of the Norway spruce stand, (iii) the trunk yield class, and (iv) the cutting regime after Wiedemann (see Schober, 1987, pp. 62–75). Item (i) has already been mentioned above: the mean needle age in Würzburg is 6.1 ± 0.6 y. Items (ii) to (iv) can be easily obtained in the field by fitting observed trunk altitude and diameter growth rates to the most appropriate 'Wiedemann scenario', or just by phoning a competent forest officer (e.g. Guttenberger Forst in Würzburg: 26–46-year-old spruce stand, yield class I–II, moderate cutting; Schörry, personal communication). The Wiedemann tables supply data for spruce stands between 20 and 120 years old. Thus, roughly 100 years, 3 cutting regimes, and 3 yield classes means that 900 different trunk growth scenarios (plus intermediates) are available to thoroughly define the best 'table representative' in the field. Today, wood production rates at many sites are about 20% higher than Wiedemann's field observations 50 years ago. This is usually attributed to the increased wet deposition of nitrogen in spruce ecosystems in recent decades. Mitscherlich (1963) compared growth parameters from numerous European experimental sites with the predictions made by the Wiedemann tables. Most parameters (stem diameter, stem volume, etc.) matched closely to the table data in central and south Europe. The maximum deviation was only about $\pm 20\%$. Larger differences between observation and table data occurred mainly in North and West Europe.

4.6.1. Nutrient cycling

Fundamentals of nutrient cycling in forest ecosystems are discussed by Ulrich (1981b). In Figure 38, detailed Mg^{2+} demand and cycling rates of Norway spruce in the Guttenberger Forst in Würzburg are estimated. The total annual growth demand is $352 \, mol \, Mg^{2+} \, ha^{-1} \, y^{-1}$. This corresponds to the highest annual Mg^{2+} growth supply from the soil in Figure 37 (moderate cutting, 35-year-old stand, yield class I). Consequently, this value ($\approx 8.5 \, kg \, Mg^{2+} \, ha^{-1} \, y^{-1}$) can be taken as an estimation of the maximum annual Mg^{2+} uptake from the soil that is consumed for growth purposes or tissue storage in Würzburg (for leaching demands, see below). About

170

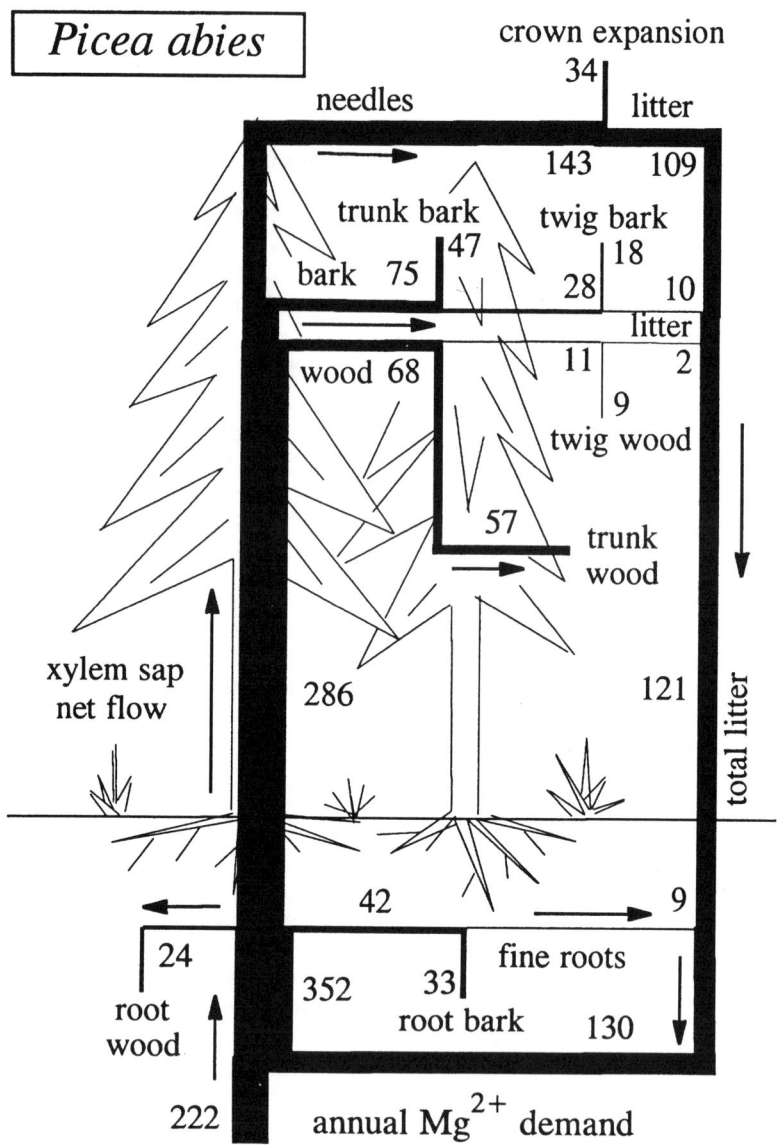

Figure 38. Mg²⁺ cycling rates of a Norway spruce stand with the following properties (example for the Würzburg situation: Guttenberger Forst): yield class I, after Wiedemann, stand age, t = 35 years, moderate cutting regime, mean needle age, NA = 6.1 years, and Mg²⁺ contents from Table 6

$286 \, mol \, Mg^{2+} ha^{-1} y^{-1}$ (81%) are consumed for trunk growth and canopy turnover. Most of the total annual Mg²⁺ transport rate within the xylem sap is recycled via the phloem sap (see section 4.2.). The total annual Mg²⁺ loss of the spruce crown via canopy litter production is $121 \, mol \, Mg^{2+} ha^{-1} y^{-1}$. Turnover of fine roots supplies an estimated additional rate of $9 \, mol \, Mg^{2+} ha^{-1} y^{-1}$. Approximately 100 times $121/352 = 34\%$ of the annual Mg^{2+} demand is therefore resupplied by microbial litter decomposition (equilibrium approach), but this rate is difficult to quantify in more detail since complete spruce litter decomposition requires roughly 20–25 years in the field (B. Ulrich, Göttingen, personal communication). Its rate in the field strongly depends on soil microbiology. Thus, there is an uncertain delay (and kinetics) of Mg^{2+} release in the field, which is usually retarded by soil podsolization and is enhanced by soil liming. The total annual Mg^{2+} uptake from the soil in Würzburg is approximately $352 \, mol \, Mg^{2+} ha^{-1} y^{-1}$. The net demand from the mineral soil accounts for at least $222 \, mol \, ha^{-1} y^{-1}$ if Mg^{2+} release after litter decomposition is as fast as litter production.

Nutrients other than Mg^{2+} must not be completely out of the scope of this communication, since Mg^{2+} supply and demand is highly dependent on other competitive or synergetic nutrients (see section 4.3.) and possibly on air pollutants (see section 4.7.). Nutrient cycling rates and growth demands of other plant nutrients besides Mg^{2+} can be determined: (i) on the basis of their tissue contents (see Tables 10, 11 and 12 for Würzburg), and (ii) the growth scenarios given in section 4.5. Thus, for any nutrient and for any growth scenario of spruce, detailed nutrient flux data can be calculated similar to Figure 38, which shows just the actual example of: (i) the Mg^{2+} demand (Würzburg) for (ii) yield class I, (iii) 35-year-old stands with (iv) moderate cutting, and (v) 6.1-year-old needles. Nutrient demand ratios change in ageing stands and vary at different sites. Thus, detailed knowledge of cycling rates of other nutrients besides Mg^{2+} is important additional information for an adequate, balanced and economic nutrition of ecosystems (see Chapter 11).

4.6.2. Canopy leaching

Canopy leaching is an important additional component of nutrient cycling. It has been disregarded so far, since the annual Mg^{2+} supply from the soil, that compensates for the precipitation-dependent loss of Mg^{2+} from the canopy, is per se not available for growth. Mg^{2+} that originates from native crowns is recycled back to the soil, where it just resupplies the additional annual Mg^{2+} demand for canopy leaching of the next year (flux equilibrium assumed). This leaching cycle transports considerable amounts of Mg^{2+}. Annual means of Mg^{2+} leaching rates from Norway spruce canopies in central Germany (Hessen) vary between 20 and $70 \, mol \, Mg^{2+} ha^{-1} y^{-1}$ (Table 13). These values are corrected means of several years per stand calculated after Balázs (1991), which consider only the real Mg^{2+} leaching rate that originates from spruce canopy tissues. Dissolved Mg^{2+} measured in the precipitation water outside the canopy must also have been in the canopy through fall. The remainder on canopy surfaces after evaporation of intercepted water is dissolved by further rain and is collected from the canopy as it falls to the ground. The ratios of the annual precipitation outside and below the crowns of all individual sites and years are factors from which the expected concentration increase in the canopy leachate can be calculated. This expected concentration increase below canopies originates

Table 10. Chemical composition of 'mean' living spruce needles (green needles, $n=200$) and fresh spruce litter (brown needles, $n=13$)

Compound	Green needles $(n=200)$ $(mmol\,kg^{-1})$	Brown needles $(n=13)$ $(mmol\,kg^{-1})$	Retranslocation $(mean \pm SD)$ $(mmol\,kg^{-1})$
C_{org} (*)	39886 ± 1532	33497 ± 530	6389 ± 1621
H_{org} (*)	64757 ± 1563	52933 ± 1679	11824 ± 2294
N_{sum} (*)	652.8 ± 121.1	441.8 ± 91.7	211.0 ± 151.9
NO_3^-	0.58 ± 0.56	3.60 ± 3.38	-3.0 ± 3.4
N_{red} (*)	652.2 ± 121.1	438.2 ± 91.9	214.0 ± 152.0
S_{sum} (*)	34.05 ± 6.20	23.94 ± 2.15	10.11 ± 6.56
SO_4^{2-}	8.89 ± 7.82	7.17 ± 2.02	1.72 ± 8.08
S_{red}	25.16 ± 9.98	16.77 ± 2.95	8.39 ± 10.41
P_{sum} (*)	33.00 ± 8.12	20.86 ± 4.54	12.14 ± 9.30
HPO_4^{2-}	13.13 ± 7.16	14.47 ± 8.66	-1.34 ± 11.24
P_{org}	19.87 ± 10.83	6.39 ± 9.77	13.48 ± 14.59
Cl^-	16.93 ± 19.18	19.74 ± 10.29	-2.81 ± 21.77
H_3BO_3	2.11 ± 0.51	2.11 ± 0.21	0.00 ± 0.55
Ca^{2+}	420.64 ± 169.68	597.32 ± 91.14	-176.68 ± 192.6
Mg^{2+}	31.08 ± 6.76	35.57 ± 5.10	-6.49 ± 8.5
K^+ (*)	122.87 ± 32.21	82.01 ± 14.12	40.86 ± 35.2
Na^+	4.46 ± 4.10	9.15 ± 3.83	-4.69 ± 5.6
Fe^{2+}	1.771 ± 0.701	2.198 ± 0.782	-0.427 ± 1.050
Mn^{2+}	0.391 ± 0.206	0.456 ± 0.060	-0.065 ± 0.215
Zn^{2+}	1.191 ± 0.565	1.225 ± 0.181	-0.034 ± 0.593
Cu^{2+}	0.043 ± 0.014	0.030 ± 0.006	0.013 ± 0.015
Cd^{2+}	0.0003 ± 0.0004	0.0052 ± 0.0090	-0.005 ± 0.009
Al^{3+}	2.860 ± 1.231	3.781 ± 1.294	-0.921 ± 1.786
H^+	0.041 ± 0.017	0.043 ± 0.015	-0.002 ± 0.023
$Inorg^+$ (mEq)	1046.2 ± 341.2	1380.2 ± 183.2	40.9
$R-NH_3^+$ (mEq)	652.2 ± 121.1	438.2 ± 91.9	214.0
$Inorg^-$ (mEq)	84.8 ± 27.6	69.0 ± 13.4	18.2
$R-COO^-$ (mEq)	1613.6 ± 363.1	1749.4 ± 205.4	236.7
Organic salts (mEq)	961.4 ± 342.3	1311.2 ± 183.7	22.7
$R-COO^-/C_{org}$ (%)	4.046 ± 0.924	5.223 ± 0.706	3.7

Values are mean needle contents from samples collected in Würzburg in all seasons of the year from 7 different trees. The third column is the relative difference of the contents of green and brown needles and thus indicates the retranslocation of nutrients before needle senescence. Data are already calibrated using the boron content of green and brown needles, since boron is immobile in the phloem sap (Jeschke, Würzburg, personal communication). The only slightly smaller boron contents in brown needles (ca. 90% of green needles) must be due to leaching from brown needles. C_{org}, H_{org} and N_{sum} were measured by elementar analysis (Heraeus), all mentioned anions by isocratic anion chromatography (HPLC), and all mentioned metal cations H_3BO_3, P_{sum} and S_{sum}, by ICP spectrophotometry. Organic N, S and P compounds (N_{red}, S_{red}, P_{org}) are differences from the total element content minus the determined inorganic anion content. The proton content H^+ is determined by pH measurements of needle homogenates. $Inorg^+$ is the sum of the equivalents of all determined inorganic cations; $R-NH_3^+$ is the sum of organic base equivalents estimated from the N_{org} content; and $Inorg^-$ equals the sum of all measured inorganic anion equivalents assuming that the phosphate content in vivo consists equally of 50% $H_2PO_4^-$ and 50% HPO_4^{2-}. The difference of all inorganic cations minus inorganic anions must be organic anions $R-COO^-$. The relative content of carboxyl groups in organic matter $R-COO^-$ per C_{org} of spruce needles equals ca. 4–5%. All analysis data after Kindermann and Slovik, unpublished

(*) indicates significant retranslocation before needle abscission

Table 11. The chemical compositions of: xylem sap gained from spruce twigs using a pressure bomb, spruce branch wood, and spruce root wood

Compound	Xylem sap ($n=200$) ($mmol\,m^{-3}$)	Twig wood ($n=50$) ($mmol\,kg^{-1}$)	Root wood ($n=3$) ($mmol\,kg^{-1}$)
C_{org}	n.d.	40992 ± 832	41098 ± 148
H_{org}	n.d.	66593 ± 1325	65592 ± 4440
N_{sum}	2417 ± 496	133.0 ± 27.4	273.0 ± 21.7
NO_3^-	12.0 ± 14	0.36 ± 0.23	0.28 ± 0.31
N_{red}	2405 ± 496	132.6 ± 27.4	272.7 ± 21.7
S_{sum}	185.7 ± 87	4.85 ± 0.81	10.20 ± 1.42
SO_4^{2-}	81.2 ± 52	0.92 ± 0.40	4.56 ± 1.84
S_{red}	104.5 ± 102	3.93 ± 0.90	5.64 ± 2.33
P_{sum}	294.4 ± 207	6.30 ± 1.84	28.63 ± 10.34
HPO_4^{2-}	292.7 ± 203	2.11 ± 0.83	17.07 ± 10.49
P_{org}	1.7 ± 290	4.19 ± 2.02	11.56 ± 14.73
Cl^-	972 ± 1084	4.91 ± 2.27	6.73 ± 4.85
H_3BO_3	n.d.	0.37 ± 0.22	0.44 ± 0.04
Ca^{2+}	1005 ± 382	42.53 ± 7.12	51.58 ± 18.76
Mg^{2+}	194 ± 89	6.43 ± 1.56	15.24 ± 1.44
K^+	2259 ± 974	20.44 ± 4.20	77.48 ± 33.53
Na^+	71 ± 41	2.72 ± 1.77	3.07 ± 0.72
Fe^{2+}	25.7 ± 61.3	0.312 ± 0.258	1.886 ± 1.574
Mn^{2+}	1.71 ± 1.39	0.116 ± 0.054	0.229 ± 0.159
Zn^{2+}	13.4 ± 12.0	0.321 ± 0.131	0.415 ± 0.146
Cu^{2+}	0.75 ± 0.65	0.059 ± 0.020	0.039 ± 0.007
Cd^{2+}	n.d.	0.0008 ± 0.0012	0.0007 ± 0.0001
Al^{3+}	2.65 ± 1.4	0.317 ± 0.273	4.66 ± 3.93
H^+	0.73 ± 1.374	n.d.	0.0019 ± 0.0009
$Inorg^+$ (mEq)	4820 ± 1258	123.6 ± 15.3	233.3 ± 51.9
$R-NH_3^+$ (mEq)	2405 ± 496	132.6 ± 27.4	272.7 ± 21.7
$Inorg^-$ (mEq)	1588 ± 1132	16.6 ± 3.7	59.1 ± 16.7
$R-COO^-$ (mEq)	5637 ± 1764	239.6 ± 31.6	446.9 ± 58.7
Organic salts (mEq)	3232 ± 1692	107.0 ± 15.7	174.2 ± 54.5
$R-COO^-/C_{org}$ (%)	n.d.	0.585 ± 0.078	1.087 ± 0.143

For details, see footnote to Table 10
n.d. = not determined

from wet deposition but not from spruce tissues. Only the observed annual Mg^{2+} leaching rate, which exceeds these expectations, originates from canopy tissues. On the basis of the stock of needles and the needle Mg^{2+} content of three stands in Hessen (Table 13), specific canopy leaching rates [$mol\,Mg^{2+}\,ha^{-1}\,y^{-1}$ (leaching) per $mol\,Mg^{2+}\,kg^{-1}$ (needle dw)] can be calculated. These specific canopy leaching rates are relative fractions of the mean Mg^{2+} content in all spruce needle age classes from the mentioned sites, which is annually leached by precipitation. The mean value of the three given sites in Table 13 is $18.1\pm7.5\%\ Mg^{2+}\,y^{-1}$. Since $(0.181\ years^{-1})^{-1}\approx5.5$ years roughly approximates the mean spruce needle age in Hessen, the annual Mg^{2+} leaching rate from the canopy equals approximately the annual Mg^{2+} demand for needle growth. On the basis of the given 'typical' 18.1% value, the annual Mg^{2+} leaching rate in Würzburg may be roughly estimated, where the mean Mg^{2+} content

Table 12. Chemical compositions of branch bark and root bark of Norway spruce growing in Würzburg

Compound	Twig bark ($n=50$) (mmol kg^{-1})	Root bark ($n=3$) (mmol kg^{-1})
C_{org}	40356±1048	38307±1591
H_{org}	66312±1028	59725±4464
N_{sum}	539.1±118.7	458.7±62.8
NO_3^-	2.30±3.73	0.44±0.33
N_{red}	536.8±118.7	458.3±62.8
S_{sum}	24.24±5.11	17.32±0.97
SO_4^{2-}	9.37±5.36	5.97±1.22
S_{red}	14.87±7.40	11.35±1.56
P_{sum}	36.50±7.23	48.71±7.14
HPO_4^{2-}	11.86±5.29	38.60±8.69
P_{org}	24.64±8.96	10.11±11.25
Cl^-	16.45±14.90	14.98±11.72
H_3BO_3	1.35±0.10	1.14±0.16
Ca^{2+}	341.24±92.75	286.47±56.36
Mg^{2+}	37.34±6.59	43.88±9.45
K^+	98.96±18.51	139.06±21.37
Na^+	11.47±16.97	3.01±0.21
Fe^{2+}	4.188±1.745	4.307±1.989
Mn^{2+}	0.530±0.177	0.581±0.317
Zn^{2+}	2.259±0.769	1.177±0.211
Cu^{2+}	0.168±0.049	0.071±0.013
Cd^{2+}	0.0005±0.0006	0.0019±0.0004
Al^{3+}	7.18±2.83	11.78±5.90
H^+	0.0073±0.0033	0.0041±0.0039
Inorg$^+$ (mEq)	903.4±187.9	850.4±117.7
R-NH$_3^+$ (mEq)	536.8±118.7	458.3±62.8
Inorg$^-$ (mEq)	92.2±21.6	100.4±16.1
R-COO$^-$ (mEq)	1348.0±223.3	1208.3±134.4
Organic salts (mEq)	811.2±189.1	750.0±118.8
R-COO$^-$/C$_{org}$ (%)	3.340±0.560	3.154±0.375

For details, see footnote to Table 10

Table 13. Mean annual Mg^{2+} leaching rates from spruce canopies ($n=5$ years, see Balázs, 1991), the estimated stock of needle dry matter (see Figure 19; yield class and site age data after Gärtner, 1987), the mean Mg^{2+} content of needles from these stands (Kindermann and Slovik, unpublished) and the calculated specific leaching rate (mean annual percentage of Mg^{2+} loss per Mg^{2+} needle content) at three sites in Hessen (central Germany)

Site	Mg^{2+} leaching (mol ha^{-1} y^{-1})	Needle stock (kg ha^{-1})	Needle content (mmol Mg^{2+} kg^{-1})	Specific leaching (fraction per year)
Königstein	47.0±33.1	12100±1200	18.2±1.4	0.213
Grebenau	20.6±32.8	11800±300	18.3±2.7	0.095
Witzenhausen	69.1±30.7	12600±2400	23.3±2.3	0.235

in spruce needles is 31.1 mmol kg^{-1} (see Table 6). The appropriate annual Mg^{2+} leaching rate, which is consistent with the Mg^{2+} flux rates given in Figure 38, is thus estimated to be 0.181 y^{-1} times 0.0311 mol Mg^{2+} kg^{-1} (needle dw) times 18450 kg (dw needles) ha^{-1} (see Figure 19) ≈ 100 mol Mg^{2+} ha^{-1} y^{-1} (≈ 2.5 kg Mg^{2+} ha^{-1} y^{-1}) in

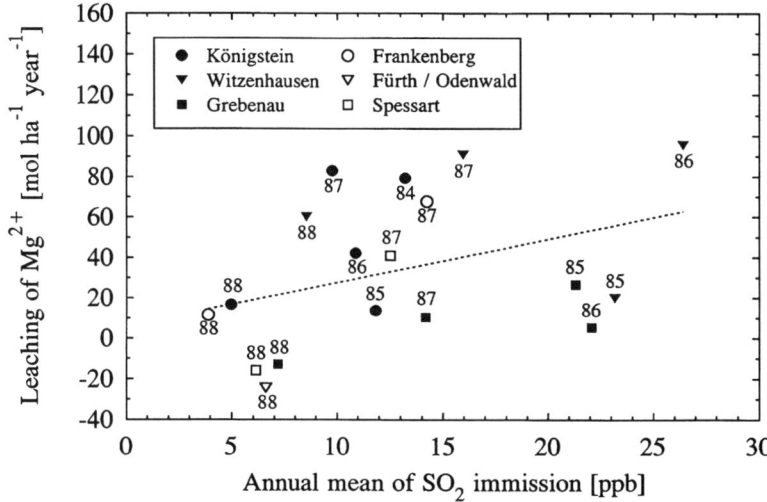

Figure 39. Leaching rates of Mg^{2+} from Norway spruce canopies at six different sites in Germany (Hessen) as measured between 1985 and 1988 in the field (recalculated after Balázs, 1991). Data are plotted versus the annual mean of SO_2 pollution that was synchronously measured within these canopies (data after Hessische Landesanstalt für Umwelt, Wiesbaden, Germany). The dotted regression line, which is a tentative estimate of an increase in Mg^{2+} leaching by $2 \, mol \, ha^{-1} \, y^{-1} \, ppb^{-1} \, SO_2$, is significant only at the 93% level

Würzburg, where canopy leaching rates were not measured. This is about two thirds of the total annual Mg^{2+} demand for needle growth ($143 \, mol \, Mg^{2+} \, ha^{-1} \, y^{-1}$) in Würzburg (see Figure 38). Since the leaching demand of Mg^{2+} adds to the annual demand of $352 \, mol \, Mg^{2+} \, ha^{-1} \, y^{-1}$ (see Figure 38), the total annual Mg^{2+} demand, which must be supplied by the soil, equals $352 + 100 \approx 450 \, mol \, Mg^{2+} \, ha^{-1} \, y^{-1}$ in Würzburg (ca. $11 \, kg \, Mg^{2+} \, ha^{-1} \, y^{-1}$). The annual Mg^{2+} uptake from the soil that compensates the canopy leaching demand, is therefore about 22% of the total annual Mg^{2+} uptake by the roots.

In Figure 39, the annual canopy leaching rate of Mg^{2+} [$mol \, ha^{-1} \, y^{-1}$] at six spruce stands in Hessen is plotted versus the measured SO_2 concentration (annual mean) in the ambient air. The dotted line represents the regression function, which is significant only at the 93% level. Conventionally speaking, there is only a hard-to-prove SO_2-dependent Mg^{2+} leaching from spruce canopies but there is a highly significant correlation (99.9% level) between SO_4^{2-} leaching rates from Norway spruce canopies and Mg^{2+} leaching rates (Figure 40). The slope of $0.213 \pm 0.048 \, mol \, (Mg^{2+}) \, mol \, (SO_4^{2-})$ indicates, that $21 \pm 5\%$ of the leached sulfate anions are neutralized by Mg^{2+} cations in the field (see section 4.7.). The intercept in Figure 40 is close to zero ($-7.6 \pm 12.0 \, mol \, Mg^{2+} \, ha^{-1} \, y^{-1}$). The soil sulfur content below spruce trees ($900 \, kg \, S \, ha^{-1}$) was three times higher than that below beech trees ($300 \, kg \, S \, ha^{-1}$) in the Solling mountains (Meiwes, 1979; Ulrich *et al.*, 1979a,b), i.e. evergreen spruce trees 'filter' much more wind- and rain-borne sulfur than the deciduous beech trees.

In winter 1973/74, the rain-borne sulfur supply per hectare below canopies of different tree species were: 13.9 kg S (*Fagus sylvatica*), 41.5 kg S (*Quercus robur*), 43.1 kg S (*Pinus sylvestris*) and 61.7 kg S (*Picea abies*) relative to 12.9 kg S outside the canopy. Thus, the mentioned species can be arranged according to their ability to 'filter' wet deposition from the air:

Fagus sylvatica < *Quercus robur* < *Pinus sylvestris* < *Picea abies*

This sequence, which does not arrange net leaching of solutes from the canopy, but wet deposition plus canopy leaching, is the same as for dry deposition of dust. Since Lampadius (1968) found hardly any difference in the 'quality' of air moving through either pine or spruce canopies before and behind the stand on dry days, there must be wet surfaces of intercepted water to strongly enable dissolution of gaseous SO_2 if effective 'sulfate filtering' by the canopy is to occur. Ulrich (1972) found that the magnitude of water imbibition of bark and dead twigs is an important factor in absorbing fog and mist, and in adsorbing hydrophilic air pollutants with the canopy (see Table 4). A high potential water inhibition of canopy tissues correlates with high water interception and thus with high wet deposition rates and with high surface adsorption of gaseous pollutants. The imbibed phloem of spruce bark increases its volume by ca. 40% (Volz, 1974).

All these findings lead to the disputed question of, from which canopy tissues does leaching Mg^{2+} mainly originate: (i) extraction from 'dry' deposited dust, (ii) from ion exchange at needle cuticles and bark surfaces, or (iii) extraction of Mg^{2+} salts from living or decomposing tissues within the spruce canopy (bark, dead needles etc.). In the first case (i), measured Mg^{2+} leaching rates would not constitute an additional annual Mg^{2+} demand for spruce stands, but, in contrast, an external atmospheric Mg^{2+} supply into spruce stands. In case (ii), there would be net leaching of cations from the canopy only if net proton/cation exchange dominated in the field, e.g. by 'acid mist' deposition. pH values of the canopy leachate below Norway spruce crowns are generally lower than outside the canopy in native precipitation water in Hessen (Balázs, 1991). The corresponding mean acid deposition rate is $564 \pm 142 \, mol \, H^+ \, ha^{-1} \, y^{-1}$ outside of spruce canopies, but it is $1493 \pm 555 \, mol \, H^+ \, ha^{-1} \, y^{-1}$ below spruce crowns ($n=23$ years from 6 spruce stands; Balázs, 1991; see Table 14). Therefore, spruce canopies are no net proton buffering targets but in fact sources of (organic) acids in the field ($638 \pm 489 \, mol \, H^+ \, ha^{-1} \, y^{-1}$). This value is not the difference of the means given above, but the mean of the individual differences. There is no net proton (H^+)/cation (Cat^+) exchange between precipitation water and native spruce canopy surfaces. Still, the possibility remains that there is Cat^+/Mg^{2+} exchange in spruce canopies, which thereby increases Mg^{2+} leaching rates. Table 14 shows that there is no ion with lower measured leaching rates than precipitation rates. Spruce canopies are net sources for all important plant nutrient cations. The precipitation/leaching balance is close to zero only for sodium, zinc and copper, i.e. atmospheric supply and precipitation-dependent loss of these cations is almost equal in the long term. Regarding all other cations, the dominant net source of canopy leaching must be extraction of soluble mineral compounds from living or

Table 14. Mean annual nutrient and acid precipitation rates and spruce canopy leaching rates in the field at Hessen (Balázs, 1991). Besides measured leaching rates, recalculated 'corrected' canopy leaching rates are also given, which estimate the contribution of the spruce canopy itself to the total canopy through-fall fluxes. For almost all analyzed compounds, spruce canopies are net sources. The net interception/net leaching balance is close to zero only for Na^+, Zn^{2+} and Cu^{2+} in the field

Ion species	Precipitation (measured) $(mol\,ha^{-1}\,y^{-1})$	Canopy leaching $(mol\,ha^{-1}\,y^{-1})$	
		Measured	Corrected
H_3O^+	564±142	1493±555	638±489
NH_4^+	289±93	507±157	75±134
K^+	88±28	560±118	427±122
Na^+	368±94	471±129	−78±126
Mg^{2+}	102±26	194±37	40±36
Ca^{2+}	267±71	571±138	167±110
Al^{3+}	12±5	57±29	38±23
Fe^{2+}	3.6±1.6	12.6±5.9	7.1±5.2
Mn^{2+}	6.4±3.1	77.8±41.0	68.0±42.1
Zn^{2+}	6.1±2.5	8.1±2.5	−1.1±3.4
Cu^{2+}	0.95±0.56	1.08±0.53	−0.47±0.62
SO_4^{2-}	175±35	486±125	221±119
NO_3^-	119±28	238±66	59±55
Cl^-	412±114	765±198	142±200

decomposing tissue surfaces (needles, litter, bark etc.) within the spruce canopy but not net ion exchange processes. Based on the real (=corrected) canopy leaching rates given in Table 14, the relative 'availability' of different cations to leak from spruce canopies can be deduced:

$$H^+ > K^+ > Ca^{2+} > NH_4^+ > Mn^{2+} > Mg^{2+} > Al^{3+} > Fe^{2+}$$

This is a 'typical' sequence based on field data from six sites in central Germany (see Balázs, 1991), which may vary at 'exceptional' sites. Relative to the leaching rates of (organic) acids and of potassium salts from spruce canopies, leaching of Mg^{2+} is not dominant in the field. Leached potassium is easily resupplied via the xylem sap. There is no net drop in the K^+ contents in potassium leaching needles (Fiedler *et al.*, 1984). Leaching of Ca^{2+}, Mn^{2+} and Al^{3+} is interpreted here as a means of eliminating an excess supply with 'waste cations'.

It is generally accepted that leaching of ions from needle and leaf surfaces almost exclusively affects the apoplasm of epidermal cells (Adams and Hutchinson, 1984; Fink, 1992; Foster, 1990; Hutchinson and Adams, 1987; Mecklenburg *et al.*, 1966; Tukey and Tukey, 1969; Tukey, 1970). Diffusion of cations across cuticles is slow. Leaching of cations from healthy plants is smaller than from deficient or declining plants, and leaching of K^+, Mg^{2+} and Ca^{2+} from ageing tissues is more pronounced than from vital young plant tissues (Arens, 1934; Helder, 1956; Tukey and Tukey, 1969; cf. Mengel *et al.*, 1987). Klemm *et al.* (1989) found that the release of Mg^{2+} from green needles grown in the Fichtelgebirge with artificial precipitation was close to zero within experiments that lasted only 5 hours. It was higher from yellow needles and maximum from brown needles, but absolute differences were not significant since there was an enormous scattering of the data. Mitterhuber *et al.* (1989) found that Mg^{2+} leaching rates from small spruce twigs grown in Würzburg was linearly 0.0061% of the water-soluble needle Mg^{2+} content per hour in vitro (10 h per day artificial rain with distilled water; see also Klemm *et al.*, 1989). Subsequently, an artificial 'acid rain' (1 mmol/L H_2SO_4, pH 2.95) was sprayed onto the same twigs. There was only a short-term increase of Mg^{2+} leaching for a few hours before the Mg^{2+} leaching rate was linearly continued with a stable rate of 0.0058% h^{-1} despite continuation of the H_2SO_4 treatment for another 50 h. Very similar results were obtained with rinsed isolated spruce needles. The Mg^{2+} leaching rate was the same in distilled water (0.010% $Mg^{2+}h^{-1}$) and in 1 mmol/L H_2SO_4 (0.011% $Mg^{2+}h^{-1}$). Mg^{2+} leaching from isolated needles was almost twice as high as from intact twigs, apparently due to wounding effects. Compared with xylem flux rates, leaching rates from the canopy are small, i.e. the leaching demand can be readily resupplied via xylem sap import of lost cations. Thus, small contents of Mg^{2+}, Ca^{2+} and Mn^{2+} in needles represent poor soils, but not increased leaching rates (Fiedler *et al.*, 1984). The mentioned Mg^{2+} leaching data after Mitterhuber *et al.* (1989) may be employed to estimate the annual Mg^{2+} leaching rate L_{Mg} [% Mg^{2+} y^{-1}] if the realistic precipitation scenario after Klemm *et al.* (1989) is applied (93 days per year of wet spruce canopies in the field):

$$L_{Mg} = 0.006\% \ Mg^{2+}h^{-1} \cdot 24\,h\,day^{-1} \cdot 93\,days\,year^{-1} = 13.4\% \ Mg^{2+}y^{-1}$$

In the field, $L_{Mg} = 18.1 \pm 7.5\% \ Mg^{2+}y^{-1}$ was measured as a mean value from 15 site-years in Hessen (see Table 13 and text above). The difference is small and it may be attributed to Mg^{2+} leaching from 'naked' bark surfaces, which adds to the Mg^{2+} leaching rate originating from needle surfaces. Von Droste zu Hülshoff (1970) determined the canopy surface in a 76-year-old Norway spruce stand close to Munich (Ebersberger Forst): total needle surface = 21.61 ha ha^{-1}; total bark surface = 4.54 ha ha^{-1} (trunk surface = 1.25 ha ha^{-1}, living twigs and branches = 2.79 ha ha^{-1}, dead twigs and branches = 0.50 ha ha^{-1}). The leached needle surface accounts for 82.6% of the total canopy surface. The bark surface percentage is 17.4%, and the bark surface per needle surface ratio equals 1.4/82.6 = 0.21, i.e. there is 1.21 times more canopy surface (0.21 units bark surface) than needle surface. The content ratio of Mg^{2+} in spruce needles and spruce bark in Würzburg equals $c_r = 37.3$ mol $Mg^{2+}kg^{-1}$ (dw bark)/31.1 mol $Mg^{2+}kg^{-1}$ (dw needles) = 1.2 mol (bark-Mg^{2+}) mol^{-1} (needle-Mg^{2+}) (see Table 6). Thus, a total annual Mg^{2+} leaching percentage of the whole spruce canopy can be estimated:

$$L_{Mg} = L_{needles} + 0.21 \cdot c_r \cdot L_{bark}$$

where $L_{needles}$ [% $Mg^{2+}y^{-1}$] is the percentage of the Mg^{2+} content in needles that is leached per year, and L_{bark} [% $Mg^{2+}y^{-1}$] is the leached percentage of the Mg^{2+} content in bark. Assuming that $L_{bark} \approx L_{needles} \approx 13.4\% \ Mg^{2+}y^{-1}$ (= extrapolated in-vitro leaching rate; see above), the total annual canopy leaching percentage L_{Mg} can be estimated to be $L_{Mg} = 16.8\% \ Mg^{2+}y^{-1}$. This independently estimated relative leaching rate is very close to the measured value of $18.1 \pm 7.5\% \ Mg^{2+}y^{-1}$ in the field.

Thus, in-vitro leaching rates seem to be good representatives of field data.

Besides extraction of Mg^{2+} from needle and bark surfaces, it was assumed that there may be Mg^{2+} loss also from the xylem sap via diffusion of Mg^{2+} across the cambium into the bast (living phloem) and rhytidoma (dead bark) tissues via the radial rays of the bark into the precipitation water moistening the spruce bark (Klemm et al., 1989). The apparent permeability coefficient of the cambium was determined to be $P_{Mg} = 9 \cdot 10^{-9}\,m\,s^{-1}$ for $MgCl_2$ (Klemm et al., 1989) after full hydration and swelling of spruce bark tissues (see legend of Table 9). There is a Mg^{2+} concentration gradient between xylem sap $[Mg^{2+}]_{xyl} \approx 0.194\,mol\,m^{-3}$ (annual mean, see Table 11) and 'typical' Mg^{2+} contents in precipitation water from central Germany of $[Mg^{2+}]_{prec} \approx 0.0115\,mol\,m^{-3}$ (annual mean from 23 years of 6 spruce stands; after Balázs, 1991). The Mg^{2+} concentration difference $\Delta[Mg^{2+}] \approx 0.180\,mol$ $Mg^{2+}\,m^{-3}$ between both is a maximum 'gradient' assuming that there is fresh precipitation water adhering all the time in statu nascendi on spruce bark surfaces. Thus, the maximum annual Mg^{2+} loss J_{Mg} [mol $Mg^{2+}\,ha^{-1}\,day^{-1}$] from the xylem sap into the spruce canopy leaching water equals:

$$J_{Mg} = P_{Mg} \cdot \Delta[Mg^{2+}] \cdot 45400\,m^2\,ha^{-1} \cdot 86400\,s\,day^{-1} \approx 6.4\,mol\,ha^{-1}\,day^{-1}$$

$P_{Mg} = 9 \cdot 10^{-9}\,m\,s^{-1}$ is the permeability coefficient of bark for Mg^{2+} and $\Delta[Mg^{2+}] \approx 0.180\,mol\,Mg^{2+}\,m^{-3}$ is a 'typical' maximum Mg^{2+} concentration difference between Norway spruce xylem sap and precipitation water. There are $45400\,m^2$ spruce bark surface per ha soil (after Von Droste zu Hülshoff, 1970) and $86400\,s/day$. Assuming again 93 days of adhering interception water in the field (after Klemm et al., 1989), the maximum total annual Mg^{2+} loss from the xylem sap by Mg^{2+} leaching from spruce canopies would be $595\,mol\,ha^{-1}\,y^{-1}$. As already mentioned, this is an overestimation. A more realistic rate is only half of 595 ($\approx 300\,mol\,ha^{-1}\,y^{-1}$) if there is a linear Mg^{2+} concentration increase in the interception water adhering to bark surfaces until concentrations, that are measured in the canopy through-fall, are achieved. The maximum annual Mg^{2+} leaching rate measured in the spruce canopy through-fall in Hessen was, exceptionally, $279\,mol\,Mg^{2+}\,ha^{-1}\,y^{-1}$ (Witzenhausen in 1986, see Balázs, 1991). The mean Mg^{2+} leaching rate in the field is only $40 \pm 36\,mol\,ha^{-1}\,y^{-1}$ (see Table 14 and Figure 40). This much smaller observed Mg^{2+} leaching rate is readily explained by extraction of Mg^{2+} from needle and bark surfaces (see above). The calculated Mg^{2+} loss from the xylem sap is therefore unrealistic and the permeability coefficient P_{Mg} must be too high.

4.6.3. Wet deposition of magnesium

Total annual wet deposition rates of Mg^{2+} via precipitation of rain, snow, fog etc. at six Norway spruce stands in Hessen (central Germany) from 1984 to 1988 are summarized in Table 15. The mean of these ombrogene Mg^{2+} deposition rates from all sites and years equals $102 \pm 26\,mol\,Mg^{2+}\,ha^{-1}\,y^{-1}$. Compared with the mean canopy leaching rate of $40 \pm 36\,mol\,Mg^{2+}\,ha^{-1}\,y^{-1}$, there is 2.5 times more Mg^{2+}

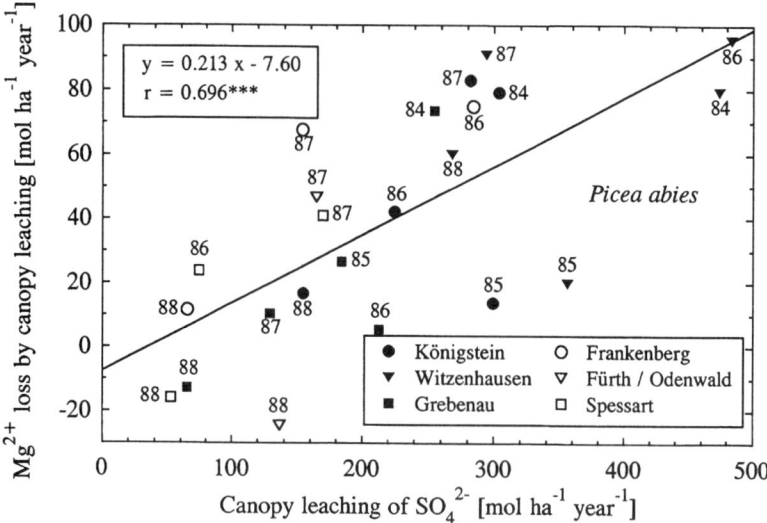

Figure 40. Mg²⁺ leaching rates of different Norway spruce sites in Hessen plotted versus the synchronously measured SO_4^{2-} leaching rates (recalculated after Balázs, 1991). The slope of the given significant linear regression function estimates that 21.3% of leaching SO_4^{2-} equivalents are electrostatically compensated by Mg²⁺ ions

Table 15. Net supply of Mg²⁺ via precipitation (wet deposition) at different spruces sites in Hessen from 1984 to 1988

Site	Wet deposition of Mg (mol ha⁻¹ y⁻¹)				
	1984	1985	1986	1987	1988
Königstein	52.2	88.6	83.6	119.9	126.4
Grebenau	42.6	75.9	89.0	96.5	86.1
Witzenhausen	87.5	109.0	130.6	119.5	107.1
Spessart	—	—	93.2	145.4	124.8
Frankenberg	—	—	77.6	116.7	99.4
Fürth/Odw.	—	—	—	130.5	136.7

After Balázs, 1991. The mean Mg²⁺ supply rate from all given sites and years equals 102 ± 26 mol ha⁻¹ y⁻¹

imported into spruce forests via precipitation than net leaching of Mg²⁺ from spruce canopies. Similar wet deposition rates were observed in the Fichtelgebirge (NE Bavaria, 700 m asl: 78 mol Mg²⁺ ha⁻¹ y⁻¹, Hantschel, 1987), in the Solling (Lower Saxony, FRG, 500 m asl: 83 mol Mg²⁺ ha⁻¹ y⁻¹; Ulrich *et al.*, 1979a,b) and in Bodenmais (east Bavaria, FRG, 950 m asl: 66 mol Mg²⁺ ha⁻¹ y⁻¹; Dunkl and Rehfuess, 1988). All these values are within the variance of the different sites and years given in Table 15. There are lower Mg²⁺ precipitation rates for example in Hubbard Brook (New Hampshire, USA, 500–800 m asl: 24 mol Mg²⁺ ha⁻¹ y⁻¹; Likens *et al.*, 1977). Compared with the annual Mg²⁺ demand of spruce stands (e.g.

Figure 38 for Würzburg: ca. $350\,\text{mol}\,\text{ha}^{-1}\,\text{y}^{-1}$ (growth demand) $+\,100\,\text{mol}\,\text{ha}^{-1}\,\text{y}^{-1}$ (leaching demand), the mean precipitation flux of ca. $100\,\text{mol}\,\text{Mg}^{2+}\,\text{ha}^{-1}\,\text{y}^{-1}$ in Germany supplies roughly 22% of the annual Mg^{2+} demand.

The Mg^{2+} concentration in precipitation water strongly depends (not only) on the mass ratio of dust compounds (condensation nuclei) per condensed water in the precipitation drops. For this and other reasons, the Mg^{2+} concentration in precipitation water varies from 0.1 to $26\,\text{mmol}\,\text{m}^{-3}$. The mean Mg^{2+} content in rain water from the Fichtelgebirge was $1.55\,\text{mmol}\,\text{m}^{-3}$ (Eiden et al., 1989). The mean Mg^{2+} content in fog water was $16-18\,\text{mmol}\,\text{m}^{-3}$. Mg^{2+} accounted for only $2.1\pm2.5\%$ of the total cation equivalents in the Fichtelgebirge (Eiden et al., 1989). In fog water, Herterich and Paffrath (1986) found $1.6-55$ (mean: 12) $\text{mmol}\,\text{Mg}^{2+}\,\text{m}^{-3}$ at the top of the Ochsenkopf in the Fichtelgebirge (Bavaria, 1020 m asl). The Mg^{2+} concentration in fog water from other sites in the Fichtelgebirge ranged from 1.9 to $45\,\text{mmol}\,\text{Mg}^{2+}\,\text{m}^{-3}$ (mean: $15\,\text{mmol}\,\text{Mg}^{2+}\,\text{m}^{-3}$; Herterich, 1987). In the Taunus (north of Frankfurt/M., FRG), it ranged from 4.1 to $135\,\text{mmol}\,\text{Mg}^{2+}\,\text{m}^{-3}$ (mean: $12\,\text{mmol}\,\text{Mg}^{2+}\,\text{m}^{-3}$; Georgii et al., 1986). Thus, the Mg^{2+} content in fog water is about one order of magnitude higher than in rain water. This corresponds very roughly to the inverse volume ratio of fog and rain droplets. The wind-dependent fog water flux rate inside the spruce canopy was only one third of the flux rate at the edges of the stand (Eiden et al., 1989). Thus, Mg^{2+} wet deposition is expected to be focused at these edges of forest canopies. The percentage of precipitation by fog increases with rising altitude of the site above sea level. Baumgartner (1958) found that 70% of the total annual precipitation was fog on the top of the Großer Falkenstein mountain (Bayerischer Wald, FRG). In the Karpatic Mountains (Poland), the fog percentage varies from 8 to 88% (altitudes ranged from 740 m to 1991 m asl; Ermich et al., 1967). Thus, cum grano salis, there is a tendency for wet deposition of plant nutrients to rise with increasing site altitude asl, at least relative to the inversely dependent annual nutrient demand of spruce stands (see section 4.5.). Figure 41 plots the annual wet deposition rate of Mg^{2+} versus the site altitude of six spruce stands (see also Table 15). The correlation is weak, but significant.

Norway spruce is one of the forest tree species which collects most interception water with the crown (Figure 42). The maximum observed specific interception of water in spruce crowns equals $0.1\,\text{L}/\text{m}^2$ of the total spruce canopy surface (Eiden et al., 1989). This strongly predisposes just evergreen spruce stands growing in high altitudes asl to collect nutrient and pollutant-rich fog water within their crowns. This makes hill and mountain spruce stands also most sensitive to high nitrogen wet deposition rates. It is well known that nitrogen-dependent growth extends the 'physiological length' of the annual vegetation period. Excess nitrogen supply reduces the achieved frost resistance of many plants in the winter (Aber et al., 1989), which can be fully induced only in dormant plant tissues. This becomes most harmful at high-altitude sites. Details of this interesting dependency of frost resistance on nitrogen wet deposition, and the striking synchronous onset of forest decline in the early 1980s in central Germany after a few separated years with extreme frost events, are out of the scope of this communication.

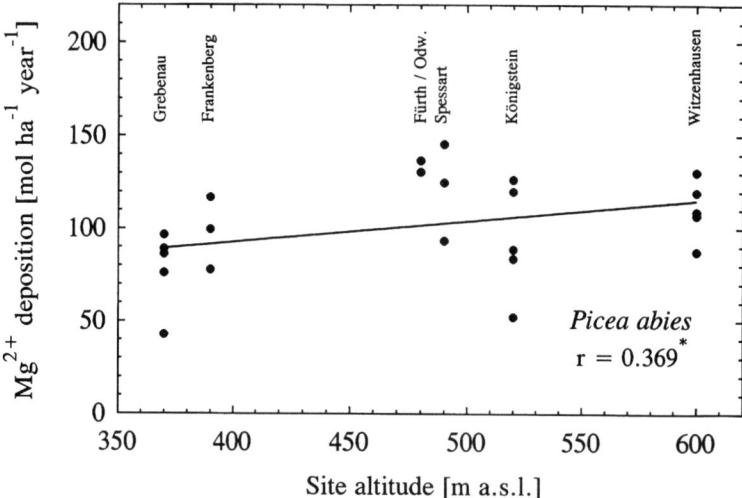

Figure 41. Mg^{2+} wet deposition at different sites in Hessen in relation to the site altitude above sea level (after Balázs, 1991). The weak correlation is significant at the 95% level

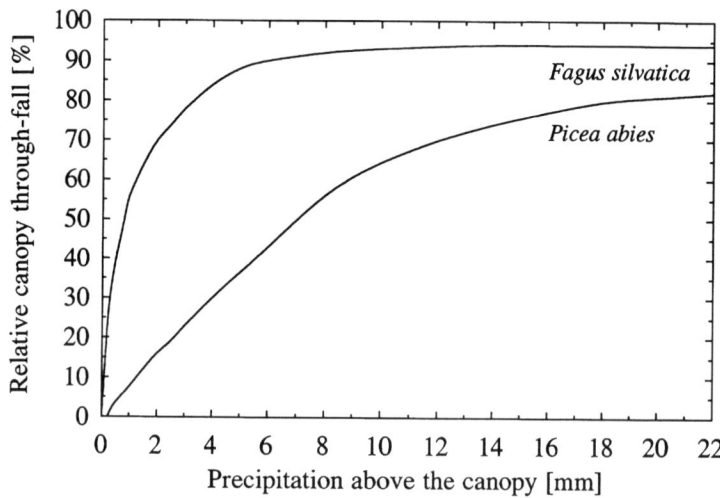

Figure 42. Percentage of the canopy through-fall of precipitation water of beech and spruce stands in relation to the precipitation intensity (after Schmidt-Vogt, 1987)

4.6.4. Dry deposition of magnesium

Dry deposition of dust within the crowns of different tree species has been investigated by Günther and Knabe (1976). The following relative sequence of dust deposition within the canopy of different tree species was found in the Ruhrgebiet, which is an industrialized area in Germany:

Fagus sylvatica < Quercus robur < Pinus sylvestris < Picea abies

Behrends (1974) confirmed that coal ash adheres better on spruce than on pine. The absolute amount of dust deposition can be estimated after Keller (1971) to be 30 mg (dust) g^{-1} (needle dw) for spruce, which corresponds to 420 kg ha^{-1} y^{-1} (for 14000 kg spruce needles ha^{-1}), 70 mg g^{-1} (leaf dw) for beech, which corresponds to 280 kg ha^{-1} y^{-1} (for 4000 kg beech leaves ha^{-1}) and 90 mg (dust) g^{-1} (leaf dw) for oak, which corresponds to 540 kg ha^{-1} y^{-1} (6000 kg oak leaves ha^{-1}). Thus, the order of species given above may vary. Meldau (1956) reported most extreme annual dust deposition rates: 32000 kg (dust) ha^{-1} y^{-1} in Norway spruce stands, 35000 kg (dust) ha^{-1} y^{-1} in Scotch Pine stands, and 68000 kg (dust) ha^{-1} y^{-1} in beech stands (summer). This corresponds to 300–700 mg (dust) cm^{-2} (leaf surface). More important than 'possible' deposition rates are actual mean dust deposition rates in central Europe at the present time. These data can be estimated from the annual man-made dust emission rate in Germany, which was reduced from 1.3 million tonnes in 1970 to 0.46 million tonnes in 1989 (natural dust sources omitted here; after Anonymous, 1992). This corresponds to an annual dust deposition rate of 52 kg ha^{-1} y^{-1} in 1970, and 18 kg ha^{-1} y^{-1} in 1989, if the annual dust emission within Germany is evenly deposited within the German state area (248 500 km^2 before reunification with the GDR). In the Fichtelgebirge, 41% of the very heterogenous particulate aerosol mass was ions and 3.2±2.5% of the mean fractional ionic charge was Mg^{2+} cations; the Mg^{2+} content ranged from approximately 7–150 mmol Mg^{2+} kg^{-1} (dw dust); the mean was ca. 27 mmol Mg^{2+} kg^{-1} (dw dust); (Eiden *et al.*, 1989). These data are sufficient to very roughly estimate an annual Mg^{2+} supply rate via dust deposition of 1.4 mol Mg^{2+} ha^{-1} y^{-1} in 1970, and 0.5 mol Mg^{2+} ha^{-1} y^{-1} in 1989. This 'mean' dry deposition rate of Mg^{2+} is by two orders of magnitude too small to explain canopy leaching rates of Mg^{2+}. If dust deposition rates of 420 kg ha^{-1} y^{-1} are assumed (after Keller, 1971; see above), the annual Mg^{2+} dry deposition would supply ca. 11 mol Mg^{2+} ha^{-1} y^{-1} in spruce forests, which is roughly only 2–3% of the annual Mg^{2+} growth and leaching demand of ca. 350 (growth) + 100 (leaching) = 450 mol ha^{-1} y^{-1}. Usually, dust deposition is an unimportant source of Mg^{2+} supply to spruce ecosystems.

4.6.5. Trunk harvesting and yield waste

Besides Mg^{2+} release via microbial decomposition of litter produced by spruce stands, there is also recycling of Mg^{2+} via decomposition of 'yield waste' (branches, stumps, roots and usually trunk bark), which remains in the stand after trunk harvesting in most cases. On the other hand, there is harvest and export of trunk wood by cutting of the stand, which corresponds to an irreversible Mg^{2+} loss. These ecophysiological Mg^{2+} flux rates can be quantified. Results deduced here will be consistent with growth data deduced in section 4.5. The Wiedemann tables supply the number n of harvested (index 'H') trees $\partial n_H/\partial t(t)$, the corresponding trunk

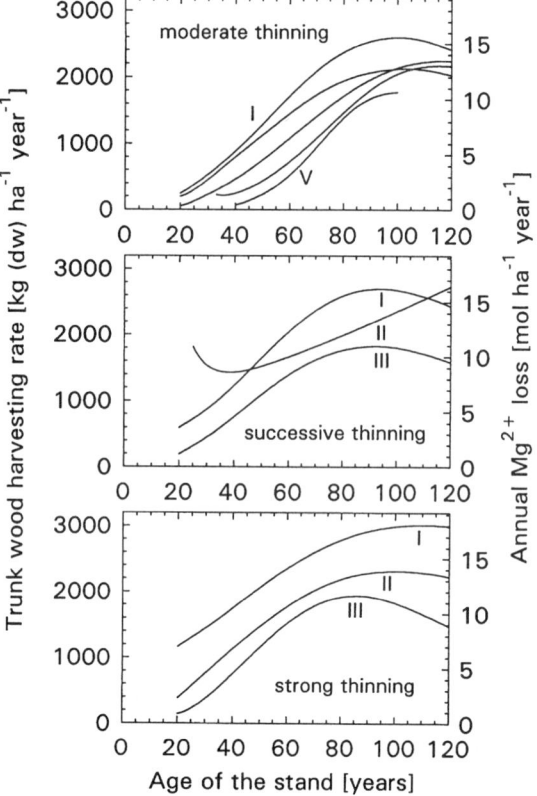

Figure 43. Trunk wood harvesting rates in relation to the age of the stand, the cutting regime and the yield class, after Wiedemann (see Schober, 1987). The right ordinate indicates the corresponding Mg^{2+} loss of the spruce stand if the Mg^{2+} content of sap wood from Würzburg (see Table 6) is used for calculations (see text)

volume harvesting rate $\partial V_H/\partial t(t)$ and the mean trunk diameter $d_H(t)$ of the harvested trees as depending on the stand age t for different yield classes and cutting regimes (see Schober, 1987, pp. 62–75, columns 11, 14 and 15). Employing the relative bark volume of spruce trunks as a function of the trunk diameter (see Figure 26 in section 4.5), the annual wood volume harvesting rates can be deduced from trunk harvesting rates. Based on the bark percentage, on the mean diameter of the harvested trunks and on their actual age, the mean annual increment of trunk wood circles $[cm\,y^{-1}]$ can be calculated. This information is necessary to get the mean specific weight of the harvested trunk wood (see Figure 33 in section 4.5.). Figure 43 shows the mean annual trunk wood harvesting rate $[kg\,(dw\,trunk\,wood)\,ha^{-1}\,y^{-1}]$ and the corresponding Mg^{2+} loss (Mg^{2+} content in spruce wood from Würzburg). Despite decreasing (mean) numbers of cut trees per year, the trunk wood yield rate per hectare increases in ageing stands for essentially all growth and cutting scenarios. The annual Mg^{2+} loss accounts for a maximum of $10-15\,mol\,Mg^{2+}ha^{-1}\,y^{-1}$. In the

Guttenberger Forst (Würzburg, see Figure 38), the mean annual Mg^{2+} loss by export of harvested trunk wood accounts for ca. $4\,mol\,Mg^{2+}\,ha^{-1}\,y^{-1}$. This is only 1% of the annual Mg^{2+} growth and leaching demand of this 35-year-old stand (yield class I, moderate cutting). This percentage strongly depends on the age of the stand, yield class etc., and must not be generalized.

Harvesting of trunks produces yield waste, which usually remains and decomposes within the spruce stand. Its full microbial mineralization requires approximately 20–25 years (B. Ulrich, Göttingen, personal communication). This is only a very rough estimation (see Chapter 8 for information on Mg^{2+} in soils). The following waste production and Mg^{2+} recycling rates are based on the assumption that the waste production rate equals the microbial waste decomposition rate in flux equilibrium. Consequently, this approach estimates realistic Mg^{2+} recycling rates only if there were no net degradation (e.g. after liming) or net accumulation of organic humus matter. This assumption is of course not fully true but it allows rough estimations. More detailed information is possible only on the basis of intensive spruce tissue decomposition studies in the field.

The trunk bark waste rate, which is calculated mutatis mutandis similar to the trunk wood production rate, is shown in Figure 44. The annual Mg^{2+} loss (if the trunk bark is exported from the forest) and the respective annual Mg^{2+} recycling rate via trunk bark decomposition approximately equals the annual Mg^{2+} loss via export of trunk wood from the stand (roughly $10\,mol\,Mg^{2+}\,ha^{-1}\,y^{-1}$). Bark dry matter contains more Mg^{2+} than wood dry matter. Much higher potential Mg^{2+} recycling rates are resupplied to surviving spruce trees via decomposition of canopy waste (needles, branch wood, branch bark). The production rate of needle waste is given in Figure 45. Table 8 supplies information on the relative composition of spruce canopies (needles:wood:bark = 32.9:49.5:17.6 [%]). Table 6 supplies data on the absolute Mg^{2+} content (needles:wood:bark = 31.1:6.4:37.3 [mmol kg^{-1}]). Thus, the mean Mg^{2+} content of 1 kg (dw) canopy waste $[Mg^{2+}]_{canopy}$ in Würzburg equals:

$$[Mg^{2+}]_{canopy} = 0.329 \cdot 31.1 + 0.495 \cdot 6.4 + 0.176 \cdot 37.3 = 20.0\,mol\,Mg^{2+}\,kg^{-1}$$

The canopy waste production rate R_{canopy} as dependent on the needle waste production rate (= gap expansion growth rate) R_{NS} (see Figure 25) equals:

$$R_{canopy} = (1.000 + 1.505 + 0.535) \cdot R_{NS}$$

The total canopy dry matter of harvested trees consists of 1.000 relative units of needles, 1.505 relative units of branch wood per unit of needle matter, and 0.535 relative units of branch bark per unit of needle matter (see Table 8). Results of canopy waste production and Mg^{2+} recycling rates are shown in Figure 45. The highest nutrient recycling rates (ca. $40\,mol\,Mg^{2+}\,ha^{-1}\,y^{-1}$) are achieved in young stands, where cutting rates are still high.

Root waste production rates (bark, wood), and the corresponding Mg^{2+} recycling rates are shown in Figure 46. The mean content $[Mg^{2+}]_{root}$ in spruce roots equals:

$$[Mg^{2+}]_{root} = 0.32 \cdot 43.9 + 0.68 \cdot 15.2 = 24.4\,mol\,Mg^{2+}\,kg^{-1}$$

186

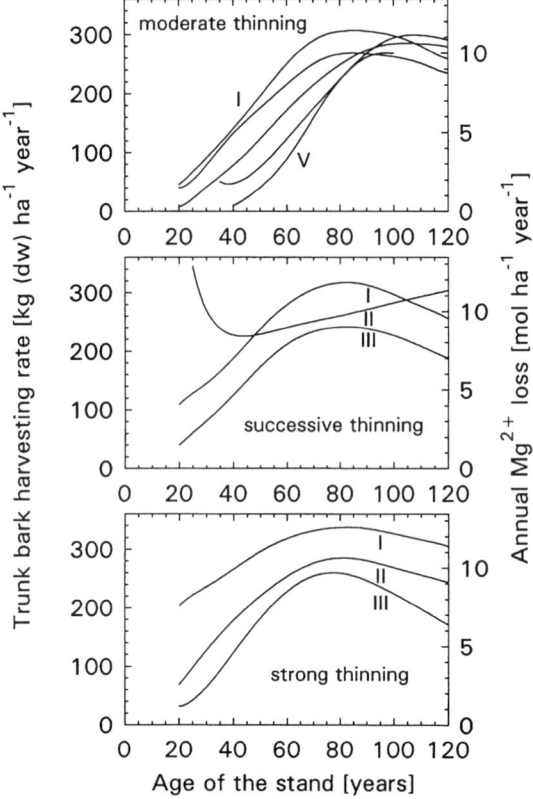

Figure 44. Trunk bark harvesting rates in relation to the age of the stand, the cutting regime and the yield class, after Wiedemann (see legend of Figure 43 for details)

where 0.32 and 0.68 are the relative root bark respective root wood fractions (see section 4.5.), and 43.9 and 15.2 mmol kg^{-1} (dw) are the corresponding Mg^{2+} contents in both root tissues (see Table 6). The stock of decomposing root matter R$_{root}$ equals 0.186 times the stock of trunk wood matter (see section 4.5.).

4.6.6. Soil depletion and Mg^{2+} fertilization

Balancing all ecological above-ground Mg^{2+} flux rates, the following situation for the Guttenberger Forst in Würzburg (35 years old, yield class I, moderate cutting, 200 m asl) can be estimated. The total annual growth demand equals ca. 352 mol Mg^{2+} ha^{-1} y^{-1} (see Figure 38). Additionally, canopy leaching must be compensated, which was estimated to be ca. 100 mol Mg^{2+} ha^{-1} y^{-1}. Thus, the total annual Mg^{2+} uptake rate of spruce roots is approximately 450 mol Mg^{2+} ha^{-1} y^{-1}.

This demand is supplied by different sources. The litter decomposition rate accounts for ca. 121 mol Mg^{2+} ha^{-1} y^{-1} (maximum; see Figure 38), and the recycled

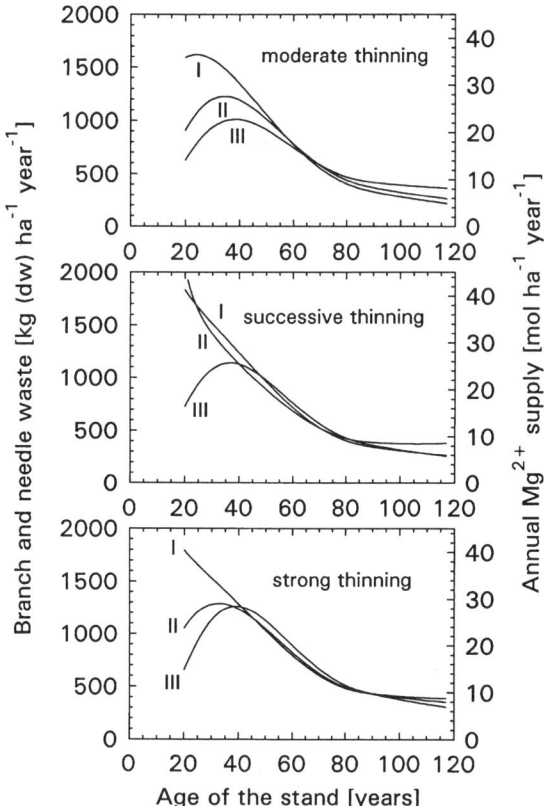

Figure 45. Branch and needle waste produced by different cutting regimes in relation to the age of the stand and the yield class, after Wiedemann. The corresponding Mg^{2+} supply rates via canopy waste decomposition (ecological flux equilibrium) is plotted at the right ordinate using the Mg^{2+} contents in needles, bark and wood of twigs and branches, after Table 6 (Würzburg situation)

canopy leaching rate for ca. $100\,mol\,Mg^{2+}ha^{-1}y^{-1}$ (flux equilibrium assumed). Recycling of Mg^{2+} via decomposition of 'harvesting waste' accounts for another 4 (trunk bark) + 34 (canopy waste) + 3 (root waste) = $41\,mol\,Mg^{2+}ha^{-1}y^{-1}$ (see Figures 44 – 46). The mean annual wet deposition of Mg^{2+} in Würzburg (200 m asl) may be (as at numerous other sites in Germany) roughly $70\,mol\,Mg^{2+}ha^{-1}y^{-1}$ (see Figure 41). The dry deposition of Mg^{2+} may be about $1 - 10\,mol\,Mg^{2+}ha^{-1}y^{-1}$. Thus, the sum of all mentioned Mg^{2+} source rates is approximately $300\,mol\,Mg^{2+}ha^{-1}y^{-1}$, and only the difference of ca. $150\,mol\,Mg^{2+}ha^{-1}y^{-1}$ is Mg^{2+} *net* supply from the soil matrix (ion exchange, soil mineral weathering); 66% of the annual Mg^{2+} demand of a 'typical' Norway spruce stand growing in the Guttenberger Forst in Würzburg is (re)supplied via atmospheric Mg^{2+} deposition and nutrient (re)cycling. Consequently, disturbance of Mg^{2+} (re)cycling rates may cause sensitive imbalance in the annual Mg^{2+} supply of spruce forests.

188

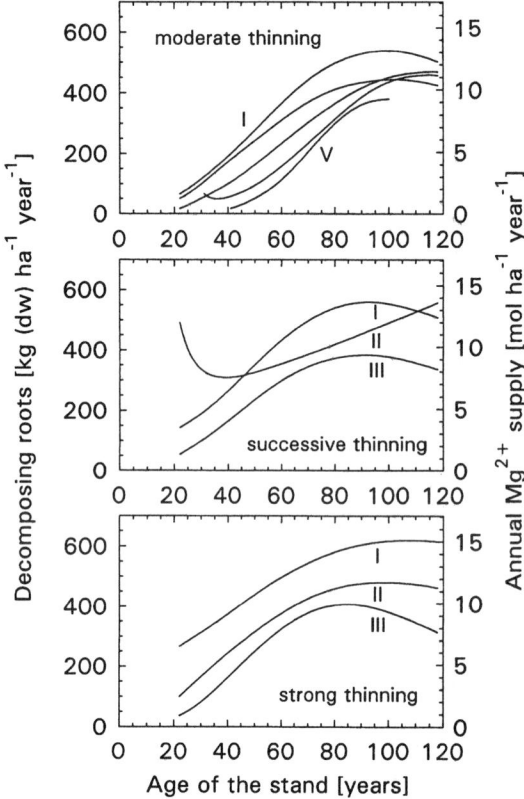

Figure 46. Total root matter waste decomposing in differently thinned Norway spruce stands in relation to the age of the stand and the yield class, after Wiedemann. The corresponding Mg^{2+} supply rates via root bark and root wood decomposition (ecological flux equilibrium) are plotted at the right ordinate using the Mg^{2+} contents in bark and wood from spruce roots in Table 6 (Würzburg)

4.7. Impact of man-made air pollutants

In this section, the fate of dry deposition of trace gases in spruce canopies and its impact on the physiology and nutrient cycling of Mg^{2+} is briefly discussed. Harmful effects on Mg^{2+} in forest ecosystems and soils via wet deposition of acid precipitation, nitrogen burden etc. are discussed by Kaupenjohann (Chapter 8, this volume). Potential causes of forest decline are discussed e.g. by Kenk (1990), Schulze (1989), Ulrich (1980, 1981a), Wentzel (1982, 1985) and Zöttl (1986).

4.7.1. *Harm due to chronic SO_2 pollution*

There are two main possible targets of trace gas action: (i) adsorption and chemical changes on canopy surfaces, and (ii) uptake of trace gases via leaf stomata and interference of these trace gases with metabolic processes within the mesophyll.

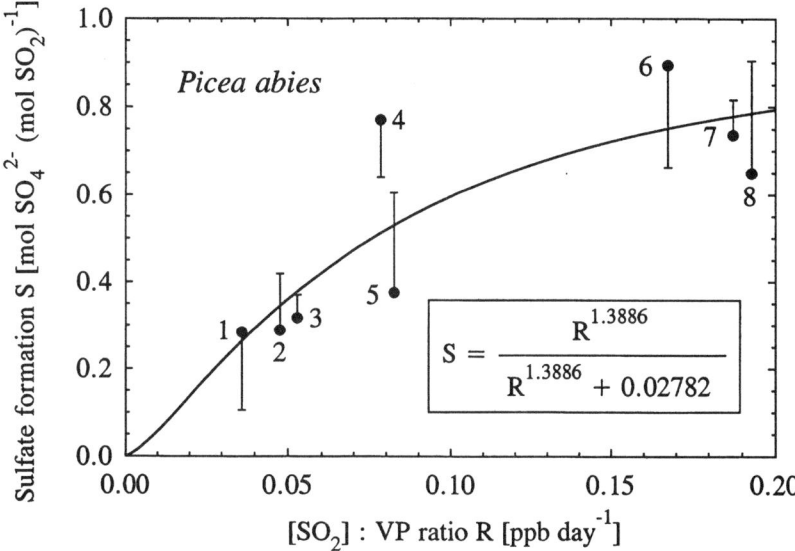

Figure 47. Relative SO_4^{2-} formation, S [mol SO_4^{2-} mol^{-1} SO_2] plotted versus the physiological SO_2 burden parameter, R = SO_2 : VP given in ppb day^{-1}. The numbers next to symbols identify the sites: 1 = Königstein (Hessen), 2 = Würzburg (Bavaria), 3 = Grebenau (Hessen), 4 = Witzenhausen (Hessen), 5 = Fichtelgebirge (Bavaria), 6 = Höckendorf (Saxony), 7 = Oberbärenburg (Saxony), 8 = Kahleberg (Saxony). The insert shows a cooperative saturation type function [0 → 1], which best approximates the data points S = f(R). The function parameters were determined by employing a Marquardt–Levenberg fitting algorithm, which minimizes the least squares (from Slovik *et al.*, 1995)

From Figure 39 (section 4.6.) an SO_2-dependent Mg^{2+} leaching rate from Norway spruce canopies of roughly 2 mol Mg^{2+} ha^{-1} y^{-1} ppb^{-1} (SO_2) can be estimated. This is only a tentative value, since a statistical significance of 93% (see Figure 39) may not be accepted. At 32 ppb SO_2, which is a typical annual mean in the Erzgebirge, a Mg^{2+} leaching rate of roughly 60 mol (leaching Mg^{2+}) ha^{-1} y^{-1} would result in the long term, and about half of it in the Fichtelgebirge at 15 ppb SO_2. This chronic impact of SO_2 on Mg^{2+} leaching is now to be compared with the impact of stomatal SO_2 uptake, which is quantified in Figure 4 (section 4.2.). Again regarding the Erzgebirge situation, an annual stomatal SO_2 uptake of ca. 15–20 mmol SO_2 kg^{-1} (needle dw) y^{-1} can be expected (Slovik *et al.*, 1995, see also Figure 4). A typical stock of needles in Norway spruce stands is 14000 ± 4000 kg (dw) ha^{-1} (Figure 19). Thus, the stomatal SO_2 uptake in the Erzgebirge is approximately 210–280 mol SO_2 ha^{-1} y^{-1} and, accordingly, about 130 ± 40 mol SO_2 ha^{-1} y^{-1} in the Fichtelgebirge. In Figure 47, the relative SO_4^{2-} formation rate [mol SO_4^{2-} mol^{-1} SO_2] is shown (after Slovik *et al.*, 1995) for different Norway spruce stands in central Europe (point 5 = Fichtelgebirge, points 6–8 = Erzgebirge). At sites with low SO_2 pollution [SO_2] and long vegetation periods VP (= small [SO_2] : VP ratios, sites 1 to 3; 3 = Würzburg), there are favourable site conditions that facilitate the metabolic reduction of stomatal SO_2 burden to organic sulfur compounds using SO_2 as a sulfur fertilizer. Additionally, the phloem removal of SO_2 assimilation compounds (mainly glutathione) from spruce needles is promoted at sites with long vegetation periods. The situation is inverse at sites (6–8 in Figure 47) with high SO_2 pollution and short vegetation periods (heavily polluted high-altitude sites). Under these conditions, SO_2 acts mainly as a burden, not just as a sulfur fertilizer (Heber *et al.*, 1987). This is the situation in the Erzgebirge, where the stomatal SO_2 burden is oxidized by 80–90% to sulfate, which accumulates in spruce needles (see also Kaiser *et al.*, 1991). At these sites, the reductive SO_2 assimilation and the phloem removal of glutathione, sulfate etc. are kinetically overburdened. Thus, in the Erzgebirge, there is a sulfate accumulation rate in spruce needles of $0.8 \, mol \, SO_4^{2-} mol^{-1} SO_2$ times $210–280 \, mol \, SO_2 \, ha^{-1} y^{-1} = 170–220 \, mol \, SO_4^{2-} ha^{-1} y^{-1}$. The corresponding value in the Fichtelgebirge equals $0.35 \, mol \, SO_4^{2-} mol^{-1} SO_2$ times $130 \, mol \, SO_2 \, ha^{-1} y^{-1} \approx 40 \, mol \, SO_4^{2-} ha^{-1} y^{-1}$ (see point 5 in Figure 47). The first oxidative SO_2 detoxification product in needle mesophyll cells is sulfuric acid, which is sequestered into needle vacuoles, where sulfate accumulates. In SO_2 fumigation experiments in closed chambers the physiological neutralization demand of the synchronously generated protons can be overburdened leading to vacuolar acidification, which can be detected by an observed decrease in the pH value of the tissue homogenate of fumigated needles (Kaiser *et al.*, 1993). In contrast to these fumigation experiments, neutralization of sulfuric acid is not kinetically overburdened in the field at lower SO_2 concentrations. This is observed even in the Erzgebirge, where parallel accumulation of sulfate and of mainly potassium (ca. 80%) is observed in ageing spruce needles (after Kaiser *et al.*, 1993). It is uncertain whether the remaining 20% is supplied by magnesium since these figures are rough estimates from the differences of the slopes of (positive or negative) cation accumulation rates in ageing needles in the Erzgebirge (ca. $32 \, ppb \, SO_2$) relative to the Würzburg situation (ca. $5–10 \, ppb \, SO_2$). The findings of Kaiser *et al.* (1993) have been confirmed by Hüve *et al.*, (1995): we compared needles from different individual spruce trees in the Erzgebirge at Höckendorf and observed that they accumulated different cations in their vacuoles (see Figure 48). Tree 1 mainly accumulated K^+ together with sulfate, tree 2 preferentially accumulated Mg^{2+}, and tree 3 accumulated mainly Mn^{2+}. These observations tentatively explain why different trees of the same species growing at the same site in the same SO_2-polluted atmosphere differ in developing deficiency symptoms on Mg^{2+}-depleted soils, e.g. in the Fichtelgebirge. We studied the SO_4^{2-}-dependent cation content in spruce needles in more detail also in Würzburg (Kindermann and Slovik, unpublished) and again confirmed findings from the Erzgebirge. The SO_4^{2-}-dependent Mg^{2+} accumulation has already been shown in Figure 18. Results of similar correlations of other metal cations plotted versus the SO_4^{2-} content analyzed in the same spruce needles from Würzburg are summarized in Table 16. There is SO_4^{2-}-dependent accumulation of potassium (ca. $0.59 \, Eq \, K^+ Eq^{-1} SO_4^{2-}$), magnesium (ca. $0.36 \, Eq \, Mg^{2+} Eq^{-1} SO_4^{2-}$) and zinc ($0.05 \, Eq \, Zn^{2+} Eq^{-1} SO_4^{2-}$) in Würzburg. Interestingly, just those cations which accumulate together with sulfate in healthy spruce needles in Würzburg are often deficient in the field at sites where 'forest decline' occurs (Hüttl, 1986, 1987; Lange *et al.*, 1989a,b). Experiments with

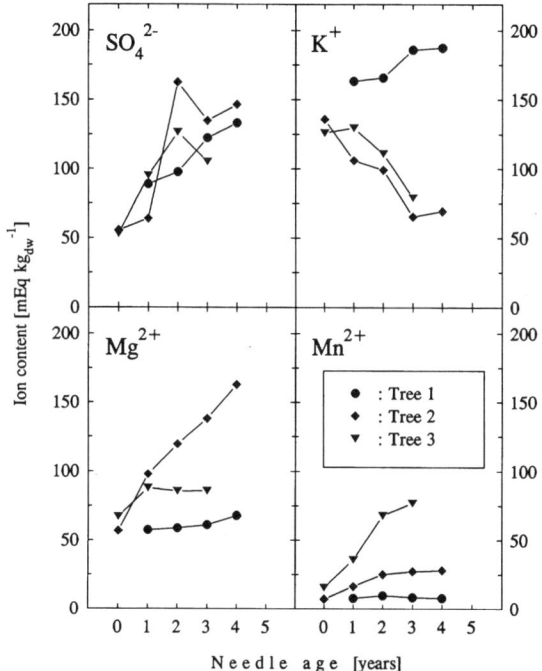

Figure 48. Ion contents of SO_4^{2-}, K^+, Mg^{2+} and Mn^{2+} in spruce needles in relation to the needle age class. Data are from Norway spruce trees growing in the Erzgebirge; see Hüve *et al.* (1995)

Table 16. SO_4^{2-}-dependent metal cation accumulation rates in healthy needles of Norway spruce trees growing in the botanical garden of Würzburg ($n=49$; Kindermann and Slovik, unpublished)

Cation	Slope (Eq Eq^{-1})	Corrected values (%)
K^+	$+0.680\pm0.300*$	58.9 ± 26.0
Mg^{2+}	$+0.415\pm0.111***$	36.0 ± 9.6
Zn^{2+}	$+0.059\pm0.006***$	5.1 ± 0.5
Mn^{2+}	$-0.011\pm0.004**$	—
Fe^{2+}	$-0.037\pm0.012**$	—
Al^{3+}	$-0.072\pm0.033*$	—
Sum	1.034 ± 0.322	100.0 ± 27.7

Data were determined by plotting the cation content versus the sulfate content of the same needle probes. The slope is determined by regression analysis (see Figure 18 for the example of Mg^{2+}). The number of asterisks indicates the statistical significance (*=95% level, **=99% level and ***=99.9% level). Besides the absolute slope ($\pm$SD), corrected percentage values are given presuming that the sum of all positive regression slopes corresponds to 100%

tonoplasts from tree species are not available but are from herbaceous plants. By means of electrophysiological patch-clamp techniques, K^+ and Mg^{2+} channels in tonoplast membranes of vacuoles from herbaceous plants have been indeed identified (Hedrich and Schröder, 1989). Also Zn^{2+} is readily sequestered into vacuoles (Brune *et al.*, 1994). Zn^{2+} deficiency of Norway spruce is associated with apex meristem damage and thus 'stork's nest symptoms' (Bergmann, 1988). In the Fichtelgebirge, besides the Mg^{2+} content in damaged needles, the Zn^{2+} content is also reduced relative to healthy control needles. In summary, the following relative order of vacuolar cation immobilization (together with SO_2-dependent SO_4^{2-}) in 'mean' spruce needles was observed:

$$K^+ > Mg^{2+} \text{ or } Mn^{2+} \text{ or } Zn^{2+}$$

Since these cations accumulate together with sulfate in vacuoles of ageing needles, they are not available for cytosolic demands. Accumulated cations in spruce needles are shed forming needle litter after their mean life-time. The SO_2-dependent accumulation of cations in spruce needles therefore induces an additional demand on these cations, which adds up to their 'background' demand in clean air, and which increases the litter loss of these cations every year as long as there is too high chronic SO_2 pollution (see Figure 47). Finally, the whole tree runs into a shortage of cations, which becomes obvious as mineral deficiency and spruce decline symptoms at sites with low cation supply (see Slovik *et al.*, 1993). Hence, the soil 'quality' defines which cation species first becomes limiting after chronic SO_2 pollution. It must be stressed here that SO_2-dependent Mg^{2+} sequestration into needle vacuoles does not occur at all sites or in all individuals.

The magnitude of this SO_2-dependent additional demand of e.g. Mg^{2+} can be estimated. In the Erzgebirge, the annual SO_4^{2-} accumulation rate in spruce needles is approximately $170-220 \, mol \, SO_4^{2-} \, ha^{-1} \, y^{-1}$ (see above). About $20-36\%$ of accumulating SO_4^{2-}-equivalents are neutralized by Mg^{2+} cations in extreme cases. In the Erzgebirge, SO_2 induces a roughly estimated additional Mg^{2+} demand of $0.20-0.36 \, mol \, Mg^{2+} \, mol^{-1} \, SO_4^{2-}$ times $170-220 \, mol \, SO_4^{2-} \, ha^{-1} \, y^{-1}$ which equals $34-79 \, mol$ (vacuolar Mg^{2+}) $ha^{-1} \, y^{-1}$. The corresponding vacuolar Mg^{2+} sequestration rate in the Fichtelgebirge would equal $0.20-0.36 \, mol \, Mg^{2+} \, mol^{-1} \, SO_4^{2-}$ times ca. $40 \, mol \, SO_4^{2-} \, ha^{-1} \, y^{-1} = 8-15 \, (\approx 10) \, mol$ (vacuolar Mg^{2+}) $ha^{-1} \, y^{-1}$. This is the SO_2-dependent Mg^{2+} demand, which must be additionally supplied from the soil (e.g. via Mg^{2+} fertilization), or by which the annual tree growth must be reduced if poor soils cannot supply this SO_2-dependent additional Mg^{2+} demand. In the Fichtelgebirge, strong Mg^{2+} deficiency symptoms can be observed at a SO_2 pollution rate of ca. 15 ppb (annual mean). In Figure 38, the annual Mg^{2+} growth demand of spruce canopies was deduced for the Würzburg situation, i.e. for a mean needle age of $6.1 \pm 0.6 \, y$ and a Mg^{2+} content in green spruce needles of $31.1 \pm 6.9 \, mol \, Mg^{2+} \, kg^{-1}$ (dw). Under these conditions, the annual Mg^{2+} demand of the spruce canopy growth (needles, branches) equals $182 \, mol \, ha^{-1} \, y^{-1}$, and the total annual growth demand of the whole trees is ca. $352 \, mol \, ha^{-1} \, y^{-1}$. In the Fichtelgebirge, the situation is different: the mean needle age NA *before* onset of spruce decline symptoms can be estimated to be 9.5 years (see Figure 22) at ca. 1000 m asl (Schneeberg, Fichtelgebirge). The Mg^{2+} content in healthy spruce needles from the Fichtelgebirge is ca. $20 \, mmol \, kg^{-1}$ (dw); see Figure 2. Thus, the needle turn-over rate was ca. $9.5/6.1 \approx 1.6$ times longer in the Fichtelgebirge before onset of damage symptoms than in Würzburg, and the

Mg^{2+} content in healthy needles from the Fichtelgebirge is only 0.64 times that in needles from Würzburg. Consequently, the annual growth demand of spruce canopies in the Fichtelgebirge is ceteris paribus only $0.64/1.6 = 0.4$ times as high as in Würzburg, i.e. only 0.4 times $182\,mol\,ha^{-1}\,y^{-1} \approx 73\,mol\,ha^{-1}\,y^{-1}$. Annually, there is a vacuolar immobilization (additional litter loss) of $10\,mol\,Mg^{2+}\,ha^{-1}\,y^{-1}$ (see above). Thus, $10\,mol\,(litter\text{-}Mg)/73\,mol\,(growth\text{-}Mg)$ times $100\% = 14\%$ of the annual Mg^{2+} growth demand of the spruce canopy is consumed to neutralize SO_2-dependent SO_4^{2-} accumulation every year. If, first, the canopy turnover rate is not reduced in spite of SO_2 pollution, then the Mg^{2+} content of spruce needles, branch bark and branch wood must decrease by 14% per year starting with an original Mg^{2+} content in needles of ca. $20\,mmol\,kg^{-1}$ (dw). Consequently, within only four or five years, the pollution of SO_2 forces the Mg^{2+} content of spruce needles, twigs and branches in the Fichtelgebirge down from $20\,mmol\,kg^{-1}$ (dw) by 4.3 times 14% of $20\,mmol\,kg^{-1}$ to only $8\,mmol\,kg^{-1}$ (dw). At these Mg^{2+} contents, severe Mg^{2+} deficiency symptoms become apparent (Figure 2). As little as 15 ppb SO_2 may induce, within a few years, severe Mg^{2+} deficiency symptoms in the Fichtelgebirge because the Mg^{2+}-depleted soils are apparently not able to resupply the SO_2-dependent additional annual Mg^{2+} demand after canopy litter decomposition. These calculations stress the possible importance of SO_2 in causing forest decline symptoms in the Fichtelgebirge on its Mg^{2+} depleted soils. Today, after the onset of Mg^{2+} deficiency, correlation between Mg^{2+} and SO_4^{2-} may be weak. It remains an open question, whether SO_2-dependent Mg^{2+} depletion had in fact occurred when canopy symptoms started to develop. Mg^{2+}-poor soils are expected to strongly predispose Norway spruce stands to chronic SO_2 damage. The only way to save Mg^{2+} is to reduce the annual growth demand of other tree parts in order to supply the SO_2-induced additional Mg^{2+} demand of the canopy. There is observed reduction in trunk growth in the Fichtelgebirge, which is in fact one of the most potent means of saving considerable amounts of Mg^{2+}. Figure 38 estimates that the Mg^{2+} demand for trunk growth is roughly 0.57 times as high as the Mg^{2+} demand of the whole canopy, which roughly equals $50\,mol\,Mg^{2+}\,ha^{-1}\,y^{-1}$ in the Fichtelgebirge. Thus, the trunk growth demand in the Fichtelgebirge equals ca. 0.57 times $50\,mol\,Mg^{2+}\,ha^{-1}\,y^{-1} = 29\,mol\,Mg^{2+}\,ha^{-1}\,y^{-1}$. A reduction of the trunk growth rate by 35% is sufficient to just 'save' $10\,mol\,Mg^{2+}$ $ha^{-1}\,y^{-1}$ for vacuolar SO_2 neutralization demands in the needles instead.

In SO_2 fumigation experiments Keller (1978) found that the CO_2 assimilation of needles was reduced by 25% and the growth in breadth of the trunk dropped. The reduction of wood production was stronger after fumigation within the growth period compared with winter fumigation (Keller, 1978). This presumably was just the consequence of the annual stomatal aperture kinetics shown in Figure 3. Reduction of assimilate production and trunk wood growth has been often observed in SO_2-polluted stands (Bucher and Keller, 1978; Grill and Härtel, 1972; Keller, 1980, 1981; Materna, 1972; Pollanschütz, 1962; Schröder, 1873; Vins, 1961). According to Wentzel (1971, 1983), the trunk height growth is more affected by air pollutants than the trunk breadth growth. With increasing spruce decline symptoms, trunk wood formation is increasingly focused at the apical trunk parts and reduced at the basal parts (Franz, 1983). This change and reduction in wood growth seems to be an active but indirect reaction of the trunk cambium to SO_2 burden, since there are no clear and evident 'damage symptoms' of wood anatomy (Athari, 1980; Eckstein et al., 1981; Grill et al., 1979; Halbwachs and Kisser, 1967; Liese et al., 1975). The sap wood area of trunks is reduced in damaged trees (Bauch, 1983), but there are no symptoms of wet heart wood formation (German: Naßkern). There are also no detectable changes in biomechanical and technologically valuable wood properties in damaged trees (Frühwald et al., 1984). 'Spruce decline' is therefore not equivalent to wood damage. Correspondingly, there are also no wood anatomical differences between trace-gas-sensitive and more tolerant trees (Greve et al., 1985).

Similar findings for air-pollution-dependent Mg^{2+} deficiency are now becoming available for deciduous trees. Bergmann (1988) defined 'new type autumn yellowing' symptoms (German: 'Neuartige Herbstfärbung') which are characterized by early chlorosis and yellowing of leaves and apical shoots of deciduous trees and shrubs before 'regular' autumnal leaf coloration. Baule (1978) found that Mg^{2+} deficiency symptoms of deciduous woody plants are apparent mainly in the autumn. These are only weak hints. We urgently need detailed information concerning the interdependency of SO_2 pollution and Mg^{2+} deficiency of deciduous trees.

In Table 16, we discussed only *accumulating* cations as a consequence of sulfate accumulation. Concerning the Mn, Fe and Al content of spruce needles, there is a significant but weak negative correlation in Würzburg. The needle content of these cations is reduced with increasing sulfate content. The sum of all SO_4^{2-}-dependent cations is $1.034 \, Eq \, Cat^+ \, Eq^{-1} \, SO_4^{2-}$. Since the difference from the ideal value of $1.000 \, Eq \, Eq^{-1}$ is not significant, there are no unidentified SO_4^{2-}-dependent cations. Concerning all other ICP-analyzed cations in Würzburg (Ca^{2+}, Na^+, Cu^{2+}), there was no significant SO_4^{2-} dependency of their needle content in Würzburg. It is surprising that there is no SO_4^{2-}-dependent accumulation of $CaSO_4$ in rinsed spruce needles despite Ca^{2+}-rich shell–lime soils in Würzburg. Of course, we found Ca^{2+} accumulation in ageing needles from Würzburg similar to that from many other sites, but this Ca^{2+} accumulation is not related to $CaSO_4$, but to a needle age-dependent accumulation of apoplasmic calcium oxalate (Fink, 1992). In Würzburg, $4.50 \pm 1.57 \, mmol \, kg^{-1}$ (needle dw) free oxalic acid can be extracted from green needles. This oxalic acid must be localized in the spruce vacuoles since the apoplasm (xylem sap) of the same spruce twigs contained only trace concentrations of oxalic acid in the μmol/L range (Kindermann and Slovik, unpublished). Figure 49 shows the maximum possible contribution of soluble Ca^{2+} to vacuolar SO_4^{2-} neutralization if there are concomitant oxalic acid anions in the vacuole that precipitate Ca^{2+} forming insoluble calcium oxalate (solubility product of calcium oxalate $K_{sol} = 2.56 \cdot 10^{-9} \, mol^2$; see Weast, 1989). The pH of the needle homogenate was 4.39 ± 0.09. The pK_a values of oxalic acids equal $pK_{a1} = 1.19$ and $pK_{a2} = 4.21$. Since the relative water content of living spruce needles is ca. 50% and since most water in leaves is vacuolar water, the numerical values of the abscissa of Figure 49 are approximately also the needle sulfate content given in $mmol \, kg^{-1}$ (dw). The

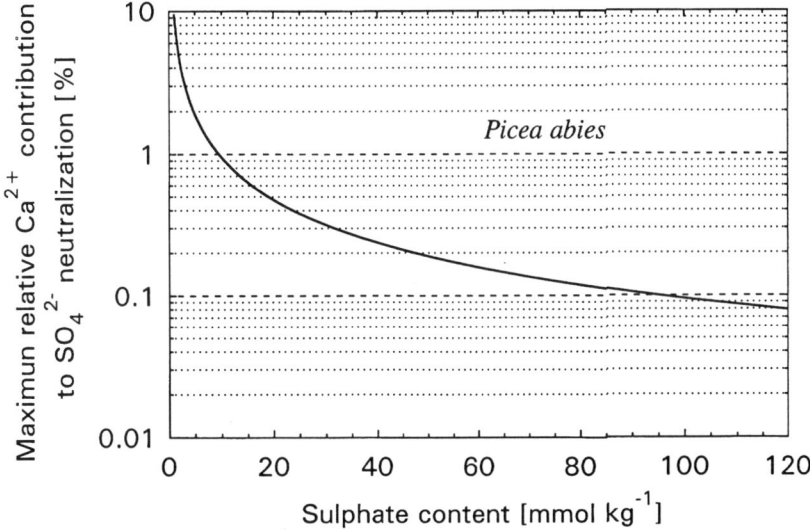

Figure 49. Maximum possible contribution of Ca^{2+} to vacuolar SO_4^{2-} neutralization in the presence of free extractable oxalic acid (4.50 ± 1.57 mmol kg^{-1} needle dw) in spruce needle vacuoles. Calculations were performed using the solubility product of calcium oxalate ($K_{sol} = 2.56 \cdot 10^{-9}$ M^2; see Weast, 1989), $pK_{a1} = 4.21$, $pK_{a2} = 1.19$. The pH of the needle homogenate (4.39 ± 0.09) estimates the pH of the vacuolar sap. An ion activity of 0.1 mol mol^{-1} was assumed for Ca^{2+} and SO_4^{2-}

maximum relative contribution of Ca^{2+} to vacuolar SO_4^{2-} neutralization drops from ca. 10% at 2 mmol SO_4^{2-} kg^{-1} to below 1% at SO_4^{2-} contents greater than 10 mmol kg^{-1} (activity coefficient γ is assumed to be 0.1 mol mol^{-1} for Ca^{2+} and SO_4^{2-} in vivo). This percentage further drops with increasing SO_4^{2-} content and it is only 0.1% at 100 mmol SO_4^{2-} kg^{-1} needle dw. In the Erzgebirge, sulfate contents beyond 100 mmol SO_4^{2-} kg^{-1} (dw) can be analyzed. In Würzburg, the SO_4^{2-} content equals 9.0 ± 2.9 mmol kg^{-1}. Gypsum crystals associated with needle surfaces and leaching of epidermal $CaSO_4$ is outside the scope of this section (see section 4.6.).

4.7.2. *Harm due to chronic HCl pollution and NaCl deposition*

Generally speaking, HCl pollution induces damage symptoms similar to those of SO_2 pollution (Berge, 1970). Both trace gases mainly act as potentially acidic air pollutants, which must be neutralized, and both produce anions (Cl^-, SO_4^{2-}) which generally occur in all plants. These anions are sequestered into vacuoles, where they may accumulate together with cations. In *Picea abies* needles from nonfumigated healthy spruce trees growing in Würzburg, there is Cl^--dependent accumulation of potassium (1.27 Eq K^+ Eq^{-1} Cl$^-$), sodium (0.13 Eq Na^+ Eq^{-1} Cl$^-$) and manganese (0.01 Eq Mn^{2+} Eq^{-1} Cl$^-$), and a significant reduction of the magnesium (-0.30 Eq Mg^{2+} Eq^{-1} Cl$^-$) and zinc content (-0.30 Eq Zn^{2+} Eq^{-1} Cl$^-$). Compared with Table 16, accumulation of Cl^- and SO_4^{2-} share (i) the dominant vacuolar potassium consumption, and (ii) the complete absence of Ca^{2+} involvement in neutralizing Cl^- or SO_4^{2-} anions. In contrast to SO_4^{2-}, with increasing Cl^- content, there is an apparent substitution of Mg^{2+} by K^+ in spruce needles in Würzburg. Again, accumulation of Cl^- anions interferes with the cellular compartmentalization with respect to source–sink relationships of Mg^{2+}, Mn^{2+} and Zn^{2+} cations. It is well known, that *Picea abies*, *Pinus sylvestris* and *Acer* spp. are sensitive to chloride burden, irrespective of the chloride source, air-borne or soil-borne (Dässler *et al.*, 1975; see Dässler, 1991). Excessive chronic chloride burden induces Cl^- toxicity symptoms with *Abies alba*, which also is SO_2 sensitive. Cl^- accumulation induces Mg^{2+} deficiency symptoms of *Picea omorica* (Serbian spruce). This type of Cl^--dependent 'Omorika decline' is known in German as 'Omorika-Sterben' (Alt *et al.*, 1981, 1982; Bergmann, 1988). Since gymnosperms are especially chloride sensitive, Mg^{2+} deficiency symptoms must be corrected by fertilization with $MgSO_4$ in the field, not by employing $MgCl_2$ fertilizers (see also Zöttl, 1990).

4.7.3. Harm due to chronic NO_x and NH_3 pollution and deposition

Oren *et al.* (1988), Oren and Schulze (1989), Schulze *et al.* (1987) and Schulze (1989) proposed that a high nitrogen supply supports canopy growth relatively more than the reduction in growth as a consequence of Mg^{2+} deficiency. According to these authors, the net effect would be a dilution of Mg^{2+} in spruce growing in the Fichtelgebirge. It is still an open question, whether a nitrogen-dependent increase in the canopy turnover rate in the past has depleted Mg^{2+} from Norway spruce crowns. Per se, this method of Mg^{2+} depletion seems possible, since the retranslocation of nitrogen via the phloem sap from senescent leaves and needles is relatively more efficient than recovery of Mg^{2+} in any plants (Bussler, 1973; see Table 10 for spruce; see section 4.3.). An important supposition of this nitrogen-dependent and time-consuming 'Mg^{2+} dilution process' is to violate and infringe 'Liebig's Law', i.e. that the minimum nutrient supply from the soil (here: Mg^{2+}) defines the actual possible growth rate. Liebig's Law was deduced from findings with herbaceous crop plants which contain hardly any 'buffering' resources of nutrients. It is extremely difficult to judge and criticize the immediate validity of Liebig's Law for spruce trees, which possess a large stock of cation buffering sap wood and several needle age classes (see section 4.3.). Since the complete Mg^{2+} demand of the new flush in the spring can be transiently supplied only from sap wood – uncoupled from the soil supply in the spring (see section 4.3.) Liebig's Law transiently seems in fact *not* to be valid immediately and under all circumstances whatsoever. It still remains valid of course in the long term, i.e. after the Mg^{2+} buffers in trees have been depleted. Today, the absolute trunk volume growth rate and the canopy growth rate are reduced in the Fichtelgebirge (relative to comparable healthy sites) as a consequence of Mg^{2+} deficiency. There is no dominant continuation of the postulated 'historical' Mg^{2+} dilution process via increased nitrogen-dependent growth now. It is no longer possible to experimentally test this dilution theory in the field, since research started too late. We can now observe results, but not the mechanisms. Under conditions of reduced nitrogen supply (and growth rates), mineral deficiency symptoms do not develop (Hällgren and Näsholm, 1988), and an imbalanced nutrient supply may indeed result in nutrient deficiencies of needles (Ingestad, 1959, 1982). Maximum growth rates of spruce seedlings can be achieved if the molar $N:Mg^{2+}$ ratio is about $28\,mol\,mol^{-1}$ (Ingestad, 1959, 1982). In Würzburg, the molar $N:Mg^{2+}$ ratio in needles is $21\,mol\,mol^{-1}$, i.e. there is sufficient Mg^{2+} supply (see Table 10). There are no visible Mg^{2+} deficiency symptoms in Würzburg. In the Fichtelgebirge, the $N:Mg^{2+}$ ratio in healthy spruce needles is $26\,mol\,N\,mol^{-1}\,Mg^{2+}$, and, in visually Mg^{2+}-deficient needles, the ratio is $71\,mol\,N\,mol^{-1}\,Mg^{2+}$ (Schulze et al., 1989). The absolute N content in spruce needles was $850\,mmol\,kg^{-1}\,(dw)$ in the Fichtelgebirge and $650\,mmol\,kg^{-1}\,(dw)$ in Würzburg.

As already shown above, the pollution of SO_2 in the Fichtelgebirge (annual mean ca. $15\,ppb$ SO_2) may be sufficient to deplete Mg^{2+} from spruce canopies over the course of a few years and to induce Mg^{2+} deficiency and forest decline symptoms in spruce stands growing on Mg^{2+}-depleted soils. The $N:Mg^{2+}$ ratio in spruce needles is therefore automatically raised if the nitrogen supply at least has remained constant. Since forest decline in the geographical neighbourhood of the Fichtelgebirge (Erzgebirge; ca. $30\,ppb$ SO_2 annual mean) is clearly attributable to SO_2 pollution, it is per se not necessary to postulate another reason for forest decline in the Fichtelgebirge. Still, the synergetic contribution of stomatal NO_x uptake and – perhaps more important – the wet deposition of nitrate and ammonium must be kept in mind. It is difficult to quantify the impact of nitrogen supply on canopy turnover, Mg^{2+} demand, Mg^{2+} depletion ('dilution') and the interaction of these factors with decline symptoms. Wet deposition of HNO_3 and nitrification of precipitated NH_4^+ both enhance soil acidification and root uptake of Mg^{2+} competes with NH_4^+ uptake. Under these conditions, the Mg^{2+} 'availability' from the soil decreases and the 'availability' of potentially toxic Mn^{2+} and various aluminum cations increases instead. If, in contrast, the nitrification of NH_4^+ is inhibited in acidic soils, then accumulating NH_4^+ inhibits the edaphone, and thus the litter decomposition (Buchner, 1985), which is the bottle neck of Mg^{2+} recycling in podsolized soils (Figure 38). Pedological details are discussed by Kaupenjohann, Chapter 8 in this volume.

4.7.4. Harm due to chronic O_3 pollution

The direct impact of ozone pollution on plant physiology and Mg^{2+} cycling is per se restricted to the above-ground organs of trees. Its potentially harmful action can be divided into: (i) physiological consequences of stomatal ozone uptake, which are out of the scope of this communication, and (ii) and effect of ozone on leaf or needle cuticles and waxes, which may be oxidized by this highly reactive oxygen radical. An increased permeability of the cuticles may be assumed. A synchronous wet deposition of acid mist would then increase Mg^{2+} leaching. This interference with nutrient cycling may then cause forest decline (e.g. spruce decline) symptoms. This hypothesis was convincingly falsified by fumigation experiments that combined chronic ozone pollution ($100 \, \mu g \, m^{-3}$ + peaks between 130 and $360 \, \mu g \, m^{-3}$) and acid mist (pH 3.0) treatments for 14 months under controlled conditions in closed chambers (Pfirrmann *et al.*, 1990). Weak pollutant-related effects were observed, but cation leaching from the needles was generally lower than observed in the field. There is no ozone-enhanced Mg^{2+} leaching from native spruce canopies in the field. There is also no increase in cation leaching by the interception of acidic rain at ambient environmental pH values of the precipitation water (Slovik, Balázs and Siegmund, unpublished). There was 'ozone'-dependent potassium and calcium leaching in the field; however, this is not caused by ozone itself, but presumably by organic acids that are produced from hydrocarbons of exhaust fumes in the presence of ozone. These data do not support the hypothesis that enhanced cation leaching from needles contributes to magnesium deficiency in mature spruce stands (Pfirrmann *et al.*, 1990; Mitterhuber *et al.*, 1989). Apparently, turn-over and repair of cuticular waxes was sufficient to cope with the action of ozone and acid mist (Barthlott, 1992) in the experiments of Pfirrmann *et al.*, 1990, but the turn-over, structure and wetability of waxes is impaired by Mg^{2+} deficiency and by Al toxicity (Schwab *et al.*, 1993a, Schwab *et al.*, 1994a,b), which may confuse the situation in the field.

4.8. Magnesium deficiency and phytopathogens

Spruce decline is often associated with secondary phytopathogenesis by fungi. *Armillaria mellea*, the most prominent example, is a facultative virulent spruce parasite, that forms hyphal rhizomorphs in the cambial zone between the cortical bast and the sap wood of basal trunk parts. The virulence of this fungus is strongly apparent in SO_2-polluted spruce forests. There is also only rare and confusing information on how Mg^{2+} deficiency is interrelated with (i) the virulence of phytopathogens and (ii) damage by phytophagous insects (see Bergmann 1988, p. 90). Generally speaking, there seems to be a tendency towards reduced virulence of fungal pathogens in well-Mg^{2+}-supplied agricultural crop plants that are infected by the fungal species, *Phytophthora infestans*, *Rhizoctonia solani* or *Puccinia graminis*. In contrast, well-Mg^{2+}-supplied crop plants tend to be affected more by phytophagous aphids and mites. Kiss and Pozsár (1977) reported that Mg^{2+} fertilization significantly reduced damage to crop plants caused by the fungi, *Botrytis* spp. or *Phytophthora* spp. (on tomato), and *Cercospora beticola* or *Pleospora betae* (on sugar beet). This was confirmed by Schwab *et al.* (1993a,b). Sufficient Mg^{2+} supply also reduced infection of sugar beet by the bacterium *Pseudomonas phaseolicola*, and reduced potato virus diseases by 73%. Information is poor concerning the interdependency of phytopathogens and Mg^{2+} supply in forest trees.

4.9. Conclusions

Magnesium is an important key element in tree nutrition. It is catalytically involved as a cofactor in most essential biochemical and regulatory processes of plant metabolism. Knowledge of the physiology of Mg^{2+} seems sufficient to explain observed Mg^{2+} deficiency symptoms in the field. Mg^{2+} is mobile in the xylem sap and the phloem sap of trees, and Mg^{2+} is recycled several times per year within the canopy. Details of Mg^{2+} transport in the phloem sap of trees are unknown. In the spring, the young flush of spruce is supplied with Mg^{2+} that originates mainly from the youngest needle age class and from the sap wood. The pool of transiently depleted Mg^{2+} in sap wood and older needles is filled up in the course of the subsequent summer by the net uptake of Mg^{2+} from the soil. There is Mg^{2+} exchange between sap wood and the xylem sap. The varying Mg^{2+} content in different tree organs and tissues depends on the supply rate from the soil and on the tree species. On the basis of commonly used yield tables, available literature information and chemical tissue analysis data, the annual growth demand of Mg^{2+} for all important tissues can be estimated for spruce stands. The actual annual Mg^{2+} demand of healthy spruce stands mainly depends on the stand age and on the growth scenario (trunk wood yield class, cutting regime, site altitude asl etc.). Annual nutrient cycling rates, including canopy leaching, trunk harvesting loss, yield waste decomposition and atmospheric wet and dry Mg^{2+} deposition have been deduced and summarized. On the basis of complete information on ecological Mg^{2+} cycling rates and balances, the impact of manufactured air pollutants on Mg^{2+} supply and tree vitality can be tentatively quantified for the example of chronic SO_2 pollution in originally healthy spruce stands. In Würzburg (healthy site), Mg^{2+} is sequestered together with SO_2-dependent SO_4^{2-} anions into vacuoles of Norway spruce needles. Thereby immobilized Mg^{2+} in needle vacuoles will add to the annual Mg^{2+} loss via needle litter production. In the long term, this process may massively compete with the annual Mg^{2+} demand. As a consequence, SO_2-dependent Mg^{2+} deficiency and reduction of the needle matter ('spruce decline') may develop. The quantification of these processes depends on meteorological, pedological, physiological and pollution data. Most SO_2-sensitive are: (i) old spruce stands at (ii) high altitudes asl, which (iii) are growing on Mg^{2+}-depleted soils. Since chronic SO_2 pollution and an additional annual Mg^{2+} demand are equivalent in terms of ecological Mg^{2+} balance, SO_2-dependent forest decline symptoms on Mg^{2+}-deficient soils can be efficiently tackled by Mg^{2+} fertilization with $MgSO_4$ or by liming. The latter facilities microbial litter decomposition and thereby indirectly increases the Mg^{2+} supply. Besides Mg^{2+}, K^+ and Zn^{2+} are also trapped in spruce vacuoles together with sulfate. In some cases, fertilization with K^+ and Zn^{2+} has alleviated forest decline symptoms (Bergmann, 1988; Hüttl 1985, 1986, 1987). In the long term, reduction of chronic SO_2 pollution is an effective means of minimizing or avoiding SO_2-dependent cation-deficiency symptoms and SO_2-dependent spruce decline. In the field, Mg^{2+}-depleted soils and 'acid rain', SO_2-dependent vacuolar Mg^{2+} immobilization, nitrogen-dependent Mg^{2+} 'dilution' via growth (NO_x pollution, NO_3^- wet deposition) and competition for Mg^{2+} uptake from the soil (wet deposition of competitive NH_3/NH_4^+) seem to act synergistically with varying hard-to-quantify relative importance. As a result, Mg^{2+} deficiency symptoms and spruce decline has developed at many Mg^{2+}-deficient sites.

200

Acknowledgements

This work has been performed within the Sonderforschungsbereich 251 of the University of Würzburg. It has been financed by Fonds of the Chemische Industrie and by the Projektgruppe Bayern zur Erforschung der Wirkung von Umweltschadstoffen (PBWU). Advice and discussions with Prof. U. Heber and Prof. O. L. Lange (Würzburg, Germany), Prof. Ch. Körner (Basel, Switzerland) and Prof. B. Ulrich (Göttingen, Germany) are gratefully acknowledged.

4.10. References

Aber JD, Nadelhoffer KJ, Steudler P, Melillo JM. 1989. Nitrogen saturation in northern forest ecosystems. Excess nitrogen from fossil fuel combustion may stress the biosphere. BioScience. 39, 378–386.

Abrahamson G. 1980. Acid precipitation, plant nutrition and forest growth. In: Drablos D, Tollan A, Eds. Ecological Impact of Acid Precipitation. pp. 58–63. Norw. Interdisciplinary Res. Program, SNSF Project As, Norway.

Adams CM, Hutchinson TC. 1984. A comparison of the ability of leaf surfaces of three species to neutralize acidic rain-drops. New Phytol. 97, 463–478.

Aikawa JK, Ed. 1981. Magnesium: Its Biologic Significance. CRC Press Inc, Boca Raton, Florida, USA.

Alt D, Zimmer R, Stock M. 1981. Omorikensterben – Chloridüberschuß – eine wesentliche Ursache für die Schäden – keine parasitäre Krankheit. Gartenbörse Gartenwelt. 81, 379–382.

Alt D, Zimmer R, Stock M. 1982. Erhebungsuntersuchungen zur Nährstoffversorgung von *Picea omorika* im Zusammenhang mit dem Omorikasterben. Pflanzenern Bodenkde. 145, 117–127.

Altherr E, Unfried J, Hradetzky, Hradetzky V. 1978. Statistische Rindenbeziehungen als Hilfsmittel zur Ausformung und Ausmessung unentrindeten Stammholzes. Teil 4: Fichte, Tanne, Douglasie und Sitka-Fichte. Mitt Forstl Versuchs- und Forschungsanst Baden-Württ. Vol. 90.

Andrussow L. 1969. Diffusion. In: Landolt-Börnstein. Zahlenwerte und Funktionen aus Physik, Chemie, Astronomie, Geophysik, Technik. In: Borchers H, Hausen H, Hellwege KH, Schäfer K, Eds. 5. Teil, Bandteil a (Schäfer K, Ed: Transportphänomene I Viskosität und Diffusion). Springer Verlag Berlin, Heidelberg, New York, pp. 513–729.

Anonymous. 1939. Holzeigenschaftstafel: Fichte. Holz Roh- Werkstoff. 2, 407–408.

Anonymous. 1986. Forschungsbeirat Waldschäden/Luftverunreinigungen der Bundesregierung und der Länder, 2. Bericht, 229 pp. Elser Druck, Mühlacker.

Anonymous. 1992. Daten zur Umwelt 1990/91. Umweltbundesamt, Red.: Fachgebiet I 1.2 "Umweltforschung/Umweltstatistik" (ed), 675 pp. Erich Schmidt Verlag, Berlin.

Arens K. 1934. Die kutikuläre Exkretion des Laubblattes. Jb wiss Bot. 80, 230–248.

Arndt U, Nobel W, Schweizer B. 1987. Bioindikatoren. Möglichkeiten, Grenzen und neue Erkenntnisse, Eugen Ulmer, Stuttgart.

Athari S. 1980. Untersuchungen über die Zuwachsentwicklung rauchgeschädigter Fichtenbestände. Diss Forstl Fak Univ Göttingen. 164 pp.

Athari S, Kramer H. 1983a. Erfassung des Holzzuwachses als Bioindikator beim Fichtensterben. Allg Forstz. 30, 767–769.

Athari S, Kramer H. 1983b. Problematik der Erfassung von umweltbedingten Zuwachsverlusten in Fichtenbeständen. Forst-Holzwirt. 38, 204–206.

Balázs Á. 1991. Niederschlagsdeposition in Waldgebieten des Landes Hessen. Ergebnisse von den Mess-Stationen der "Waldökosystemstudie Hessen". Forschungsberichte Hessische Forstliche Versuchsanstalt, Hessisches Ministerium für Landesentwicklung, Wohnen, Landwirtschaft, Forsten und Naturschutz, Ed. Vol. 11. 168 pp. Hann. Münden, Germany.

Barber J, Mills J, Nicolson J. 1974. Studies with cation specific ionophores show that within the intact chloroplast Mg^{2+} acts as the main exchange cation for H^+ pumping. FEBS Lett. 49(1), 106–110.

Barthlott W. 1992. Die Selbstreinigungsfähigkeit pflanzlicher Oberflächen durch Epicuticularwachse: Ihre ökologische Bedeutung als Abwehrstrategie gegen Pathogene, Ihre Störung durch Umwelteinflüsse, Möglichkeiten technischer Anwendung. In: Rheinische Friedrichs-Wilhelms-Universität, Ed. Klima und Umweltforschung und der Universität Bonn. pp. 117–120.

Bauch J. 1983. Biologische Veränderungen in Stamm und Wurzeln umweltbelasteter Waldbäume. In: SO₂ und die Folgen. GSF-Bericht A3, pp. 49–57. München–Neuherberg.

Baule H, Fricker C. 1967. Die Düngung von Waldbäumen. BLV, München.

Baule H. 1984. Zusammenhänge zwischen Nährstoffversorgung und Walderkrankungen. Allg Forstztg. 30, 775–778.

Baule H. 1978. Grundlagen der Forstpflanzenernährung und -düngung. Wirtschafts- und Forstverlag Euting KG.

Baumeister W, Ernst W. 1978. Mineralstoffe und Pflanzenwachstum. G. Fischer, Stuttgart.

Baumgartner A. 1958. Nebel und Nebelniederschlag als Standortsfaktoren am Großen Falkenstein (Bayerischer Wald). Forstw. Cbl. 77, 257–272.

Behrends C. 1974. Vergleichende Untersuchungen zum artspezifischen Staubfangvermögen von *Pinus sylvestris* und *Picea abies*. Diploma thesis, Sekt Forstwirtschaft, Tharandt, TU Dresden.

Ben-Hayyim G. 1978. Mg²⁺ translocation across the thylakoid membrane: studies using the ionophore A 23187. Eur. J. Biochem. 83, 99–104.

Benecke P. 1978. Der Wasserumsatz eines Buchen- und Fichtenwaldökosystems im Hochsolling. Habilitationsschrift, Univ. of Göttingen, Germany.

Bennet JH, Hill AC. 1975. Interactions of air pollutants with canopies of vegetation. In: Mudd JB, Kozlowski TT, Eds. Responses of Plants to Air Pollution. pp. 273–306. Academic Press, New York.

Berge H. 1970. Immissionsschäden (Gas-, Rauch- und Staubschäden). In: Sorauer P, Ed. Handbuch der Pflanzenkrankheiten, Bd I, 7. Aufl., 4. Lieferung, pp. 1–169. Parey, Berlin.

Bergmann W, Ed. 1988. Ernährungsstörungen von Kulturpflanzen. Entstehung, visuelle und analytische Diagnose. Gustav Fischer Verlag Jena, 2. Auflage. 762 pp.

Bergmann L, Rennenberg H. 1993. Glutathione metabolism in plants. In: De Kok LJ, Stulen I, Rennenberg H, Brunold C, Rauser WE, Eds. Sulfur Nutrition and Assimilation in Higher Plants. Regulatory and Environmental Aspects. pp. 109–123. SPB Academic Publishing bv, The Hague, The Netherlands.

Bernhart A. 1965. Fischfeuchtigkeit und Schwindverhalten von Fichtenholz. Forstw. Cbl. 84, 347–356.

Bernhart A. 1967. Fragen über die Rohdichte beim Fichtenholz. Allg Forstztg. 78, 264–267.

Bertog H. 1895. Untersuchung über den Wuchs und das Holz der Weißtanne und Fichte. Forst Naturwiss Z. 4, 97–112 and 177–216.

Beyschlag W, Wedler M, Lange OL, Heber U. 1987. Einfluß einer Magnesium-düngung auf Photosynthese und Transpiration von Fichten an einem Magnesium-Mangelstandort im Fichtelgebirge. Allg Forst Zeitschrift. 27/28/29, 738–741.

Blossfeld O, Haasemann W, Wonka R. 1963. Rindendicken und Rindenanteile von dünnem Fichten- und Kiefernholz. Holztechnologie. 4, 163–169.

Bosch C, Pfannkuch E, Baum U, Rehfuess KE. 1983. Über die Erkrankung der Fichte (*Picea abies* Karst.) in den Hochlagen des Bayerischen Waldes. Forstw. Cbl. 102, 167–181.

Böttcher HD. 1987. Leitfähigkeitsuntersuchungen an unterschiedlich geschädigten Fichten der Arten *Picea abies* (Karst.) sowie *Picea jezoensis* (Carr.) im Hessischen Forstamt Kaufungen. Forschungsberichte Hessische Forstliche Versuchsanstalt. Vol. 4, pp. 79–91. Hann Münden, Germany.

Boxman D. 1988. Effects of excess nitrogen on the nutritional state of trees. In: Nilsson J, Ed. Critical Loads for Sulphur and Nitrogen. UN-ECE Workshop, Sweden.

Brune A, Urbach W, Dietz K-J. 1994. Compartmentation and transport of zinc in barley primary leaves as basic mechanisms involved in zinc tolerance. Plant Cell Environ. 17, 153–162.

Buchanan BB. 1980. Role of light in the regulation of chloroplast enzymes. Ann. Rev. Plant Physiol. 31, 341–374.

Bucher JB, Keller T. 1978. Einwirkungen niedriger SO₂-Konzentrationen im mehrwöchigen Begasungsversuch auf Waldbäume. VDI-Berichte Nr. 314, 237–242.

Buchner A. 1985. Nadelverfärbungen die auf Nährstoffmangel beruhen. Forst- Holzwirt. 40, 279–285.

Buchner A, Isermann K. 1984. Wie sind Waldschadensursachen aus der Sicht der Pflanzenernährung zu beurteilen? Allg Forstztg. 39, 781–785.

Burger H. 1927. Die Lebensdauer der Fichtennadeln. Schweiz Z. Forstwes. 78, 372–375.

Burger H. 1937. Holz, Blattmenge und Zuwachs (III). Nadelmenge und Zuwachs bei Föhren und Fichten verschiedener Herkunft. Mitt Schweiz Anst forstl Versuchswes. 20, 101–114.

Burger H. 1939a. Der Kronenaufbau gleichalteriger Nadelholzbestände. Mitt Schweiz Anst forstl Versuchswes. 21, 5–57.

Burger H. 1939b. Baumkrone und Zuwachs in zwei hiebsreifen Fichtenbeständen. Mitt Schweiz Anst forstl Versuchswes. 21, 147–176.

202

Burger H. 1951. Aufastung, Entnadelung und Zuwachs bei jungen Fichten und Föhren. Mitt Forstl Bundes-Versuchsanst Mariabrunn. 47, 8–16.

Burger H. 1952. Holz, Blattmenge und Zuwachs (XII). Fichten im Plenterwald. Mitt Schweiz Anst forstl Versuchswes. 28, 109–156.

Burger H. 1953. Holz, Blattmenge und Zuwachs (XIII). Fichten im gleichalterigen Hochwald. Mitt Schweiz Anst forstl Versuchswes. 29, 38–130.

Bussler W. 1973. The dependence of the development of deficiency symptoms from physiological function of a nutrient. Curso Intern. de Fertilidad de Suelos y Nutr. Vegetal, pp. 1–13. Madrid.

Bussler W. 1979. Mangelerscheinungen an höheren Pflanzen. II. Mangel an Hauptnährstoffen. Z. Pflanzenkrankh. Pflanzenschutz. 86, 43–62.

Campbell WH. 1989. Structure and regulation of nitrate reductase in higher plants. In: Wray JL, Kinghorn JR, Eds. Molecular and Genetic Aspects of Nitrate Assimilation. pp. 125–154. Oxford University Press.

Clijsters J. 1971. Blattfall bei 'Golden Delicious'. Obstbau-Weinbau, Lana. 7/8, 222–223.

Dambrine E, Carisey N, Pollier B, et al. 1992. Dynamique des éléments minéraux dans la sève xylémique d'épicéas de 30 ans. Ann. Sci. For. 49, 489–510.

Dambrine E, Martin F, Carisey N, Granier A, Hällgren JE, Bishop K. 1995. Xylem sap composition: a tool for investigating mineral uptake and cycling in adult spruce. Plant Soil. [In press].

D'Ans J, Lax E. 1949. Taschenbuch für Chemiker und Physiker. Springer Verlag, Berlin.

Dässler HG, Börtitz S, Liessner A. 1975. Zur Phytotoxizität von Auftausalzen. Arch Natursch Landschaftsf. 15, 69–76.

Dässler HG. 1991. Einfluß von Luftverunreinigungen auf die Vegetation, 4., überarbeitete Auflage. G Fischer, Jena, 266pp.

De Kok LJ, Stulen I. 1993. Role of glutathione in plants under oxidative stress. In: De Kok LJ, Stulen I, Rennenberg H, Brunold C, Rauser WE, Eds. Sulfur nutrition and assimilation in higher plants. Regulatory and environmental aspects, pp. 125–138. SPB Academic Publishing bv, The Hague, The Netherlands.

Dietrich H. 1963. Untersuchungen zum Nährstoffkreislauf von Fichtenbeständen im Osterzgebirge. Arch. Forstwes. 12, 1116–1136.

Dilley RA, Vernon LP. 1965. Ion and water transport processes related to the light-dependent shrinkage of spinach chloroplasts. Arch. Biochem. Biophys. 111, 365–375.

Dittrich APM, Yin Z-H, Slovik S, Heber U. 1991a. Wirkung von Schadgasen auf Blätter. In: Reuther M, Kirchner M, Rösel K, Projektgruppe Bayern zur Erforschung der Wirkung von Umweltschadstoffen (PBWU), Eds. Proceedings 2. Statusseminar der PBWU zum Forschungsschwerpunkt "Waldschäden", 4–6. Februar 1991, GSF-Forschungszentrum München-Neuherberg, GSF-Bericht 26/91, 439–451.

Dittrich APM, Slovik S, Heber U, Kaiser W. 1991b. Proton accumulation is mainly responsible for reduced viability of spruce growing in poor soils and exposed to SO_2. In: Research on Forest Decline in Eastern Central Europe and Bavaria, Tagung der Projektgruppe Bayern zur Erforschung der Wirkung von Umweltschadstoffen (PBWU), Eds. Passau, 1990.

Dunkl I, Rehfuess KE. 1988. Stoffdeposition mit dem Freiland- und dem Bestandesniederschlag an fünf Waldstandorten in Südbayern. Forstliche Forschungsberichte 86. pp. 27–81. Lehrstuhl für Bodenkunde, Univ. Munich.

Eckstein D, Greve U, Frühwald A. 1981. Anatomische und mechanisch-technologische Untersuchungen am Holz einer SO_2-geschädigten Fichte und Tanne. Holz Roh- Werkst. 39, 477–487.

Eh H. 1961. Untersuchungen über die Rindenstärke der Fichte in einigen Wuchsbezirken des Württembergischen Oberschwabens. Allg Forst- Jagdztg. 132, 104–109.

Eichhorn J, Ackerbauer E. 1987. Nadelkoeffizient und Kronenschadstufe als Vitalitätsweiser zur Beurteilung des Gesundheitszustandes von Fichten (Picea abies Karst). Forschungsberichte Hessische Forstliche Versuchsanstalt, Vol. 4, pp. 7–48. Hann Münden.

Eiden R, Förster J, Peters K, Trautner F, Herterich R, Gietl G. 1989. Air pollution and deposition. In: Schulze E-D, Lange O-L, Oren R, Eds. Forest Decline and Air Pollution. A Study of Spruce (Picea abies) on Acid Soils. Ecological Series Vol. 77, pp. 57–103. Springer, Berlin.

Eidmann FE. 1943. Untersuchungen über die Wurzelatmung und Transpiration unserer Holzarten. Schriftenr Akad. Dt. Forstwiss. 5, 1–143.

Eidmann FE, Schwenke HJ. 1967. Beiträge zur Stoffproduktion, Transpiration und Wurzelatmung einiger wichtiger Baumarten. Forstwiss. Forsch. 23, 1–46.

Ermich K, Bednarz Z, Feliksik E. 1967. Wstepne Badania Nad Osadzami Mgly w Karpackim Obszarze Lesnym (Polish with English and Russian summary). Introducing investigations on fog deposition in forests of the Karpatic Mountains. Zeszyt. Warschau. 16, 123–144.

Eschrich W, Fromm J, Essiamah S. 1988. Mineral partitioning in the phloem during autumn senescence of beech leaves. Trees. 2, 73–83.

Fankhausen F, Schumacher R, Stadler W. 1976. Blattspritzungen zur Verhüung des vorzeitigen Blattfalles bei 'Golden Delicious'. Schweiz Z. Obst- und Weinbau Wädenswil. 10, 211–214.

Farquhar GD, Firth PM, Wetselaar R, Weir B. 1980. On the gaseous exchange of ammonia between leaves and the environment: Determination of the ammonia compensation point. Plant Physiol. 66, 710–714.

Fiedler HJ, Höhne H. 1965. Vorkommen und Gehalt der Makronährstoffe in Waldbäumen. Wissenschaftl. Zeitschr. Techn. Univ. Dresden. 14, 989–999.

Fiedler HJ, Hofmann W, Ilgen G. 1984. Die Ernährung der Fichte mit Mengen-elementen in den Hoch- und Kammlagen des Thüriger Waldes. Proc Mengen- und Spurenelemente, pp. 52–61. Arbeitst Karl-Marx-Univ., Leipzig.

Fink S. 1992. Histologische und histochemische Untersuchungen zur Nährstoffdynamik in Waldbäumen im Hinblick auf die Neuartigen Waldschäden. Projekt Europäisches Forschungszentrum für Maßnahmen zur Luftreinhaltung, Ed. KfK-PEF Bericht 98, 88 pp.

Flurry P. 1987. Einfluß der Berindung auf die Kubierung des Schaftholzes. Mitt Schweiz Centralanst. Forstl. Versuchswes. 5, 203–255.

Foster JR. 1990. Influence of pH and plant nutrient status on ion fluxes between tomato plants and simulated acid mists. New Phytol. 116, 475–485.

Foy CD, Chaney RL, White MC. 1978. The physiology of metal toxicity in plants. Ann. Rev. Plant Physiol. 29, 511–566.

Franz F. 1983. Auswirkungen der Walderkrankungen auf Struktur und Wuchsleistung von Fichtebeständen. Forstw. Cbl. 102, 186–200.

Frühwald A, Bauch J, Göttsche-Kühn H. 1984. Biologische und technologische Eigenschaften von Fichtenholz aus Waldschadensgebieten. Holz Roh- Werkst. 42, 441–449.

Gärtner EJ. 1987. Beobachtungseinrichtungen des hessischen Untersuchungs-programmes "Waldbelastung durch Immissionen – WdI" (Konzeption und Aufbau). Forschungsberichte Hessische Forstliche Versuchsanstalt. Vol. 1, 110 pp. Hann Münden, Germany.

Georgii HW, Grosch S, Schmitt G. 1986. Feststellung der Schadstoffbelastung von Waldgebieten in der Bundesrepublik Deutschland durch trockene und nasse Deposition. Inst. Meteorol. Geophys. Final Reports Part A, pp. 1–247. Univ. Frankfurt/Main.

Gimmler H, Schäfer G, Heber U. 1974. Low permeability of the chloroplast envelope towards cation. In: Avron M, Ed. Proceedings of the 3rd International Congress on Photosynthesis, pp. 1381–1392. Elsevier, Amsterdam.

Glavac V, Koenies H, Jocheim M, Ebben U. 1989. Mineralstoffe im Xylemsaft der Buche und ihre jahreszeitlichen Konzentrationsveränderungen entlang der Stammhöhe. Angew Botanik. 63, 471–486.

Glavac V, Koenies H, Ebben U. 1990. Seasonal variations in mineral concentrations in the trunk xylem sap of beech in a 42 year old beech forest stand. New Phytol. 116, 47–54.

Godbold DL, Fritz E, Hüttermann A. 1988. Aluminium toxicity and forest decline. Proc. Natl. Acad. Sci. USA. 85, 3888–3892.

Godbold DL. 1991. Aluminium decreases root growth and Ca and Mg uptake in *Picea abies* seedlings. In: Wright RJ, Baligar VC, Murrmann RP, Eds. Plant–Soil Interactions, pp. 747–753. Kluwer, Dordrecht.

Godde D. 1991. Turnover of the D-1 reaction center polypeptide from photosystem II in intact spruce needles and spinach leaves. Z. Naturforschung. 46c, 245–251.

Godde D. 1992. Photosynthesis in relation to modern forest decline; content and turnover of the D-1 reaction centre polypeptide of photosystem 2 in spruce trees (*Picea abies*). Photosynthetica. 27(1–2), 217–230.

Godde D, Buchhold J. 1992. Effect of long term fumigation with ozone on the turnover of the D-1 reaction center polypeptide of photosystem II in spruce (Picea abies). Physiol. Plantarum. 86, 568–574.

Greve U, Eckstein D, Scholz F, Schweingruber FH. 1985. Holzbiologische Untersuchungen an Fichtenklonen unterschiedlicher Empfindlichkeit gegen eine HF-Begasung. Angew Bot. Vol. 59.

Grill D, Härtel O. 1972. Zellphysiologische und biochemische Untersuchungen an SO_2-begasten Fichtennadeln. Resistenz und Pufferkapazität. Mitt. Forstl. Bundes-Versuchsanstalt Wien. 97/II, 367–386.

Grill D, Liegl E, Windisch E. 1979. Holzanatomische Untersuchungen an abgasbelasteten Bäumen. Phytopath. Z. 94, 335–342.

Grimme H. 1981. Wirkung einer Mg-Düngung im Gefäßversuch bei verschiedenen durch K- und Kalkdüngung veränderten Al-Konzentrationen in der Bodenlösung. Kali-Briefe. 15, 761–772.

Grimme H. 1982. The effect of Al on Mg uptake and yield of oats. Proceedings of the 9th Intern. Plant Nutr. Coll. Vol. 1, pp. 198–204. Warwick University, England.

Grimme H. 1983. Aluminium induced magnesium deficiency in oats. Pflanzenern u. Bodenk. 146, 666–676.

Grimme H. 1984. Aluminium tolerance of soybean plants as related to magnesium nutrition. Proc. VIth Intern. Coll. Optim. Plant Nutrition, pp. 242–249. Montpellier.

Gruber F. 1990. Verzweigungssystem, Benadelung und Nadelfall der Fichte (*Picea abies*). Contributiones biologiae arborum. Vol. 3, 136 pp. Birkhäuser, Basel.

Günther KH, Knabe W. 1976. Messung der Schwefel- und Säureniederschläge im Ruhrgebiet in der Zeit von Juli 1973 bis März 1975. Schriftenr Landesamt f Immissions- und Bodennutzungsschutz d Landes NRW, Heft 39, pp. 36–44. Essen.

Günther T. 1981. Biochemistry and pathobiochemistry of magnesium. Magnesium-Bull. 3, H 1a, 91–101.

Hager H. 1975. Kohlendioxid-Konzentrationen, -Flüsse und -Bilanzen. Univ München Meteorol. Inst. Wiss Mitt. 26 PhD thesis, Univ. Munich.

Hahlin M. 1973. Der Effekt der Kaliumdüngung hängt vom Verhältnis K:Mg im Boden ab. Växt Pressen, SUPRA, informationsavd, Fack, H 4, 6–7, Sweden.

Hakkila P. 1970. Weight and composition of the branches of large Scots pine and Norway spruce trees. Commun. Inst. Forest Fenn. 67, 37 pp.

Halbwachs G, Kisser J. 1967. Durch Rauchimmissionen bedingter Zwergwuchs bei Fichte und Birke. Cbl. Forstwes. 84, 156–173.

Hällgren JE, Näsholm T. 1988. Critical loads for nitrogen. Effects on forest canopies. In: Nilsson J, Grennfelt P, Eds. Critical loads for sulfur and nitrogen, pp. 323–342. Nordic council of Ministers and UN-ECE. ISBN 91-7996-096-0.

Hantschel R. 1987. Wasser- und Elementbilanz von geschädigten, gedüngten Fichtenökosystemen im Fichtelgebirge unter Berücksichtigung von physikalischer und chemischer Bodenheterogenität. PhD thesis, Univ. Bayreuth.

Hartig R. 1885. Das Holz der deutschen Nadelwaldbäume, Berlin.

Hartig R. 1892. Die Verschiedenheit in der Qualität und im anatomischen Bau des Fichtenholzes. Forstl-Naturwiss Z. 1, 209–233.

Hartig R. 1896. Wachstumsuntersuchungen an Fichten. Forstl-Naturwiss Z. 5, 1–15 and 33–45.

Hartung W, Slovik S. 1991. Physicochemical properties of plant growth regulators and plant tissue determine their distribution and redistribution: stomatal regulation by abscisic acid in leaves. Tansley Review No 35, New Phytol. 119, 361–382.

Hauhs M, Wright RF. 1986. Regional pattern of acid deposition and forest decline along a cross section through Europe. Water Air Soil Pollution. 31, 463–474.

Heath RL. 1975. Ozone. In: Mudd JB, Kozlowski TT, Eds. Responses of Plants to Air Pollution, Academic Press, New York, San Francisco, London.

Heber U, Laisk A, Pfanz H, Lange O-L. 1987. Wann ist SO_2 Nährstoff und wann Schadstoff? Ein Beitrag zum Waldschadensproblem. Allg. Forst. Zeitschrift. 27/28/29, 700–705.

Hecht-Buchholz C, Jorns CA, Keil P. 1987. Effect of excess aluminium and manganese on Norway spruce seedlings as related to magnesium nutrition during Al stress. J. Plant Nutr. 10, 1103–1110.

Hedrich R, Schröder RI. 1989. The physiology of ion channels and electrogenic pumps in higher plants. Ann. Rev. Plant Physiol. Mol. Biol. 40, 539–569.

Heil K. 1970. Grünästung an Fichte. Allg. Forstz. 25, 256–257.

Heinsdorf D. 1963. Beitrag über die Beziehungen zwischen dem Gehalt an Makronährstoffen N, P, K, Mg in Boden und Nadeln und der Wuchsleistung von Kiefernkulturen in Mittelbrandenburg. Albrecht-Thaer-Archiv. 7, 331–353.

Helder RJ. 1956. The loss of substances by cells and tissues. In: Handbuch der Pflanzenphysiologie, Vol. 2, pp. 468–488. Springer, Berlin, Göttingen, Heidelberg.

Herterich R. 1987. Eigung und Anwendung von Rotating Arm Collectoren zur Bestimmung von Nebeleigenschaften – Flüssigwassergehalt, ionische Spurenstoffe, Wasserstoffperoxid. Masters thesis, Unvi. Bayreuth.

Herterich R, Paffrath D. 1986. Untersuchung großräumiger Schadstoffbelastung im Zusammenhang mit den Waldschäden in Bayern – Meßergebnisse des Schadstoffgehalts in Nebelwasserproben aus dem Fichtelgebirge. Deutsche Forschungs- und Versuchsanstalt für Luft- und Raumfahrt, IB553, 86/23, Oberpfaffenhofen.

Hildebrandt G, Ed. 1954. Untersuchungen an Fichtenbeständen über Zuwachs und Ertrag an reiner Holzsubstanz. Dtsch. Verl. Wissenschaften. 133 pp.

Hind G, Nakatani HY, Izawa S. 1974. Light-dependent redistributions of ions in suspensions of chloroplast thylakoid membranes. Proc. Natl. Acad. Sci. USA, 71, 1484–1488.

Hocking D, Hocking MB. 1977. Equilibrium solubility of trace atmospheric sulphur dioxide in water and its bearing on air pollution injury to plants. Environ. Pollution. 13, 57–64.

Hoeß P. 1986. Nährstoffgehalte der Fichtennadeln (*Picea abies*) im Verlauf ihrer Altersentwicklung: Untersuchungen an einem geschädigten und einem ungeschädigtem Baum im Fichtelgebirge. Diploma thesis, Univ. Würzburg, FRG.

Hoffmann J. 1958. Untersuchungen über die Rindenstärke der Fichte auf verschiedenen Standorten im südöstlichen Thüringer Wald. Wiss Z. TU Dresden. 7, 361–368.

Höhne H. 1964a. Über den Einfluß des Baumalters auf das Gewicht und den Elementgehalt 1 – 4jähriger Nadeln der Fichte. Arch. Forstwes. 13, 247–265.

Höhne H. 1964b. Untersuchungen über die jahrezeitlichen Veränderungen des Gewichtes und Elementgehaltes von Fichtennadeln in jüngeren Beständen des Osterzgebirges. Arch. Forstwes. 13, 747–774.

Höhne H. 1964c. Der Einfluß der soziologischen Stellung der Fichte auf das Gewicht und den Elementgehalt ihrer Nadeln. Arch. Forstwes. 13, 833–842.

Hunger W. 1964. Düngungsdiagnosen und Düngungsergebnisse in älteren Fichtenbeständen des Erzgebirges und Vogtlandes. PhD thesis, Univ. Dresden.

Hutchinson TC, Adams CM. 1987. Comparative abilities of leaf surfaces to neutralize acidic raindrops. I. The influence of calcium nutrition and charcoal-filtered air. New Phytol. 106, 169–183.

Hüttl RF. 1985. "Neuartige" Waldschäden und Nährelementversorgung von Fichtenbeständen (*Picea abies* Karst.) in Südwestdeutschland. Freiburger Bodenkd. 16, 1–195.

Hüttl RF. 1986. Neuartige Waldschäden und Ernährungszustand von Fichtenbeständen in Südwestdeutschland am Beispiel Oberschwaben. Kali-Briefe. 17, 1–7.

Hüttl RF. 1987. Neuartige Waldschäden und Nährelementversorgung von Fichtenbeständen (*Picea abies* Karst.) in Südwestdeutschland. Freiburger Bodenk Abh H 16 d Inst f Bodenk u Waldernährungslehre, Freiburg i. Br.

Hüve K, Dittrich A, Kindermann G, Slovik S, Heber U. 1995. Detoxification of SO_2 in conifers differing in SO_2-tolerance. A comparison of *Picea abies*, *Picea pungens* and *Pinus sylvestris*. Planta. [In press].

Ingestad T. 1958. Studies on manganese deficiency in a forest stand. Medd Skogsforskn Inst. Stockholm. 48, 1–20.

Ingestad T. 1959. Studies on the nutrition of forest tree seedlings. II. Mineral nutrition of spruce. Physiol. Plantarum (Copenhagen). 12, 568–593.

Ingestad T. 1962. Macro element nutrition of pine, spruce and birch seedlings in nutrient solutions. Medd Skogsforskn Inst. Stockholm. 51, 1–150.

Ingestad T. 1982. Relative addition rates and external concentrations: driving variables used in plant nutrition research. Plant Cell Environ. 5, 443–453.

Isermann K. 1985. Diagnose und Therapie der "neuartigen Waldschäden" aus der Sicht der Waldernährung. VDI-Berichte. 560, 897–920.

Jensen RG, Bassham JA. 1967. Photosynthesis by isolated chloroplasts. Light activation of the carboxylation reaction. Biochim. Biophys. Acta. 153, 227–234.

Jentschke G, Schlegel H, Godbold DL. 1991. The effect of aluminium on uptake and distribution of magnesium and calcium in roots of mycorrhizal Norway spruce seedlings. Physiol. Plantarum. 82, 266–270.

Jeschke WD, Pate JS. 1991. Ionic interactions of petiole and lamina during the life of a leaf of castor bean (*Ricinus communis* L.) under moderately saline conditions. J. Exp. Bot. (Oxford). Vol. 42, No. 241, 1051–1064.

Jeschke WD, Atkins CA, Pate JS. 1985. Ion circulation via phloem and xylem between root and shoot of nodulated white lupin. J. Plant Physiol. 117, 319–330.

Jeschke WD, Pate JS, Atkins CA. 1987. Partitioning of K^+, Na^+ Mg^{++}, and Ca^{++} through xylem and phloem to component organs of nodulated white lupin under mild salinity. J. Plant Physiol. 128, 77–93.

Jorns A, Hecht-Buchholz C. 1985. Aluminiuminduzierter Magnesium- und Calciummangel im

206

Laborvesuch bei Fichtensämlingen. Allg. Forstz. 46, 1248–1252.

Jung J, Dressel J. 1977. Über die Wirkung von "Stickstoffmagnesia" auf zwei an Magnesium unterschiedlich vararmten Böden (Lysimeterversuche). Acker- Pflanzenbau. 114, 268–279.

Kaiser WM, Dittrich APM, Heber U. 1991. Sulfatakkumulation in Fichtennadeln als Folge von SO_2-Belastung. In: Reuther M, Kirchner M, Rösel K, Projektgruppe Bayern zur Erforschung der Wirkung von Umweltschad-stoffen (PBWU), Eds. Proceedings 2. Statusseminar der PBWU zum Forschungs-schwerpunkt "Waldschäden", 4–6. Februar 1991, GSF-Bericht 26/91, pp. 425–437, GSF-Forschungszentrum, München-Neuherberg.

Kaiser WM, Spill D, Brendle-Behnisch E. 1992. Adenine nucleotides are apparently involved in the light-dark modulation of spinach-leaf nitrate reductase. Planta. 186, 236–240.

Kaiser WM, Dittrich APM, Heber U. 1993. Sulfate concentrations in Norway Spruce needles in relation to atmospheric SO_2: a comparison of trees from various forests in Germany with trees fumigated with SO_2 in growth chambers. Tree Physiol. 12, 1–13.

Kandler O, Miller W, Ostner R. 1987. Dynamik der "akuten Vergilbung" der Fichte. Allg. Forsztg. 27/28/29, 715–723.

Kaupenjohann M, Hantschel R, Horn W, Zech W. 1985. Nährstoffversorgung gedüngter, unterschiedlich geschädigter Fichten auf immissionsbelasteten Stand-orten in NO-Bayern. Mitt. Dtsch. Bodenkundl. Ges. 43, 969–974.

Kazda M, Weilgony P. 1988. Seasonal dynamics of major cations in xylem sap and needles of Norway spruce in stands with different soil solution chemistry. Plant Soil. 110, 91–100.

Keller T. 1971. Die Bedeutung des Waldes für den Umweltschutz. Schweiz Z. Forstwes. 122, 600–613.

Keller T. 1972. Gaseous exchange of forest trees in relation to some edaphic factors. Photosynthetica. 6, 197–206.

Keller T. 1978. Der Einfluß einer SO_2 Begasung zu verschiedenen Jahreszeiten auf CO_2-Aufnahme und Jahrringbreite der Fichte. Schweiz Z. Forstwes. 129, 381–393.

Keller T. 1980. The effect of a continuous springtime fumigation with SO_2 on CO_2 uptake and structure of the annual ring. Can. J. Forest Res. 10, 1–6.

Keller T. 1981. Folgen einer winterlichen SO_2-Belastung für die Fichte. Gartenbauwiss. 46, 170–178.

Keller T, Häsler R. 1986. The influence of a prolonged SO_2-fumigation on the stomatal reaction of spruce. Eur. J. Forest Pathol. 16, 110–115.

Keller T, Häsler R. 1987. Some effects of long-term ozone fumigation on Norway spruce. I. Gas exchange and stomatal response. Trees. 1, 129–133.

Kelley GJ, Latzko E, Gibbs M. 1976. Regulatory aspects of photosynthetic carbon metabolism. Annu. Rev. Plant Physiol. 27, 181–205.

Kenk G. 1990. Effects of air pollution on forest growth in southwest Germany – Hunting for a phantom? In: Proc., Div. 2, XIX IUFRO World Congress, 5–11, Aug. 1990, Montreal, CDN, pp. 388–395.

Kerner H, Gross E, Koch W. 1977. Structure of the assimilation system of a dominating spruce tree (*Picea abies* [L.] Karst.) of closed stand: computation of needle surface area by means of a variable geometric needle model. Flora. 166, 449–459.

Kindermann G, Slovik S, Urbach W, Heber U. 1992. The base content of the xylem sap restricts the long-term SO_2-resistance of leaves. Phyton (Austria). 32, 63–68.

Kirkby EA, Mengel K. 1976. The role of magnesium in plant nutrition. Pflanzenern Bodenkde H. 2, 209–222.

Kirwald E, Ed. 1950. Forstlicher Wasserhaushalt und Forstschutz gegen Wasserschäden, Stuttgart-Ludwigsburg.

Kiss SA. 1981. Das Magnesium und dessen Rolle in den Pflanzen. Magnesium-Bull. 3, H1a, 6–12.

Kiss SA, Pozsár BI. 1977. Resistance increased by magnesium nutrition. Acta Agron. Acad. Sci. Hungaricae. 26, 156–163.

Klemm O, Kuhn U, Beck E et al. 1989. Leaching and uptake of ions through above-ground Norway spruce tree parts. In: Schulze E-D, Lange O-L, Oren R, Eds. Forest Decline and Air Pollution. A Study of Spruce (*Picea abies*) on Acid Soils. Ecological Series, Vol. 77, pp. 210–237, Springer, Berlin.

Koltzenburg C. 1981. Mechanisch-biologischer Schälschutz an Fichte, Auswirkungen auf Holz und Rinde. Schriften Forstl Fak Univ Göttingen. Vol. 73, 120 pp.

Korell U. 1972. Über Rindendicken der Fichte. Beitr Forstwirtsch. 115, 54–55.

Körner C, Perterer J, Altrichter C, Meusburger A, Slovik S, Zöschg M. 1995. Ein einfaches Modell zur Berechnung der jährlichen Schadgasaufnahme von Fichten- und Kiefernadeln. Allg Forst Jagdzeitschrift. [In press].

Korte H. 1985. Ionenaustauschereigenschaften des Holzgewebes von Tanne (*Abies alba* Mill). Diploma thesis, Univ. of Hamburg, 59 pp.

Köstner BMM. 1989. Jahresverlauf der Chloroplastenpigmente von ungeschädigten und chlorotischen Fichten an einem Waldschadenstandort im Fichtelgebirge in Abhängigkeit von Alter und Mineralstoffgehalt der Nadeln. PhD thesis, Univ. Würzburg.

Köstner B, Czygan F-C, Lange O-L. 1990. An analysis of needle yellowing in healthy and chlorotic Norway Spruce (*Picea abies*) in a forest decline area of the Fichtelgebirge (NE Bavaria). I. Annual time-course changes in chloroplast pigments for five different needle age classes. Trees. 4, 55–67.

Kozlowski TT. 1971. Growth and Development of Trees. Vol. I. Academic Press, New York, London.

Kramer H. 1976. Grünästung und Düngung bei Fichte. Allg. Forst- Jagdzeitung. 147, 25–33.

Krause GH. 1977. Light induced movement of Mg^{2+} ions in intact chloroplasts. Spectroscopic determination with eriochrome blue SE. Biochim. Biophys. Acta. 460, 500–510.

Kruis A, May A. 1962. Lösungsgleichwichte von Gasen in Flüssigkeiten. In: Bartels J, Borchers H, Hausen H, Hellwege KH, Schäfer K, Schmidt E, Eds. Landolt-Börnstein. Zahlenwerte und Funktionen aus Physik, Chemie, Astronomie, Geophysik, Technik, Vol. 2: Eigenschaften der Materie in ihren Aggregatzuständen. Bandteil 2b (Schäfer K, Lax E, Eds.): Gleichgewichte außer Schmelzgleichgewichten. Lösungsgleichgewichte I, pp. 1–210. Springer Verlag Berlin, Göttingen, Heidelberg.

Kuhn AJ, Bauch J, Schröder WH. 1995. Monitoring uptake and contents of Mg, Ca and K in Norway spruce as influenced by pH and Al, using microprobe analysis and stable isotope labelling. Plant Soil. [In press].

Ladefoged K. 1946. De enkelte kronedeles produktionsmaessige betyduing hos rødgran. Forstl. Forsøgsv. Danm. 16, 365–400.

Ladefoged F. 1963. Transpiration of forest trees in closed stands. Physiol. Plant. 16, 378–414.

Laisk A, Kull O, Moldau H. 1989. Ozone concentration in leaf intercellular spaces is close to zero. Plant Physiol. 90, 1163–1167.

Lampadius F. 1968. Die Bedeutung der SO$_2$-Filterung des Waldes im Blickfeld der forstlichen Rauchschadenstherapie. Z. Th. Dresden. 17, 503–511.

Lange O-L, Gebel J, Zellner H, Schramel P. 1986. Photosynthesekapazität und Magnesiumgehalte verschiedener Nadeljahrgänge bei der Fichte in Waldschadensgebieten des Fichtelgebirges. In: Führ F, Ganser S, Kloster G, Prinz B, Stüttgen E, Eds. Statusseminar im Auftrage der Interministeriellen Arbeitsgruppe (IMA) "Waldschäden/Luftverunreinigungen". pp. 127–147, Jülich.

Lange O-L, Zellner H, Gebel J, Schramel P, Köstner B, Czygan F-C. 1987. Photosynthetic capacity, chloroplast pigments, and mineral content of the previous year's spruce needles with and without the new flush: analysis of the forest-decline phenomenon of needle bleaching. Oecologia. 73, 351–357.

Lange O-L, Heber U, Schulze E-D, Ziegler H. 1989a. Atmospheric pollution and plant metabolism. In: Schulze E-D, Lange O-L, Oren R, Eds. Forest Decline and Air Pollution. A Study of Spruce (*Picea abies*) on Acid Soils. Ecological Studies. Vol. 77, pp. 238–276. Springer, Berlin.

Lange O-L, Weikert RM, Wedler M, Gebel J, Heber U. 1989b. Photosynthese und Nährstoffversorgung von Fichten aus einem Waldschadensgebiet auf basenarmen Untergrund. Allg. Forstzeitschr. 3/89, 55–64.

Langner W. 1932. Die Wasserverteilung im Stammholz der Fichte und ihre Veränderungen. Botan. Archiv. 34, 1–80.

Lanzl A, Führer G, Lippert M, Kaiser WM. 1989. Sulfat- und Nitratgehalte in Nadeln und Blättern als Indikatoren für Belastung mit SO$_2$ und NO$_2$. In: Reuther M, Kirchner M, Rösel K, Projektgruppe Bayern zur Erforschung der Wirkung von Umwelt-schadstoffen (PBWU), Eds. Proceedings 1. Statusseminar der PBWU zum Forschungsprojekt "Waldschäden", GSF-Bericht 6/89. pp. 185–194. GSF-Forschungszentrum, München-Neuherberg.

Leinert S. 1962. Die Bedeutung der Rohwichte für die Beurteilung des Zuwachses bei Fichte. PhD thesis, Nat-Math Fak. Univ. Freiburg i. Br., 158 pp.

Leyton L. 1948. Mineral nutrient relationships of forest trees. Forestry Abstracts, Vol. IX(4), 399–408.

Lichtenthaler HK, Buschmann C. 1984. Das Waldsterben aus botanischer Sicht – Verlauf, Ursachen und Maßnahmen, G Braun GmbH, Karlsruhe.

Liese W, Dujesiefken D. 1986. Das Holz der Fichte. In: Schmidt-Vogt H, Ed. Die Fichte. Ein Handbuch in zwei Bänden. Vol. II/1 Wachstum, Züchtung, Boden, Umwelt, Holz, pp. 373–443. Paul Parey Verlag, Hamburg und Berlin.

Liese W, Parameswaran N. 1971. Über die Rindenanatomie starkborkiger Fichten. Forstw. Cbl. 90, 370–375.

Liese W, Schneider M, Eckstein D. 1975. Histometrische Untersuchungen am Holz einer rauchgeschädigten Fichte. Eur. J. For. Path. 5, 152–161.

208

Likens GE, Bormann FH, Pierce RS, Eaton JS, Johnson NM, Eds. 1977. Biogeochemistry of a Forested Ecosystem. Springer, Berlin, Heidelberg, New York.

Lilley R McC, Holborow K, Walker DA. 1974. Magnesium activation of photosynthetic CO_2-fixation in a reconstituted chloroplast system New Phytol. 73, 657–662.

Linder S, Flower-Ellis J. 1992. Environmental and physiological constraints to forest yield. In: Teller A, Mathy P, Jeffers JNR, Eds. Responses of Forest Ecosystems to Environmental Changes. pp. 149–164. Elsevier Applied Science, London.

Lippert M. 1988. Photosynthesekapazität, Chlorophyllgehalt und Nährstoffgehalte von Nadeln geschädigter und ungeschädigter Fichten mit und ohne Konkurrenz des diesjährigen Triebs: Experimenteller Ansatz zur Analyse der Waldschäden im Fichtelgebirge, Diploma thesis, Univ. Würzburg.

Lütz C, Steiger A, Godde D. 1992. Influence of air pollutants and nutrient deficiency on D-1 protein content and photosynthesis in young spruce trees. Physiol. Plantarum. 85, 611–617.

Lyr H, Polster H, Fiedler HJ, Eds. 1967. Gehölzphysiologie. Gustav Fischer Verlag, Jena.

MacKintosh C, Cohen P. 1989. Identification of high levels of type 1 and type 2A protein phosphatases in higher plants. Biochem. J. 262, 335–339.

MacKintosh RW, MacKintosh C. 1993. Regulation of plant metabolism by reversible protein (serine/threonine) phosphorylation. In: Battey NH, Dickinson HG, Hetherington AM, Eds. Post-translational Modifications in Plants, Society for Experimental Biology Seminar Series 53, pp. 197–212. Cambridge University Press, Cambridge.

MacKintosh C, Coggins J, Cohen P. 1991. Plant protein phosphatases. Subcellular distribution, detection of protein phosphatase 2C and identification of protein phosphatase 2A as the major quinate dehydrogenase phosphatase. Biochem. J. 273, 733–738.

Mansfield TA, Ed. 1976. Effects of Air Pollution on Plants. Cambridge University Press, New York.

Mantinger H. 1974. Blattfall bei Golden Delicious. Landwrit (Bozen). 28, 463–466.

Marschner H. 1986. Mineral Nutrition of Higher Plants. Academic Press, San Diego, 674 pp.

Marschner H. 1991. Mechanisms of adaptation of plants to acid soils. Plant Soil. 134, 1–20.

Marschner H. 1992. Bodenversauerung und Magnesiumernährung der Pflanzen. In: Glatzel G, Jandl R, Sieghardt M, Hager H, Eds. Magnesiummangel in Mitteleuropäischen Waldökosystemen. Ergebnisse eines Symposiums in Salzburg am 8. und 9. April 1991. Forstliche Schriftenreihe der Universität für Bodenkultur (Wien) Vol. 5, 15 pp.

Materna J. 1962. Auswertung von Düngungsversuchen in rauchgeschädigten Fichtenbeständen. Wissensch. Zeitschrift Techn. Univ. Dresden. 11(3), 589–593.

Materna J. 1972. Einfluß niedriger SO_2-Konzentrationen auf die Fichte. Mitt. Forstl. Bundes-Versuchsanstalt. 97/I, pp. 219–232, Wien.

Mecklenburg RA, Tukey HB Jr, Morgan JV. 1966. A mechanism for the leaching of calcium from foliage. Plant Physiol. 41, 610–613.

Meiwes KI. 1979. Der Schwefelhaushalt eines Buchenwald- und Fichtenwald-Ökosystems. Göttinger Bodenkundliche Berichte. Vol. 60.

Meldau R. 1956. Handbuch der Staubtechnik I, Düsseldorf.

Mengel K. 1984. Ernährung und Stoffwechsel der Pflanze. 6. Auflage, Gustav Fischer Verlag, Jena.

Mengel K, Lutz HJ, Breininger MT. 1987. Auswaschung von Nährstoffen aus jungen intakten Fichten (Picea abies). Z. Pflanzenern Boden. 150, 61–68.

Mette H-J, Korell U, Eds. 1989. Richtzahlen und Tabellen für die Forstwirtschaft. Deutscher Landwirtschaftsverlag, Berlin.

Meyer H. 1956. Über Zusammenhänge zwischen Assimilationsmasse und Holzzuwachs in Fichtenbeständen. Arch. Forstwes. 5, 1–7.

Meyer H. 1959. Der Einfluß von Kronenverkürzungen an Fichten der II. Altersklasse auf deren Höhen und Stärkenwachstum. Arch. Forstwes. 8, 812–849.

Meyer H. 1968. Auswertungsergebnis über Kronenkürzungen an Fichten nach Ablauf einer Zehnjahresperiode. Arch. Forstwes. 17, 781–796.

Meyer J, Schneider BU, Werk KS, Oren R, Schulze E-D. 1988. Performance of two Picea abies (L.) Karst. stands at different stages of decline. V. Root tip and ectomycorrhiza development and their relation to above-ground and soil nutrients. Oecologia (Berlin). 77, 7–13.

Michael G. 1941. Über die Aufnahme und Verteilung des Magnesium und dessen Rolle in der höheren grünen Pflanze. Bodenkde Pflanzenern. 25, 65–120.

Michal G. 1978. Biochemical Pathways. Boehringer GmbH, Mannheim.

Mies E, Zöttl HW. 1985. Zeitliche Änderung der Chlorophyll- und Elementgehalte in den Nadeln eines gelb-chlorotischen Fichtenbestandes. Forstw. Cbl. 104, 1–8.

Mitscherlich G. 1963. Das Wachstum der Fichte in Europa. Allg. Forst. Jagdzeitung. 134, 29–45, 61–72, 93–110, 125–140.

Mitscherlich G. 1970. Wald, Wachstum und Umwelt, Vol. 1. J Sauerländer, Frankfurt/Main.

Mitscherlich G. 1975. Wald, Wachstum und Umwelt, Vol. 3. Boden, Luft und Produktion, J Sauerländer, Frankfurt/Main.

Mitscherlich G, Von Gadow K. 1968. Über den Zuwachsverlust bei der Ästung von Waldbäumen. Allg. Forst- Jagdzeitung. 139, 175–184.

Mitterhuber E, Pfanz H, Kaiser WM. 1989. Leaching of solutes by the action of acidic rain: a comparison of efflux from twigs and single needles of *Picea abies* (L.) Karst. Plant Cell Environ. 12, 93–100.

Møller CM. 1946. Untersuchungen über Laubmenge, Stoffverlust und Stoffproduktion des Waldes. Forstl. Forsøgsv. Danm. 17, 1–287.

Moosmayer HU. 1984. Erkenntnisse über die Walderkrankung. Dargestellt an Projekten der Forstlichen Versuchs- und Forschungsanstalt Baden-Württemberg. Forstw. Cbl. 103, 1–16.

Mothes K, Baudisch W. 1958. Untersuchungen über die Reversibilität der Ausbleichung grüner Blätter. Flora. 146, 521–531.

Münch E. 1928. Winterschäden an Fichte und anderen Gehölzen. Thar forstl Jb. 79, 276.

Nagel D. 1968. Untersuchungen über die Form und Formentwicklung des Fichtenschaftes. PhD thesis, Univ. Freiburg i. Br.

Nebe W. 1963. Über die Beurteilung der Düngebedürftigkeit von Mittelgebirgsstandorten durch Blattanalysen. Arch. Forstwes. 12, 1024–1052.

Nebe W. 1967. Zur Manganernährung der Fichte. Arch. Forstwes. 16, 109–118.

Németh K, Grimme H. 1974. Einfluß einer Düngung auf die Aufnahme nicht gedüngter Nährstoffe im Gefäßversuch. Z. Pflanzenern Bodenk. 137, 203–213.

Nobel PS, Ed. 1983. Biophysical Plant Physiology and Ecology. WH Freeman and Company, New York.

Olischläger K. 1969. Untersuchungen über den Wertzuwachs von Fichten nach Ästung. PhD thesis, Univ. Göttingen, Hann Münden.

Oren R, Schulze ED. 1989. Nutritional disharmony and forest decline: a conceptual model. In: Schulze E-D, Lange O-L, Oren R, Eds. Forest Decline and Air Pollution. A Study of Spruce (*Picea abies*) on Acid Soils. Ecological Studies, Vol. 77, pp. 425–443. Springer, Berlin.

Oren R, Zimmermann R. 1989. CO_2 assimilation and the carbon balance of healthy and declining Norway spruce stands. In: Schulze E-D, Lange O-L, Oren R, Eds. Forest Decline and Air Pollution. A Study of Spruce (*Picea abies*) on Acid Soils. Ecological Studies, Vol. 77, pp. 352–369. Springer, Berlin.

Oren R, Schulze E-D, Maryssek R, Zimmermann R. 1986. Estimating photosynthetic rates and annual carbon gain in conifers from leaf weight and leaf biomass. Oecologia. 70, 187–193.

Oren R, Schulze E-D, Werk KS, Meyer J. 1988. Performance of two *Picea abies* (L.) Karst. stands at different stages of decline. VII. Nutrient relations and growth. Oecologia (Berlin). 77, 163–173.

Osonubi O, Oren R, Werk KS, Schulze E-D, Heilmeier H. 1988. Performance of two *Picea abies* (L.) Karst. stands at different stages of decline. IV. Xylem sap concentrations of magnesium, calcium, potassium and nitrogen. Oecologia. 77, 1–6.

Östlin E. 1963. Barkuppgifter för tall, gran, björk m. fl. Del 1 u. 2. Rapp. Uppsatser Inst. Skogstaxering Nr 5 u. 6, Stockholm.

Palaniyandi R, Smith CB. 1978. Growth and nutrient interrelationships in snap beans as affected by several sources of potassium and magnesium. J. Am. Soc. Hort. Sci. 103, 109–113.

Palaniyandi R, Smith CB. 1979. Effects of nitrogen sources on growth responses and magnesium and manganese leaf concentrations of snap beans. Com. Soil Sci. Plant Anal. 10, 869–881.

Parameswaran N, Fink S, Liese W. 1985. Feinstrukturelle Untersuchungen an Nadeln geschädigter Tannen und Fichten aus Waldschadensgebieten im Schwarzwald. Eur. J. For. Pathol. 15, 168–182.

Parker DR, Kinraide TB, Zelazny LW. 1989. On the phytotoxicity of polynuclear hydroxy-aluminium complexes. Soil Sci. Soc. Am. J. 53, 789–796.

Paulus W, Bresinsky A. 1989. Soil fungi and other microorganisms. In: Schulze E-D, Lange O-L, Oren R, Eds. Forest Decline and Air Pollution. A Study of Spruce (*Picea abies*) on Acid Soils. Ecological Studies, Vol. 77, pp. 110–120. Springer, Berlin.

Pechmann H. 1958. Über die Heilungsaussichten bei hagelbeschädigten Waldbeständen. Forstw. Cbl. 77, 357–373.

Perterer J, Körner C. 1990. Das Problem der Bezugsgröße bei physiologisch-ökologischen Untersuchungen an Koniferennadeln. Forstw. Cbl. 109, 220–241.

Pfeiffer S. 1987. Netto-Photosynthese und Transpiration einer gedüngten und einer geschädigten Fichte bei natürlicher externer CO_2-Konzentration am Freiland-standort im Fichtelgebirge. Diploma thesis, Univ. Würzburg.

Pfirrmann T, Runkel K-H, Schramel P, Eisenmann T. 1990. Mineral and nutrient supply, content and leaching in Norway spruce exposed for 14 months to ozone and acid mist. Environ. Poll. 64, 229–253.

Pfluger R. 1973. Investigations on ion fluxes of chloroplasts with an intact envelope. Z. Naturforschung C. 28, 779–780.

Pisek A, Cartellieri E. 1939. Zur Kenntnis des Wasserhaushalts der Pflanzen. Jb. wiss Botanik. 88, 22.

Pisek A, Cartellieri E. 1941. Zur Kenntnis des Wasserhaushalts der Pflanzen. Jb. wiss Botanik. 90, 255–291.

Pisek A, Tranquillini W. 1951. Transpiration und Wasserhaushalt der Fichte (*Picea excelsa*) bei zunnehmender Luft- und Bodentrockenheit. Physiol. Plant. 4, 1–27.

Pollanschütz J. 1962. Rauchschadensfeststellung unter besonderer Berücksichtigung von Bohrkern-analysen. Allg. Forstztg. 73, Beibl Inform Dienst Forstl Bundesversuchsanstalt Mariabrunn. 55, 1–4.

Poller S. 1968. Zur physikalisch-mechanischen Beschaffenheit von Fichtenholz auf verschiedenen Standorten. Arch. Forstwes. 17, 733–752.

Polster H, Ed. 1950. Die physiologischen Grundlagen der Stofferzeugung im Walde. Untersuchungen über Assimilation, Respiration und Transpiration unserer Hauptholzarten, München.

Polster H. 1954. Gesichertes und Ungesichertes über den Wasserhaushalt des Waldes. Forst Jagd. 4, 256–258.

Portis AR Jr. 1981. Evidence of a low stromal Mg^{2+} concentration in intact chloroplasts in the dark. 1. Studies with the ionophore A 23187. Plant Physiol. 67, 985–989.

Portis AR Jr, Heldt HW. 1976. Light-dependent changes of the Mg^{2+} concentration in the stroma in relation to the Mg^{2+} dependency of CO_2 fixation in intact chloroplasts. Biochim. Biophys. Acta. 449, 434–446.

Portis AR Jr, Chon CA, Mosbach A, Heldt HW. 1977. Fructose- and sedoheptulose bisphosphatase. The sites of a possible control of CO_2 fixation by light-dependent changes of the stromal Mg^{2+} concentration. Biochim. Biophys. Acta. 461, 313–325.

Quast P. 1981. Welchen Beitrag können bestimmte Schorffungizide zur Spurenelement-versorgung von Apfelanlagen und zur Verhinderung von Blattflecken leisten? Mitt Obstbauversuchsr des Alten Landes. 36, 144–153.

Rabe R, Kreeb KH. 1980. Bioindication of air pollution by chlorophyll destruction in plant leaves. Oikos. 34, 163–167.

Rauterberg E, Miraftabi B. 1970. Nährstoffmangeldiagnose durch Untersuchung der zuerst von Mangelerscheinungen befallenen Pflanzenteile mit einfachen Schnell-methoden. Pflanzenern Bodenkde. 125, 156–168.

Raven JA. 1985. pH regulation in plants. Sci. Prog. Oxf. 69, 495–509.

Reemtsma JB. 1964. Untersuchungen an Fichte und anderen Nadelbaumarten über den Nährstoffgehalt der lebenden Nadeljahrgänge und der Streu. PhD thesis, Univ. Göttingen, Hann. Münden.

Rennenberg H. 1984. The fate of excess sulfur in higher plants. Ann. Rev. Plant Physiol. 35, 121–153.

Rennenberg H, Brunold CH. 1994. Significance of Glutathione Metabolism in Plants Under Stress. Progess in Botany, Vol. 55, pp. 142–156. Springer, Berlin.

Schimansky C. 1991. Einfluß von Aluminium und Protonen auf die Aufnahme und den Transport von Magnesium (^{28}Mg) bei Gerste (*Hordeum vulgare*). Z. Pflanzenernähr. Bodenkd. 154, 1–4.

Schmidt-Vogt H. 1949. Die Verzweigungstypen der Fichte und ihre Bedeutung für die forstliche Pflanzenzüchtung. PhD thesis, Univ. Munich. Summary (1952) in: Z. Forstgenetik Forstpflan-zenzüchtung. 1, 81–91.

Schmidt-Vogt H. 1953. Kronen- und Zuwachsuntersuchungen an Fichten des bayerischen Alpenvorlandes. Forstw. Cbl. 72, 276–286.

Schmidt-Vogt H, Ed. 1986. Die Fichte. Ein Handbuch in zwei Bänden. Vol. II/1 Wachstum, Züchtung, Boden, Umwelt, Holz, Paul Parey Verlag, Hamburg and Berlin. 563 pp.

Schneider BU, Meyer J, Schulze E-D, Zech W. 1989. Root and Mycorrhizal Development in Healthy and Declining Norway Spruce stands. In: Schulze E-D, Lange O-L, Oren R, Eds. Forest Decline and Air Pollution. A Study of Spruce (*Picea abies*) on Acid Soils. Ecological Studies, Vol. 77, pp. 370–391. Springer, Berlin.

Schober R, Ed. 1987. Ertragstafeln wichtiger Baumarten. 3. neubearb. und erweiterte Auflage. 166 pp. J Sauerländer, Frankfurt/Main.

Schöpfer W. 1961. Beiträge zur Erfassung des Assimilationsapparates der Fichte. Schriftenreihe

Landesforstverwaltung Baden-Württemberg, Vol 10.

Schröder J. 1873. Die Einwirkung der schwefligen Säure auf die Pflanzen. Thar. forstl. Jahrb. 23, 217–267.

Schubert A. 1939. Untersuchungen über den Transpirationsstrom der Nadelhölzer und den Wasserbedarf von Fichte und Lärche. Thar. Forstl. Jb. 90, 821–883.

Schulze E-D. 1989. Air pollution and forest decline in a spruce (Picea abies) forest. Science. 244, 776–783.

Schulze E-D, Oren R, Zimmermann R. 1987. Die Wirkung von Immissionen auf 30-jährige Fichten in mittleren Höhenlagen des Fichtelgebirges auf Phyllit. Allg. Forstztg. 27/28/29, 725–730.

Schulze E-D, Oren R, Lange O-L. 1989. Nutritional relations of trees in healthy and declining Norway spruce stands. In: Schulze E-D, Lange O-L, Oren R, Eds. Forest Decline and Air Pollution. A Study of Spruce (Picea abies) on Acid Soils. Ecological Studies, Vol. 77, pp. 399–417. Springer, Berlin.

Schumacher F. 1976. Mögliche Ursachen des vorzeitigen Blattfalles bei Golden Delicious. Schweiz Obst- Weinbau Wädenswil. 12, 260–264.

Schwab M, Noga G, Barthlott W. 1993a. Einfluß eines Mg- und Ca-Mangels auf Synthese, chemische Zusammensetzung und mikromorphologische Ausbildung von epicuticulären Wachsen bei Kohlrabiblättern. Angew. Botanik. 67, 172–179.

Schwab M, Noga G, Barthlott W. 1993b. Einfluß einer unzureichenden Mg- und Ca-Versorgung auf die Anfälligkeit von Kohlrabi gegenüber Botrytis cinerea. Angew. Botanik. 67, 180–185.

Schwab M, Noga G, Barthlott W. 1995a. Einfluß hoher Aluminium-Konzentrationen auf die Feinstruktur der epikutikulären Wachse von Fichtensämlingen. Forstw. Cbl. [In press].

Schwab M, Noga G, Barthlott W. 1995b. Einfluß eines Mg- und Ca-Mangels auf die Mikromorphologie der epicuticulären Wachse und die Benetzbarkeit von Fichtennadeln. Z. Pflanzenern Bodenk. [In press].

Schweingruber FH. 1980. Jahrringe als Klimatologische Datenquelle. In: Oeschger H, Messerli B, Svilar M, Eds. Das Klima – Analysen und Modelle, Geschichte und Zukunft. pp. 246–256. Springer, Berlin, Heidelberg, New York.

Slovik S, Hartung W. 1992a. Compartmental distribution and redistribution of abscisic acid in intact leaves. II. Model analysis. Planta. 187, 26–36.

Slovik S, Hartung W. 1992b. Compartmental distribution and redistribution of abscisic acid in intact leaves. III. Analysis of the stress-signal chain. Planta. 187, 37–47.

Slovik S, Hartung W. 1992c. Stress-induced redistribution kinetics of ABA in leaves: Model considerations. In: Karssen CM, Van Loon LC, Vreugdenhil D, Eds. Progess in Plant Growth Regulation. Proceedings of the 14th International Conference on Plant Growth Substances, Amsterdam, 21–26 July, 1991, Kluwer Academic Publishers, Dordrecht, Boston, London. Curr. Plant Sci. Biotechnol. Agricult. 13, 464–473.

Slovik S, Kaiser WM, Körner C, Kindermann G, Heber U. 1992a. Quantifizierung der physiologischen Kausalkette von SO_2-Immissionsschäden. (I) Ableitung von SO_2-Immissionsgrenzwerten für akute Schäden an Fichte. Allg. Forst. Zeitschrift. 15, 800–805.

Slovik S, Heber U, Kaiser W, Kindermann G, Körner C. 1992b. Quantifizierung der physiologischen Kausalkette von SO_2-Immissionsschäden. (II) Ableitung von SO_2-Immissionsgrenzwerten für chronische Schäden an Fichte. Allg. Forst. Zeitschrift. 17, 913–920.

Slovik S, Baier M, Hartung W. 1992c. Compartmental distribution and redistribution of abscisic acid in intact leaves. I. Mathematical formulation. Planta. 187, 14–25.

Slovik S, Kindermann G, Heber U. 1993. Die physiologische Kausalkette chronischer SO_2-Immissionsschäden und daraus ableitbare Immissionsgrenzwerte für Fichten. In: Tesche M, Feiler S, Eds. Air Pollution and Interactions between Organisms in Forest Ecosytems. Proceedings of the 15th International Meeting of Specialists in Air Pollution Effects on Forest Ecosystems (IUFRO Centennial, 9–11 Sept. 1992), pp. 104–108. Dresden.

Slovik S, Siegmund A, Kindermann G, Riebeling R, Balázs Á. 1995. Stomatal SO_2 uptake and sulfate accumulation of Norway spruce stands (Picea abies) in Central Europe. Plant Soil. 169, 405–419.

Sonn SW, Ed. 1960. Der Einfluß des Waldes auf die Böden. G Fischer, Jena.

Strebel O. 1961. Nadelanalytische Untersuchungen an Fichten – Altbeständen sehr guter Wuchsleistung im Bayerischen Alpenvorland. Forstw. Cbl. 80, 344–352.

Thaler M. 1991. Lichtabhängige Änderung cytoplasmatischer Ionenaktivitäten bei Eremosphaera viridis. Untersuchungen mit ionensensitiven Mikroelektroden. PhD thesis, Univ. of Würzburg.

Themlitz R. 1958. Ein Beitrag zur Düngung in forstlichen Pflanzgärten. Beobachtungen zum Kali-Kalk-Antagonismus bei jungen Nadelholzpflanzen. Kali-Briefe, Fachgebiet 6, Folge 1, 10.

212

Theophrastos Eresios. ≈ 300 BC. Περὶ φυτικῶν ιστορι ῶν α' - ϑ' (Inquiry into Plants). In Warmington EH, Ed. Loeb Classical Library, Vol. 70 and Vol. 79. Harvard University Press, Cambridge, Massachusetts (Greek with English translation by Hort AF).

Thomasius H. 1964. Über die Abhängigkeit des Wachstums der Waldbäume von Zeit und Umwelt. Dt. Akad. Landwirt Wiss Berlin. 66, 123–149.

Tischner R, Peuke A, Godbold DL, Feig R, Merg G, Hüttermann A. 1988. The effect of NO_2-fumigation on aseptically grown spruce seedlings. J. Plant Physiol. 133, 243–246.

Trendelenburg R. 1934. Untersuchungen über das Raumgewicht der Nadelhölzer. Thar. Forstl. Jahrb. 85, 649–747.

Trendelenburg R. 1936. Aufbau und Eigenschaften des Fichtenholzes und anderer Zellstoffhölzer. Papier-Fabrikant. 34, 389–435.

Tukey HB Jr, Tukey HB. 1969. The leaching of materials from leaves. In: Scharrer, Linser, Eds. Handbuch der Pflanzenernährung und Düngung. pp. 585–594. Springer, Wien, New York.

Tukey HB Jr. 1970. The leaching of substances from plants. Annu. Rev. Plant Physiol. 21, 305–324.

Ulrich B. 1972. Filterfunktion von Wald-Ökosystemen für organische Luftverun-reinigungen. In: Ellenberg H, Ed. Ökosystemforschung, Springer, Berlin.

Ulrich B. 1980. Die Wälder in Mitteleuropa: Meßergebnisse ihrer Umweltbelastung, Theorie über Gefährdung, Prognose ihrer Entwicklung. Allg. Forstzeitschr. 35, 1198–1202.

Ulrich B. 1981a. Eine ökosystemare Hypothese über die Ursachen des Tannensterbens (Abies alba Mill.). Forstw. Cbl. 100, 228–236.

Ulrich B. 1981b. Theoretische Betrachtung des Ionenkreislaufs in Waldökosystemen. Pflanzenern Bodenkunde. 144, 647–659.

Ulrich B, Mayer R, Khanna PH. 1979a. Deposition von Luftverunreinigungen und ihre Auswirkungen in Waldökosystemen im Solling. Schriften aus d Forstl Fak Univ. Göttingen und der Niedersächs Forstl Versuchsanstalt, Vol. 58.

Ulrich B, Mayer R, Khanna PH. 1979b. Fracht an chemischen Elementen in den Niederschlägen im Solling. Z. Pflanzenernähr Bodenkd. 142, 601–615.

Vanselow K. 1951. Krone und Zuwachs der Fichte in gleichaltrigen Reinbeständen. Forstw. Cbl. 70, 705–719.

Veloso D, Guyma RW, Oskarsson M, Veech RL. 1973. The concentrations of free and bound magnesium in rat tissues. J. Biol. Chem. 248, 4811–4819.

Vins B. 1961. Störung der Jahresringbildung durch Rauchschäden. Naturwiss. 48, 484–485.

Volz KR. 1974. Untersuchungen über die Eigenschaften der Rinde von Fichte, Kiefer und Buche und ihre Eignung als Rohstoff für Flachpreßplatten. PhD thesis, Univ. Göttingen, 227 pp.

Von Droste Zu Hülshoff B. 1969. Struktur und Biomasse eines Fichtenbestandes auf Grund einer Dimensionsanalyse an oberirdischen Baumorganen. PhD thesis, Univ. Munich.

Von Droste Zu Hülshoff B. 1970. Über die Kronenstruktur in einem älteren Fichtenbestand. Allg. Forst-Jagdzeitung. 141, 253–256.

Von Guttenberg A, Ed. 1915. Wachstum und Ertrag der Fichte im Hochgebirge. Wien, Leipzig.

Vuokila Y. 1968. Karsiminen ja kasvu (finn. m. engl. Zusammenfassung: Pruning and increment). Comm. Inst. Forest Fenn. 48, 1–138.

Vyskot M a Kolektiv. 1981. Ceskoslovenské pralesy (with Russian and English summary). Czechoslovak Virgin Forests, Praha.

Wachter H. 1985. Zur Lebensdauer der Fichtennadeln in einigen Waldgebieten Nordrhein-Westfalens. Forst- Holzwirt. 16, 420–425.

Walker DA. 1976. Regulatory mechanisms in photosynthetic carbon metabolism. In: Horecker BL, Ed. Current Topics Cellular Regulation, Vol. 11, pp. 203–241. Academic Press, New York.

Wandt R. 1937. Die Eigenschaften "stamm- und kronenbürtigen" Holzes. Mitt Forstwirtsch Forstwiss 8, 343–369.

Weast RC. 1989. Handbook of Chemistry and Physics, 70th edn. Chemical Rubber Co. Cleveland, CRC Press, Boca Raton, Florida.

Wedler M. 1986. Einfluß einer Magnesiumdüngung auf Photosynthese und Wasser- haushalt geschädigter und ungeschädigter Fichten an einem Magnesium-Mangel-standort im Fichtelgebirge. Diploma thesis, Univ. Würzburg.

Wehrmann J. 1959a. Die Mineralstoffernährung von Kiefernbeständen (Pinus sylvestris) in Bayern. Forstw. Cbl. 78, 129–149.

Wehrmann J. 1959b. Methodische Untersuchungen zur Durchführung von Nadel-analysen. Forstw. Cbl. 78, 77–79.

Wehrmann J. 1963. Möglichkeiten und Grenzen der Blattanalyse in der Forstwirtschaft. Landwirtschaftl Forschung. 16, 130–145.

Weigl J, Ziegler H. 1962. Die räumliche Verteilung von ^{35}S und die Art der markierten Verbindungen in Spinatblättern nach Begasung mit $^{35}SO_2$. Planta. 58, 435–447.

Weikert RM. 1986. Die photosynthetische Leistungsfähigkeit verschieden stark geschädigter Fichten – Vergleich zwischen einem "belasteten" und "unbelastetem" Standort. Diploma thesis, Univ. Würzburg.

Weikert RM, Wedler M, Lippert M, Schramel P, Lange O-L. 1989. Photosynthetic performance, chloroplast pigments, and mineral content of various needle age classes of spruce (Picea abies) with and without the new flush: an experimental approach for analysing forest decline phenomena. Trees. 3, 161–172.

Wellburn AR. 1990. Why are atmospheric oxides of nitrogen usually phytotoxic and not alternative fertilizers? Tansley Review No. 24, New Phytologist. 115, 395–429.

Welte E, Werner W. 1959. Über die physiologische Funktion des Mg in der Pflanze. Landwirtschaftl. Forschung. Vol. 13 (special issue).

Wentzel KF. 1971. Habitus-Änderung der Waldbäume durch Luftverunreinigung. Forstarchiv. 42, 164–172.

Wentzel KF. 1982. Ursachen des Waldsterbens in Mitteleuropa. Allg. Forstzeitung. 45, 1365–1368.

Wentzel KF. 1983. Höhenzuwachs-Analysen zur Diagnose von Immissionswirkungen. Allg. Forstzeitung. 38, 342.

Wentzel KF. 1985. Hypothesen und Theorien zum Waldsterben. Forstarchiv. 56, 51–56.

Werk KS, Oren R, Schulze E-D, Zimmermann R, Meyer J. 1988. Performance of two Picea abies L. (Karst.) stands at different stages of decline. III Canopy transpiration of green trees. Oecologia. 76, 519–524.

Wichmann W. 1976. Ermittlung von Grenzwerten der Pflanzenanalyse zur Kennzeichnung der Magnesium-Versorgung von Getreide in Schleswig-Holstein. PhD Thesis, Univ. Kiel.

Wiedemann E, Ed. 1923. Zuwachsrückgang und Wuchsstockungen der Fichte in den mittleren und unteren Höhenlagen der Sächsischen Staatsforsten, 2nd edn. University of Dresden, Forest Academy of Tharandt.

Wiedemann E, Ed. 1938. Ertragstafeln für Buche (1931), Fichte (1936), Douglasie (1937), Hannover.

Wiedemann E, 1939. Untersuchungen der Preußischen Versuchsanstalt über Ertragstafeln. Mitt Forstwirtschaft Forstwirtschaft Forstwiss. 10, 401–438.

Wiedemann E, Schober R. 1957. Ertragstafeln wichtiger Holzarten bei verschiedener Durchforstung, Hannover.

Wiersma JH. 1963. A new method of dealing with results of provenance tests. Silvae Genet. 12, 200–205.

Wild A. 1988. Licht als Streßfaktor bei Waldbäumen. Naturwiss. Rundschau. 41, 93–96.

Wilhelm E, Battino R, Wilcock RJ. 1977. Low-pressure solubility of gases in liquid water. Chem. Rev. 77, 219–262.

Wilhelmi T. 1968. Zur Zuwachsleistung der Fichte nach Grünästung. Forst- Holzwirt. 23, 157–161.

Winner WE, Mooney HA. 1980. Ecology of SO_2 resistance: I. Effects of fumigations on gas exchange of deciduous and evergreen shrubs. Oecologia (Berlin). 44, 290–295.

Winter K, Winkelmann E, Königer M. 1989. Zur Rolle der Photoinhibition bei der Wirkung von Luftschadstoffen auf den Photosyntheseapparat. In: Reuther M, Kirchner M, Rösel K, Projektgruppe Bayern zur Erforschung der Wirkung von Umwelt-schadstoffen (PBWU), Eds. Proceedings 1. Statusseminar der PBWU zum Forschungsprojekt "Waldschäden", GSF-Bericht 6/89, pp. 207–214. GSF-Forschungszentrum, München–Neuherberg.

Woodrow IE, Murphy DJ, Latzko E. 1984. Regulation of stromal sedoheptulose-1,7-bisphosphatase activity by pH and Mg^{2+} concentration. J. Biol. Chem. 259, 3791–3795.

Yin Z-H. 1990. Durch Licht oder Luftschadstoffe induzierte pH-Änderungen in verschiedenen Kompartimenten der Blätter höherer Pflanzen. PhD thesis, Univ. Würzburg, 123 pp.

Zech W. 1968. Kalkhaltige Böden als Nährsubstrat für Koniferen. PhD thesis, Ludwigs-Maximilians – Univ. Munich.

Zech W, Popp E. 1983. Magnesiummangel, einer der Gründe für das Fichten- und Tannensterben in NO-Bayern. Forstw. Cbl. 102, 50–55.

Zederbauer E. 1916. Beiträge zur Biologie der Bäume. II. Lebensdauer der Blätter. Cbl. ges Forstwes. 42, 339–341.

Zieger E, Ed. 1960. Technologie der Holzentrindung. Leipzig, Fachbuchverlag, 334 pp.

Ziegler H. 1982. Flüssigkeitsströme in Pflanzen. In: Hoppe W, Lohmann W, Markl H, Ziegler H, Eds.

Biophysik, 2nd edn. pp. 652–663. Springer, Berlin, Heidelberg, New York.

Zöttl H. 1983. Zur Frage der toxischen Wirkung von Aluminium auf Pflanzen. Allg. Forst. Zeitg. 38(8), 206–208.

Zöttl H. 1986. Possible causes of forest damage in Germany. CONCAWE Rep. 86/61, 55–70.

Zöttl H. 1990. Ernährung und Düngung der Fichte. Forstw. Cbl. 109, 130–137.

Zöttl H, Mies E. 1983. Nährelementversorgung und Schadstoffbelastung von Fichtenökosystemen im Südschwarzwald unter Immissionseinfluß. Mitt. Dtsch. Bodenkundl. Ges. 38, 429–434.

5
Influence of magnesium supply on tree growth

K. MAKKONEN-SPIECKER and H. SPIECKER

5.1. Introduction

After analyzing nutrient contents of the soils and needles in Norway spruce stands in the northern Black Forest in the 1960s, Moll (1965) noted that the magnesium content of the sites with well-growing stands on higher variegated sandstone was 40–300 times as high as that of the poorly growing stands. He described the role of magnesium 'as an essential activator within the nutrient cycles, especially in the roots' and drew attention to its stimulating effects on leaf-substance production.

Since the 1970s, magnesium deficiency has become very common in the Black Forest, and also in other parts of Germany and Europe as well. In the areas with 'montal yellowing', foliar nutrition status has been investigated extensively. There are several publications describing the effects of magnesium fertilization on soil and on magnesium uptake by trees. A summary of these findings has been compiled by Hüttl (1991), who also revealed long-lasting effects of magnesium fertilization. However, the influence of magnesium deficiency or magnesium supply on tree growth has been examined rather rarely. This paper summarizes the results of research devoted to this question. It does not include investigations into the visual or physiological effects of Mg deficiency on trees.

5.2. Effects of magnesium on the growth of young trees

Investigations into the effects of magnesium deficiency on the growth of young trees are mainly carried out as pot experiments with different soil substrates. There are not many field trials with small trees concerning this problem. With the exception of Lamb (1977), who found a negative correlation between foliar magnesium and growth (height and basal area) of young planted *Eucalyptus deglupta*, a fast-growing tree species, in Papua New Guinea, all other here mentioned studies showed neutral or positive growth effects of magnesium. Differing results could be due to the different tree ages and species used or could be of a methodological nature.

Ericsson (1993) observed in experiments with birch (*Betula pendula*) seedlings that their root growth, but not the shoot growth, was reduced when magnesium (and K or Mn) deficiency restricted plant development, whereas the opposite pattern was found when N, P or S acted as growth-limiting elements. Gonzáles Cascón *et al.* (1990) also described root growth-reducing effects of magnesium deficiency on silver fir (*Albies alba*) seedlings.

In a pot experiment, Austrian pine (*Pinus nigra*) and Scots pine (*P. sylvestris*)

R. F. Hüttl & W. Schaaf (eds): Magnesium Deficiency in Forest Ecosystems, 215–226.
© 1997 *Kluwer Academic Publishers. Printed in Great Britain.*

seedlings were grown for two years at various levels of magnesium in the form of $MgSO_4$ (Fiedler *et al.*, 1991): Magnesium treatment increased magnesium content of the needles but had little effect on growth. Dumbroff (1965) came to similar conclusions in a study of slash pine (*Pinus elliottii*) seedlings grown in sand culture: Differences of 6–78 ppm magnesium in the nutrient solution did not have any growth effect, but did affect the magnesium content of the needles.

Hunter *et al.* (1986) treated 6-year-old *Pinus radiata* stands in an area with extreme magnesium deficiency in New Zealand with a mixture of ground dolomite ($CaCO_3$*$MgCO_3$) and Epsom salts to supply 100 kg Mg ha^{-1}. Recovery in tree appearance and growth was slow, but 2 years after treatment a strong response was observed. Over a 5-year period, trees treated with dolomite and Epsom salts had a 66% greater height growth and a 45% greater diameter growth than untreated trees. Biomass analysis 5 years after fertilization showed further that treated trees had taken up 29 kg Mg ha^{-1} more than untreated trees.

Küppers *et al.* (1985) investigated the effect of various levels of magnesium fertilization (in the form of $MgSO_4$) on young Scots pines (*Pinus sylvestris*), growing on Mg-deficient soils derived from quaternary gravels in Northern Bavaria. There were differences in the annual height growth of the pines due to different magnesium content during the first 2 years (Figure 1) but after 7 years the total height of the pines and mean length increment of lateral branches were only reduced noticeably by very low magnesium needle contents of under 0.3 mg g^{-1} dry weight (Figures 2a and b). In contrast, the needle dry weight, and the length and the surface area of the needles were related to increases in foliar magnesium (Figure 2c and d).

The same tendency was observed by Strebel (1960) in a young Norway spruce (*Picea abies*) plantation. No reduction in growth could be registered at magnesium levels in needles of 0.3–0.5 mg g^{-1}; the spruces displayed a medium growth rate.

Sauter (1991) examined the influence of sulfate-, carbonate- and silicate-bound magnesium fertilizers on nutritional status and growth in a pot experiment with 3-year-old cloned Norway spruces and in two field experiments with 10–12-year-old spruces on sandy and loamy soil. In the pot experiment, the root biomass production was increased clearly by fertilization in carbonate form but not by silicate and sulfate-bound fertilization. This last result contradicts those of Gonzáles Cascón *et al.* (1990) showing root-growth promoting effects on silver fir (*Abies alba*) seedlings through fertilization with Mg sulfate, and the results of Schneider and Zech (1990/91) dealing with Norway spruces. These contradictions could be due to the differences in the magnesium contents of the needles, which were deficient in the plants investigated by Gonzáles Cascón *et al.* (1990) and Schneider and Zech (1990/91) but normal in the case of Sauter (1991).

Sauter (1991) could not discover any fertilizer effects on the height growth or the root collar diameter growth. The magnesium contents of the needles were all at normal levels but showed differences between treatments. In his field experiments, fertilization had the same effects but they were delayed and not significant.

In a pot experiment, Makkonen-Spiecker and Evers (1993) investigated the responses to water stress of young cloned Norway spruces (*Picea abies*) with

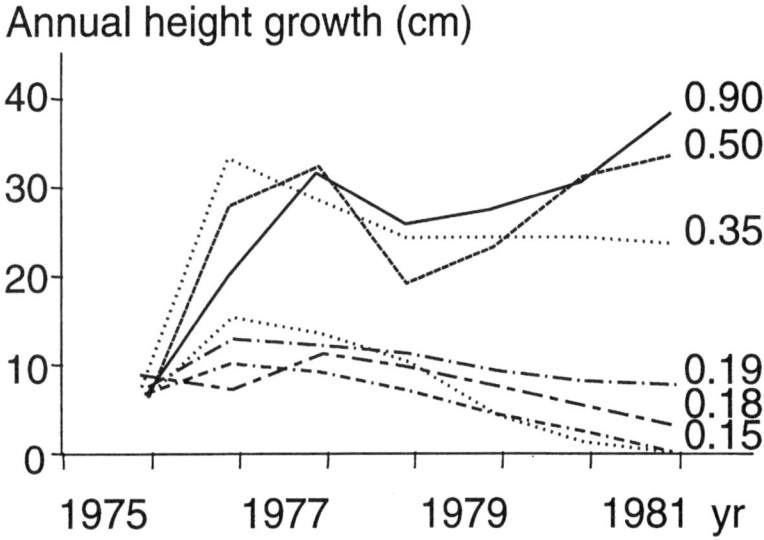

Figure 1. Annual height growth of pine seedlings at different Mg-nutrition levels. Mg contents of the needles (mg g^{-1}) were measured at the end of August 1981 (Küppers *et al.*, 1985, modified)

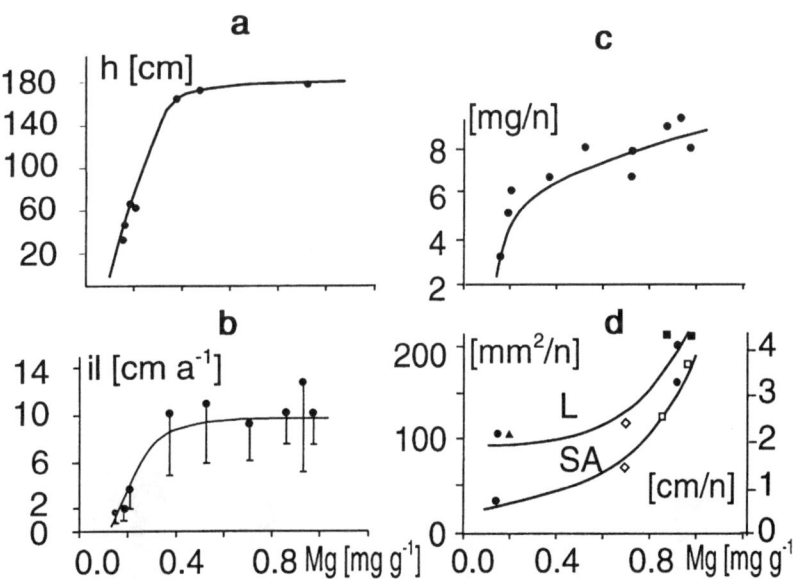

Figure 2. Total height (a), mean length increment of lateral branches (b), needle dry weight (c), needle surface area (SA, d) and needle length (L, d) of 7-year-old pines in relation to the Mg content of the needles (Küppers *et al.*, 1985, modified)

218

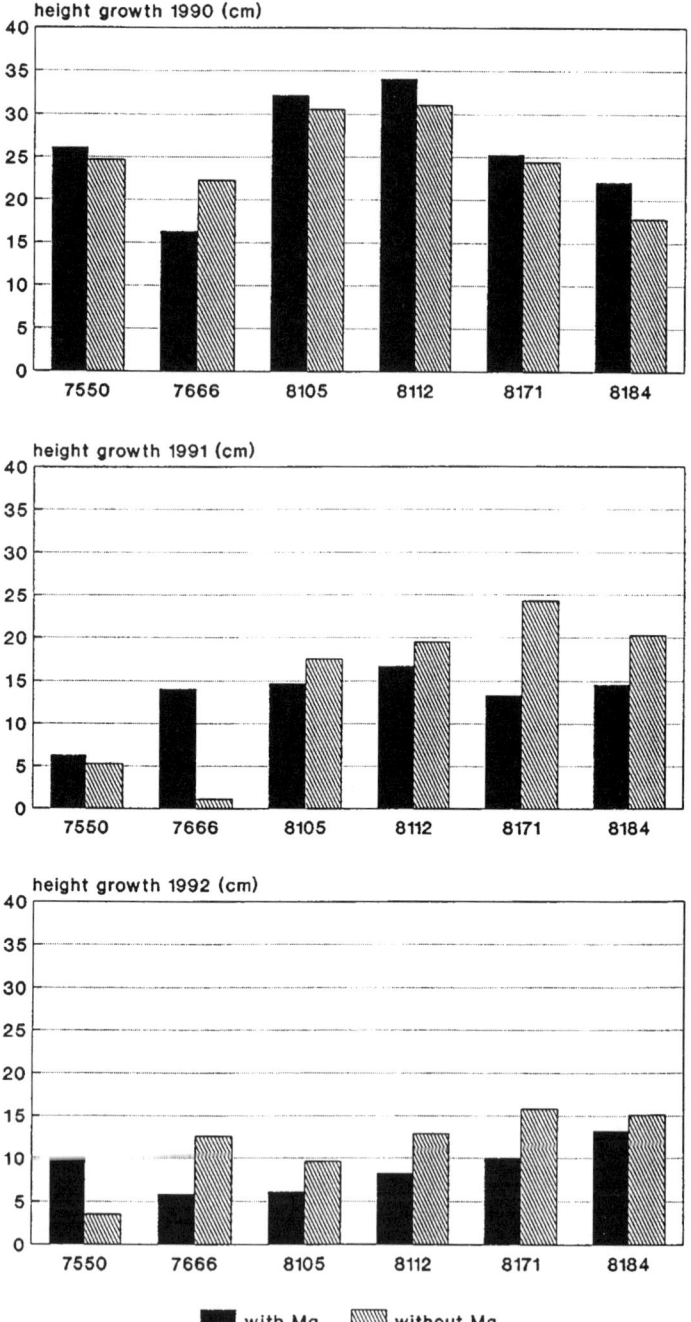

Figure 3. Mean height growth of young Norway spruces of six clones at different Mg-nutrition levels in the years 1990–1992 (Makkonen-Spiecker and Evers, 1993, modified)

Table 1. Magnesium contents in the youngest needles of young cloned Norway spruces treated with (+Mg) and without (−Mg) magnesium (according to Makkonen-Spiecker and Evers, 1993)

Clone	Mg (mg g⁻¹ dry weight)		
	1990	1991	1992
7550			
+Mg	0.90	2.94	1.78
−Mg	0.25	0.41	0.57
7666			
+Mg	0.99	2.49	1.29
−Mg	0.27	0.48	0.52
8105			
+Mg	0.59	1.51	1.00
−Mg	0.26	0.40	0.22
8112			
+Mg	0.73	1.59	1.08
−Mg	0.34	0.36	0.26
8171			
+Mg	0.77	1.84	1.05
−Mg	0.23	0.38	0.23
8184			
+Mg	0.51	1.38	0.88
−Mg	0.23	0.28	0.25

different magnesium nutrition. Figure 3 shows the terminal shoot growth of trees watered with and without magnesium treatment in the years 1990–1992; in Table 1, the respective magnesium contents of the needles are listed. The growth of the trees treated without magnesium and having very low magnesium contents, compared with the trees with magnesium treatment, was, with one exception, slightly reduced in 1990 but this was compensated for by a slightly increased growth in the following years. Altogether, the differences between the treatments were not significant. The growth differences were mainly due to the different nitrogen contents of the needles; low nitrogen content of the needles in 1992 reduced growth. The growth of lateral branches reacted similarly.

Furthermore, the experiments revealed that the number of lateral branches in the first whorl was reduced by magnesium deficiency (Figure 4), mainly because not all of the existing buds flashed. The number of needles was also reduced considerably (Figure 5), whereas the other investigated needle parameters (needle length and needle dry weight) were more influenced by genetics than by magnesium supply (Makkonen-Spiecker and Evers, 1993).

On the other hand, Uebel and Trillmich (1974), in experiments with one-year-old pine seedlings (*Pinus sylvestris*), found that magnesium fertilization increased the number of needles considerably, whereas the dry weight of the plants was reduced. The water content of the needles was increased through magnesium (and K) ferti-lization. The magnesium levels were not deficient in these investigations.

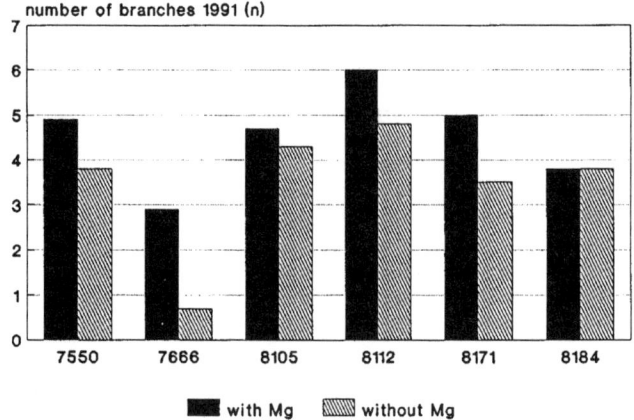

Figure 4. Mean number of branches in the first whorl 1991 of young Norway spruces of six clones at different Mg-nutrition levels (Makkonen-Spiecker and Evers, 1993, modified)

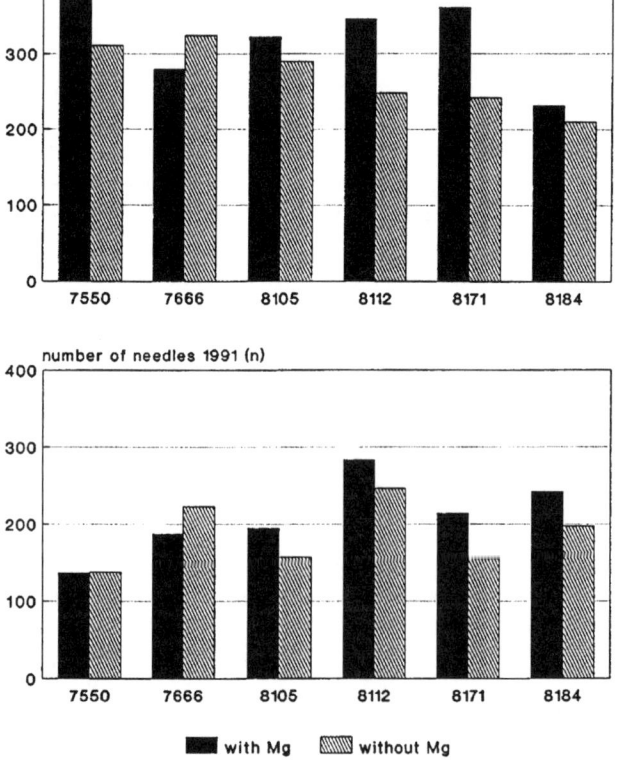

Figure 5. Mean number of needles on the branches (first whorl, 1990 and 1991) of young Norway spruces of six clones at different Mg-nutrition levels (Makkonen-Spiecker and Evers, 1993, modified)

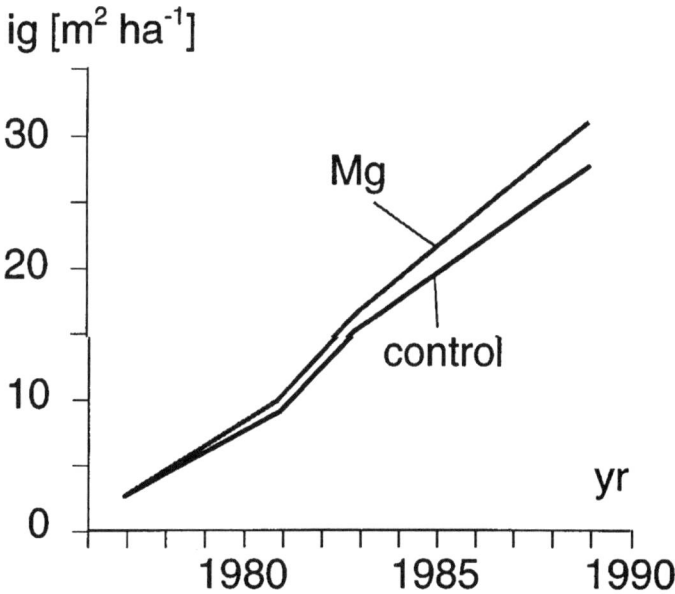

Figure 6. Basal area growth of Norway spruces with visible magnesium deficiency symptoms (control) and after treatment with Mg slag (Mg, 80 kg ha⁻¹, 1978) (Monchaux and Thivolle-Cazat, 1990/91, modified)

In a further experiment, the effects of several types of magnesium fertilizer on a 9-year-old plantation of Norway spruce with visible magnesium deficiency symptoms were investigated in the Vosgian mountains in France (Monchaux and Thivolle-Cazat, 1990/91): Five magnesium fertilizers (2 forms of slag, 2 forms of limestone, and Epsom salts) at two rates of magnesium (40 and 80 kg ha⁻¹) were compared with a control treatment. The highest positive result was obtained by treatment with basic Mg slag at 80 kg ha⁻¹; the circumference growth and the basal area growth were slightly but not significantly higher than the control (Figure 6). The appearance of the trees improved but there was no lasting effect on tree growth. These results were verified in another experiment at the same place (Barneoud and Jover, 1978, cited by Monchaux and Thivolle-Cazat, 1990/91).

5.3. Effects of magnesium fertilization on mature stands: field experiments

Fertilization with $MgSO_4$ and calcined dolomite increased root growth in the organic surface layer of Mg-deficient Norway spruces through improved magnesium nutrition of the shoot, which, according to Schneider and Zech (1990/91), implies that magnesium supply had been a growth-limiting factor because they had not observed such a growth reaction when spruces were sufficiently supplied with magnesium.

Altherr and Evers (1975) also concluded that an increase in growth of 48-year-old Norway spruces on medium variegated sandstone in the southern Odenwald 14 years after fertilizing could have been caused by magnesium, which had been an accompanying element of other fertilizer applications. They observed a similar development with 41-year-old beeches (*Fagus sylvatica*) in the same area (Altherr and Evers, 1974).

Will (1961) came to similar conclusions in experiments with magnesium-deficient pine seedlings growing in pumice soil nurseries in New Zealand: The application of a low rate of $(NH_4)_2SO_4$ increased the uptake of magnesium, as evidenced by increased foliage content and growth. This suggests that it could be the nutritional disharmony that reduces tree vigor rather than one single element, in this case magnesium (cf. Feger, 1993). This kind of nutritional disharmony might have been the reason why Hunger and Fielder (1965) found a positive correlation between magnesium nutrition and the growth rate of Norway spruce stands on some soils but not on others.

To confirm earlier observations on the effect of magnesium on beech growth, Altherr and Evers fertilized one plot which had received CaP-fertilization (1958 and 1962) and another unfertilized plot with $MgSO_4$ (1970 and 1972: 200 kg ha^{-1}); a third plot was left unfertilized (Altherr and Evers, 1977). Six years after the first fertilization, the diameter was increased by 11.4% (CaPMg) and 29.7% (Mg), the magnesium content being higher than in the control plot. This surprisingly high magnesium fertilizing effect was possibly due to the method used to investigate the growth parameters (cf. Spiecker, 1992).

The results of the investigations of Hildebrand and Schöpfer (1993) are exceptional as well: In a forest-decline study, they found that magnesium contents < 0.54 mg g^{-1} in the 3-year-old needles influenced the radial increment negatively, whereas even extreme low magnesium contents did not reduce height growth. On the other hand, in contrast to all earlier cited investigations, they did not find any correlation between the nitrogen nutrition of these trees and their growth. It is rather unlikely that variation of nitrogen content in the needles has no effect whereas variation of magnesium has an effect on growth (see also Katzensteiner and Glatzel, Chapter 6 in this book).

After examining 69 Norway spruce stands in Bavaria, Strebel (1960) could not find any correlation between the magnesium content in the needles and the class of height growth (Figure 7; Strebel, 1960). Schmidt-Vogt and Makkonen-Spiecker (1986) also found no correlation between growth rate of Norway spruce and magnesium contents in the needles which were medium to high.

The effects of magnesium fertilization in the form of $MgSO_4$ on the vitality and nutrition of European beech (*Fagus sylvatica*) with extreme magnesium deficiency (0.23 mg Mg g^{-1} needle dry weight) were investigated in the southern Black Forest (Ende and Zöttl, 1990/91): Shoot growth was stimulated within the first growth period after treatment, the best results being obtained with the lowest dose (500 kg ha^{-1}).

In the research project ARINUS (a watershed study of nutrient cycling in spruce eco-systems of the Black Forest; see Zöttl and Feger, 1990), the effects of magnesium

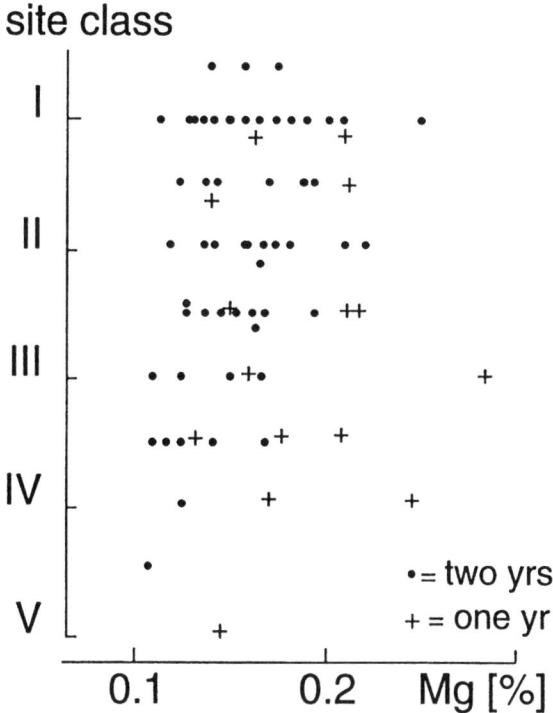

Figure 7. Relationship between the site class and the Mg content in the needles of Norway spruce stands in Bavaria (Strebel, 1960, modified)

fertilization on the growth of about 50-year-old spruces were investigated by Mäkinen (1996). The spruces were fertilized with 750 kg $MgSO_4$ (120 kg Mg ha^{-1}) in 1988, the foliar magnesium content being about 0.8 mg g^{-1} in the youngest needles and considerably lower in older needles (Feger *et al.*, 1990). Development of height growth and diameter growth are shown in Figures 8 and 9: Differences in height growth and diameter growth between the fertilized and non-fertilized trees are similar before and after treatment; no fertilization effect is evident. This confirms the earlier cited results that a growth effect occurs only when the Mg content is extremely low before fertilization.

Truong-Dinh-Phu (1979) could not register any magnesium fertilizing effect on white spruce (*Picea glauca*) in Canada either: In a 36-year-old plantation, 336 kg hydrated lime, 115 kg N and 93 kg K, with or without 22 kg Mg ha^{-1}, produced, over the following 10-year period, significantly higher basal area increments. Without N and K, dolomite or hydrated lime alone or together with Mg had no marked beneficial effects on the growth of white spruce 5–10 years after treatment (see also Spiecker, 1991).

224

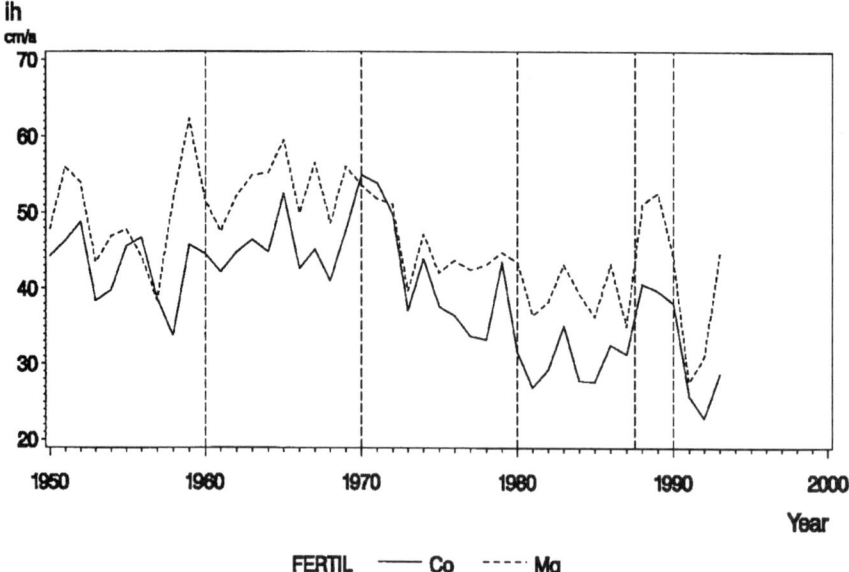

Figure 8. Development of the height increment on two Norway spruce plots in the Black Forest: control plot (Co) and Mg-fertilized (1988) plot (Mg) (Mäkinen, 1996)

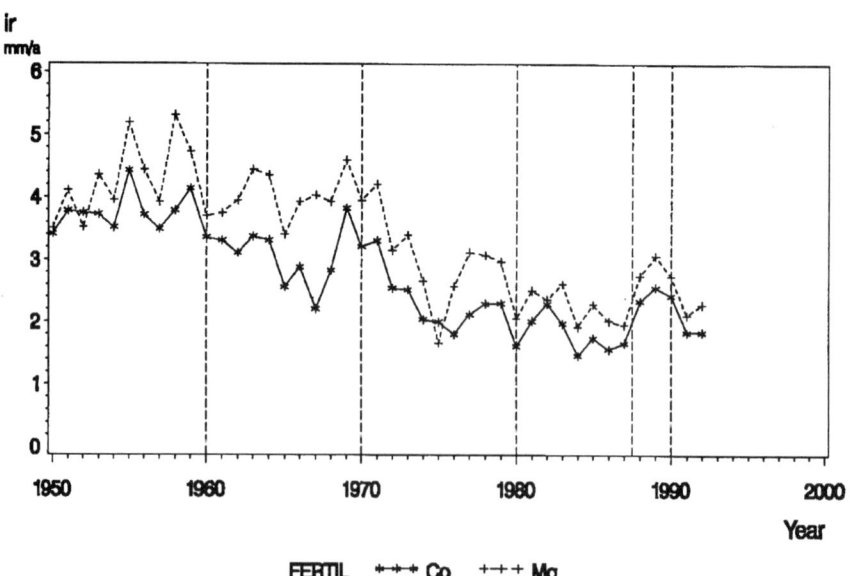

Figure 9. Development of the radial increment on two Norway spruce plots in the Black Forest: control plot (Co) and Mg-fertilized (1988) plot (Mg) (Mäkinen, 1996)

5.4. Conclusions

Reactions to magnesium supply are different between young and mature trees as well as between different species. It has become clear that these differences can be due to the different methods used: pot/field experiments, site differences and use of different fertilizers. This should be taken into account when results are compared. Investigations show that magnesium fertilization increases the magnesium content in the foliage of trees, associated with a disappearance of the visible magnesium-deficiency symptoms, but has little effect on growth (see also Katzensteiner and Glatzel, Chapter 6 in this book). The following conclusions can be drawn from research on the fertilizing effects of magnesium on tree growth:

1. It might be nutrient imbalances that reduce tree vigor rather than one single nutrient. However, whenever magnesium deficiency restricts plant development, root growth is affected first (see Raspe, Chapter 10 in this book).

2. Very low foliar magnesium content (below $0.3\,mg\,g^{-1}$ dry weight) can create a reduction of shoot growth. Mg deficiency in conifers may also cause a reduction in the number of needles.

3. Positive effects of magnesium fertilization on tree growth occur only when the foliar magnesium content before fertilization is extremely low.

4. To some extent, the effects of magnesium are influenced by genetics.

5.5. References

Altherr E, Evers FH. 1974. Unerwarteter Düngungserfolg bei Magnesiummangel in einem jungen Buchenbestand auf mittlerem Butsandstein des Odenwaldes. Allg. Forst- u. J.-Ztg. 145, 121–125.

Altherr E, Evers FH. 1975. Magnesium-Düngungseffekt in einem Fichtenbestand des Buntsandstein-Odenwaldes. Allg. Forst- u. J.-Ztg. 146, 217– 225.

Altherr E, Evers FH. 1977. Nachweis eines Magnesium-Düngungseffekts in einem Buchenbestand auf mittlerem Buntsandstein des Odenwaldes. Allg. Forst- u. J.-Ztg. 148, 45– 48.

Barneoud C, Jover L. 1978. Carance magnésienne sur épicéa. Commun. Annales de recherchers sylvicoles AFOCEL, 443– 466.

Dumbroff EB. 1965. Mineral nutrition of slash pine: effects of nitrogen, phosphorus, and magnesium and their interactions on growth and other physiological processes. Abstract of thesis, in Diss. Abstr. 25 (10).

Ende HP, Zöttl HW. 1990/91. Effects of magnesium fertilizer on the vitality and nutrition of a European beech (*Fagus sylvatica* L.) stand in the Southern Black Forest of West Germany. Water Air Soil Pollution. 54: 561–566.

Ericsson T. 1993. Growth and dry matter allocation of seedlings in relation to nutrient availability. CEC/IUFRO Symposium, Nutrient Uptake and Cycling in Forest Ecosystem, Halmstad, Sweden, June 7–10, 1993.

Feger K-H. 1993. Bedeutung von ökosysteminternen Umsätzen und Nutzungseingriffen für den Stoffhaushalt von Waldlandschaften. Freiburger Bodenk. Abhandl. 31, 237 S.

Feger K-H, Zöttl HW, Brahmer G. 1990. Projekt ARINUS: IV. Auswirkungen der Kieseritdüngung. Project ARINUS: IV. Effects of Kieserite Fertilization. KfK-PEF. 61, 21–35.

Fiedler HJ, Heinze M, Ngo-Van-Vien. 1991. Effect of lime and magnesium on growth and nutrition of Austrian pine and Scots pine seedlings. Mengen- und Spurenelemente: 11 Arbeitstagung, 12 and 13 Dezember 1991 in Leipzig (Eds. Anke, M. *et al.*), 172–179, Jena/Germany; Friedrich Schiller Universität.

González Cascón MR, Alcubilla M, Rehfuess KE. 1990. Wirkungen von Magnesium- und Calcium-

226

Sulfat und -Carbonat auf Sproß- und Wurzelentwicklung junger Weißtannen (Abies alba Mill.) im Topfversuch mit sauren Böden. Allg. Forst- u. J.-Ztg. 161, 21–28.

Hildebrand EE, Schöpfer W. 1993. Ergebnisse der Belastungsinventur Baden-Württemberg 1988. Mitt. Forstl. Versuchs- u. Forschungsanstalt Baden-Württ. 172, 157S.

Hunger W, Fiedler HJ. 1965. Düngungsdiagnosen für ältere Fichtenbestände des Erzgebirges und Vogtlandes. Arch. Forstwes. 14, 963–986.

Hunter IR, Prince JM, Graham JD, Nicholson GM. 1986. Growth and nutrition of Pinus radiata on rhyolitic tephra as affected by magnesium fertilizer. NZ J. Forestry Sci. 16:2, 152–165.

Hüttl R. 1991. Die Nährelementversorgung geschädigter Wälder in Europa und Nordamerika. Freiburger Bodenkundl. Abhandl. 28, 440S.

Küppers M, Zech W, Schulze E-D, Beck E. 1985. CO_2-Assimilation, Transpiration und Wachstum von Pinus silvestris L. bei unterschiedlicher Magnesium versorgung. Forstw. Cbl. 104, 23–36.

Lamb D. 1977. Relationships between growth and foliar nutrient concentrations in Eucalyptus deglupta. Plant Soil. 47:2, 495–508.

Mäkinen H. 1996. Wachstum von Fichten auf den ARINUS-Flächen. Auswirkungen der Revitalisierungsdüngungen und witterungsbedingte Zuwachsvariation im Südschwarzwald. Diss. Freiburg 1996, 141 p.

Makkonen-Spiecker K, Evers FH. 1993. Untersuchungen zur Reaktionsweise junger Klonfichten (Picea abies (L.) Karst.) auf Trockenstreß und Magnesiummangel. KfK-PEF. 114, 81 S.

Moll W. 1965. Nährstoffversorgung von Fichtenbeständen im Nordschwarzwald. Schriftenr. Forstl. Abt. Albert-Ludwigs-Univ. Freiburg, Bd. 4, Forstl. Hochschulwoche. 1964. 252–265.

Monchaux P, Thivolle-Cazat A. 1990/91. Correction of magnesium deficiency in a young stand of Norway spruce. Water Air Soil Pollution. 54: 595–605.

Sauter U. 1991. Versuche zur Wirkung von sulfatisch, carbonatisch und silikatisch gebundenem Magnesium auf Ernährungszustand und Wachstum junger Fichten, chemischen Bodenzustand und Sickerwasserbefrachtung. (Effect of sulfate-, carbonate- and silicate-bound magnesium on nutritional status and growth of young Norway spruce, soil chemistry and percolation-water loading). Forstl. Forschungsber. München. No. 114, 442 p.

Schmidt-Vogt H, Makkonen-Spiecker K. 1986. Unterschiede im Nährelementspiegel von Nadeln verschiedener Fichtenherkünfte (Picea abies (L.) Karst.). Allg. Forst- u. J.-Ztg. 157, 8, 145–152.

Schneider BU, Zech W. 1990/91. The influence of Mg fertilization on growth and mineral contents of fine roots in (Picea abies (Karst) L.) stands at different stages of decline in NE-Bavaria. Water Air Soil Pollution. 54: 469–476.

Spiecker H. 1991. Liming, nitrogen and phosphorus fertilization and annual volume increment of Norway spruce stands on long-term permanent plots in Southwestern Germany. Fertilizer Res. 27: 87–93.

Spiecker H. 1992. Which trees represent stand growth? In: Bartholin ThS, Berglund BE, Eckstein D, Schweingruber FH, Eggertsson O, Eds. Tree rings and environment. Proceedings of the International Dendrochronological Symposium, Ystad, South Sweden, 3–9 Sept. 1990, Lundqua Report 34, 1992, 308–312.

Strebel O. 1960. Mineralstoffernährung und Wuchsleistung von Fichtenbeständen (Picea abies) in Bayern. Forstw. Cbl. 79, 17–42.

Truong-Dinh-Phu. 1979. Effet du chaulage sur la croissance de l'épinette blanche plantée à Grand Mère. Can. J. Forest Res. 9: 3, 305–310.

Uebel E, Trillmich H-D. 1974. Ergebnisse eines Gefäßversuches zur Prüfung des Einflusses von N-, K- und Mg-Gaben zu einem Sandbodenmaterial auf Bodeleben und Wachstum von Kiefernsämlingen. Pedobiologia Bd. 14, 41–50.

Will GM. 1961. Magnesium deficiency in pine seedlings growing in pumice soil nurseries. NZ J. Agric. Res. 4 (1/2), 151–160.

Zöttl HW, Feger K-H. 1990. Waldökosystemforschung in den ARINUS-Versuchsgebieten Schluchsee und Villingen. Forest ecosystem research at the ARINUS sites Schluchsee and Villingen. KfK/PEF. 61, 11–20.

6
Causes of magnesium deficiency in forest ecosystems

K. KATZENSTEINER and G. GLATZEL

6.1. Introduction

Magnesium is an element essential both for plants and animals and magnesium deficiency has been demonstrated to cause health problems even in humans. Magnesium deficiency in forest ecosystems is usually studied on the basis of individual forest trees or stands of trees. This approach is very limited and can only be attributed to methodological difficulties in studying magnesium nutrition in soil micro-organisms.

The flowchart (Figure 1) shows a number of factors and pathways recognized as causes of the development of Mg deficiencies in forest trees.

Magnesium deficiency in vascular plants is the result of uptake rates which are insufficient for the magnesium demand caused by plant growth and magnesium losses due to leaching or shedding of biomass. Uptake rates of magnesium which are too low to maintain physiologically optimal magnesium levels in the plant cells can be caused by a number of factors. Low magnesium concentrations in the soil due to a parent material low in magnesium content, or as a result of prolonged or intensive leaching, must be considered a primary cause of magnesium deficiency in forest ecosystems. An important step is magnesium uptake into the plant. Aside from physical and chemical constraints in ion exchange and transport in the soil solution, competition of soil organisms with mycorrhiza and tree roots for magnesium may decrease magnesium availability to the tree. Most important is the functioning of the root system. Limitation of spatial access to the soil volume and poor efficiency due to root damage from abiotic and biotic agents can limit plant uptake despite satisfactory magnesium concentrations in the soil solution.

Magnesium losses from the tree symplast, which exceed uptake rates, are another important cause of magnesium deficiency. Leaching of magnesium from the canopy, losses in flowering and fructification, litter fall, parasitism by fungi or higher plants and the action of insects and other herbivores must be considered in the magnesium balance of a tree.

A very important, yet frequently overlooked, aspect is the relation of growth to uptake. Uptake of nutrients must follow growth, which has genetically determined limits, in order not to develop deficiencies (Chapin, 1988). The interplay between nutrient levels in the cells and restriction of growth differs among elements. It is rather tight for nitrogen, showing a good correlation between supply and growth, and less tight for magnesium. While plants growing on nutrient-poor bogs have evolved to stay small even in an environment which provides ample light, water and

227

R. F. Hüttl & W. Schaaf (eds): Magnesium Deficiency in Forest Ecosystems, 227–251.
© 1997 *Kluwer Academic Publishers. Printed in Great Britain.*

228

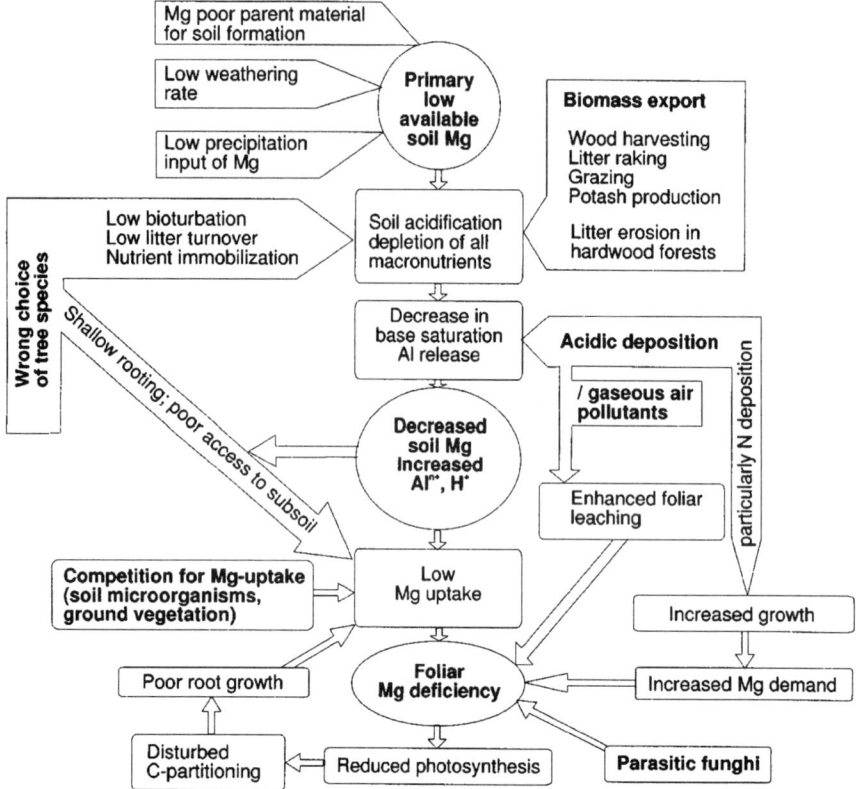

Figure 1. Causes of Mg deficiencies in forest ecosystems (modified after Roberts *et al.*, 1989)

warmth and thus avoid deficiencies, many forest tree species, in particular conifers, easily outgrow their magnesium supply and develop conspicuous deficiency symptoms. Thus, on the very same site, the forester's choice of tree species or even cultivar can decide the development of magnesium deficiency.

6.2. Primary magnesium deficiency

Magnesium content of parent materials for soil formation varies widely (Scheffer-Schachtschabel, 1989), depending on the amount of magnesium-containing minerals, such as muscovite (0–2.4% MgO), biotite (0.3–28% MgO), pyroxene (10–18% MgO), amphibole (3–25% MgO) and olivine (38–47% MgO). Table 1 lists the magnesium contents of common rock types as cited in the literature. Sandstone, quartzite, granites and rhyolites are rather low in magnesium.

Weathering liberates magnesium from the crystal lattice of minerals and makes it available to plants. Physical and chemical weathering depend both on climate and parent mineral. Weathering stability of silicate materials can be ranked in the order

Table 1. Average chemical composition of common rock types – percentage contents

Plutonic rocks

	Peridotites ($n=23$)	Gabbros ($n=160$)	Diorites ($n=50$)	Monzonites ($n=46$)	Granodirorites ($n=137$)	Granites ($n=172$)
SiO_2	43.54	48.36	51.86	55.36	66.88	72.08
Al_2O_3	3.99	16.84	16.4	16.58	15.66	13.86
Fe_2O_3	2.51	2.55	2.73	2.57	1.33	0.86
FeO	9.84	7.92	6.97	4.58	2.59	1.67
MgO	34.02	8.06	6.12	3.67	1.57	0.52
CaO	3.46	11.07	8.4	6.76	3.56	1.33
K_2O	0.25	0.56	1.33	4.68	3.07	5.46
Al:Mg	0.05	0.83	1.06	1.79	3.94	10.54

Volcanic rocks

	Basalts ($n=137$)	Andesites ($n=49$)	Dacites ($n=50$)	Rhyolites ($n=22$)
SiO_2	50.83	54.2	63.58	73.66
Al_2O_3	14.07	17.17	16.67	13.45
Fe_2O_3	2.88	3.48	2.24	1.25
FeO	9.05	5.49	3	0.75
MgO	6.34	4.36	2.12	0.32
CaO	10.42	7.92	5.53	1.13
K_2O	0.82	1.11	1.4	5.35
Al:Mg	0.88	1.56	3.11	16.61

Sedimentary rocks

	Buntsandstein** (colored sandstone)	Sandstones ($n=253$)	Greywackes ($n=61$)	Shales ($n=277$)	Tillites ($n=68$)	Limestones ($n=93$)	Dolomites** ($n=95$)
SiO_2	85.46	78.7	66.7	58.9	58.9	6.9	1.76
Al_2O_3	6.74	4.8	13.5	16.7	15.9	1.7	0.39
Fe_2O_3	1.94	1.1	1.6	2.8	3.3	0.98	0.29
FeO	0.14	0.3	3.5	3.7	3.7	1.3	
MgO	0.45	1.2	2.1	2.6	3.3	0.97	21.75
CaO	0.58	5.5	2.5	2.2	3.2	47.6	29.02
K_2O	2.89	1.3	2	3.6	3.9	0.57	0.28
Al:Mg	5.99	1.58	2.54	2.54	1.9	0.69	0.01

Wedepohl, 1969; **Fiedler and Hunger, 1970

olivine < garnet < pyroxene < amphibole < biotite < plagioclase < orthoclase < muscovite < quartz (Scheffer-Schachtschabel, 1989). In cold climates, physical fragmentation of rocks leads to increased weathering of micas as compared with feldspars. Coarse-grained granites show fast physical but slow chemical weathering, and resulting soils have a high content of primary silicate minerals as compared with soils on fine-grained basic magmatites, where small grains are susceptible to intensive chemical weathering. The increase in exchange sites associated with clay formation leads to a better buffering capacity and resistance against acidification. Mineralogical proper-ties of the rocks and chemical and physical environment determine which clay minerals are formed and thus cation exchange capacity and magnesium supply. Tomlinson (1990/91) points out that the molar ratios of Al to Ca and Mg as well as Ca+Mg to K of the parent material indicate which element is susceptible to depletion by weathering and leaching in various rock types. In general, granites are rather susceptible to depletion of magnesium.

Forest soils which provide very little magnesium to plants can be found on parent material with low magnesium content and high resistance to weathering. Magnesium deficiency in pine, spruce and beech stands on quartz sands, sandstone (e.g. Buntsandstein in Germany) or glacial outwash (Adirondacs) has been repeatedly described since the turn of the century (Altherr and Evers, 1974; Baule and Fricker, 1967; Becker-Dillingen, 1940; Möller, 1904; Stone 1953). It can be assumed that magnesium deficiency in sensitive tree species can arise on such sites without the intervention of man. In such cases the term 'primary magnesium deficiency' seems appropriate.

Antagonisms or competition of magnesium with other elements such as calcium for absorption and uptake sites (Marschner, 1986; Mengel, 1968) can inhibit sufficient uptake of magnesium, causing deficiency. This effect has to be considered in fertilization. In natural forest ecosystems, it is rare because soils on calcareous bedrock usually contain sufficient magnesium to prevent antagonism. In very degraded soils, increased availability of manganese, aluminium or iron may cause antagonism with magnesium (Marschner, 1991).

Erosion of biomass on exposed ridges or hilltops, where storms carry away leaf litter, can aggravate the effect of poor magnesium supply in the soil. Differences in soil properties, caused by lateral nutrient transfer in a hardwood system in the Vienna Woods (Halmschlager, 1987), may serve as an example. Erosion of leaf litter (narrow C/N ratio) and retention of heavier woody debris (wide C/N ratio) on a wind-exposed ridge led to marked decrease of nutrients on the windward side and to extreme accumulation on the leeward side (Figure 2).

In cool and wet climates, formation of peat soils can be another reason for low magnesium availability in the root zone, because the mineral soil is isolated from the roots by waterlogged layers of peat and plant nutrients are diluted in the accumulating peat. Yet, despite low magnesium concentrations in the soil solution, woody plants rarely develop magnesium-deficiency symptoms on peat because concurrent low nitrogen availability checks growth.

Altogether, primary magnesium deficiency does not seem to be very common under normal deposition regimes of air constituents (i.e. in the absence of acid deposition from natural or anthropogenic sources). Magnesium imported from dust and lateral biomass transfer, and magnesium retention in living organisms tend to build up and maintain magnesium levels sufficient for ecosystem function on all but the very poorest sites.

Deposition is a major source of magnesium in forest ecosystems. In coastal regions, the magnesium cycle is dominated by sea salt spray. A comparison of magnesium input in a north–south gradient from N Germany to Austria shows the different deposition patterns of magnesium (Table 2).

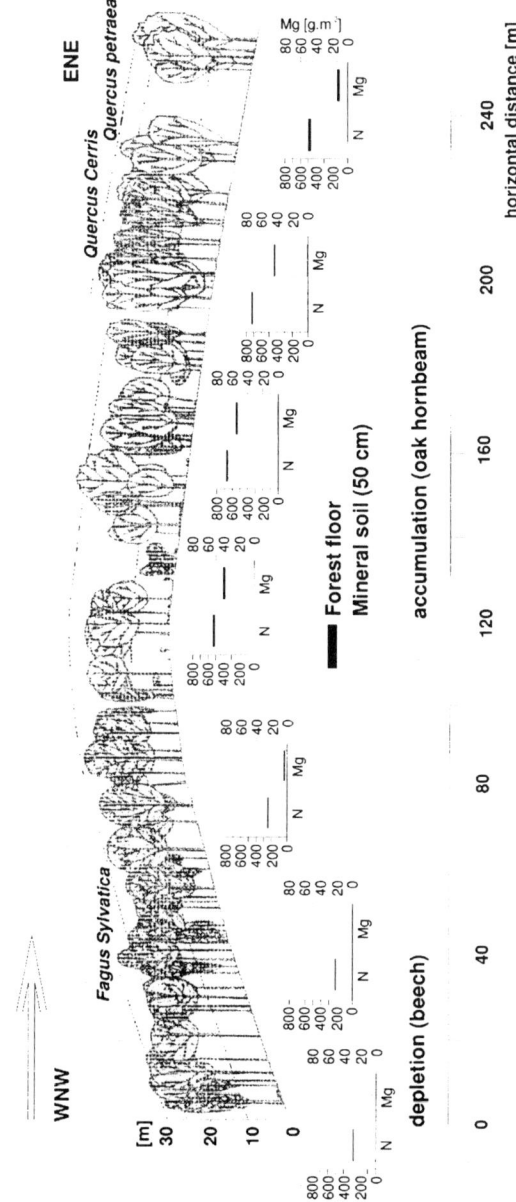

Figure 2. Effect of lateral biomass transport on nutrient stores of soils in the Vienna Woods. N: total; Mg: total in forest floor, BaCl₂ - exchangeable in mineral soil; Halmschlager, 1987

Table 2. Nutrient input and output in European forest ecosystems

Location, dates and reference	H	NH$_4$-N	NO$_3$-N	SO$_4$-S	K	Ca	Mg
Wingst, 1984–1986, Büttner (1992)							
Bulk deposition	0.2–0.3	11.3–14.0	5.9–6.3	12.9–16.20	3.7–3.9	3.7–6.4	2.1–3.6
Throughfall Norway spruce	0.0–0.4	33.4–50.6	7.2–10.3	45.4–61.6	25.7–29.8	11.3–19.9	6.1–9.2
Output	0.1–0.2	0.2–2.6	9.1–15.4	32.2–57.6	1.8–6.4	3.5–8.8	2.4–4.2
Solling, 1969–1985, Matzner (1988)							
Bulk deposition	0.6–1.3	9.0–15.6	6.1–11.0	19.6–27.2	2.4–5.6	6.5–21.8	1.3–3.9
Throughfall + stemflow beech	0.79–1.72	9.0–24.8	8.3–20.9	39.1–66.0	18.1–40.0	18.4–32.1	3.2–4.9
Output beech	0.2–1.0	0.0–0.5	0.1–4.3	14.8–79.6	2.0–7.4	4.0–34.5	1.7–5.8
Throughfall Norway spruce	2.1–5.0	8.7–19.4	10.2–20.4	54.0–107.6	20.0–41.6	19.6–41.6	3.0–6.1
Output Norway spruce	0.2–0.8	0.1–0.3	2.5–28.7	22.2–225.0	1.9–7.7	7.4–31.1	2.7–11.2
Harz, 1983, Hauhs (1985)							
Bulk deposition	0.6–0.9	8.7–12.5	6.8–10.4	15.4–22.4	2.7–5.3	6.4–12.4	1.2–2.3
Throughfall Norway spruce	0.6–2.1	2.5–21.5	1.9–22.8	13.4–52.1	0.9–11.7	1.7–27.4	0.4–5.1
Output Norway spruce	0.5	<0.4	2.4	34.5	3.7	8.7	2.8
Black Forest 1988–1990, Feger (1993)							
Bulk deposition	0.1–0.3	4.7–7.3	4.0–5.8	6.7–8.9	1.2–3.2	4.3–5.3	0.7–1.0
Throughfall Norway spruce	0.3	2.1–5.1	4.1–7.4	10.0–12.3	10.7–14.2	6.5–8.0	1.2–1.6
Output Norway spruce	0.1–0.2	<0.1–0.3	<0.1–6.9	13.1–34.8	3.6–11.5	1.1–12.1	0.9–3.5
Bavarian Forest 1980–1984, Hüeser and Rehfuess (1988)							
Bulk deposition	0.3–0.5	9.3–16.5	6.4–13.3	20.5–34.8	2.8–14.9	5.7–12.5	1.1–2.2
Throughfall Norway spruce	1.0–1.2	7.0–16.4	9.1–19.5	39.1–58.2	10.5–31.6	10.5–18.4	2.4–3.6
Bohemian Forest 1986–1987, Glatzel et al. (1988)							
Bulk deposition	0.03–0.2	8.7–10.2	5.4–7.3	14.4–15.2	2.5–2.6	6.5–7.4	0.7–1.3
Throughfall Norway spruce	0.3–0.7	9.0–13.0	5.0–11.0	18.0–30.0	11.0–16.0	7.0–12.0	1.0–2.0
Northern calcareous Alps, 1992–1995, Berger and Glatzel (1992)							
Bulk deposition	0.3	9.4	7.7	10.0	0.3	8.7	2.5
Throughfall Norway spruce	0.2–0.3	4.0–4.4	7.1–8.0	7.8–7.9	7.6–1.1	13.3–15.6	4.5–4.6
Gleinalm (Styria), 1992–1993 (unpublished)							
Bulk deposition	0.1	4.0–4.8	3.6	9.3–11.2	5.5–6.5	5.6–7.0	1.4–1.5
Throughfall Norway spruce	0.2	2.0–3.0	3.0–3.3	12.0–17.6	7.7–7.8	8.4–9.6	4.4–5.2

Bold: areas with pronounced Mg-deficiencies; all units are kg ha^{-1} y^{-1}

While Mg input with bulk deposition can reach values of $4 \, \text{kg} \, \text{Mg} \, \text{ha}^{-1} \text{y}^{-1}$, in N Germany (Solling area), the deposition rates in southern Germany and Austria lie between 0.5 and $2.2 \, \text{kg} \, \text{ha}^{-1} \text{y}^{-1}$ with exception of the northern limestone Alps. Dust from dolomitic rock formations and scree rises bulk deposition to $2.5 \, \text{kg} \, \text{ha}^{-1} \text{y}^{-1}$ in this region. In coastal areas, dry and occult deposition of sea salt aerosols can add considerable amounts of magnesium. In the Wingst area in NW Germany, total magnesium deposition is between 4.0 and $7.6 \, \text{kg} \, \text{ha}^{-1} \text{y}^{-1}$ (Büttner, 1992). It has to be noted that the reduction of dust emissions from industrial sources has reduced deposition of basic cations, especially Ca and Mg during recent decades. In the Bavarian station, Grafrath, for example, calcium and magnesium bulk deposition declined by nearly 50% during the decade 1970–1980 (Hüser and Rehfuess, 1988). Novel agricultural methods, which try to avoid having bare soil, may further decrease dust deposition, while the use of dolomitic grit on winter roads may increase magnesium deposition in the vicinity of major roads and cities.

6.3. Secondary magnesium deficiency

Secondary magnesium deficiency may be defined as a condition wherein net loss of magnesium leads to a decline of the pool of magnesium available to plants in a forest ecosystem and causes deficiency response in trees. Critical dilution of magnesium in plant tissues caused by excessive growth of introduced species as well as effects of parasitic fungi in individual trees will be included here.

6.3.1. Harvesting of biomass from forest ecosystems

In the early history of mankind, forests were primarily sources of food. This pattern changed when permanent settlements and seafaring evolved and wood was needed as a construction material. After metals came into use, wood continued to be the only source of energy for thermal use for a long time. But the use of forests as sources of nitrogen and minerals for food production has prevailed into this century in central Europe and is still common in many parts of the world. As early as 1840, Justus von Liebig had pointed out that food requires reasonable amounts of nitrogen and minerals while wood consists of little more than carbon, oxygen and hydrogen. In pre-industrial and early industrial times, forests were also sources of raw materials for many technical processes: bark was harvested for tanning of leather, and biomass was burned to yield potash, just to mention two widespread usages.

The cycles of magnesium and other nutrient elements were interrupted by these land use practices because most were extractive, with very little or no return to the forest ecosystem. Based on historic documents and modern data on magnesium content in various biomass fractions, the impact of historic land use on magnesium relationships in forest ecosystems can be estimated.

6.3.2. Historic land use

An excellent source of detailed description and quantitative data on land use in the Alps during the 19th century is Wessely's book *Die Österreichischen Alpenländer und ihre Forste (The Austrian Alpine Provinces and their Forests)* published in 1853. Litter raking and lopping of branches in forest stands yielded 1.47 million cart loads annually in Tyrol and Vorarlberg from a total forest area of 614000 ha. Assuming that only 60% of the forest area was accessible, this amounts to an annual harvest of about 27 m^3 per hectare or about 2.5–3.5 tons of dry biomass per hectare. Assuming a magnesium content of 0.1%, the annual export of magnesium was of the order of about 3 kg per hectare. Kreutzer (1972) calculated nutrient losses due to litter raking without lopping for Scotch pine stands (*Pinus sylvestris*) in Bavaria. He estimated that about 100–150 kg magnesium were lost per hectare over a 100-year period.

When forests were eventually felled for timber, agriculture was practised on the clear-cut areas for some time. According to Wessely, the following practice was widespread. Immediately after felling and burning of slash, field crops (oats, rye, turnips) were cultivated for a period of 2–3 years on the clear-cut area. In the third year, tree seeds (usually spruce) were sown between field crops. The area was then grazed for up to ten years by cattle, sheep or goats until the herbaceous vegetation was shaded out by the regenerated tree stand. Intensive herding was usually prescribed by the land owner to prevent excessive damage to the regeneration. When the stands reached an age of 80–100 years, they were open enough to allow some grazing. It is difficult to estimate nutrient exports from intermittent cultivation and grazing. Based on yield data by Rösch (1992) and data on metabolism and assimilation of food by domestic animals (Kirchgeßner, 1987), nutrient loss due to grazing for a 100-year rotation period can be estimated as 100–1400 kg N, 17–210 kg K, 70–900 kg Ca, 7–90 kg P, and 2–30 kg Mg ha^{-1} 100y^{-1}.

The effects of litter raking and grazing on nutrient stores are reflected in today's soil chemistry, even though these practices were abandoned long ago. The Tyrolean Forest Soil Inventory (Stöhr *et al.*, 1988) shows that soils where litter raking or grazing is documented have an average base saturation of 30% while soils from less intensively used forests have a base saturation of more than 42%. Because better sites were preferred for litter raking, grazing and lopping, these figures actually underestimate the impact.

Additional nutrient losses resulted from potash and charcoal production. These losses are at least of the same magnitude as losses resulting from whole tree harvesting as compared with harvesting of stem wood only.

Because all these land use practices removed an excess of basic cations, they decreased the acid-neutralizing capacity (ANC) of the soil and, in turn, created conditions for retarded mineralization of organic matter and increased leaching of metals due to formation of soluble organic complexes. Figure 3 (Glatzel, 1991) compares the contribution of various processes to the consumption of ANC. The tremendous effect of litter raking compared with other influences is obvious.

ACIDIFICATION OF FOREST SOILS IN AUSTRIA

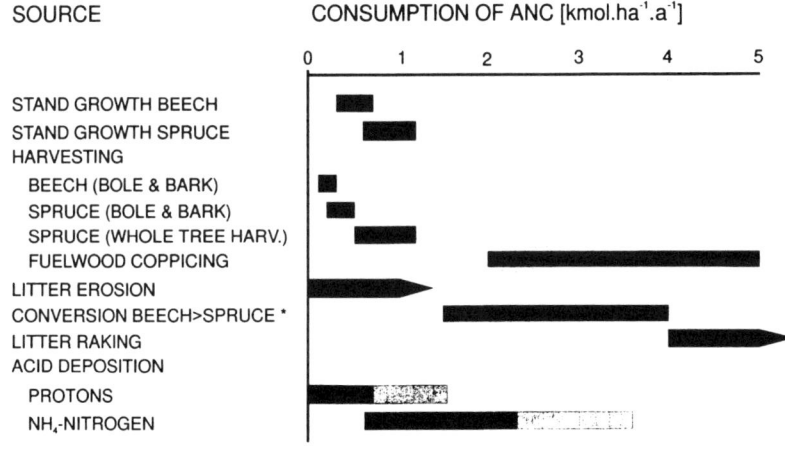

Figure 3. Contribution of various processes to the depletion of acid buffering capacity of forest soils in Austria. Figures for the conversion of beech to spruce and for litter raking are peak rates which decline as the system reaches equilibrium or declines in productivity (Glatzel, 1991)

6.3.3. Contemporary forestry

Since the last century, commercial forestry has tried to ban litter raking, grazing and other destructive practices from forests. Except for grazing in alpine areas, biomass harvesting in forests is nowadays limited to timber and fuel wood.

A large database exists on the magnesium content of the wood and bark of common timber trees. Multiplying these figures by the amount of wood extracted gives estimates of magnesium losses from harvesting in contemporary forestry. As an example, Figure 4 shows nutrient stores in a chronological sequence of Norway spruce stands (*Picea abies*) of Upper Austria (site index ~12) (Bauer, 1989). After canopy closure, magnesium content in the foliage remains constant, while accumulation in wood and bark increases with stand age and almost equals the exchangeable pool in the mineral soil of this site at the age of 100 y.

The harvest method, i.e. whole-tree harvesting versus conventional harvesting of boles only, has a significant influence on nutrient losses. Whole tree harvesting, especially with high-lead cable logging, may double the loss of ANC compared with conventional stem-only harvesting (Figure 3).

Because of the difference between species, the difference in site-specific yield class and the difference in harvesting methods, magnesium export with harvested wood may vary widely. Magnesium losses of between 0.3 and 0.8 kg ha^{-1} y^{-1} can be assumed to be a fair estimate for central European spruce forests, when boles with

236

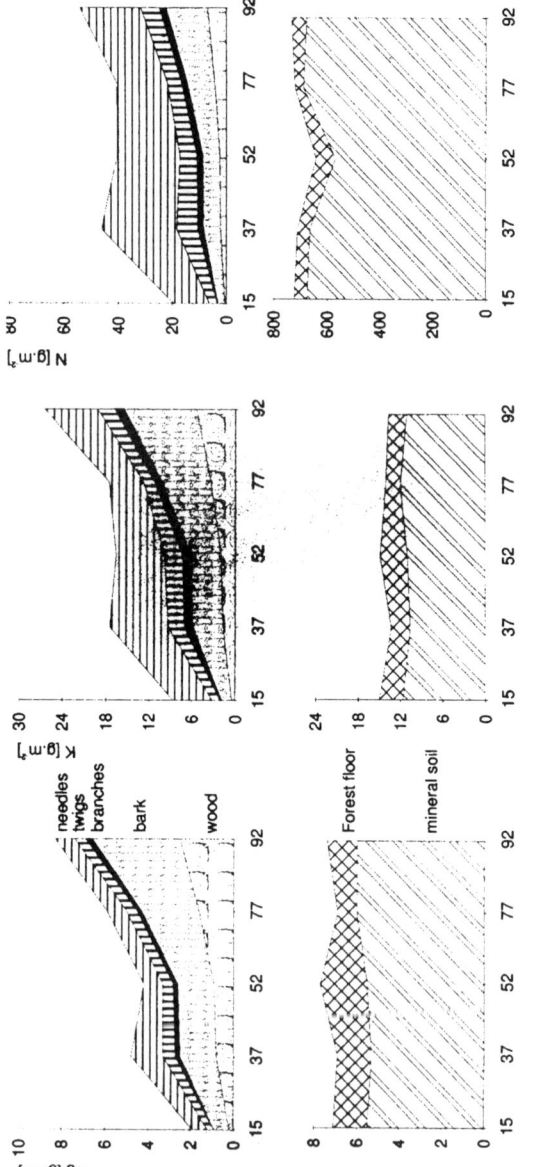

Figure 4. Nutrient stores in a chronosequence of Norway spruce stands. N: total in above ground biomass, forest floor and mineral soil; K, Mg: total in above-ground biomass, BaCl$_2$-exchangeable in mineral soil; Bauer, 1989

bark are harvested. This almost amounts to an order of magnitude less than the combined losses through land use at early industrial times.

Clear-cutting entails additional losses of magnesium because mineralization of organic matter continues while uptake by tree roots is absent. Obviously, losses are most severe on sites with high precipitation and sandy, water-permeable soils of poor absorption capacity. Research by Mann et al. (1988) and Tiedemann et al. (1988) in the NW USA shows that post-harvest nutrient losses are elevated for the first 3 years after felling but are considered to be minor in relation to harvest removals for most elements. Ca and K are considered possible exceptions by Mann et al. (1988). Unfor-tunately, magnesium was not investigated in their studies. Hornbeck et al. (1990) estimate total harvesting losses of magnesium at three forests in New England to be between 35 and 52 kg ha^{-1}. The increase in the hydrological output during the first three years after felling was 12–16 kg ha^{-1}, which amounts to 20–30% of the total harvesting loss.

A special case is the conversion of deciduous broadleaf forests to coniferous forests. Coniferous forests tie up nutrients and acid neutralizing capacity in the much denser evergreen canopy and in the humus layer, originating from the slow decomposition of conifer needles which evolved to resist the attack of herbivores and fungi in the canopy for several years. Consequently, the mineral soil is depleted of nutrients and acidifies (Chen and Glatzel, 1988; Glatzel, 1991; Nilsson et al., 1982). The fixation of magnesium in the biomass, and the dilution caused by the rapid growth of conifers, may trigger deficiency on poor sites (Miller, 1995).

Kreutzer et al. (1986) found an increased leaching of NO$_3^-$ and base cations under a spruce plantation following a hardwood forest compared with an adjacent beech stand. They attributed this difference to decomposition of the humus pool in soil horizons currently not rooted by spruce.

6.3.4. Magnesium losses due to enhanced leaching

Since the beginning of the 1980s, cases of magnesium deficiency have been described in areas where granite, gneiss, phyllite, rhyolite or other silicatic rocks form the parent material (Fichtelgebirge: Zech and Popp, 1983; Black Forest: Zöttl and Hüttl, 1986; Bavarian Forest: Bosch et al., 1983; Bohemian Forest: Glatzel et al., 1987; Vosges: Becker, 1991; Appalachians: McLaughlin et al., 1990; Hendershot et al., 1992; Ardennes: Weissen et al., 1990). Increased leaching due to acid deposition is thought to have contributed significantly to this situation.

Leaching of magnesium by percolating soil water depends on the availability of mobile anions and organic complexing agents. Uptake of magnesium from lower soil horizons by plant roots and subsequent deposition on the soil surface by litter fall from the canopy, as well as bioturbation, are counteracting mechanisms.

In neutral or weakly acidic soils, bicarbonate ions from respiratory processes in the soil are abundant and leaching of readily available calcium and magnesium ions is pronounced. The degree of hardness of ground and spring water is largely the result of this process.

In moderate-to-medium-level acidic soils, essential mineral elements are retained in amounts necessary for ecosystem functioning because mobile anions are scarce and competition of organisms for uptake is fierce. If mobile anions are introduced into such a system or generated inside the system from precursors, leaching losses will occur. Loss of sinks for uptake has the same effect as demonstrated in many small watershed studies by clear-cutting or herbicide treatment (Bormann and Likens, 1979). An interesting case of internal anion generation occurs in stands of trees capable of dinitrogen fixation with the aid of symbiotic micro-organisms. Excess nitrogen compounds are oxidized to nitric acid which acidifies the soil and drives leaching of base cations (Van Miegroet and Cole, 1985).

In pre-industrial times, the prevalent mobile anion deposited from the atmosphere was sulfate, originating from the oxidation of reduced sulfur compounds emitted by the oceans or sulfur dioxide from volcanism (Birks, 1994; Officer et al., 1987). Impact of asteroids, according to model calculations, created sufficient temperatures to oxidize atmospheric nitrogen and may have caused catastrophes of nitrate acidification and leaching in the earth's history (Griffis and Chapman, 1990). When coal and oil became the dominant sources of thermal energy, emission of sulfur dioxide increased dramatically, and sulfate from man-made sources became an important constituent of atmospheric deposition, causing acidification of forest soils (Rusnov, 1919). The invention of industrial processes for conversion of atmospheric dinitrogen into ammonia and nitric acid for use as fertilizer for crops and raw material for military explosives increased nitrogen availability in the atmosphere. The oxidation of traces of dinitrogen in many combustion processes contributes further, and in many areas nitrate from deposition or from the oxidation of deposited ammonium nowadays exceeds sulfate as the predominant mobile anion in forest soils. Ulrich (1983) developed a comprehensive theory on the effects of acid deposition on central European forest ecosystems.

Table 2 lists data on deposition and leaching for selected central European forest ecosystems. Deposition of free H^+ covers a range from 0 up to $1.3 \, kmol \, ha^{-1} \, a^{-1}$ in the open, while $5 \, kmol \, ha^{-1} \, a^{-1}$ has been recorded in throughfall of Norway spruce because of increased dry and droplet deposition in the canopy. The forester's choice of tree species is a major factor influencing deposition patterns. In beech stands of the Solling, H^+ fluxes below the canopy were less than 50% of the amounts measured in an adjacent Norway spruce stand.

A number of forest soil inventories have proved that a significant decline in pH values has taken place throughout central and northern Europe during recent decades (i.e. Falkengren-Grerup, 1987; Hallbäcken and Tamm, 1986; Stöhr, 1984). This decline is not only attributable to the direct input of H^+ to forests. High rates of nitrogen deposition potentially increase acidification in forest ecosystems. Van Breemen et al. (1982, 1983) described the acidifying effect of $(NH_4)_2SO_4$ in throughfall of forest ecosystems due to H^+ generation via uptake or nitrification of NH_4^+. Especially in areas with intensive animal husbandry, acidification due to deposition of reduced N compounds may exceed acidification due to H^+ load. Additionally 'climatic acidification pulses' due to humus disintegration and increased nitrification rates following warm dry years can add to the total acid load of the soil. Such acidification pulses with increased leaching of NO_3^- accompanied by magnesium and other cations have been frequently described (e.g. in stands of the Austrian Bohemian Forest, Kazda and Katzensteiner, 1993).

Adsorption energy of basic cations on the exchange complex of soils ranks from $Na < K < Mg < Ca < Al$. Potassium is retained better in forest ecosystems than this ranking suggests because it is stored in illites and 3-layer minerals due to specific fixation and is actively taken up by plants and thus, kept in the biochemical cycle. Therefore, magnesium, the next easily-exchangeable element, which shows no specific absorption on mineral soil particles, is the most easily leachable macro-nutrient. Despite this, Matzner (1988) found no reduction of the extremely low exchangeable magnesium pools in the Solling sites in the period 1969–1985, though the total magnesium balance of the soil was negative. Magnesium is released at rates of $3.3 – 4.4 \, kg \, ha^{-1} \, a^{-1}$ from silicate weathering while exchangeable magnesium contents remain at a minimum value sustained by biochemical cycling. Hildebrand (1986) also found almost no change in base cation content of heavily acidified soils at a base saturation of 5% in the past 15 years, while H^+ and Fe^{3+} increased and Al^{3+} decreased significantly due to acidic deposition. However, long-term acidification experiments by Abrahamsen et al. (1994) showed increased leaching of calcium and magnesium from soils due to increased acid load; base saturation and soil pH were severely reduced. Reduction of magnesium in the soil has had the most critical effects on nutrition and even on growth of trees.

6.4. Magnesium supply and tree nutrition

For agricultural soils, a content of exchangeable magnesium in excess of $20 – 30 \, mg \, kg^{-1}$ seems to be sufficient for crops (Scheffer-Schachtschabel, 1989). Liu and Trüby (1989) found severe magnesium deficiency in Norway spruce (*Picea abies*) below values of $20 \, mg \, kg^{-1}$ Mg (NH_4Cl-exchangeable; content of topsoil 0 –10 cm weighted by 0.7, subsoil 20–30 cm weighted by 0.3) in forest soils of SW Germany. Meiwes et al. (1984) classified soils where magnesium saturation was less than 1–2% of CEC_{eff} as sensitive to acidification. Katzensteiner (1994) found a relationship between pH value, Mg, K, Ca and Mn in % of the CEC_{eff} and the magnesium content of Norway spruce needles (Equation 1). The correlation between absolute amounts of exchangeable magnesium in the soils and Mg content of needles was weak.

$$Mg_{current \, needles(\% \, of \, DM)} = -0.076 + 0.045 pH(H_2O) + 0.078 \, Mg_{\% \, of \, the \, exch. \, complex} \quad (1)$$

$$-0.011 \, K_{\% \, of \, the \, exch. \, c.} -0.011 \, Mn_{\% of \, the \, exch. \, c.} -0.011 \, Ca_{\% \, of \, the \, exch. \, c.}$$

$$r^2 = 0.58^{***}, \, n = 50$$

Kaupenjohann et al. (1987) has shown that magnesium concentration in percolate from undisturbed soil was closely correlated with magnesium content of needles, while correlation with extracts from sieved soil samples was poor. This supports the theory of dominant base cation depletion of aggregate surfaces (Hildebrand, 1986) in biologically inactive soils.

6.4.1. Interaction with other elements: The role of nitrogen

Interaction between nutrient elements is well known from fertilizer trials in agriculture (Marschner, 1986; Mengel, 1968). Ion competition for absorption sites, uptake and enzyme action in plants is usually called antagonism. In the case of borderline magnesium supply, liming with pure calcium carbonate can induce magnesium deficiency. The same is true for fertilizing with potassium salts without addition of magnesium. For this reason, many fertilizers are multi-element fertilizers which contain both potassium and magnesium. For the liming of forests with low magnesium supply, dolomitic lime is usually recommended.

Interactions with nitrogen are much more complex. Input of nitrogen compounds influences nutrition of forests directly as a nutrient and indirectly via soil acidification. Nitrogen is the major constituent of the biological systems in cells and thus tightly coupled with growth. Additional nitrogen enhances growth and dilutes elements with less-close coupling to critical levels (Glatzel et al., 1987; Katzensteiner et al., 1992; Roelofs et al., 1985; Schulze, 1989). Houdijk and Roelofs (1993) found a significantly negative correlation between magnesium content of conifer needles and ammonium and ammonium/cation ratios in deposition in a regional study in the Netherlands. Investigations of Nebe (1991) and Hippelli and Branse (1992) showed a clear trend towards increasing N contents and decreasing magnesium content in needles of Scotch pine and Norway spruce stands in former Eastern Germany.

The effect of widespread nitrogen deposition on forest ecosystems is probably one of the causes of the observed growth enhancement of Central European forests in recent decades (Kauppi et al., 1992; Eckmüllner, 1988; Kenk and Fischer, 1988; Sterba, 1992). Increased growth consequently leads to an increased demand for base cations. When this demand is not met, deficiency results (Schulze, 1989).

6.4.2. Direct effects of gaseous pollutants and acid rain on nutritional status of trees

One of the early hypotheses on causes of the new type of forest decline in Germany was the effect of O_3 and photo-oxidants on leaf and needle surfaces of forest trees (i.e. Prinz et al., 1982). A number of controlled experiments validated this hypothesis. Krause et al. (1985) described a significant increase of base cation leaching from Norway spruce seedlings treated with O_3 in combination with acidic mist. Guderian et al. (1985) induces chlorotic symptoms on Norway spruce seedlings with combined fumigation of O_3 and SO_2 especially in plants with insufficient Mg and Ca supply but could not decrease magnesium content in the plants; on the contrary, magnesium content in the needles even increased. Bosch et al., 1986 detected only a minor effect of O_3 but a significant effect of acidic mist and a combination of both on leaching of calcium and magnesium, especially of unfertilized seedlings. While magnesium concentration in needles from magnesium-fertilized trees showed no treatment effect – probably due to increased turnover rates – unfertilized trees had a lower magnesium content in the combined O_3 + acidic mist treatment. Skeffington and Roberts (1985), however, were not able to produce needle yellowing or magnesium deficiency in Scots pine and Norway spruce with O_3 and acidic mist. Again, needle concentrations increased due to O_3 treatment. Seufert (1988) found enhanced leaching of base cations from leaves and needles in open-top chamber experiments after SO_2 fumigation of Norway spruce (*Picea abies*), European beech (*Betula pubescens*) and silver fir (*Abies alba*) combined with an acid rain treatment (pH 4.0). Combined fumigation with SO_2 and O_3 further increased base cation leaching. The author concludes that O_3 improves oxidation of dissolved SO_2 on the needle surface. Simple O_3 fumigation did not increase leaching of calcium and magnesium. The increased leaching of base cations at pH 4.0 compared with pH 5 coincides with findings of the other authors. To sum up, it seems that direct effects of gaseous air pollutants and acidic mist on the canopy surface only plays a role in magnesium deficiency in combination with other stress factors, especially poor supply of base cations from the soil.

6.4.3. *Parasitism*

Parasitic fungi or parasitic vascular plants may influence mineral nutrition of infected trees. Singh and Bhurke (1974) and Tomiczek (1990) observed a significant decrease of foliar magnesium content in trees suffering from infections with *Armellaria* spp. and other fungi which cause root or stem rot. Infected trees exhibited yellowing and symptoms typical for magnesium deficiency, while the colour of the foliage of nearby healthy trees was normal (Tomiczek, 1990). Another important factor may be the infection of twigs and needles with pathogenic fungi. Such infections are frequently observed in areas where Mg deficiencies are described (Neumüller, 1994). The interrelationship between fungal attack and Mg nutrition is not yet understood.

Also, mistletoes may deprive trees of nutrients which are usually retranslocated from the foliage into the branches because they act as terminal sinks for minerals transported both via the xylem and the phloem (Glatzel, 1982).

Neglecting these effects may lead to erroneous interpretation of ecosystem nutrient status from foliar analysis.

6.5. FIW-case study, Bohemian Forest

To visualize the effect of historic land use and current input of air pollutants on magnesium nutrition of forests, results from the FIW-case study of the Schlaegl Monastery in the Bohemian Forest/Upper Austria (Führer and Neuhuber, 1994) will be stated briefly. The factors recognized as causing agents for the development of Mg deficiencies in Figure 1 and the preceding paragraphs fit well for the forests of this study area.

242

6.5.1. The study area (Katzensteiner and Glatzel, 1994)

The FIW-research area is located in the Austrian part of the Bohemian Forest ($48°39'$N/$14°3'$E – $48°46'$N/$13°47'$E). Elevations range between 500 and 1379 m. Bedrock materials are Mg- and Ca-poor granites and gneisses (i.e. paragneiss: 0.55% CaO, 2.52% MgO; orthogneiss: 2.4% CaO, 1.4% MgO; Eisgarner-Granite: 0.91% CaO, 0.60% MgO; Sulzberg-Granite: 0.61% CaO, 0.60% MgO). Soils had developed mainly on prepleistocene periglacial deposits with low contents of easily weatherable primary silicates. Soil types range from histosols and gleysols to cambisols and podzols.

Average annual precipitation is rather high (880–1100 mm on valley sites, increasing with elevation) and air temperature is low (6.7–7.8°C). Fog water deposition contributes to the total precipitation especially on exposed ridges. Leaching of elements and formation of peat soils is frequent under these conditions.

Hence only a little arable land surrounds the forests and no major dust emissions occur; in the vicinity, in the input of Mg- and Ca-containing dust is currently low (bulk deposition of Ca: 0.7–6.4 kg ha^{-1}y^{-1}; Mg: 0.7–1.3 kg ha^{-1}y^{-1}).

Due to the parent material, soil genesis, and climate, soil Mg supply on a number of sites was probably already low before the intervention of man. Primarily low available soil Mg contents can be postulated.

6.5.2. Historic land use (Scholl and Katzensteiner, 1994)

Biomass exports The colonization of the forest area started in the 12th century. In the first phase of settlement, wood demand was covered by clearing of woodland. Crop fields and meadows were established, even at high altitudes. Hunting and gathering of mushrooms and fruits were minor influences on the virgin forests. Until the end of the 15th century, forests were pushed back to current limits. It can be assumed that, until 1500, no major nutritional disturbances due to human impact occurred in the remaining forests. It is even likely that input of ash from the clearing and subsequent burning of woodland led to an enrichment of base cations.

From 1500 to 1767, uncontrolled extensive forestry took place. Only valuable timber in accessible areas close to settlements was used in a form of selective tree harvesting. Species composition of remaining stands was hardly changed and nutrient losses with timber were small. Pasturing of cattle in forests was an additional extensive land use.

A very disturbing influence on the forests, however, was the practice of litter raking. Litter raking was very intensive, especially in the areas close to settlements. As the rental of the rights for litter raking was part of the income of the foresters, no limits were set by them. Mostly the whole forest floor was removed. Unfortunately, no records about the amounts withdrawn are available for this time. In the calculation of nutrient losses, removal of nutrients stored currently on the forest floor of 60–100-year-old forest stands (Katzensteiner, 1992) was chosen as a lower limit. A monastery report on the forest status, published in 1864, attributed a number of forest health problems (failing of natural regeneration of broadleaves trees, bark beetle infestations (water stress?)) to litter raking. From this date onwards, litter raking was restricted to old growth stands, the intervals of litter raking were extended to three years and only small amounts of litter were allowed to be removed.

The second major influence on forests was the large-scale exploitation from 1767 onwards, when fuelwood was produced for the city of Vienna 250 km distant. A technical masterpiece at this time was the Schwarzenberg logging flume with a length of 34 km, crossing the land divide to the next river, where the wood was drifted down to the Danube and transported to Vienna. The remaining wood of small dimensions and slash were used for potash production for the local glass industry. The nutrient export from this multiple land use, therefore, was comparable to whole-tree harvesting.

The use of sites as meadows for some decades also led to severe nutrient depletion. Even if an extremely low yield of 1000 kg ha^{-1} of hay (extensively used *Nardus strica* meadows (Klapp, 1971)) with a low Mg content of 1.5 mg g^{-1} dry matter is calculated, the annual Mg export rate is at least double the export from conventional forestry.

The nutrient losses due to these biomass exports dramatically reduced the nutrient pools in the ecosystem (Figure 5). In the site survey from 1965–1970 (Institut f. Forstliche Standortslehre, 1971), a large part of the mapped forest, especially in easily accessible areas, was classified as 'degraded'.

Choice of tree species As natural regeneration of the huge clearings (20 ha and more) failed, cultural measures were applied from 1838 onwards. Oat was grown together with Norway spruce and harvested as a field crop until the spruce seedlings were established. As a reaction to the bad condition of the forest – failing regeneration and dominance of young age classes – seedbeds were established in the forests from 1875 onwards and clearings were replanted mainly with Norway spruce. Also a number of meadows were afforested at this time. These meadows were greatly acidified due to the high rate of nutrient export over the centuries. Establishment of regeneration at such sites was a severe problem. Since then, a sustained forestry has been established and, from 1930 onwards, silvicultural methods favoring natural regeneration have been applied.

The natural forest vegetation types in the Bohemian Forest are mainly *Abies–Fagus*, *Abies* and partly *Picea* associations. During the period of the exploitation, at least 30% of the produced fuel wood consisted of hardwood. The ratio of silver fir to Norway spruce in the softwoods was not reported. A considerable amount of silver fir can be expected from the natural vegetation types. According to the site map, the stocking goal for the investigated district would be 89% conifer–deciduous tree mixtures, 4% spruce–fir mixtures and 7% pure Norway spruce stands. Currently, pure Norway spruce stands cover 46% of the area – despite efforts to promote mixed stands.

All the negative effects of Norway spruce monocultures on nutrient cycling described previously, like shallow rooting, low biological activity in soils and

244

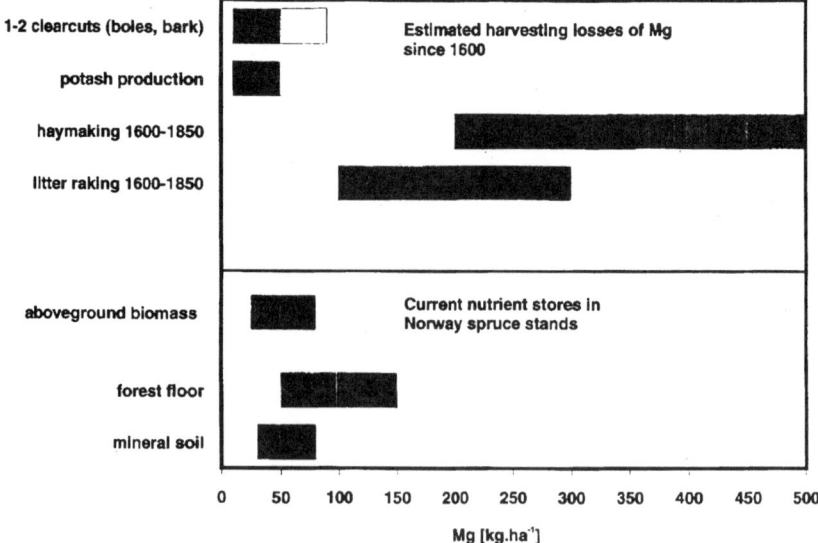

Figure 5. Nutrient pools and estimated nutrient exports from selected Norway spruce stands in the Bohemian Forest

therefore nutrient immobilization in the forest floor (Figure 6), can be observed. Subsequent nutrient losses with seepage water due to humus decomposition in deeper soil horizons are probable. The effects of nutrient accumulation in growing trees and forest floor on soil acidification have been described (Chen and Glatzel, 1988; Nilsson *et al.*, 1982). High nutrient demand in the early stages of growth, when the main foliage biomass is formed, may already lead to nutrient stress (Miller, 1995). Stands at higher elevations in the Bohemian Forest are frequently damaged by wet snowfall and hoar-frosts. The loss of the upper part of the crown depletes the nutrient pool of the tree. The tree has to form new foliage without retranslocating nutrients from old needles and may again run into nutrient stress. Miller (1995) gives examples of transient nutrient stress in forests due to loss of foliage by thinning or insect defoliation. In the same context, the loss of older needles as a reaction to drought may be seen. A number of dry growth periods since the early eighties have probably contributed to the observed needle losses in the investigated area as postulated as a main trigger for the development of forest decline symptoms throughout central Europe (Rehfuess, 1989). Drought and disturbed C-partitioning may lead to damage to the fine root system, resulting in limited capability for nutrient uptake.

6.5.3. Input of air pollutants

The input of air pollutants in the Bohemian Forest depends on the exposure of the sites to the prevailing winds. Deposition measurements along a NW–SW transect showed that, on the lee side of the mountain range, fluxes of air pollutants are much lower than on the plateau and on the SW slope (throughfall in 1987: 14kg ha^{-1} N_{total} and 19kg ha^{-1} SO_4^2-S on the lee side; 24kg ha^{-1} N_{total} and 30kg ha^{-1} SO_4^2-S on the west slope. Besides the acidification due to mineral acids, the input of nitrogen exceeds by far the demand of the spruce stands in the area. Enhanced nitrification, especially after warm dry periods, leads to nitrate leaching from the subsoil accompanied by cations, on the sites with highest N-input rates. Additionally, high Al concentrations in the soil solution occur (Kazda and Katzensteiner, 1993). An unbalanced ratio of N to Mg and Ca in Norway spruce needles from the most polluted regions of the Bohemian Forest also indicate nitrogen saturation (Katzensteiner *et al.*, 1992).

Biochemical, ultrastructural and physiological investigations (Grill *et al.*, 1994) suggest an additional effect of gaseous air pollutants directly on the needles. While ambient SO_2 and NO_x air concentrations are rather low throughout the year, apart from episodic SO_2 peaks in later winter, O_3 concentrations in the summer period are very high (average for the vegetation periods 1985–1988: 0.081mg m^{-3}; maximal value over half an hour: 0.309mg m^{-3}) and may contribute to the observed enhanced formation of toxic radicals, low chlorophyl content and structural changes of thylakoid membranes. These disturbances in the photosynthetic active part of the plant contributes to an unbalanced C-partitioning of the tree, reducing root growth and thus uptake of water and nutrients.

6.5.4. *Soil status, tree nutrition and vigor*

The investigation of the mineral nutrient status of 50 Norway spruce stands situated along two elevational transects (Katzensteiner, 1992) clearly shows the Ca and Mg depletion of soils at higher altitudes. The areas with lowest exchangeable soil Mg contents are the same as the areas which receive the highest input of air pollutants. As an example, in Figure 6, the chemical soil condition from an afforested meadow at the Baerenstein plateau (close to site S2) is compared with that of a soil form from a NE-exposed site, close to the valley bottom. As can be seen from these figures, both soils are heavily acidified in the upper horizons. Probably as a result of its former use as a meadow and the reduced humus mineralization due to temporary waterlogging, the soil from the plateau, developed on a granite substrate, shows a rather high humus content and thus cation-exchange capacity. The exchange sites are, however, occupied mainly by Al and Fe ions. Almost no base cations remain in the lower soil horizons. Mn is already totally depleted. On this site, a large amount of nutrients is immobilized in the forest floor. Most of the fine roots are concentrated in the forest floor and the upper 20 cm of the mineral soil. Stands developed on this and similar sites frequently show reduced tree vigor and poor Ca and Mg nutrition. The soil from the valley, developed on paragneiss, also shows acidified upper horizons due to heavy litter raking in the past. Cation exchange capacity is less than in the soil from the plateau but, in the subsoil, there is a sufficient satur-

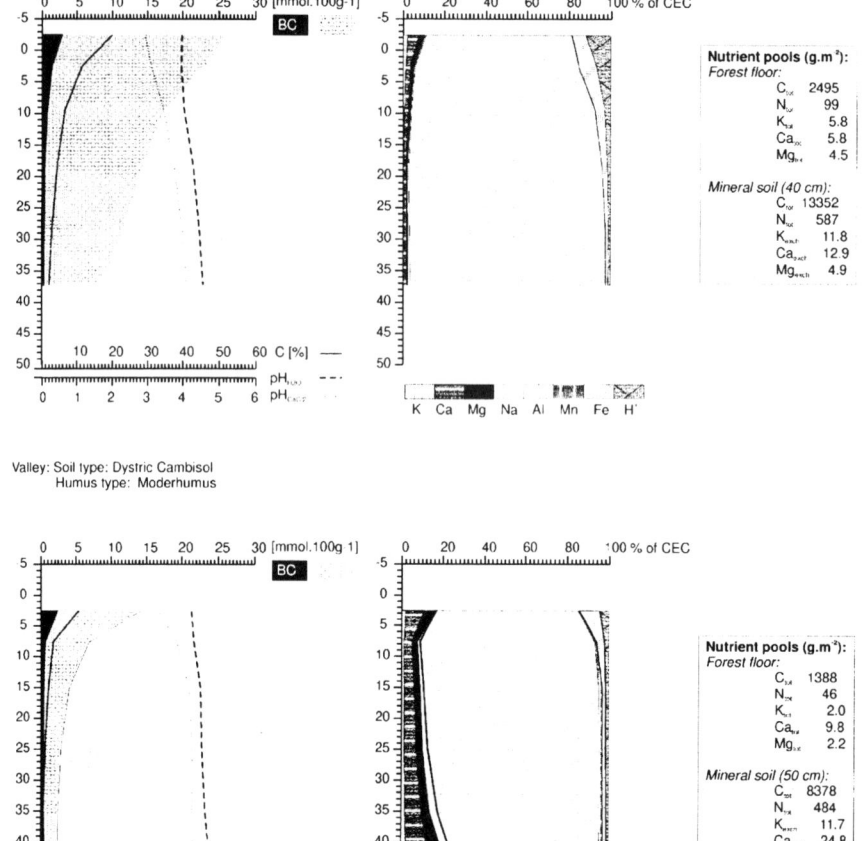

Plateau: Soil type: Stagni-Gleyic Podzol
Humus type: Hydromorphic Rawhumus

Nutrient pools (g.m⁻²):
Forest floor:

Mineral soil (40 cm):

Valley: Soil type: Dystric Cambisol
Humus type: Moderhumus

Nutrient pools (g.m⁻²):
Forest floor:

Mineral soil (50 cm):

Figure 6. Nutrient pools and chemical soil characteristics for two selected Norway spruce stands in the Bohemian Forest

ation of exchangeable base cations. On such sites, forest stands exhibit 'normal' nutrition and vitality.

The dependence of Mg nutrition on the soil status was described in paragraph 6.4. Another close correlation has been found between the intensity of needle loss in Norway spruce and the Mg content of current years' needles (Figure 7, after Katzensteiner, 1992). Mg contents below a limit of $0.8 \, mg \, g^{-1}$ are associated with severe deficiency symptoms and are in accordance with deficiency limits for Norway spruce set by other authors (Hüttl *et al.*, 1986).

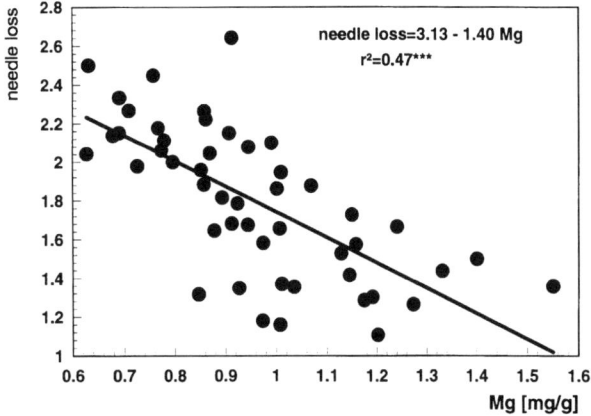

Figure 7. Relationship between needle losses of Norway spruce stands and Mg contents of current needles. Needle-loss class: 1, < 15%, 2, 15–30%; 3, 30–60%; 4, >60%. After Katzensteiner *et al.*, 1992

Hence, the differentiation between soil changes (low supply and uptake) and (enhanced leaching from the canopy of the observed Mg-deficiency symptoms is not possible. It is evident, however, that abandoned meadows afforested with Norway spruce show the worst nutritional status.

How the simultaneous occurrence of fungal diseases on twigs and needles of Norway spruce (Neumüller, 1994) relates to Mg deficiency has yet to be examined experimentally.

All of the described stress factors, either single or together, influence forest nutrition. Mg deficiency causes the most vivid symptoms in a complex forest deterioration.

6.6. Conclusions

Despite the fact that specific retention sites for magnesium do not exist in the mineral soil, forest ecosystems are capable of accumulating and maintaining magnesium levels for ecosystem functioning by means of tight biochemical cycling on most sites under natural conditions. Excessive biomass harvesting in past land use has removed large amounts of magnesium. Despite this, magnesium deficiency is the exception because other nutrients, in particular nitrogen, were harvested in equal proportions, which caused a reduction in growth but prevented critical dilution. Deposition of mobile anions and their precursors from natural sources and air pollution creates conditions for leaching of base cations, in particular when biochemical cycling is weakened by destruction of sinks or poor root vitality. Nitrogen deposition enhances growth, and, consequently, magnesium may be critically diluted in plant organs. Planting of rapidly growing tree species on sites with limited magnesium supply can cause the same effect. It can be stated that deposition of acidic air pollutants with a high nitrogen content in forest ecosystems must be considered to be the leading cause of the marked increase in magnesium deficiency in forest ecosystems during the last few decades.

248

6.7. References

Abrahamsen G, Stuanes AO, Tveite B, Eds. 1994. Long-term Experiments with Acid Rain in Norwegian Forest Ecosystems. Springer-Verlag, New York, Berlin, Heidelberg, 342 p.

Altherr E, Evers FH. 1974. Unerwarteter Düngungserfolg bei Magnesiummangel in einem jungen Buchenbestand auf mittlerem Buntsandstein des Odenwaldes. Allg. Forst. Jagdztg. 145, 121–125.

Bauer H. 1989. Nährstoffvorräte von Fichtenbeständen auf einer Standortseinheit im Kobernaußerwald. Diploma Thesis, BOKU Vienna, 90 p.

Baule H, Fricker C. 1967. Die Düngung von Waldbäumen. BLV Landwirtschaft GmbH, München, Basel, Wien, 259 p.

Becker M. 1991. Impact of climate, soil and silviculture on forest growth and health. DEFORPA 2nd Report, 23–38.

Becker-Dillingen. 1940. Die Magnesiafrage im Waldbau. Forstarchiv. 16, 88–92.

Berger TW, Glatzel G. 1992. Eintrag und Umsatz landzeitwirksamer Luftschadstoffe in Waldökosystemen der Nordtiroler Kalkalpen. Report BMLF-Project 56.810/18-VA2/92, 34 p.

Birks HJB. 1994. Did Icelandic volcanic eruptions influence the post-glacial vegetational history of the British Isles? Trends Ecol. Evol. 9, 312–313.

Bormann FH, Likens GE. 1979. Pattern and Processes in a Forested Ecosystem. Springer-Verlag, New York, Heidelberg, Berlin, 253 p.

Bosch C, Pfannkuch E, Baum U, Rehfuess KE. 1983. Über die Erkrankung der Fichte (*Picea abies* Karst.) in den Hochlagen des Bayerischen Waldes. Forstw. Cbl. 102, 167–181.

Bosch C, Pfannkuch E, Rehfuess KE, Runkel KH, Schramel P, Senser M. 1986. Einfluß einer Düngung mit Magnesium und Calzium, von Ozon und saurem Nebel auf Frosthärte, Ernährungszustand und Biomasseproduktion junger Fichten (Picea abies [L.] Karst). Forstw. Cbl. 105, 218–229.

Büttner G. 1992. Stoffeinträge und ihre Auswirkungen in Fichtenôkosystemen in nordwestdeutschen Küstenraum. Ber. des Forschungszentrums Waldökosysteme, Reihe A, 84, Göttingen, 192 p.

Chapin FS. 1988. Ecological aspects of plant mineral nutrition. In Tinker B, Läuchli A, Eds. Advances in Plant Nutrition 3, 161–192.

Chen C, Glatzel G. 1988. Vergleich des Bodenzustandes unter Buche und Fichte im Wienerwald. In: FIW-Symposium 1988 Waldsterben in Österreich, Theorien, Tendenzen, Therapien. Ed. Führer E, Neuhuber F, 253–254. BMWF Wien.

Eckmüllner O. 1988. Zuwachsuntersuchungen an Fichte im Zusammenhang mit neuartigan Waldschäden. Dissertation BOKU Vienna, 129 pp.

Falkengren Grerup U. 1987. Long-term changes in pH of forest soils in Southern Sweden. Environ. Pollut. 43, 79–90.

Feger KH. 1993. Bedeutung von ökosysteminternen Umsätzen und Nutzungseingriffen für den Stoffhaushalt von Waldlandschaften. Freib. Bodenk. Abh. 31, 237 p.

Fiedler HJ, Hunger W. 1970. Geologische Grundlagen der Bodenkunde und Standortslehre. Steinkopff, Dresden, 382 p.

Führer E, Neuhuber F, Eds. 1994. Status diagnosis and rehabilitation concepts for impact forest sites in the Bohemian Massif. Forstl. Schriftenr. Univ. Bodenkultur Wien, 7, 304 p.

Glatzel G. 1982. Ökophysiologische Untersuchungen zum Mineralstoff- und Wasserhaushalt mistelbefallener Eichenbestände und daraus abgeleitete Hinweise für Bekämpfungsmaßnahmen. In: Mayer H, Ed. Die Eichemistel im Weinviertel. Forschungsber. Univ. BOKU, Wien.

Glatzel G, Kazda M, Grill D, Halbwachs G, Katzensteiner K. 1987. Ernährungsstörungen bei Fichte als Komplexwirkung von Nadelschäden und erhöhter Stickstoffdeposition – Ein Wirkungsmechanismus des Waldersterbens? Allg. Forst. Jagdztg. 158, 91–97.

Glatzel G, Katzensteiner K, Kazda M, Kühnert M, Markart G, Stöhr D. 1988. Deposition langzeitwirksamer Luftschadstoffe in Wäldern und Einfluß auf den Ionenhaushalt. Report BMWF-Project 36.036/2-23/85, Vienna, 47 p.

Glatzel G. 1991. The impact of historic land use and modern forestry on nutrient relations of Central European forest ecosystems. Fer. Res. 27, 1–8.

Griffis K, Chapman DJ. 1990. Modeling Cretaceous–Tertiary boundary events with extant photosynthetic plankton: Effects of impact-related acid rain. Lethaia. 23, 379–383.

Grill D, Tausz M et al. 1994. Die physiologische und biochemische Bioindikation und ihre Anwendung am Beispiel der Fallstudie Schöneben. Forstl. Schriftenr. Univ. f Bodenkultur Wien. 7, 123–146.

Guderian R, Küppers K, Six R. 1985. Wirkungen von Ozon, Schwefeldioxid und Stickstoffdioxid auf Fichte und Pappel bei unterschiedlicher Versorgung mit Magnesium und Kalzium sowie auf de Blattflechte Hypogymnia physodes. VDI-Berichte. 560, 657–701.

Hallbäcken L, Tamm CO. 1986. Changes in soil acidity from 1927 to 1982–1984 in a forest area of South-West-Sweden. Scand. J. For. Res. 1, 219–232.

Halmschlager E. 1987. Bodeneigenschaften entlang eines Querprofils über einen windexponierten Rücken mit starker Streuverfrachtung in einem Laubwaldbestand des Wienerwaldes. Diploma Thesis, BOKU Vienna, 148 p.

Hauhs M. 1985. Wasser- und Stoffhaushalt im Einzugsgebiet der Langen Bramke/Harz. Berichte des Forschungszentrums Waldökosysteme/Waldsterben, Reihe A, 17, Göttingen, 206 p.

Hendershot WH, Courchenese F, Schemenauer RS. 1992. Soil acidification along a topographic gradient on Roundtop Mountain, Quebec, Canada. Water Air Soil Pollut. 61, 235–242.

Hildebrand EE. 1986. Zustand und Entwicklung der Austauschereigenschaften von Mineralböden aus Standorten mit erkrankten Waldbeständen. Forstw. Cbl. 105, 60–67.

Hippeli P, Branse C. 1992. Veränderung der Nährelementkonzentration in den Nadeln mittelalter Kiefernbestände auf pleistozänen Sandstandorten Brandenburgs in den Jahren 1964 bis 1988. Forstw. Cbl. 111, 44–66.

Hornbeck JW, Smith CT, Martin QW, Tritton LM, Pierce RS. 1990. Effects of intensive harvesting on nutrient capitals of three forest types in New England. For. Ecol. Manage. 30, 55–64.

Houdijk ALFM, Roelofs JGM. 1993. The effects of atmospheric nitrogen deposition and soil chemistry on the nutritional status of Pseudotsuga menziesii, Pinus nigra and Pinus sylvestris. Environ. Pollut. 80, 79–84.

Hüser R, Rehfuess KE. 1988. Stoffdeposition durch Niederschläge in ost und südbayerischen Waldbeständen. Forstl. Forschungsber. München 86, 153 p.

Hüttl RF. 1986. Forest Fertilisation: Results from Germany, France and the Nordic Countries. The Fertiliser Proceedings 250, 40 p.

Institut für Forstliche Standortslehre. 1971. Standortskartierung der Reviere nördlich der Großen Mühl des Forstbetriebes Schlägl. Revier Obernhof. Vienna, 115 p.

Katzensteiner K. 1992. Mineralstoffernährung, Bodenzustand und Baumvitalität in Fichtenwaldökosystemen des Böhmerwaldes. FIW-Forschungsber. 1992/1, 195 p.

Katzensteiner K, Glatzel G, Kazda M. 1992. Nitrogen induced nutritional imbalances – a contributing factor to Norway spruce decline in the Bohemian Forest (Austria). For. Ecol. Manage. 51, 29–42.

Katzensteiner K. 1994. Mineralstoffernährung und Bodenzustand in Fichtenwaldökosystemen des Böhmerwaldes (Oberösterreich). Forstl. Schriftenr., Univ. Bodenkultur Wien, 7, 57–66.

Katzensteiner K, Glatzel G. 1994. Das FIW Fallstudiengebiet Böhmerwald. Forstl. Schriftenr., Univ. f Bodenkultur Wien, 7, 29–44.

Kaupenjohann M, Hantschel R, Horn R, Zech W. 1987. Bodenextrakte zur chemischen Kennzeichnung der Nährstoffversorgung unterschiedlich säurebelasteter Fichtenstandorte. Mitt. d. Deutschen Bodenkundlichen Gesellschaft 55/II, 607–612.

Kazda M, Katzensteiner K. 1993. Factors influencing the soil solution chemistry in Norway spruce stands in the Bohemian Forest, Austria. Agric. Ecosys. Environ. 47, 135–145.

Kauppi PE, Mielikaeinen K, Kuusela K. 1992. Biomass and carbon budgets of European forests, 1971 to 1990. Science. 256, 70–71.

Kenk G, Fischer H. 1988. Evidence of nitrogen fertilization in forests of Germany. Environ. Pollut. 54, 199–218.

Kirchgeßner M. 1987. Tierernährung. 7. Aufl, DLG-Verlag Frankfurt/Main, 533 p.

Klapp E. 1971. Wiesen und Weiden. Verlag Paul Parey, Berlin and Hamburg, 620 p.

Krause GHM, Jung KD, Prinz B. 1985. Experimentelle Untersuchungen zur Aufklärung der neuartigen Waldschäden in der Bundesrepublik Deutschland. VDI Berichte. 560, 627–656.

Kreutzer K, Deschu E, Hösl G. 1985. Vergleichende Untersuchungen über den Einfluß von Fichte (Picea abies [L.] Karst) und Buche [Fagus sylvatica L.] auf die Sickerwasserqualität. Forstw. Centralbl. 105, 364–371.

Kreutzer K. 1972. Über den Einflu der Streunutzung auf den Stickstoffhaushalt von Kiefernbeständen. Forstw. Centralbl. 91, 263–270.

Liu JC, Trüby P. 1989. Bodenanalytische Diagnose von K- und Magnesium Mangel in Fichtenbeständen (Picea abies Karst.). Zeitschrift f. Pflanzenernaehrung und Bodenkunde. 152, 307–311.

Mann LK, Johnson DW, West DC et al. 1988. Effects of whole tree and stem only clearcutting on postharvest hydrologic nutrient losses, nutrient capital, and regrowth. For. Sci. 34, 412–428.

Marschner H. 1986. Mineral Nutrition of Higher Plants. Academic Press, London, San Diego, New York, Boston, Sydney, Tokyo, 674 p.

Marschner H. 1991. Mechanisms of adaption of plants to acid soils. Plant Soil. 134, 1–20.

Matzner E. 1988. Der Stoffumsatz zweier Waldökosysteme im Solling. Berichte des Forschungszentrums Waldökosysteme, Reihe A, 40, Göttingen, 217 p.

McLaughlin SB, Andersen CP, Edwards NT, Roy WK, Layton PA. 1990. Seasonal patterns of photosynthesis and respiraton of red spruce saplings from two elevations in declining southern Appalachian stands. Can. J. For. Res. 20, 485–495.

Meiwes KJ, König N, Khanna PK, Prenzel J, Ulrich B. 1984. Chemische Untersuchungsverfahren für Mineralboden, Auflagehumus und Wurzeln zur Charakterisierung und Bewertung der Versauerung von Waldböden. Berichte des Forschungszentrums Waldökosysteme, Göttingen, 7, 1–67.

Mengel K. 1968. Ernährung und Stoffwechsel der Pflanze. Fischer Verlag, Jena, 436 p.

Miller HG. 1995. The influence of stand development on nutrient demand, growth and allocation. Plant Soil. 168–169, 225–232.

Möller A. 1904. Karenzerscheinungen bei der Kiefer. Zeitschrift für Forst- und Jagdwesen. 36, 745–756.

Nebe W. 1991. Veränderung der Stickstoff und Magnesiumversorgung immissionsbelasteter älterer Fichtenbestände in ostdeutschen Mittelgebirgen. Forstw. Cbl. 110, 4–2.

Neumüller A. 1994. Beteiligung von Pilzen am Zweig- und Aststerben der Fichte im Revier Sonnenwald (Böhmerwald). Forstl. Schriftenr., Univ. Bodenkultur Wien, 7, 171–190.

Nilsson SI, Miller HG, Miller J. 1982. Forest growth as a possible cause of soil and water acidification: an examination of the concepts. Oikos. 39, 40–49.

Officer CB, Hallam A, Drake CL, Devine JD. 1987. Late Cretaceous and paroxysmal Cretaceous/Tertiary extinctions. Nature. 326, 143–148.

Prinz B, Krause GHM, Stratmann H. 1982. Waldschäden in der Bundesrepublik Deutschland. LIS-Berichte 28, 154 p.

Rehfuess KE. 1989. Acidic deposition – extent and impact on forest soils, nutrition, growth and disease phenomena in Central Europe. Water Air Soil Pollut. 48, 1–20.

Roberts TM, Skeffington RA, Blank LW. 1989. Causes of type 1 spruce decline in Europe. Forestry. 62, 179–222.

Roelofs JGM, Kempers AJ, Houdijk ALFM, Jansen J. 1985. The effect of air-borne ammonium sulphate on Pinus nigra var. maritima in the Netherlands. Plant Soil. 84, 45–56.

Rösch K. 1992. Einfluß der Beweidung auf die Vegetation des Bergwaldes. Nationalpark Berchtesgaden, Forschungsbericht. 26, 156 S.

Rusnov P. 1919. Die Entkalkung des Bodens durch den Einfluß SO_2-hältiger Rauchgase. Mitt. forstl. Versuchsanst. Mariabrunn, Wien.

Scheffer-Schachtschabel. 1989. Lehrbuch der Bodenkunde, 12. Aufl. Enke, Stuttgart, 491 S.

Scholl T, Katzensteiner K. 1995. Historische Landnutzung im Böhmerwald. Forstl. Schriftenr., Univ. Bodenkultur Wien. 7, 45–56.

Schulze ED. 1989. Air pollution and forest decline in a spruce (Picea abies) forest. Science. 244, 776–783.

Seufert G. 1988. Untersuchungen zum Einfluß von Luftverunreinigungen auf den wassergebundenen Stofftransport in Modellökosystemen mit jungen Waldbäumen. Berichte des Forschungszentrums Waldökosysteme, Reihe A, 44, Göttingen, 258 p.

Singh P, Bhurke ND. 1974. Influence of Armillaria root rot on the foliar nutrients and growth of some coniferous species. Eur. J. For. Path. 4, 20–26.

Skeffington R, Roberts M. 1985. Effect of ozone and acidic mist on Scots pine and Norway spruce – an experimental study. VDI-Berichte. 560, 747–760.

Sterba H. 1992. Determining parameters of competition models under changing environmental conditions. In: Franke J, Roeder A, Eds. Mathematical Modelling of Forest Ecosystems. J.D. Sauerländer, Frankfurt/Main.

Stöhr D. 1984. Waldbodenversauerung in Österreich. Veränderung der pH-Werte von Waldböden während der letzten Dezennien. Wien, 165 p.

Stöhr D, Partl H, Luxner M. 1988. Bericht über den Zustand der Tiroler Böden. Amt der Tiroler Landesregierung, Innsbruck, 198 p.

Stone EL. 1953. Magnesium deficiency of some northeastern pines. Soil Sci. Soc. Proc. 17, 297–300.

Tiedemann AR, Quigley TM, Anderson TD. 1988. Effects of timber harvest on stream chemistry and dissolved nutrient losses in Northeastern Oregon. For. Sci. 34, 344–358.

Tomiczek C. 1990. Wurzel- und Stammfäulen – Eine mögliche Ursache von Nährelementmängeln in Fichtennadeln. Mitt. Forstl. Bundesversuchsanstalt, Wien 163/III, 71–94.

Tomlinson GH. 1990/91. Nutrient disturbances in forest trees and the nature of the forest decline in Quebec and Germany. Water Air Soil Pollut. 54, 61–74.

Ulrich B. 1983. A concept of forest ecosystem stability and of acid deposition as driving force for destabilization. In: Ulrich B, Pankrath HD, Eds. Effects of Accumulation of Air Pollutants in Forest Ecosystems. Reidel Publishing Company, Dordrecht, Boston, London. 1–29.

Van Breemen N, Burrough PA, Velthorst EJ et al. 1982. Soil acidification from atmospheric ammonium sulfate in forest canopy throughfall. Nature. 299, 548–550.

Van Breeman N, Mulder J, Driscoll CT. 1983. Acidification and alkalinization of soils. Plant Soil. 75, 283–308.

Van Miegroet H, Cole DW. 1985. The impact of nitrification on soil acidification and cation leaching in a red alder forest. J. Environ. Qual. 13, 585–590.

Von Liebig J. 1840. Die Organische Chemie in ihrer Anwendung auf Agricultur und Physiologie. Verlag von Friedrich Vieweg und Sohn, Braunschweig, 352 p.

Wedepohl KH. 1969. Handbook of Geochemistry. Springer-Verlag, Berlin–Heidelberg–New York, 442 p.

Weissen F, Hambuckers A, Van Praag HJ, Ramackle J. 1990. A decennial control of N-cycle in the Belgian Ardenne forest ecosystems. Plant Soil. 128, 59–66.

Wessely J. 1853. Die Österreichischen Alpenländer und ihre Forste. Wilhelm Braumüller, K.K. Hofbuchhändler, Wien, 808 p.

Zech W, Popp E. 1983. Magnesiummangel, einer der Gründe für das Fichten- und Tannensterben in NO-Bayern. Forstw. Cbl. 102, 50–55.

Zöttl W, Hüttl RF. 1986. Nutrient supply and forest decline in Southwest Germany. Water Air Soil Pollut. 31, 449–462.

Part III
Recuperation of magnesium deficiency
through fertilization

7
Soil chemistry

S. AUGUSTIN, M. MINDRUP and K.J. MEIWES

7.1. Introduction

The main retention mechanism of magnesium in soils is cation exchange on negatively charged adsorption sites. Two kinds of charges can be distinguished:

1. *Permanent charge* which originates from isomorphic substitution within the
   crystal lattice of minerals. Cation adsorption at these sites is predominantly
   caused by electrostatic forces (Coulomb forces).

2. The primary source of *pH-dependent charge* is the deprotonation of functional
   groups on the surface of soil solids.

Mg is less effective bound than other bi- and trivalent cations, like Ca and Al, as
adsorption series show (Sposito, 1989). Even in soils where the total magnesium
content in the parent material is higher than the calcium content, the corresponding
exchangeable fractions show a reversed order (Hantschel, 1987), reflecting the
higher replaceability of Mg^{2+} from exchange sites. Al^{3+} becomes more effective in
exchanging Mg^{2+} as the base saturation decreases (Hildebrand, 1991). The low
selectivity of exchange sites for Mg^{2+} retention is one reason why acid soils are poor
in adsorbed magnesium. When fertilizing forest stands with magnesium, special
attention should be given to optimizing Mg^{2+} retention in the soil.

In this chapter, different forest magnesium fertilizers are described in relation to
chemical composition and dissolution rates. The retention and transport of Mg^{2+}
applied to the soil are described as a function of the chemical properties of the
fertilizers.

7.2. Magnesium fertilizers

*7.2.1. Chemical composition of magnesium fertilizer compounds and magnesium
contents*

The forest magnesium fertilizers listed in Table 1 differ with respect to chemical
composition, Mg^{2+} content, and solubility. They originate from igneous, metamorphic and sedimentary rocks, salt deposits and from ore-smelting processes.
Rock powder consists mainly of basalt, granite, diabase, or gabbro. In 14 different
types of basalt powder analyzed by von Fragstein *et al.* (1988), the content of Mg^{2+}
extracted with 1 N HCl was $2.5 \pm 0.9\%$. This is very low when compared with other

R. F. Hüttl & W. Schaaf (eds): Magnesium Deficiency in Forest Ecosystems, 255–273.
© 1997 *Kluwer Academic Publishers. Printed in Great Britain.*

magnesium fertilizers. In carbonate forms of magnesium fertilizers, the contents are fairly high. However, it is lower in dolomite than in magnesite because of the Ca^{2+} content in dolomite. The formula for slag lime from Table 7.2.2–1 characterizes the magnesium component as Ackermanit, a mineral of the melilite group. With respect to the contents of Mg^{2+}, Ca^{2+}, Na^+, and Al^{3+}, the melilites have a wide range of chemical compositions which depend on the way the ore was smelted (Scharpenseel and Beckmann, 1964):

$$X_2 \, Y \, Z_2 \, O_7 \tag{1}$$

where: X = Ca, Na
       Y = Mg, Al, (FeII, FeIII, Zn)
       Z = Si

In 28 samples of slag lime, Munk and Rex (1992) found a mean magnesium content of 4.7% ($\pm 0.4\%$), which is low when compared with magnesium carbonates and sulfates. Kieserite and epsomite are sulfatic salts with relatively high Mg^{2+} contents. However, MgO, which can be obtained by heating magnesium carbonate, and $Mg(OH)_2$, which is a common precipitation product from sea water, have the highest Mg^{2+} contents. All magnesium fertilizers, except the sulfate forms, alter the acid–base status of acid forest soils directly because they contain a basic component.

MgCl and $MgNO_3$ are not recommended as forest fertilizers. Chloride may interfere with Cl^--sensitive species. As magnesium deficiency occurs mainly in areas of central Europe with high N deposition rates, addition of N fertilizer is not recommendable.

7.2.2. Dissolution of magnesium fertilizers

Chemical reactions   In order to change magnesium supply in the soil, Mg^{2+} in the fertilizer has to be dissolved. The amount of magnesium which can be dissolved in a water–salt system is characterized by the solubility product (K_{sp}), which is the product of the molar concentrations of the reactants: K_{sp} ($K_{AB} = [A^+] \cdot [B^-]$). The magnesium concentrations given in Table 1 are concentrations at chemical equilibrium, derived from the solubility products of the different Mg^{2+}-containing substances used as fertilizers. To regard the dependency of the ion activities from the concentrations, a more precise calculation with the Debey–Hückel and Davies equations can be made (Davies, 1962).

In a closed system with water (Table 7.2.2–1) there are high solubilities for $MgSO_4$ as compared with carbonate- and silicate-containing magnesium fertilizers. In the open system of a soil, there are, in addition to H_2O, two other main chemical solvents involved in the dissolution of magnesium carbonates and silicates: H_2CO_3 and H^+ ions (Sverdrup and Warfvinge, 1987). H^+ ions may originate from atmospheric deposition or exchange sites, while H_2CO_3 and organic acids originate mainly from biological processes in the soil.

As an example of dissolution reactions of a carbonatic magnesium fertilizer, the

Table 7.2.2–1. Names, formulas, Mg^{2+} contents and solubilities of magnesium salts and magnesium minerals, analysed by various authors

Name	Formula	Mg content (%)	Mg solubility in water conc. in mmol·L^{-1}	Solubility product (pK_{SP})
Rock powder (like basalt)	$Mg_{(1.7)}Fe_{(0.3)}Si_{(0.3)}O_6$[d] $Mg_{(2.0)}Fe_{(0.03)}Si_{(1.97)}O_6$[e]	6[b]	0.04–0.5[c]	—
Slag lime	$Ca_2MgSi_2O_7$	5[a]	—	
Forsterite	Mg_2SiO_4	34	$0.067 \cdot 10^{-3}$	28.11
Dolomite	$CaMg(CO_3)_2$	13	0.038[c]	17.09
Magnesite	$MgCO_3$	29	0.076[c]	8.24
Epsomite	$MgSO_4 \cdot 7H_2O$	10	127.3[c]	2.14
Kieserite	$MgSO_4 \cdot H_2O$	17	4943**[f]	
Magnesium oxide (Periclase)	MgO	60	0.150*[f]	
Magnesium hydroxide (Brucite)	$Mg(OH)_2$	42	0.091[c]	11.41

Different magnesium silicates: [a]Munk and Rex (1992), [b]Wedepohl (1978); solubility in water: [c]Plummer *et al.* (1986); formulas for enstatite as example for most Mg^{2+}-containing minerals in basalt, from [d]Luce *et al.* (1972) and [e]Schott *et al.* (1981), both in Sverdrup (1990), page 106); [f]Weast *et al.* (1986). *reaction with H_2O to $Mg(OH)_2$, see Brucite; ** = solubility in hot water; pK_{SP} = $-\log K$ of the solubility product

reactions for dolomite are as follows (Busenberg and Plummer, 1982):

Reaction with water

$$CaMg(CO_3)_2(s) + 2H_2O \Rightarrow Ca^{2+} + Mg^{2+} + 2HCO_3^- + 2OH^- \tag{2}$$

Reaction with carbonic acid:

$$CaMg(CO_3)_2(s) + 2H_2CO_3^\circ \Rightarrow Ca^{2+} + Mg^{2+} + 4HCO_3^- \tag{3}$$

Reaction with acids from atmospheric deposition or exchange sites:

$$CaMg(CO_3)_2(s) + 2H^+ \Rightarrow Ca^{2+} + Mg^{2+} + 2HCO_3^- \tag{4}$$

By analogy with the chemical reactions for the carbonate magnesium forms, the dissolution reactions of slag lime can be described as follows:

Reaction with water:

$$Ca_2MgSi_2O_7 + 3H_2O \Rightarrow 2Ca^{2+} + Mg^{2+} + 6OH^- + 2SiO_2 \tag{5}$$

Reaction with carbonic acid:

$$Ca_2MgSi_2O_7 + 3H_2CO_3^\circ \Rightarrow 2Ca^{2+} + Mg^{2+} + 2SiO_2 + 3HCO_3^- + 3OH^- \tag{6}$$

Reaction with H^+ ions from atmospheric deposition or exchange acidity:

$$Ca_2MgSi_2O_7 + 3H^+ \Rightarrow 2Ca^{2+} + Mg^{2+} + 2SiO_2 + OH^- \tag{7}$$

The weathering reactions for silicate minerals, which may be used as magnesium fertilizer, are analogous to reactions (5) to (7).

Magnesium oxide reacts with water:

$$MgO + H_2O \Rightarrow Mg(OH)_2 \Leftrightarrow MgOH^+ + OH^- \Leftrightarrow Mg^{2+} + 2OH^- \tag{8}$$

$Mg(OH)_2$ dissolves in water and reacts with carbonic acid and protons.

In the soil, reactions (2) to (4) or (5) to (7) occur simultaneously. In contrast to the dissolution of carbonates, the reaction of silicates with carbonic acid takes place only at CO_2 partial pressures (pCO_2) higher than 0.3 atm (Sverdrup, 1990).

Dissolution kinetics Depending on the chemical reactions of the dissolution processes of different magnesium fertilizers, the state of chemical equilibrium is reached at different velocities. This can be described by the kinetics of the dissolution reaction. Little attention is given to the dissolution kinetics of $MgSO_4$ because its dissolution rate is very high in respect of magnesium fertilization of forests. The same may be true for MgO and $Mg(OH)_2$. In contrast, the dissolution rates of magnesium silicates and carbonates in soils are low; thus there is a special interest to look at their dissolution rates, with the aim to optimize liming programs and estimate weathering rates (Sverdrup and Warfvinge, 1987; Sverdrup, 1990).

Until a chemical equilibrium is achieved, the dissolution rate ($-dm/dt$, with m = mass and t = time) is always directly proportional to the sum of the inner and outer surface areas (A) of the fertilizer, which are in contact with the bulk solution:

$$-dm/dt \sim A \tag{9}$$

The dissolution proceeds with a forward reaction rate (F) of dissolving fertilizer and a backward rate (B) of precipitating substance from the bulk solution. The dissolution rate is the difference between the two processes:

$$-dm/dt \sim F - B \tag{10}$$

Combining equations (10) and (11), the dissolution rate can be described as a function of the rate of the forward and backward reactions, the surface area and the kinetic constant, K, which is characteristic for each kind of chemical dissolution reaction:

$$-dm/dt = K * (F - B) * A \tag{11}$$

As described above, H_2O, H_2CO_3 and H^+ ions are involved in silicate and carbonate dissolution. The rate equation can be written as follows:

$$-dm/dt = A * (K1 * [H_2O] + K2 * [H_2CO_3]^m + K3 * [H^+]^n - K4 * [B]^z) \tag{12}$$

K1 – K3 are the kinetic constants of the reactions with H_2O (K1), H_2CO_3 (K2) and H^+ (K3), while K4 is the backward reaction constant. The letters, n, m and z,

Table 7.2.2–2. Coefficients of dissolution rates of dolomite and enstatite (Busenberg and Plummer, 1982; Sverdrup, 1990), given as negative logarithm: $pK = -\log K$

Mineral	pK1 (H$_2$O)	pK2 (H$_2$CO$_3$)	m	pK3 (H$^+$)	n	pK4
Dolomite	10.5	9.2	0.5	6.8	0.5	7.2
Enstatite		> 14.6	0.5	10.9	0.6	

indicate the order of dissolution reaction. Since changes in the activity of H$_2$O are negligible, the order of reaction with water is zero. The backward rate depends on the activities of the reaction products with its specific reaction order z and the constant K4. When acid forest soils are fertilized with magnesium carbonate, a backward reaction does not occur as long as HCO$_3^-$ from magnesium carbonate fertilizer is converted to H$_2$CO$_3$ and then to CO$_2$ and H$_2$O.

Table 7.2.2–2 gives the values of the coefficients in Equation (12) for the dissolution of dolomite and of enstatite, as an example of a magnesium-rich mineral in basalt powder. From pK2 and pK3, it can be seen that the dissolution reaction of dolomite with H$_2$CO$_3$ is less rapid than with protons. For enstatite, the rate of dissolution due to H$_2$CO$_3$ is less than that of dolomite. The effect of H$^+$ ions is greater on dissolution rates of dolomite than on the rates of enstatite (see pK3). As the order of reaction indicates, increasing H$^+$ ion activity accelerates the enstatite dissolution to a greater extent than that of dolomite (see n). A ten-fold increase in H$^+$ ion activity, which corresponds to a change in pH from 4 to 3, raises the dissolution rate of enstatite four times ($10^{0.6} = 3.98$), whereas dolomite dissolution increases only three times ($10^{0.5} = 3.18$). The enstatite dissolution rate can be affected also by OH$^-$ ion activity (pK[OH$^-$] = 9.6;. Sverdrup, 1990), but it is reasonable to neglect this influence in acid forest soils.

Dissolution rates in the soil The data in Table 2 characterize the chemical reaction rates of the dissolution process under well-defined conditions with respect to the surface area, and the activities of H$_2$O, H$_2$CO$_3$, and H$^+$. In the soil environment, the chemical, biological and physical milieu is different from laboratory standard conditions. The factors determining dissolution rates in the soil vary in time and space. A change in temperature leads to an increase or decrease in the reaction rates, and the soil solution is charged with CO$_2$ due to respiration of microorganisms and roots. Organic acids, originating from biological processes, act as very effective catalysts in weathering of silicates. If organic acids are present, the process of magnesium release from a mineral framework can be accelerated up to 100 times (Huang and Keller, 1972). There are many possible sinks and sources for reactants and reaction products. Diffusion and mass flow lead to constantly changing amounts of ions present in the soil solution in the vicinity of the fertilizer particles.

Since the ambient conditions in the soil are difficult to describe, only a little is known about the dissolution processes of slowly dissolving magnesium fertilizers like carbonates and silicates under field conditions. Model calculations for dolomite dissolution under conditions of chemical equilibrium estimate a magnesium dissolution rate of $120 \, kg \, ha^{-1} y^{-1}$ (assumptions: $1000 \, mm \, y^{-1}$ water flux, pH 3 in soil solution, and a pCO_2 of 0.3 atm, cation exchange not considered; Prenzel, 1985). In a field experiment in southern Germany, Kreutzer et al. (1991) found $220 \, kg \, Mg \, ha^{-1} y^{-1}$ dissolved in the first year after the application of $4 \, t \, ha^{-1}$ of dolomite. This corresponds to $1.7 \, t \, ha^{-1}$ of dolomite. After four years, 90% of the applied dolomite was dissolved (Kreutzer et al., 1991).

When magnesium fertilizer dissolves, small particles disappear first and the surface area of the remaining particles decreases. It was shown in Equation (10) that the surface area is an important factor in determining the dissolution rate. Thus, the reduction of surface area is one reason for the exponential decrease of dolomite dissolution found by Kreutzer et al. (1991).

The dissolution rate of silicate rock powder is lower than that of dolomite. This is deduced from a two-year pot experiment, where equivalent amounts of Mg^{2+} as dolomite and basalt powder of the same particle size distribution were applied (Sauter, 1991). After the application, the amounts of exchangeable Mg^{2+} were 2–3 times greater than those of basalt powder (see Figures 7.3.2–1 and 7.3.2–2, layers 2 and 3). This relationship may characterize the differences of dissolution rates of dolomite and basalt powder.

With regard to the practical aspects of magnesium fertilization in forests, the fertilizers can be grouped into 4 classes according to their dissolution rates:

$$MgSO_4 > MgO/Mg(OH)_2 > \text{slag lime} \cong \text{dolomite} \cong \text{magnesite} > \text{basalt}$$

This sequence of the four groups of magnesium fertilizers is partially deduced from studies of dissolution and weathering of minerals and different types of lime (Sverdrup and Warfvinge, 1987; Sverdrup, 1990), partly from comparative experiments with soils (Sauter, 1991; Schaaf and Zech, 1993) and also from studies of the nutritional responses of trees after fertilization (Fiedler et al., 1988).

From Equation (9), it can be expected that the dissolution rates can be manipulated by changing the particle surface area by pulverization. Hildebrand and Schack-Kirchner (1990) showed, in a short-term laboratory experiment, that magnesium concentrations in seepage water under the top organic layer were higher when dolomite with particle size diameters $<60 \, \mu m$ was applied, compared with $60–200 \, \mu m$. Under field conditions, the extent of the significance of the surface area for the slowly dissolving magnesium fertilizers is not known quantitatively.

With regard to the matter balance of a forest ecosystem, the dissolution of magnesium fertilizers can be seen as an input into the system, like weathering of minerals, because it is, at least partly, an irreversible process (Ulrich, 1988). In the remainder of this chapter, which factors determine Mg^{2+} retention and its transport in the soil will be discussed.

7.3. Retention and transport of magnesium

The retention of Mg^{2+} applied to acid forest soils depends on the magnesium concentration in the soil solution, the amount of exchange sites in the soil, and the kind of anion applied with the fertilizer.

7.3.1. Effect of magnesium concentration

The relationship between the solid and the solution phase can be described quantitatively with exchange models. The *Gapon equation* is the most common exchange model used (Bolt, 1967):

$$K^G_{Mg/Al} \cdot \frac{\gamma Mg^{1/z\,Mg}}{\gamma Al^{1/z\,Al}} = \frac{\beta_{Mg}}{\beta_{Al}} \tag{13}$$

β_{Mg} and β_{Al} are the exchangeable states of Mg^{2+} and Al^{3+}, γ_{Mg} and γ_{Al} are the activities in solution, and z is the valence. The coefficient $K^G_{Mg/Al}$ gives information on the adsorption selectivity of the cations considered. In Equation 13, Mg^{2+} and Al^{3+} are given as an example because exchange of Al^{3+} for Mg^{2+} is an important reaction if acid soils are fertilized with magnesium.

Depending on the conditions of fertilizer dissolution, the application of magnesium fertilizers leads to a more or less rapid increase of the Mg^{2+} concentration in the soil solution. The Gapon equation (13) implies that an increase in the Mg^{2+} solute concentration raises the possibility of becoming adsorbed on the exchanger. Calculations of the Mg^{2+} adsorption according to the Gapon equation shows that an increase of magnesium concentration in solution phase from $100\,\mu mol_c \cdot L^{-1}$ to $200\,\mu mol_c \cdot L^{-1}$ leads to an enhancement of the magnesium equivalent fraction of the CEC (effective cation exchange capacity in $\mu mol_c \cdot g^{-1}$, NH_4Cl extractable cations) from 0.03 to 0.11, a concentration increase up to $500\,\mu mol_c \cdot L^{-1}$ to 0.48 and a 10-fold increase of Mg^{2+} concentration to an equivalent fraction of 0.93. The assumptions for this calculation were: $K^G_{Mg/Al} = 0.004$ (Hildebrand, 1986), $CEC = 100\,\mu mol_c \cdot g^{-1}$, an Al^{3+} equivalent fraction of CEC of 0.9, and in the soil solution there is enough Mg^{2+} available for the exchange. Since an increase in Mg^{2+} concentration in the solution phase results in adsorption, a decrease in concentration leads to a desorption.

From the dissolution characteristics, it is expected that the application of kieserite $(MgSO_4 \cdot H_2O)$ or epsomite $(MgSO_4 \cdot 7H_2O)$ will result in a rapid enhancement of magnesium concentrations in solution for a short time, whereas the increase after application of dolomite is less pronounced, but lasting. Under field conditions, the application of magnesium as $MgSO_4 \cdot H_2O$ ($750\,kg \cdot ha^{-1}$ kieserite $= 132\,kg \cdot ha^{-1}$ Mg) raised the magnesium concentration in solution under the forest floor immediately from about $0.03\,mmol_c \cdot L^{-1}$ to $10\,mmol_c \cdot L^{-1}$. After 3–4 months, it dropped to $\sim 0.01-0.10\,mmol_c \cdot L^{-1}$ (Feger, 1992), and remained on this level for the next 2 years. In a similar experiment with dolomite ($4t \cdot ha^{-1} = 520\,kg$ Mg $\cdot ha^{-1}$), the magnesium concentration in the humus lysimeter was $\sim 1-2\,mmol_c \cdot L^{-1}$ after 2–5 years

(Kreutzer *et al.*, 1991), which was still twice the concentration of the control. In both studies, there was a seasonal variability of magnesium concentration in the solution phase in the top organic layer. After the dolomite application, the Mg^{2+} fluctuations were more pronounced, with the highest concentrations found in autumn and winter, showing a possible influence of the temperature on concentrations of dissolved CO_2 in the soil solution and its effect on dolomite dissolution.

7.3.2. *Effect of CEC*

Retention of magnesium in upper soil layers and downward transport to deeper horizons lower Mg^{2+} concentrations in the soil solution in deeper soil layers. A concentration gradient with depth is formed with decreasing possibilities for exchange alongside (see Equation 13) and therefore decreasing amounts of adsorbed Mg^{2+}. The gradient changes with time. These relationships were shown by Ponette *et al.* (1993) who studied the levels of exchangeable Mg^{2+} in columns of two homogenized soils over a period of 32 days. The soils differed in their levels of CEC and organic carbon (C_{org}). In general, the amount of Mg^{2+} retained in the upper layers was greater in the soil with the higher CEC than in the soil with the lower CEC. When $MgSO_4 \cdot H_2O$ was applied as dissolved kieserite at the beginning of the experiment, more Mg^{2+} was leached with seepage water from the soil with the lower CEC. In the treatment with $MgSO_4 \cdot H_2O$, the gradients of exchangeable Mg^{2+} contents with depth were the same after 8 days as well as after 32 days from the beginning of the experiment. This indicates that adsorption of Mg^{2+} occurred rapidly and that, during the course of the experiment, there was no redistribution of adsorbed magnesium throughout the soil profile. In the treatment with dolomite (irrigation with dissolved dolomite during the whole course of the experiment), the gradients of exchangeable Mg^{2+} contents with depth changed from day 8 to day 32 due to the continuous addition of magnesium. The changes of Mg_{ex} contents were found even at 5–6 cm depth in the soil with the lower content of C_{org}, while it was restricted to the upper 3 cm in the other soil. This also indicates a higher content of pH-variable charge in the soil with the high C_{org} content. The effect of CEC on Mg^{2+} retention and the gradient of exchangeable Mg^{2+} with depth after application of $MgSO_4 \cdot H_2O$ were also demonstrated by Sauter (1991) at the end of a pot experiment lasting two years. In a loamy soil with a higher CEC, more magnesium was adsorbed, the gradient with depth was more pronounced, and the Mg^{2+} losses through seepage water were lower than in the sandy soil (see Figures 7.3.2–1 and 7.3.2–2 and Table 7.3.2–1).

7.3.3. *Effects of anions on magnesium retention*

As was shown earlier, magnesium retention depends largely on both Mg^{2+} concentration in the soil solution and the cation exchange capacity of the solid phase. Another important factor is the fate of the accompanying anion of the fertilizer.

Beside the binding of cations in organic complexes and precipitation of salts,

Table 7.3.2–1. pH (KCl), sum of exchangeable (NH$_4$Cl-extractable) cations (CEC in $\mu mol_c \cdot g^{-1}$), Ma and Mb equivalent fractions of CEC (Ma=Al+Mn+Fe+H; Mb=Ca+Mg+K) in different soil layers: 1=organic layer, 2=humic mineral soil, 3=mineral soil (adapted from Sauter, 1991)

Param.	Sandy soil						Loamy soil					
	Kies.	Carb.	Basalt	Dol.	Initial	Contr.	Kies.	Carb.	Basalt	Dol.	Initial	Contr.
pH 1	3.5	5.0	4.3	6.0	3.5	3.2	3.0	3.9	3.8	4.8	3.0	2.8
2	3.3	3.7	3.5	4.0	3.1	3.1	3.0	3.2	3.2	3.3	2.8	2.9
3	3.9	3.9	4.0	4.0	4.0	3.9	3.8	3.8	3.8	3.9	4.0	3.8
CEC 1	247.4	383.8	277.3	507.8	284.5	164.1	301.5	429.2	341.1	616.5	301.5	231.4
2	59.9	68.3	70.4	86.0	77.8	55.1	150.5	136.7	133.5	153.7	181.5	138.4
3	20.0	21.3	17.4	17.9	23.8	17.0	53.8	49.6	46.3	50.2	54.0	46.6
Mb %1	87.0	96.0	92.4	98.5	77.0	68.9	74.5	97.1	82.9	98.9	71.4	47.5
2	50.8	77.3	57.2	89.7	35.0	31.2	31.4	46.6	35.8	61.8	13.0	12.4
3	10.2	14.0	9.5	14.4	13.0	8.5	8.6	7.4	5.8	7.2	4.4	5.2
Ma %1	12.9	4.1	7.7	1,6	23.0	31.1	25.5	2.8	17.1	1.1	28.6	52.5
2	49.1	22.7	42.5	10,2	65.0	68.6	68.5	53.4	64.2	38.0	87.0	87.5
3	89.0	85.9	90.8	86,0	87.0	91.2	91.5	92.3	94.2	92.8	95.4	94.6

Kies. = Kieserite (MgSO$_4 \cdot$H$_2$O), Carb. = Hydromagnesite (4 MgCO$_3 \cdot$ Mg(OH)$_2 \cdot$ 4 H$_2$O), Basalt=basalt powder, and Dol. = dolomite (Ca(Mg)CO$_3$); each 10 kmol·ha^{-1} Mg (corresponding to 243 kg Mg·ha^{-1}); Contr. = control (treatment effects), Initial = Initial chemical condition of the soils

cation exchange is the main mechanism to remove cations from soil solution. For anions, there are more mechanisms based on different principles. Chloride and nitrate are not bound at a negatively charged soil surface. Nitrate in soil solution is regulated entirely by biological processes. Phosphate is strongly influenced by specific adsorption and precipitation reactions. Bicarbonate and sulfate are affected by both biological and inorganic chemical reactions. Regarding the retention mechanisms of cations and anions, it can be concluded that in the soil solution the amount of anions determines the amount of cations because of electro-chemical neutrality in soil solution (Johnson and Cole, 1980). This has special implications for acid soil fertilization using fertilizers with different anionic compounds.

Effect of sulfate The sulfate from sulfatic fertilizers can be removed from the soil solution by various mechanisms. Sulfate uptake by organisms and sulfate reduction to H$_2$S are thought to be negligible because the amount of sulfate added by fertilizer exceeds by far the possible rates of uptake or formation of H$_2$S. Another mechanism is the formation of uncharged MgSO$_4^{\circ}$ complexes in soil solution, provided the soil solution near the surface of fertilizer particles is in equilibrium with solid MgSO$_4$. About 25% of the magnesium sulfate in this state exists as neutral MgSO$_4^{\circ}$, which cannot be adsorbed (Buchter *et al.*, 1993). At magnesium concentration of less than 0.1 mmol L^{-1} the amount of MgSO$_4^{\circ}$ in the solution remains insignificantly low (Lindsay, 1979). As the concentration of Mg in the soil solutions collected below the surface organic layer which received MgSO$_4$ fertilizer did not usually exceed the value of 0.1 mmol L^{-1} (Feger 1992), for most practical instances, this mechanism of MgSO$_4^{\circ}$ formation is expected to be of minor signifance.

Another mechanism to remove SO$_4^{2-}$ from the soil solution with subsequent Mg^{2+} retention is the specific adsorption (Hingston *et al.*, 1967) or precipitation of a basic

Al-hydroxysulfate (Prenzel, 1982), which may affect the adsorption of added magnesium. In an adsorption experiment with Bs horizon material (10–20 cm) of a podzol on phyllit with a pH_{CaCl_2} of 3.8, Mg^{2+} adsorption increased from 6 to 15 $\mu mol_c \cdot g^{-1}$ soil, when sulfate addition was increased but magnesium addition remained constant (3 mmol $Mg \cdot L^{-1}$, 0–40 μmol $SO_4 \cdot L^{-1}$ soil:solution=1:10; Kaupenjohann and Zech, 1989). The Gapon coefficient ($K^G_{Mg/Al}$) rose from ~0.07 to 0.2 at the same time. The addition of nitrate did not affect Mg^{2+} adsorption. As nitrate, unlike sulfate, cannot be adsorbed on soil surfaces, this led to the conclusion that the reduction of sulfate concentration in soil solution causes increased Mg^{2+} retention. Since further additions of sulfate did not increase magnesium retention, it can be assumed that the sulfate retention capacity was exhausted.

The most likely mechanisms causing this retention are the following (Equation 14):

$$X\text{-}OH + SO_4^{2-} H^+ \Rightarrow XOH_2^+ \text{-}SO_4^{2-} \tag{14}$$

In this case, magnesium might have changed cations retained at existing adsorption sites or at adsorption sites created during the process of sulfate retention (Zhang and Sparks, 1990). The neutral site X-OH is protonated and sulfate is bound, creating a negative charge where Mg^{2+} is adsorbed. The increase in CEC due to sulfate adsorption may be significant. In the B2 horizon of a podzol, Wiklander (1976) found increased CEC from about 40 to 60 $\mu mol_c \cdot g^{-1}$ after addition of 0.2 mol $\cdot L^{-1}$ K_2SO_4 to a soil sample compared with addition of 0.2 mol $\cdot L^{-1}$ KCl.

Assuming precipitation of an aluminium sulfate solid phase, Kaupenjohann and Zech (1989) proposed the formation of $Al(OH)SO_4(s)$ according to Equation 15:

$$XAl\text{-}OH + MgSO_4 \Rightarrow XMg + Al(OH)SO_4(s) \downarrow \tag{15}$$

When assessing their experiment, it is necessary to consider that the positive effect was achieved in a batch experiment with a high input of magnesium and sulfate, and with soil material from a Bs horizon of a podzol where sulfate adsorption was more likely to occur due to the availability of aluminium compounds and sesquioxides.

However, the importance of sulfate retention for Mg^{2+} adsorption is thought to be small in magnesium fertilizing. Data from pot (see Figures 7.3.2–1 and 7.3.2–2) and field experiments indicate that Mg^{2+} retention occurs mainly in the top organic layer and upper mineral soil. Sulfate is thought to be retained preferentially in lower soil layers and not in upper horizons rich in organic matter. In several surveys of sulfur forms in soils, it was found that sulfate contents were low in upper soil layers with high C_{org} contents, as compared with soil samples with lower C_{org} contents (Singh and Johnson, 1986; Fischer, 1989; MacDonald and Hart, 1990). Sulfate adsorption experiments also indicate a reduced retention in soil layers with high organic carbon contents (Meiwes et al., 1980; Kaiser and Kaupenjohann, 1991). To test the exact influence of sulfate adsorption on Mg^{2+} retention, both magnesium and sulfate retention must be measured throughout a soil profile.

Shortly after $MgSO_4$ application, the rate of sulfate transport with seepage water is high. This implies that magnesium losses are higher than after application of

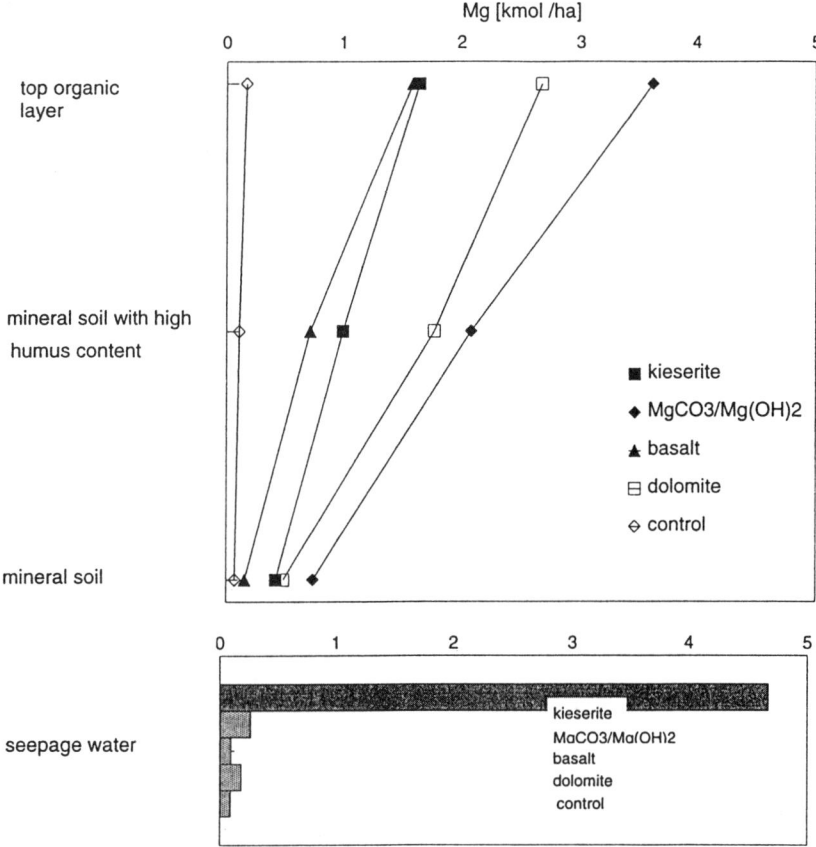

Figure 7.3.2–1. Amounts of exchangeable magnesium (NH$_4$Cl-extractable) in sandy soil layers and magnesium in the seepage water in a pot experiment lasting two years. 10 kmol Mg·ha^{-1} was mixed with the top organic layer of a sandy soil (adapted from Sauter, 1991)

carbonate or silicate magnesium fertilizers (see for example Figures 7.3.2–1 and 7.3.2–2). In pot and field experiments of two years duration (Sauter, 1991; Feger, 1992), 45–75% of the magnesium applied was still present. In a study with sandy soil, Sauter (1991) found that, two years after fertilization, only 10% of the sulfate was present in the soil while 45% of the magnesium was still adsorbed. It is therefore supposed that, during the first months after MgSO$_4$ fertilization, the sulfate losses and consequently magnesium losses were very high. Later on, the transport of the remaining sulfate did not affect Mg^{2+} transport (see also section 7.3.2.) because SO$_4^{2-}$ and magnesium retention occur at different soil depths or because leaching of sulfate is nearly complete in the initial stages. These are the reasons why Mg^{2+} losses are thought to be lower in the later stages than immediately after fertilization.

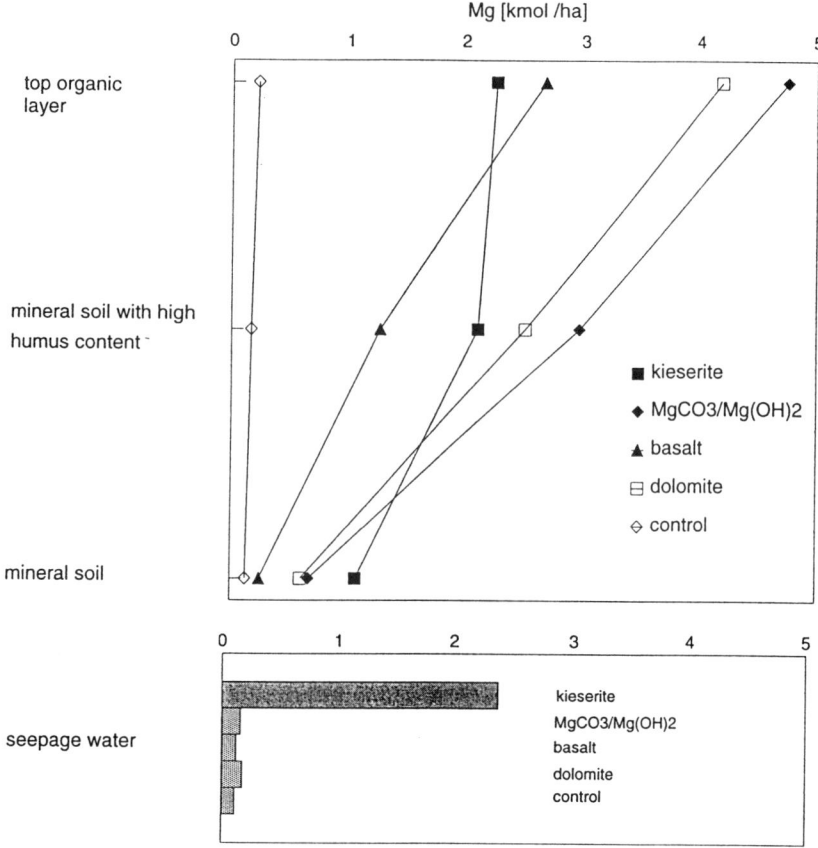

Figure 7.3.2–2   Amounts of exchangeable magnesium (NH$_4$Cl-extractable) in loamy soil layers and magnesium in the seepage water in a pot experiment lasting two years. 10 kmol Mg · ha^{-1} was mixed with the top organic layer of a loamy soil (adapted from Sauter, 1991)

In order to predict the duration of magnesium retention from short-term experiments, one has to consider the transport patterns of sulfate and magnesium as explained above. Long-term experiments are needed to obtain information about the sustainability of retention of magnesium applied as MgSO$_4$ fertilizer.

Although MgSO$_4$ is a neutral salt, its application has an effect on the acid–base status of the soil at different depths. If, for example in Sauter's experiment with sandy soil, 90% of the sulfate was leached within two years and 45% of the Mg^{2+} was retained, 45% of the sulfate must leave the system with another cation. Since Mg^{2+} exchanges H$^+$, Al^{3+}, and Ca^{2+} ions, these ions, and therefore acidity, are transported to deeper soil layers or even leave the soil profile with seepage water (Hildebrand, 1992). Since the level of adsorbed sulfate is correlated with aluminium extracted by dithionite citrate, ammonium oxalate, sodium pyrophosphate or KCl

(MacDonald and Hart, 1990) in loamy soils with high levels of these Al^{3+} fractions, the leaching of acidity due to $MgSO_4$ fertilization is expected to be lower than in sandy soils poor in extractable Al^{3+}.

Effect of basic anionic components of magnesium fertilizers on magnesium retention
If Mg^{2+} is applied with carbonate, HCO_3^- is the anion formed in the dissolution process (see Equation 2). Its concentration in soil solution is determined by the CO_2 $-H_2O-MgCO_3$ system which implies different chemical reactions and biological processes. At a given CO_2 partial pressure, the HCO_3^- concentration in acid forest soils is regulated by protonation HCO_3^- to form $H_2CO_3^*$ (Equation 16).

$$H_2CO_3^* \Leftrightarrow H^+ + HCO_3^- \qquad pK_a = 6.35 \qquad (16)$$

The pK_a value is a measure of the acid strength and indicates the degree of dissociation of an acid under given conditions. If the $pH_{(solution)}$ is equal to the pK_a, 50% of the acid is dissociated. At pH 5.35, 90% of the carbonate species are converted to H_2CO_3 and, at pH 4.35, no HCO_3^- is present in the soil solution. This means that as long as the pH is higher than 4.35, magnesium can be transported due to the presence of HCO_3^- anions. During the transport of HCO_3^--charged soil solution along a gradient of increasing soil acidity, the HCO_3^- concentration in soil solution is reduced by formation of H_2CO_3. Because of the electrochemical neutrality, this decrease in HCO_3^- concentration results in increased Mg^{2+} adsorption. This was shown by Ponette *et al.* (1993) who found increasing pH values (higher than pH 4.35) from the 8th to the 32nd day of their experiment and increasing magnesium adsorption in lower soil layers when a dolomite-saturated solution was applied on their soil columns.

The sites for Mg^{2+} adsorption are provided by deprotonation of functional groups of mainly soil organic matter if the fertilizer is applied to the forest floor. With regard to the application of magnesium carbonate, the most important functional groups are carboxyl groups because their pK_a values range from 3.0 to 5.0 (Perdue, 1985) and the pH values in the top organic layer should be increased from <3.0 to about 3.0–5.0 by the application of carbonates. Furthermore, in soil organic matter, carboxyl groups are quantitatively important (Schnitzer and Gupta, 1965). The deprotonation of the functional groups results in an increase in pH-variable charge and therefore in an increase in effective CEC (obtained by analysis with unbuffered salt solutions). The increase in CEC is linearly correlated with the amount of bases added to the soil (Nätscher, 1987). This is an important secondary effect of the application of magnesium as carbonate. However, it is of special importance in sandy soils where the cation-exchange sites of the organic matter determine to a large extent the cation-binding capacity of the soil because of the very low CEC contents in the mineral soil (see Figure 7.3.2–1 or 7.3.2–2). The short-term Mg^{2+} retention is restricted to a smaller soil layer if magnesium is applied as carbonate than as sulfate because of the high CEC in the top organic layer and the increase in CEC due to the pH-variable charge and the different transport mechanisms of sulfate and bicarbonate.

Input/output balances of fertilized soil columns show that a greater amount of Mg^{2+} is retained when applied as carbonate than as sulfate (Sauter, 1991; Ponette *et al.*, 1993). Furthermore, there are experimental results which indicate an immobilization of magnesium when applied as carbonate. After the addition of magnesium as sulfate to a top soil sample of an ultisol (hapludult), Grove *et al.* (1981) found more NH_4Cl-extractable Mg^{2+} than in $MgCO_3$-treated soil samples. At pH 4.2 and a magnesium addition level of $1.19\,mmol_c \cdot 100\,g^{-1}$, the difference was about 10%. Grove *et al.* (1981) proposed that a likely fixation mechanism was the adsorption and possible solid diffusion into newly precipitated and amorphous hydroxy Al polymers, created when acid soils are limed.

If Mg silicates act as fertilizers on acid forest soils, the reduction in anion concentration due to protonation of $H_3SiO_4^-$ may not influence magnesium transport in soil solution because of the protonation of $H_3SiO_4^-$.

$$H_4SiO_4 \Leftrightarrow H^+ + H_3SiO_4^- \qquad pK_a = 9.71 \qquad (17)$$

As long as the pH value is > 4.35, the anion HCO_3^- is present in soil solution and can facilitate magnesium transport. If MgO or $Mg(OH)_2$ are applied, the role of HCO_3^- for magnesium transport remains unchanged.

With respect to the efficiency of retention of applied Mg^{2+}, there may be differences between the basic magnesium fertilizers according to their calcium contents since calcium is more selectively adsorbed than magnesium. In a field experiment in southern Germany, four years after the application of $4\,t \cdot ha^{-1}$ dolomite, 20% of the applied Mg^{2+} had been transported into the mineral soil while the corresponding value for Ca^{2+} was 10% (Kreutzer *et al.*, 1991). The molar ratio of Ca:Mg of dolomite is 1, of slag lime about 4 and of basalt powder about 7–8. Disregarding different dissolution rates of the fertilizers concerned, and assuming a congruent dissolution of calcium and magnesium, Mg^{2+} retention in upper soil layers is thought to decrease with increasing Ca:Mg ratios in the fertilizer applied.

7.4. Conclusions

As explained above, forest magnesium fertilizers differ in Mg^{2+} contents, chemical composition and dissolution rates. Based on these differences, magnesium fertilizer management has to select the most appropriate magnesium fertilizer for various forest sites. The requirements of the forests depend on the timelag of tree response to nutrient supply, the length of the fertilizer effect, and the Mg^{2+} distribution in different soil layers.

Four groups of magnesium fertilizers have been distinguished:

1. Highly soluble magnesium sulfates with high Mg^{2+} contents,
2. Readily soluble magnesium oxides and hydroxides with high Mg^{2+} contents,
3. Magnesium carbonates and silicates (slag lime) with low dissolution rates and with either high or low Mg^{2+} contents, and
4. Pulverized silicate rocks with low dissolution rates and Mg^{2+} contents.

With respect to the anionic component, there are two groups:

1. Magnesium fertilizers with a basic anion i.e. carbonates, silicates, hydroxides and oxides, and
2. Neutral salts, like magnesium sulfates.

The properties of the different fertilizers can be combined by mixing different components. The advantage of such mixtures lies in the combination of fast- and slow-dissolving components. In this way, the velocity of dissolution can be controlled to some extent. A high Mg^{2+} retention combined with a high dissolution rate may be obtained by mixing magnesium carbonate with magnesium oxide as done by Fiedler *et al.* (1988). Using this combination, more satisfactory and sustainable results on plant nutrition were achieved when compared either with highly soluble salts or with dolomite alone. A rapid magnesium transport to deeper soil layers and a good retention in the upper soil can be achieved by combining magnesium carbonate or silicate with magnesium sulfate.

The importance of magnesium fertilization can be deduced from nutrient budget studies. Since these studies are very expensive to undertake and therefore not performed very frequently, they can give only some general information on the need of magnesium fertilization. More detailed information, which may be deduced from soil chemical data, is necessary. Studying forest dieback in the early 1980s, Ulrich *et al.* (1984) proposed a scheme of soil chemical conditions from which the need for forest fertilization could be assessed (see also Meiwes *et al.*, 1986). With regard to magnesium nutrition, soils with a Mg^{2+} equivalent fraction of CEC less than 0.01 were considered to have a very low elasticity, and less than 0.02 a low elasticity. Magnesium fertilization was considered to be essential for sites of the first group and was also recommended for soils with low elasticity. These recommendations, based on experience with soil chemistry and nutrition of trees, were given because there was a great need to counteract magnesium depletion preceding forest dieback.

A more adequate data set underlying such recommendation would be the relationship between the nutritional status of the trees (i.e. magnesium contents of leaves/ needles) and chemical soil parameters. In a survey of 44 sites in southern Germany, Liu and Trüby (1989) found a correlation between magnesium contents in spruce needles and the levels of exchangeable Mg^{2+} in soils, while Hüttl (1991) found no such correlation. Since, in Germany, actually corresponding data sets are being collected in a forest soil survey, more reliable recommendations, based on soil chemical data, for assessing the need for magnesium fertilization can be expected in the future.

The decision about the type of magnesium fertilizer to be applied depends on the objective of the fertilization and the soil chemical status of the forest site. Application of $MgSO_4$ only improves magnesium nutrition of the trees. There is a possibility of mobilization of exchange acidity in the organic layer of a forest soil with subsequent transport in the mineral soil. This is of considerable importance in soils with high levels of base-neutralizing capacity, for example raw humus forms (Hildebrand, 1991). On the other hand, mitigating soil acidification problems combined with improvement of Mg^{2+} nutrition suggests the use of basic magnesium fertilizers.

In magnesium fertilization experiments with $MgSO_4$, the rate of application ranges from $80-240 kg \cdot ha^{-1}$ Mg^{2+} (Hüttl, 1991; Sauter, 1991; Feger, 1992; Kaupenjohann, 1992). However, since no quantitative relationship between cation-exchange capacity and Mg^{2+} losses is known, only a rough application rate can be recommended for $MgSO_4$ fertilizers. From the point of view of efficient magnesium use, $50-100 kg$ $Mg \cdot ha^{-1}$ may be considered appropriate for sandy soils, where magnesium leaching losses are expected to be high. On loamy soils, up to $150 kg$ $Mg \cdot ha^{-1}$ may be applied. However, when the monetary costs of the application are important, the amount of magnesium recommended may be different. Using a basic magnesium fertilizer, the amount of fertilizer applied may depend mainly on the aim of deacidification wanted, i.e. liming the soil. In liming programs, as they exist for instance in Germany, not more than $3 t \cdot ha^{-1}$ lime are applied. Since a minimum Mg^{2+} content of 4.3% ($= 15\%$ $MgCO_3$) is required, the magnesium dose should be not less than $130 kg$ $Mg \cdot ha^{-1}$. In the case of dolomite application, which is very common, $400 kg$ $Mg \cdot ha^{-1}$ is used.

Apart from these practical problems of magnesium fertilization, more detailed knowledge of Mg^{2+} retention and Mg losses after the application of $MgSO_4$ is necessary. This refers to the role of sulfate retention and mobility in upper soil layers with high humus contents. Another problem requiring further study is the retention of Mg^{2+} from basic fertilizers in the top organic layer. Since the extraction of magnesium by a salt solution characterizes Mg^{2+} from adsorption sites as well as from fertilizer dissolution, new extraction techniques are required.

The problems of Mg^{2+} deficiency and magnesium fertilization in forests are recent. Since the majority of magnesium fertilization experiments were started during the last decade, only short-term effects are well documented. They include mainly the first $3-5$ years of the experiments. The long-term effects are expected to be described in the future. With respect to Mg^{2+} retention in the soil, magnesium losses in the first and second decades after fertilization are of special interest. From this, a more reliable recommendation of magnesium application rates is expected for the different types of soils. A model should be made for deciding magnesium fertilization, including the main components of Mg^{2+} balance, like mineral weathering, uptake of the stand and magnesium export by harvesting, as well as atmospheric input of Mg^{2+} and output with seepage water. A recent ecosystem model shows, for example, that small absolute amount of Mg^{2+} combined with a relatively high base saturation on the exchanger, but with low equivalent fraction of Mg^{2+} could be of more lasting advantage for a forest stand than high initial magnesium concentrations on the exchanger (Van Oene, 1992).

There is a tendency for decreasing atmospheric deposition of sulfate and acidity, while nitrogen deposition remains at a high level (Meesenburg et al., 1995; Matzner and Meiwes, 1994). Therefore, in the future, forests may be stressed less by acidification and more by imbalances of nutrients due to previous base cation leaching and actual nitrogen eutrophication. In this case, magnesium fertilization may provide a balanced nutrition rather than compensate for Mg^{2+} deficiency induced by intense acidification. Magnesium fertilization should be applied in combination with other nutrients more often than is presently the case.

Acknowledgement

We thank Dr Samuel Essiamah for the language review of the manuscript.

7.5. References

Bolt GH. 1967. Cation-exchange equations used in soil science – A review. Neth. J. Agric. Sci. 15, 81–103.

Buchter B, Schulin R, Flühler H. 1993. Influence of anion background on transport of calcium and magnesium. Water Air Soil Pollut. 68, 257–273.

Busenberg E, Plummer LN. 1982. The kinetics of dissolution of dolomite in $CO_2 - H_2O$ systems at 1,5 to 65°C and 0 to 1 atm pCO_2. Am. J. Sci. 282, 45–78.

Davies CW. 1962. Ion Association. Butterworth, London.

Feger KH. 1992. Bilanzierung von Stoffflüssen in magnesiumgedüngten Fichtenökosystemen im Schwarzwald. In: G Glatzel, R Jandl, M Sieghardt, H Heger, eds. Magnesiummangel in Mitteleuropäischen Waldökosystemen. Forstl. Schriftenr. Univ. f. Bodenkultur, Vol. 5.

Fiedler HJ, Leube F, Nebe W. 1988. Erste Ergebnisse einer Düngung mit MgO-haltigem dolomitischen Kalk zur Minderung von Immissionsschäden in Fichtenbeständen. Forst u. Holz. 43, 398–400.

Fischer M. 1989. Schwefel-Vorräte und -Bindungsformen süddeutscher Waldböden in Abhängigkeit von Gestein und atmogener Schwefel-Deposition. Forstl. Forschungsberichte München, Vol. 100.

Grove JH, Sumner ME, Syers JK. 1981. Effect of lime on exchangeable magnesium in variable surface charge soils. Soil Sci. Soc. Am. J. 45, 497–500.

Hantschel R. 1987. Wasser- und Elementbilanz von geschädigten, gedüngten Fichtenökosystemen im Fichtelgebirge unter Berücksichtigung von physikalischer und chemischer Bodenheterogenität. Bayreuther Bodenkundl. Berichte, Vol. 3.

Hildebrand EE. 1986. Zustand und Entwicklung der Austauschereigenschaften von Mineralböden aus Standorten mit erkrankten Waldbeständen. Forstwiss. Cbl. 105, 60–76.

Hildebrand EE. 1991. Die chemische Untersuchung ungestört gelagerter Waldbodenproben. KfK-PEF, Vol. 85.

Hildebrand EE. 1992. Sorption und Transport von gedüngtem Magnesium in Waldböden. In: G Glatzel, R Jandl, M Sieghardt, H Heger, Eds. Magnesiummangel in mitteleuropäischen Waldökosystemen. Forstl. Schriftenr. Univ. f. Bodenkultur, Vol. 5.

Hildebrand EE, Schack-Kirchner H. 1990. Der Einfluß der Korngröße oberflächig ausgebrachter Dolomite auf Lösungsverhalten und vertikale Verteilungstiefe. Forst u. Holz. 45, 139–142.

Hingston FJ, Atkinson RJ, Posner AM, Quirk JP. 1967. Specific adsorption of anions. Nature. 215. 1459–1461.

Huang WH, Keller WD. 1972. Organic acids as agents of chemical weathering of silicate minerals. Nature. 239, 149–151.

Hüttl RF. 1991. Die Nährelementversorgung geschädigter Wälder in Europa und Nordamerika. Freiburger Bodenkundl. Abh. 28.

Johnson DW, Cole DW. 1980. Anion mobility in soils: Relevance of nutrient transport from forest ecosystems. Environ. Int. 3, 79–90.

Kaiser K, Kaupenjohann M. 1991. Salz- und Säureeffekte auf die chemische Zusammensetzung von Bodenlösungen und auf die Sorptionseigenschaften immissionsbelasteter Standorte. Mitteilgn. Dt. Bodenkundl. Gessellsch. 66/I, 345–348.

Kaupenjohann M. 1992. Mehrjährige Erfahrungen mit der Magnesiumdüngung in Waldökosystemen des Fichtelgebirges. In: G Glatzel, R Jandl, M Sieghardt, H Heger, Eds. Magnesiummangel in Mitteleuropäischen Waldökosystemen. Forstl. Schriftenr. Univ. f. Bodenkultur, Vol. 5.

Kaupenjohann M, Zech W. 1989. Dynamik von Dünger-Magnesium in sauren Waldböden: Einfluß des begleitenden Anions. KfK-PEF. 55, 143–151.

Kreutzer K, Göttlein A, Pröbstle P. 1991. Dynamik und chemische Auswirkung der Auflösung von Dolomitkalk unter Fichte. In: K Kreutzer, A Göttlein, Eds. Ökosystemforschung Höglwald. Forstwiss. Forschung. 39, 186–204.

Lindsay WL. 1979. Chemical Equilibria in Soils. John Wiley and Sons, New York.

Liu J-C, Trüby P. 1989. Bodenanalytische Diagnose von K- und Mg-Mangel in Fichtenbeständen (Picea abies Karst.). Z. Pflanzenernähr. Bodenk. 152, 307–311.

Luce RW, Bartlett RW, Parks GA. 1972. Dissolution kinetics of magnesium silicates. Geochim. Cosmochim. Acta. 36, 35–50.

MacDonald NW, Hart JB Jr. 1990. Relating sulfate adsorption to soil properties in Michigan forest soils. Soil Soc. Am. J. 54, 238–245.

Matzner E, Meiwes KJ. 1994. Long-term development of elemental fluxes with bulk precipitation and throughfall in two German forests. J. Environ. Qual. 23, 162–166.

Meesenburg H, Meiwes KJ, Rademacher P. 1995. Long term trends in atmospheric deposition and seepage ouput in northwest German forest ecosystems. Water Air Soil Pollut. 85, 611–616.

Meiwes KJ, Khanna PK, Ulrich B. 1980. Retention of sulphate by an acid brown earth and its relationship with the atmospheric input of sulphur to forest vegetation. Z. Pflanzenern. Bodenk. 143, 402–411.

Meiwes KJ, Khanna PK, Ulrich B. 1986. Parameters for describing soil acidification and their relevance to the stability of forest ecosystems. Forest Ecol. Manage. 15, 161–179.

Munk H, Rex M. 1992. Zur Mobilität des Magnesiums in Böden und Düngemitteln. Allg. Forst-Zeitschr. 48, 796–799.

Nätscher L. 1987. Art, Menge und Wirkungsweise von Puffersubstanzen in Auflagehorizonten forstlich genutzter Böden des Fichtelgebirges. Ph.D. thesis, University of Munich.

Perdue EM. 1985. Acid functional groups of humic substances. In: GR Aiken, DM McKnight, RL Wershaw, P MacCarthy, Eds. Humic Substances in Soil, Sediment, and Water. John Wiley and Sons, New York.

Plummer LN, Jones BF, Truesdell AH. 1986. WATEQF: a fortran version of WATEQ, a computer program for calculating chemical equilibrium of natural waters. US Geol. Survey Water Res. Investig. Vol. 83, Reston, Virginia.

Ponette Q, Duffey JE, Weissen F, van Praag HJ. 1993. Downward effects of dolomite and kieserite on two acid soils differing in their organic carbon content. Commun. Soil Sci. Plant Anal. 24, 1439–1452.

Prenzel J. 1982. Ein bodenchemisches Gleichgewichtsmodell mit Kationenaustausch und Aluminium-hydroxosulfat. Göttinger Bodenkundl. Ber. 72.

Prenzel J. 1985. Die maximale Löslichkeit von oberflächlich ausgebrachtem Kalk. Allg. Forstzeitsch. 40, 1142.

Sauter U. 1991. Versuche zur Wirkung von sulfatisch, carbonatisch und silikatisch gebundenem Magnesium auf Ernährungszustand und Wachstum junger Fichten, chemischen Bodenzustand und Sickerwasserbefrachtung. Forstl. Forschungsber. München, Vol. 114.

Schaaf W, Zech W. 1993. Düngung mit gebranntem Magnesit und Magnesiumhydroxid zur Standortsmelioration in einem stark geschädigten Fichtenökosystem. Z. Pflanzenern. Bodenk. 156, 357–364.

Scharpenseel HW, Beckmann H. 1964. Studien über Kalkwirkungen und beobachtete Sonderwirkungen des Hüttenkalks (Hochofenschlacke) unter Verwendung von C14-Huminsäure und Fe55- sowie Ca45-Hüttenkalk. Arbeitsgemeinschaft Hüttenkalk e.V. (ed.) Düsseldorf, Breite Str. 27, Germany.

Schnitzer M, Gupta UC. 1965. Determination of acidity in soil organic matter. Soil. Sci. Soc. Am. Proc. 29, 274–277.

Schott J, Berner RA, Sjöberg L. 1981. Mechanism of pyroxenes and amphibole weathering: I. Experimental studies of iron free minerals. Geochim. Cosmochim. Acta. 45, 2123–2137.

Singh BR, Johnson DW. 1986. Sulfate content and adsorption in soils of two forest watersheds in southern Norway. Water Air Soil Pollut. 31, 847–856.

Sposito G. 1989. The chemistry of soils. Oxford University Press, New York.

Sverdrup H. 1990. The kinetics of base cation release due to chemical weathering. Lund University Press.

Sverdrup H, Warfvinge P. 1987. Upplösning av kalksten och andra neutralisationsmedel i mark. Statens Naturvardsverk, Rapport 3311, Solna, Sweden

Ulrich B. 1988. Ökochemische Kennwerte des Bodens. Z. Pflanzenern. Bodenk. 151, 171–176.

Ulrich B, Meiwes KJ, König N, Khanna PK. 1984. Untersuchungsverfahren und Kriterien zur Bewertung der Versauerung und ihrer Folgen in Waldböden. Forst. Holzwirt. 39, 278–286.

Van Oene H. 1992. Acid deposition and forest nutrient imbalances: A modelling approach. Water Air Soil Pollut. 63, 33–50.

von Fragstein W, Pertl W, Vogtmann H. 1988. Verwitterungsverhalten silikatischer Gesteinsmehle unter Laborbedingungen. Zeitschr. Pflanzern. Bodenk. 151, 141–146.

Weast RC, Astle MJ, Beyer WH. 1986. CRC Handbook of Chemistry and Physics. CRC Press, Inc. Boca Raton, Florida.

Wedepohl KH. 1978. Handbook of Geochemistry. Springer Verlag, Berlin, Heidelberg, New York.

Wiklander L. 1976. The influence of anions on adsorption and leaching of cations in soils. Grundförbättring. 27, 125–135.

Zhang PC, Sparks DL. 1990. Kinetics and mechanisms of sulfate adsorption/desorption on goethite using pressure-jump relaxation. Soil. Sci. Soc. Am. J. 54, 1266–1273.

8
Tree nutrition

M. KAUPENJOHANN

8.1. Introduction

Plant nutrition in general deals with nutrient supply of plants and aims to promote both quantity and quality of yields. Thus, a major focus of plant nutrition is the nutrient–growth relationship (Mengel and Kirkby, 1987, p. 247). In detail, aspects like nutrient availability, fertilization, nutrient uptake, nutrient assimilation and functioning in plants are considered (Bussler and Marschner, 1975).

Tree nutrition as part of plant nutrition has some special aspects which are different from nutrition of annual plants, resulting from the persistence of a forest through decades up to centuries. The major differences relate to plant internal cycling of nutrients and mycorrhiza development. Furthermore, it must be borne in mind that forest soils develop relatively stable structures since mixing through tillage as in agriculture does not occur. Soil structure, however, may significantly affect nutrient supply of trees.

Forests can be regarded as fairly independent of external nutrient supply because of tree internal cycling of nutrients (Van den Driessche, 1984). Evergreen species in particular have developed very efficient internal nutrient-conserving mechanisms which adapt them even to low fertility soils (Miller, 1981). However, the contribution of internal supply of a nutrient relative to the total amount of nutrient required depends on the mobility of the element in the plant (Mengel and Kirkby, 1987), as well as on plant species and soil properties. Switzer and Nelson (1972) have shown that, in a loblolly pine forest (*Pinus taeda* L.), about two thirds of the P, 20% of the K and traces of the Ca requirements were met by internal translocation. Magnesium cycling contributed 24% of the total amount needed.

Mycorrhiza, which is the symbiosis between fungi and plant roots, is much more obvious in natural plant communities and trees than in annual agricultural crops. Many woody species even require mycorrhizae obligatorily because their relatively coarse and unbranched root systems, which often lack root hairs, are not able to obtain sufficient amounts of mineral nutrients from the soil (e.g. St. John, 1980). A major result of mycorrhiza formation is that nutrient exploratory geometry of the root system is highly improved by the presence of external fungal hyphae which acquire nutrients also for the host (Clarkson, 1985). Thus, mycorrhizae add especially to the supply of those nutrients which are relatively immobile in the soil like P and some trace elements e.g. Cu and Zn (for a review, see Koide, 1991). Nitrogen is also principally adsorbed by fungal hyphae; however, Ames *et al.* (1983) found no evidence for improved N status of the host as a result of mycorrhiza infection.

R. F. Hüttl & W. Schaaf (eds): Magnesium Deficiency in Forest Ecosystems, 275–296.
© 1997 *Kluwer Academic Publishers. Printed in Great Britain.*

Studies on the acquisition of Mg by mycorrhiza are not available to the knowledge of the author. It may be speculated, however, that mycorrhiza formation becomes more important as the soil solution concentration of Mg decreases. For further information on mycorrhiza, see Chapter 10, section 10.5.

An effect of soil structure on nutrient availability in forest soils has recently been demonstrated. Investigations of soil aggregates from acid forest soils, for example, show that the surface of aggregates is generally more acid than their inner part (Kayser *et al.*, 1994). Thus, the base saturation of the cation exchange complex at the surface of soil aggregates and soil pores is relatively low compared with that of the whole soil (Horn, 1987). Soil solution obtained from naturally structured forest soil cores contained less base cations than solutions extracted from sieved samples (Hantschel *et al.*, 1988; Hildebrand, 1986). Since roots prefer macropores for growth, it is assumed that this kind of soil chemical heterogeneity may also affect tree nutrition (Kaupenjohann *et al.*, 1987).

These internal forest ecosystem properties have to be considered when discussing tree nutrition in general and Mg nutrition in particular. In addition, the following sections define the amount of Mg needed for forest nutrition, applying the concept of nutrient requirements. Once the quantity of Mg needed is established, the means of meeting Mg requirements will be presented. As has already been stated, one of the means is internal nutrient cycling. Others are atmospheric deposition, litter decomposition, weathering and finally fertilization. In the next section, however, general concepts of nutrient uptake will be summarized, since a basic knowledge of uptake is needed to understand Mg fertilizer effects.

8.2. Magnesium uptake

Although plant nutrients in general can be supplied via both leaves and roots, canopy uptake of Mg from atmospheric input does not generally play a significant role in Mg nutrition of trees. On the contrary, Mg is more likely to be leached from the canopy than absorbed in the natural environment. Thus, only root uptake of Mg will be considered here. For the purpose of this chapter, the physiological scale will not be taken into account. The major focus is on more macroscopic aspects of root uptake of Mg, beginning with uptake models.

In general, total nutrient uptake is mechanistically modelled as the sum of mass flow, diffusion, and root interception of ions. Root interception accounts for the quantity of nutrients to which the roots 'move' by root growth. The contribution of root interception to total nutrient supply is low, however, in comparison with mass flow and diffusion. Mass flow transports nutrients via water moving to the tree roots. The mechanism is convection. The quantity of nutrients being directed to the tree root is given by the product of transpiration and the concentration of nutrients in the soil solution.

If the amount of nutrients delivered to the tree root by this mechanism is greater than tree demand, nutrients accumulate at the root surface. On the other hand, if uptake of nutrients is faster than transport to the roots by mass flow, a concentration gradient of nutrient ions is established between the bulk soil solution and the tree root surface. Ions follow this gradient and move towards the roots. The driving mechanism is diffusion. This mechanism contributes to nutrient supply more, the lower the concentration of nutrients in the soil solution is in relation to tree demand of nutrients.

Several computer models were developed on the basis of mass flow and diffusion theory. The Barber–Cushman mechanistic nutrient uptake model (BCM) has been used extensively for nutrient uptake investigations in agronomic research (Barber, 1984). However, only recently have attempts been made to apply this model to woody species. Gillespie and Pope (1990) successfully modeled the P uptake of black locust seedlings (*Robinia pseudoacacia* L.) by modification of the BCM considering pH changes in the soil as a result of acidification of the rhizosphere. Van Rees *et al.* (1990) modeled the K uptake of slash pine seedlings (*Pinus elliottii* Engelm. var. elliottii) from low-fertility soils. They compared the BCM and the Baldwin–Nye–Tinker model (Baldwin *et al.*, 1973) and showed that both models can effectively model K fertilization in nurseries. Thus the diffusion/mass-flow supply theory seems to be valid, not only for annual species, but also for trees, and models could help in studying nutrient uptake from forest soils.

Magnesium uptake by forest species has not yet been modeled. Recently, Kelly and Barber (1991) started a first attempt using loblolly pine seedlings. The authors defined the model parameters needed in a pot study. Modeling Mg uptake obviously presents a more complicated situation with a plant like loblolly pine than with an annual species. One of the major results was that the values of c_{min} (the concentration of Mg at the root surface where influx = efflux; thus, net influx of Mg = 0) were very different at different physiological stages of the tree seedlings. During growth flush, c_{min} approached zero. Thus, Mg was withdrawn from the solution until the solution was almost depleted while, during non-flush periods, the uptake rate appeared to be greatly reduced. Thus, for appropriate modeling of Mg uptake, different values of the Michaelis–Menten kinetic parameters must be considered, e.g. growth stages of the trees.

Although complete models of Mg uptake from the soil are not yet available for trees, the above cited results are helpful when considering soil and root effects on Mg availability. Current observations and computer simulations show that root geometry is of minor significance for the interception of mobile nutrients as nitrate. All models, however, indicate that root growth and extension into unexploited volumes of the soil are of great importance in acquiring nutrients that diffuse slowly in the soil (Nye and Tinker, 1977).

These model results suggest that root geometry becomes increasingly important for sufficient Mg acquisition in soils with low Mg concentrations. Environmental impacts which reduce root growth will thus reduce Mg uptake, especially on low-fertility soils. It has been hypothesized that root growth is reduced by soil acidification (e.g. Murach, 1984; Ulrich, 1989). Besides acid deposition, roots themselves can reduce rhizosphere pH (Marschner *et al.*, 1986, 1987). Further, root growth may also be reduced by O_3 impact on tree canopies. Several recent studies have reported on decreased partitioning of dry matter to roots in response to elevated O_3 as a result of reduced C allocation to the roots (e.g. Edwards *et al.*, 1992). The effects of root properties on efficient nutrient acquisition by plants are reviewed in detail by Clarkson (1985).

Antagonistic effects of soil solution cations on the uptake of Mg are of particular interest for interpreting the results of Mg fertilization trials. Many studies have shown that K and NH_4 can decrease Mg uptake (e.g. Barber, 1984; Mengel and Kirkby, 1987). Magnesium uptake depends also on Ca levels (Barber, 1984). Furthermore, Al also reduces the uptake of Mg (Mengel and Kirkby, 1987). Grimme (1983) demonstrated Al-induced Mg deficiency on *Avena sativa*. Many solution culture experiments have been conducted with tree seedlings during the last decade, showing that Al can also reduce the Mg uptake of trees (e.g. Jorns and Hecht-Buchholz, 1985; Stienen and Bauch, 1988; Göransson and Eldhuset, 1987; Junga, 1984). Experiments with seedlings have shown that Al affects Mg nutrition before growth reduction is observed (Ilvesniemi, 1992).

Ion ratios obtained from soil solutions, however, should be interpreted carefully. With respect to Al effects studied in solution cultures, for example, Kaupenjohann (1989) has pointed out the importance of ion speciation. In soils, it should be considered that roots will significantly alter the rhizosphere solution relative to the bulk soil solution. Plants can significantly decrease the root surface concentration of a nutrient by active uptake. Detailed discussion of Mg uptake physiology is contained in Chapter 4, section 4.3.1.

8.3. Magnesium requirement

This section deals with the quantities of Mg needed for tree growth. Although plant nutrition has a history of more than 100 years, it has to be stated that the definition of nutrient requirement is still vague and thus not satisfying. Therefore, it is necessary to clearly define the term nutrient requirement as it is used in this text.

A definition of the optimum nutrient requirement of tree seedlings has been presented by Ingestad (1970, 1971, 1974). Accordingly, maximum growth is achieved when the following criteria are satisfied:

1. All mineral nutrients are present in the plant in optimum portions,
2. The NH_4/NO_3 ratio in the soil solution is optimal, and
3. The ionic strength of the solution is optimal.

Optimum Mg weight portions of trees have been observed by solution culture of seedlings for a number of species (Table 1). If all nutrients in the plants are taken up from the root medium, criterion 1 should be accomplished if the nutrient ratios in the solution are similar to those required in plant tissue (Ingestad, 1976).

In natural environments, however, optimum growth conditions for a species can hardly ever be achieved. The stability of ecosystems is based on balanced competition between individuals. Further, the requirement for Mg is altered by other growth factors, like water, light and other nutrients. Thus, the concept of optimum

Table 1. Optimum requirement of Mg nutrition (proportion by weight relative to N = 100) of seedlings of various tree species (Ingestad, 1970, 1971, 1974, 1986; Jia and Ingestad, 1984)

Pinus sylvestris	6
Pinus nigra	4
Picea abies	5
Picea sitchensis	4
Betula verrucosa	8.5
Larix leptolepis	8.5
Pseudotsuga menziesii	5
Tsuga heterophylla	5
Populus simonii	7
Paulownia tomentosa	9

nutrient requirement cannot be directly transferred from the laboratory pot experiment to the on-site growing forest. However, the concept of optimum nutrient requirement yields valuable data for estimating the genetic potential of a particular species.

Under growth conditions of a forest ecosystem, the quantity of a nutrient used for annual growth of all plant parts has been termed the nutrient requirement (Switzer and Nelson, 1972). The quantity of Mg which is leached from the canopy should be added to the total Mg requirement. This article follows the proposals of Switzer and Nelson (1972) to estimate the Mg requirements of forests.

Before summarizing literature data on Mg requirements, however, it is of interest to first establish the Mg pools in forest ecosystems. Literature data range from 24 kg Mg ha^{-1} (Switzer and Nelson, 1972) in a 20-year-old Pinus taeda stand to an extremely high value of 780 kg Mg ha^{-1} in a natural oak forest, reported by Bazilevich and Shitikova (1989). Although this value was excluded, Figure 1 shows that deciduous old and natural forests contain relatively high Mg pools. The overall average stock calculated from investigations at 16 sites (for references see Figure 1) is about 80 kg Mg ha^{-1}. Excluding another high value (328 kg Mg ha^{-1}) reported for a tropical rain forest ecosystem (Bazilevich and Shitikova, 1989) reduces this average to about 60 kg Mg ha^{-1}.

The annual Mg requirements range from 3.3 to 29.8 kg Mg ha^{-1}. Young deciduous forests require the highest portion of the Mg pools, followed by young conifers, old deciduous forests, and finally old conifers (20%, 15%, 11%, and 9%, respectively). Roughly 50% of the Mg uptake is required for above-ground biomass production, 35% is used for root production and the remaining 15% to substitute for Mg leached from the canopy (Table 2).

On the average, 1.61 kg Mg ha^{-1} y^{-1} is used for wood production and thus stored in the forests. Grenzius (1984) reported that 30–80-year-old pines incorporated 1.8–8 kg Mg ha^{-1} y^{-1} in wood. The literature data summarized in this article vary from 0.1 to 6.8 kg Mg ha^{-1} y^{-1} (Table 2).

Despite some uncertainty due to inconsistency of the data base, the results summarized in Figure 1 and Table 2 are helpful to evaluate some general function-

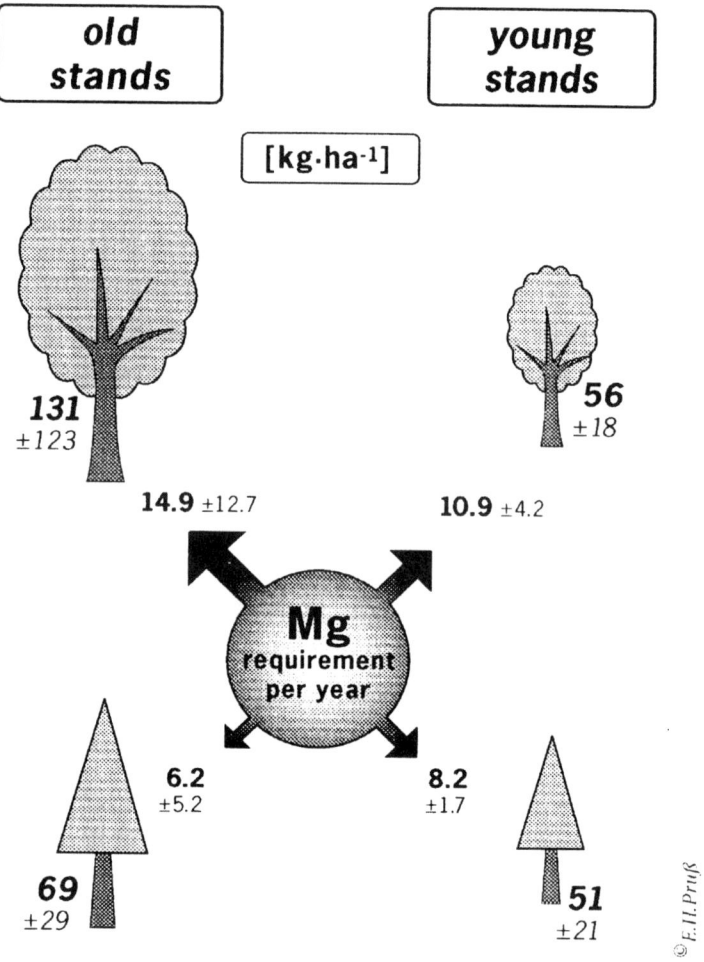

Figure 1. Total Mg contents and annual Mg requirements (± standard deviation) of different forest ecosystems. 'Old stands' are older than 70 years; data are collected from studies cited in Table 2 and from Crow *et al.*, 1991; Feger *et al.*, 1991; Johnson *et al.*, 1988; Knoepp and Swank 1994; Pehl *et al.*, 1984; Stevens *et al.*, 1989

ing of ecosystems and to judge roughly the Mg needs of forests. However, some critical comments on the concept of nutrient requirement as used here are necessary. The data summarized reflect the Mg use under natural conditions, but do not indicate whether Mg supply is sufficient for optimal growth. Le Goaster *et al.* (1990), in a 30-year-old spruce stand, found 43.3 kg Mg in healthy and 36.7 kg Mg ha^{-1} in yellow, Mg-deficient trees. The authors found higher current Mg uptake for above-ground biomass production of a healthy spruce stand than in a declining Mg-deficient stand (4.2 vs 3.2 kg ha^{-1}). Similar results have been reported by Oren *et al.* (1988). Thus, the concept of nutrient requirements as used here underestimates the needs of Mg at deficient sites.

8.4. Means of meeting Mg requirements

The nutrients required for annual growth are supplied by several naturally occurring sources, such as internal retranslocation from senescent leaves and needles to growing tissue prior to shedding, atmospheric deposition, canopy uptake of nutrients, litter mineralization and soil weathering. Switzer and Nelson (1972) expressed the means of meeting nutrient requirements in terms of a biochemical, a biogeochemical and a geochemical cycle. This approach is followed when analyzing literature data for this review. However, the geochemical cycle is split up into deposition input and soil weathering of nutrients.

Under conditions of Mg-poor soils, high Mg demand from rapid-growing forest plantations, and external input of acids or N, a shortage of Mg may result which can not be overcome by natural Mg-delivering processes. In such circumstances, Mg fertilization may be viewed as a further means of meeting the Mg requirements of a forest. Thus, this section is divided into Natural supply and Fertilizer supply of Mg.

8.4.1. Natural supply

Biogeochemical cycle   The biogeochemical cycle (① in Figure 2) accounts for nutrients supplied by litter production and decomposition and return of leached nutrients to the soil. This cycle supplies the major portion of Mg to forests. In old deciduous forests, 77% of the Mg requirement is biogeochemically supplied, while the portion is about 10% lower in young coniferous stands.

Magnesium leaching from different forest canopies ranges from about 0.5 to 5 kg ha^{-1} y^{-1} (Table 2). Leaching may be enhanced by acid precipitation (e.g. Mengel *et al.*, 1987, 1990; Potter, 1990). Leonardi and Flückiger (1989) showed that Mg and Ca leaching especially are increased by acid fog.

Litter fall returns from about 1–7 kg Mg ha^{-1} y^{-1} to the soil (Table 2). The release of Mg from decomposing foliar litter is relatively fast compared with P, N and Ca, but slower than K release (Blair, 1988). About 50% of total litter Mg of a mixed hardwood forest (dominant species: *Quercus, Carya* and *Acer* spp.) were released within one year. Magnesium release from litter has been modeled using first-order kinetics (Olson, 1963, cited in Blair, 1988).

Biochemical cycle   The internal retranslocation of nutrients in the plant is termed the biochemical cycle (② in Figure 2). Prior to leaf and needle abscission, considerable quantities of nutrients are mobilized and transferred into growing tissue. In the study by Switzer and Nelson (1972), this kind of internal transfer contributed 24% of the Mg requirements of 20-year-old *P. taeda*. Le Goaster *et al.* (1990) reported similar results for a healthy 30-year-old spruce stand. A Mg-deficient stand, however, recycled about 50% of the Mg contained in the annual

Table 2. Mg contents and Mg requirements of different forests

Location[1]	Dominant species	Age (years)	Content living biomass[2] (kg/ha)	Requirement (kg ha^{-1} y^{-1}) Above ground[3]	Below ground[4]	Leaching[5]	Litter fall[6]	Accretion[7] (kg ha^{-1} y^{-1})	References
South USA	*Pinus taeca* L. (plantations)	20	24[a]	6.3	n.d.	1.0	3.9[a]	1.4[a]	Switzer and Nelson, 1972
Southern Appalachians, USA	Several *Quercus* spp	Natural	136.5[a]	12.5	10.2	2.3	17.1[b]	3.3	Monk and Day, 1985
Valday Hills, Russia	*Picea* spp	Natural (steady state not yet reached)	120[c]		8	6.4[c]	0.8	0.8	Bazilevich and Shitikova, 1989
Oka-Don plain, Russia	*Quercus* spp.	Natural	780[c]		20	11.2[a]	n.d.	n.d.	Bazilevich and Shitikova, 1989
Amazonas Basin	Tropical rainforest	Natural	328[c]		28	1.8[a]	n.d.	n.d.	Bazilevich and Shitikova, 1989
East Netherlands[a]	*Quercus* spp. and *Betula* spp.	55	n.d.	10.5–15.6	n.d.	4.5–5.2	7.0–8.8	0.4–1.6	Pape *et al.*, 1989
Wisconsin, USA	*Quercus macrocarpa* Michx.	56	n.d.		8.8[a]	1.1[b]	4.5[d]	3.2[b]	Bockheim and Leide, 1991
Wisconsin, USA	*Pinus banksiana* Lamb.	51	n.d.		9.6[a]	1.1[b]	1.7[d]	6.8[b]	Bockheim and Leide, 1991
North Carolina, USA	*Picea rubens* Sarg.	Old growth	52[a]	3.3	n.d.	1.9	1.3	0.1[c]	Johnson *et al.*, 1991
North Carolina, USA	*Picea rubens* Sarg.	Old growth	49[a]	3.4	n.d.	1.9	1.3	0.2[c]	Johnson *et al.*, 1991

Table 2. Continued

Location[1]	Dominant species	Age (years)	Content living biomass[2] (kg/ha)	Requirement (kg ha^{-1} y^{-1})				Accretion[7] (kg ha^{-1} y^{-1})	References
				Above ground[3]	Below ground[4]	Leaching[5]	Litter fall[6]		
Colorado, USA	*Picea engelmannii* Parry	200–500 Old growth	52[a]	2.2	0.7	0.4	1.3[c]	1.2	Arthur and Fahey, 1992
New York, USA	*Acer rubrum* L.	75–100	60.8[b]	5.8	n.d.	1.0	4.4	0.4	Foster et al., 1992
Ontario, Canada	*Acer rubrum* L.	Uneven Old growth	60.8[b]	3.8	n.d.	1.0	2.4	0.4	Foster et al., 1992
Bavaria, D[b]	*Picea abies*	30	66	4.4	3.9	0.4	3.1/3.7[c]	1.7[d]	Schulze et al., 1989
Bavaria, D[c]	*Picea abies*	30	35	2.8	3.5	0.6	1.4/3.7[c]	1.1[d]	Schulze et al., 1989

n.d., not determined;
[1], [a]4 plots, [b]healthy, [c]declining, Mg-deficient;
[2], [a]including roots, [b]without roots, [c]not explained;
[3], in general: above-ground requirement = uptake = cycled + retained, above-ground requirement + below-ground requirement = litter fall + leaching + accretion, [a]data only for dominant species;
[4], [a]data only for dominant species
[5], [a]nutrient leaching from crowns and stems by precipitation', not clearly explained, [b]data only for dominant species;
[6], [a]total above-ground requirement – leaching – retained, [b]including root and other litter and arthropod consumption, [c]litter fall and mortality, [d]foliar litterfall and fine-root turnover for dominant species only, [e]above-ground/below-ground litter;
[7], [a]'retained', [b]perennial production, only dominant species, [c]wood increment, [d]requirement for stem wood production

284

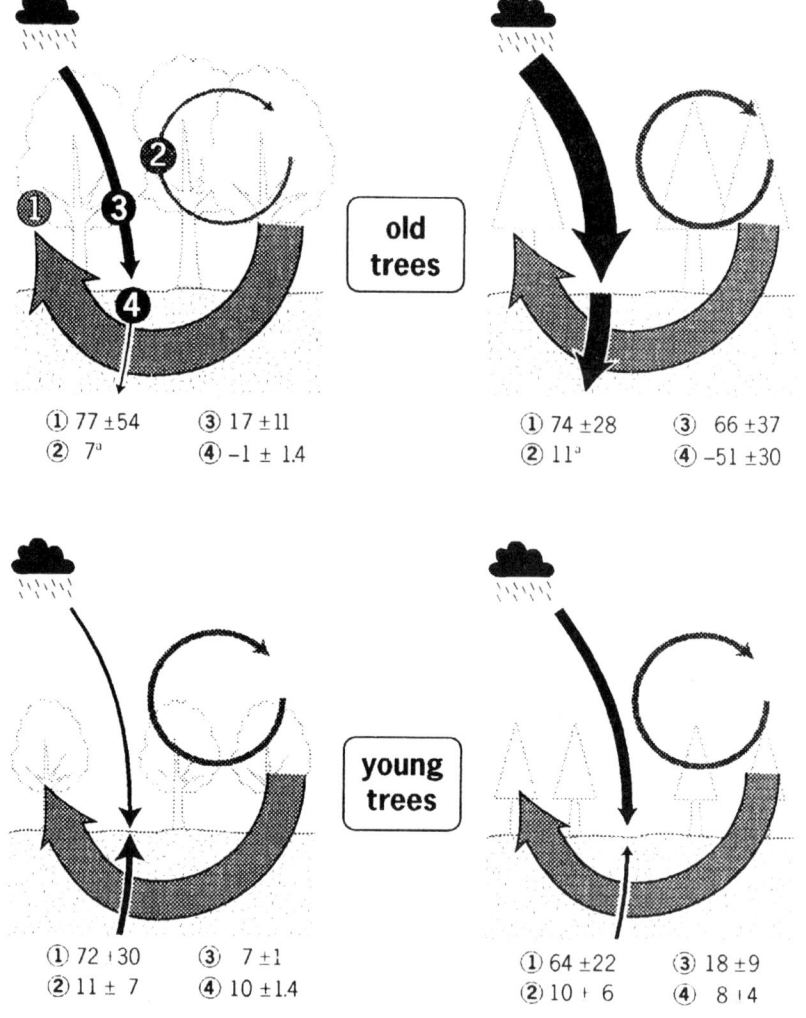

① 77 ±54	③ 17 ±11
② 7[a]	④ −1 ± 1.4

① 74 ±28	③ 66 ±37
② 11[a]	④ −51 ±30

① 72 ±30	③ 7 ±1
② 11 ± 7	④ 10 ±1.4

① 64 ±22	③ 18 ±9
② 10 ± 6	④ 8 ±4

Figure 2. Means of meeting Mg requirements (all data in % of the total requirements ± standard deviation) of different forest stands. [a]Only one study cited; data are collected from the studies cited in Figure 1

above-ground biomass production from old tissue. Yet, in the study by Oren *et al.* (1988), recycling of Mg was negligible in a healthy 30-year-old spruce stand. A declining stand, however, in agreement with the findings of the previous authors, recycled some Mg (8%). Differences between deciduous and coniferous trees are not reflected by the data.

As depicted from eight studies (for references, see Figure 2), the average contribution of retranslocation to Mg requirement of forests is about 10%, which is a much lower rate than reported for P retranslocation (60%) and much higher than for Ca retranslocation (traces, Switzer and Nelson, 1972).

Geochemical cycle The geochemical cycle summarizes Mg supply via atmosperic deposition (② in Figure 2) and weathering of soil minerals (④ in Figure 2). Deposition data are available from many forest ecosystem studies, most of which were started since the beginning of the eighties. Magnesium input into forests ranges from 0.5 to 8.8 kg ha^{-1} y^{-1} (15 studies; for references, see Figure 2). Old coniferous stands receive the highest rates while young deciduous forests gain the lowest inputs of Mg. These differences are related to canopy structure and have been discussed elsewhere (Kaupenjohann, 1989, p. 54). Old coniferous stands may receive more than 50% of the Mg requirement from the atmosphere, while young deciduous forests gain only up to 10%.

The contribution of soil weathering to Mg requirements is calculated as the difference between Mg requirement and Mg supply via cycles ① through ③. Thus, cycle ④ in Figure 2 is negative if the geo- and the biochemical cycles plus Mg deposition supply more Mg than required. In this case, the soil acts as a sink for Mg, rather than as a Mg source, which occurs in old forest stands. Young forests, by contrast, need Mg release from the soil to meet their Mg requirements.

8.4.2. Magnesium fertilization

Fertilization is a tool in the hand of man to balance a difference between nutrient requirements and the amount of nutrients supplied by natural means. Several substances, mainly Mg salts, are commercially available to be used as Mg fertilizers in forests. Beside Mg fertilizers, other fertilizers containing Mg in addition to other nutrients are used in forestry. With respect to solubility and soil solution effects, three different types of Mg salts can be distinguished, on the basis of the corresponding anions, as neutral salts of Mg, e.g. chlorides and sulfates, silicates and carbonates. These products have frequently been used in forestry.

Older forest fertilization aimed primarily at increasing stand productivity. Since N and P were the major growth-limiting mineral nutrients, fertilization was often restricted to NP application; sometimes K was added. Nevertheless, there are some older fertilizer trials on trees, which include Mg. These, however, mainly contain Mg in combination with other nutrients. Often Mg was added to the forest during amelioration treatments using dolomitic lime. Kaupenjohann and Zech (1987) and recently Hüttl and Zöttl (1993) have reviewed the results of former liming trials.

Direct Mg fertilization is a rather new strategy. Starting after the first observations of low Mg status in 'new-type declining forests', several diagnostic Mg fertilization experiments have been established since the early eighties. In contrast to the major goal of former forest fertilization, these Mg additions aimed at testing whether Mg could cure declining trees. The major results of these experiments will be summarized in this section with respect to effects on tree nutritional status. First, however, the main outcome of older fertilization trials will be highlighted.

In a stand which was treated with lime, P, N, K and Mg in 1959, Kenk *et al.* (1984) observed 63% healthy trees, whereas, among the unfertilized trees, only 23% were healthy individuals. The authors found correlations between exchangeable Mg in the soil and Mg needle concentrations as well as growth of spruce and suggested that Mg may play a key role for growth in that experiment. In a second treatment with Ca, P, and N only, the proportion of healthy trees was 47%.

Gussone and Zöttl (1975) studied the effect of application time of NPKMg on yield response of young spruce. The authors reported that nutritional imbalances were quickly corrected by fertilization. Early spring application resulted in same-year growth response while autumn application of fertilizer induced increased growth in the following year. The major growth effect was attributed to N; however, since the NPKMg treatment yielded higher growth than addition of N only, K and possibly Mg growth effects are suggested.

Kaupenjohann and Zech (1987) have summarized the results of an evaluation of older fertilization experiments with respect to the ability to prevent forests from decline. The major result is that those forests which were supplied with nutrients, like Mg and K, were significantly protected. Former exclusive lime application did not reduce the occurrence of 'new-type forest decline'.

In summary, the critical evaluation of older fertilization and liming trials suggests that, whenever Mg was included, the occurrence of advanced stages of 'new-type forest decline' was less and Mg deficiencies were reduced (Zöttl, 1990; Hüttl and Zöttl, 1993).

Sauter (1991) conducted pot experiments to evaluate the response of three-year-old spruce seedlings to Mg sulfate, carbonate and silicate fertilizers. Magnesium sulfate increased the Mg needle concentrations within one vegetation period after fertilizer application. Response to carbonate was a little slower and even silicate-bound Mg tended to increase the Mg needle concentrations after two vegetation periods (Table 3).

Table 3. Effects of different Mg fertilizers, applied in April 1985, on the Mg concentrations in needles ($mg\,g^{-1}$) of three-year-old Norway spruce saplings in a pot experiment (Sauter, 1991, pp. 95, 98, 106)

Treatment	Control	MgSO$_4$	MgSO$_4$	Dolomite	Basalt	Basalt
Mg rate ($kg\,ha^{-1}$)	0	243	122	243	243	486
Needle sampling						
Jan. 1986	0.83	1.94	1.18	1.25	0.79	0.85
Nov.–Dec. 1986[a]	0.70	1.77	1.46	1.82	1.29	1.21
Nov.–Dec. 1986[b]	0.84	1.89	1.52	1.54	1.03	0.98

[a]current-year needles; [b]all older needles

These results are fully supported by the field observations of many authors addressing the question of whether 'new-type forest decline' can be cured by fertilization. Magnesium sulfate fertilizers have been applied to Mg-deficient forests in varying amounts and in some cases combined with K and P fertilization (e.g. Zech and Popp, 1983; Hüttl, 1985, 1988, 1989; Bosch, 1986; Hüttl and Fink, 1988; Sauter, 1989, 1991; Kaupenjohann *et al.*, 1987, 1994).

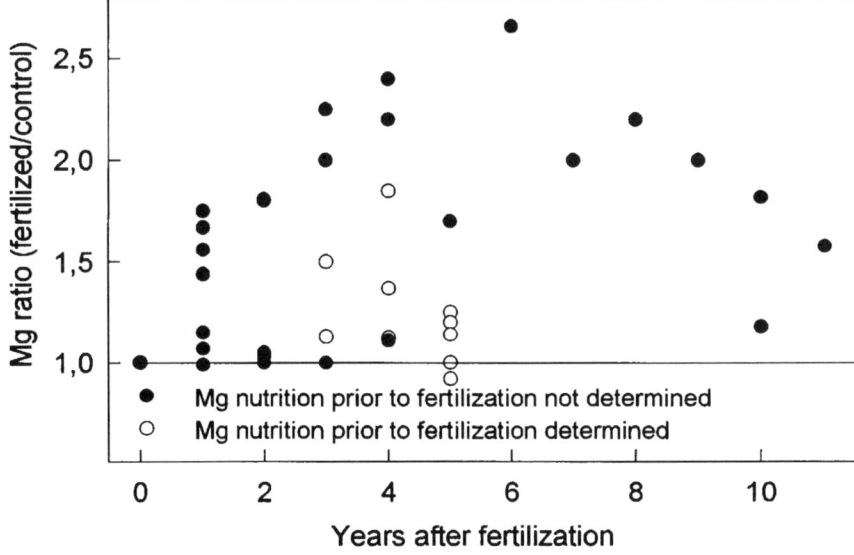

Figure 3. Effect of MgSO$_4$ fertilization on the ratio of Mg concentration in current-year spruce needles of fertilized and unfertilized (control) trees normalized to one in the year of fertilization. Data collected from Hüttl, 1988, 1989; Kaupenjohann, 1989; Kenk *et al.*, 1984; Sauter, 1989, 1991

The response of Mg needle concentrations in current-year tissue to MgSO$_4$ fertilization in various published studies is summarized in Figure 3. In order to construct Figure 3, the ratios of the Mg concentrations in the needles of fertilized trees to the corresponding values in control trees (Mg ratio) are plotted against time after fertilization. Differences between fertilized and control plots prior to fertilization are accounted for by dividing all Mg ratios by the Mg ratio prior to fertilization. This, however, was not always possible because not all authors reported the Mg nutrition in the experimental plots prior to fertilization. In these cases, it was assumed that both fertilized and unfertilized plots were supplied with similar levels of Mg.

Although – in accordance with the pot study results of Sauter (1991) – significant increases in the Mg needle concentrations were recorded in the fall after a spring application of Mg, trees at a few sites did not respond to Mg fertilization. Even years after Mg addition, these sites showed only slight, if any, increases in Mg needle concentrations, while, at the responding sites, the Mg fertilizer effects were lasting.

Liu (1988, p. 96) found positive Mg fertilizer effects only when the Mg needle concentrations of current-year foliage of spruce were less than $0.8\,\text{mg}\,\text{g}^{-1}$. Further analysis of the Mg fertilization effect on the N needle concentrations shows that those stands where MgSO$_4$ application increased Mg needle concentrations, generally showed decreased N needle concentrations (Figure 4). This is not a result of growth-induced dilution of N, since the concentrations of all other minerals in the

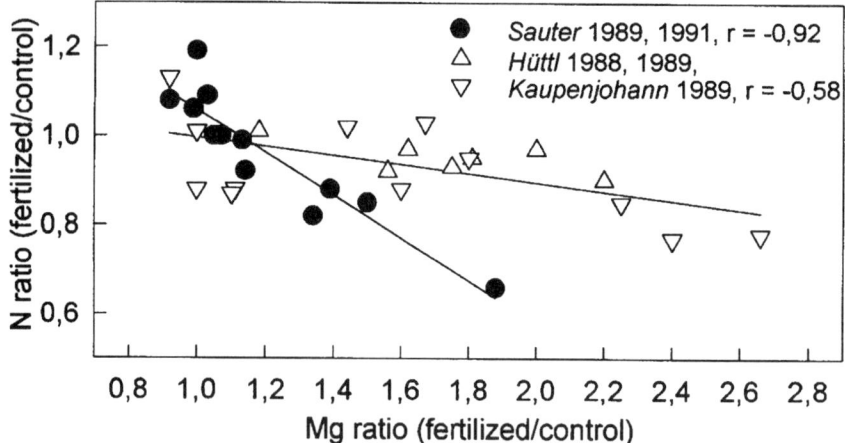

Figure 4. Ratios of Mg concentrations in current-year spruce needles of MgSO₄-fertilized and unfertilized (control) trees plotted vs. the N needle ratios of fertilized and control trees

needles were not simultaneously reduced.

It is possible that fertilizer Mg reduced the uptake of NH_4. This suggestion is in line with the assumption of Liu (1988, p. 145) who reported that Mg deficiency is not caused mainly by low Mg concentrations, but rather by high NH_4 concentrations in the soil.

Different Mg fertilizers, like MgO and $MgCO_3$ were used by Sauter (1989, 1991), Kaupenjohann *et al.* (1987), Kaupenjohann (1989), Dreyer *et al.* (1994) and others. Schaaf (1992) studied the effects of $Mg(OH_2)$ and MgO on a severely Mg-deficient Norway spruce stand. Sauter (1991) also used silicate-bound Mg in a field study. Almost all studies showed increased Mg needle concentrations after Mg application. However, the response was not as rapid as that observed after $MgSO_4$ application.

The Mg fertilizer effects on tree nutrition as summarized above can be illustrated by the example of a fertilization trial at Oberwarmensteinach, Fichtelgebirge NE-Bavaria. A 30-year-old Mg-deficient Norway spruce stand on Spodosols received $MgSO_4$ (160 kg Mg ha⁻¹) and $MgO*CaCO_3$ (1800 kg Mg ha⁻¹). The experimental plots were established in May 1983 and needle analyses have been conducted every year in autumn since that time. Further details about site and fertilization are given by Hantschel (1987) and Kaupenjohann (1989).

Only one vegetation period after $MgSO_4$ fertilization, the Mg concentrations of current-year needles reached the threshold value for sufficient growth and were almost twice as high as in the control trees. The Mg concentrations of the fertilized trees have remained high over the last 11 years; however, fluctuations in the control trees' Mg needle values are reflected (Figure 5). This may express the overall seasonal effects on Mg nutrition.

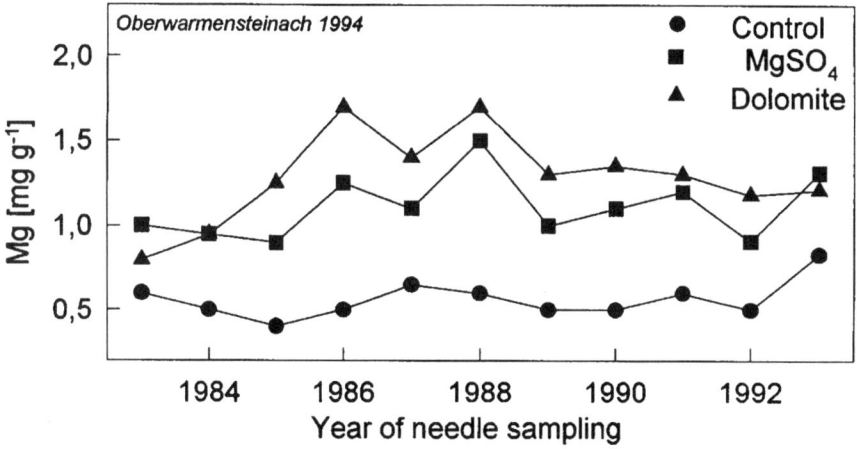

Figure 5. Current-year spruce needle concentrations of Mg after fertilization in May 1983 with MgSO$_4$ (160 kg ha^{-1}) and dolomite (1800 kg Mg ha^{-1}) at Oberwarmensteinach, Fichtelgebirge

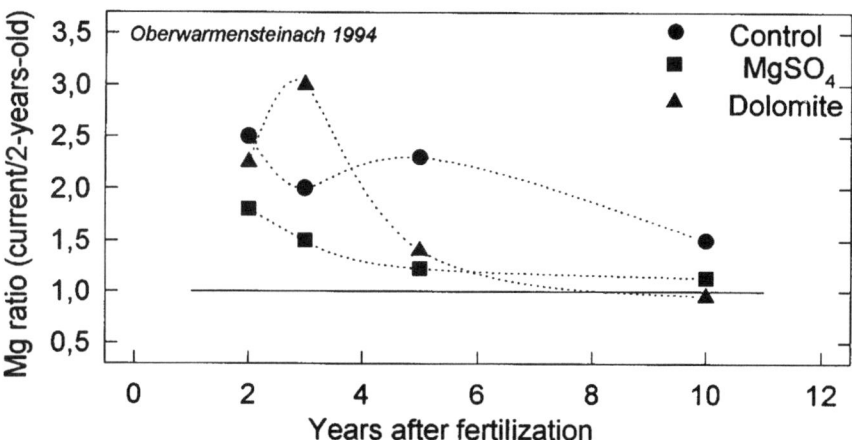

Figure 6. Ratios of the Mg concentrations in current-year and two-year-old spruce needles after fertilization with MgSO$_4$ (160 kg ha^{-1}) and dolomite (1800 kg Mg ha^{-1}) at Oberwarmensteinach, Fichtelgebirge

Needle response to dolomite application was a little slow compared with MgSO$_4$. Two vegetation periods after application, the current needle values of the MgSO$_4$ treatment were reached. However, older needles contained still less Mg compared with the MgSO$_4$ treatment (Figure 6). Since year 5 after fertilization, dolomite application has been superior with respect to Mg needle concentrations in both current-year and older needles compared with MgSO$_4$ fertilization. Eleven years after dolomite application, two-year-old needles showed even higher Mg concentrations than current-year tissue (Figure 6).

290

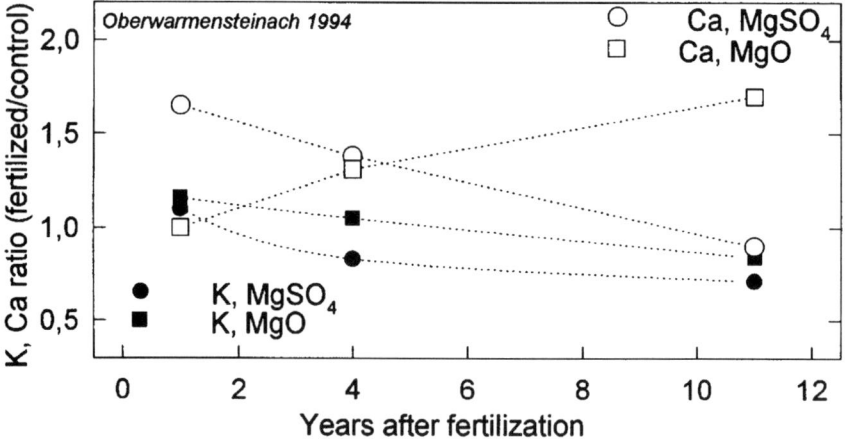

Figure 7. Effects of MgSO₄ and dolomite application on K and Ca nutrition expressed by the ratios of K and Ca in current-year needles of fertilized and unfertilized (control) trees at Oberwarensteinach, Fichtelgebirge

Thus, in summary, it is concluded from the literature review and the results of the Oberwarmensteinach fertilizer trial that a rapid improvement in the nutritional status of Mg-deficient forests is possible by Mg fertilization. The results of studies dating back to the beginning of the eighties also suggests that these effects are lasting.

Several studies on MgSO₄ effects on tree nutrition have suggested that, besides Mg, Ca nutrition was also improved after MgSO₄ application (e.g. Kaupenjohann *et al.*, 1987; Sauter, 1991). The evaluation of the Oberwarmensteinach needle analytical data showed that K nutrition is improved after MgSO₄ fertilization as well. This effect, however, lasted only for a short time after fertilizer application. By the third year after fertilization, the K needle values were already below those of the control trees (Figure 7).

These results are not yet explained sufficiently. The study of Kelly and Barber (1991) combined with the hypothesis of Seggewiss and Jungk (1988) may help to understand these data. The following hypothesis is proposed: According to Kelly and Barber (1991), trees may have significantly decreased Mg concentrations at the root surface, especially during periods of growth flush. Magnesium sulfate application, which, in all cited experiments, was carried out in spring at or shortly before growth flush, would increase the Ca and K concentrations in the soil solution due to exchange processes at the cation-exchange complex. The Ca, K/Mg ratios in the soil solution would decrease due to high Mg input. At the root surface, however, these ratios may even increase as a function of Mg absorption by the roots. Seggewiss and Jungk (1988) proposed that this mechanism could explain the beneficial effects of K fertilization on the Mg nutrition of plants. Figure 8 illustrates the hypothesis using soil solution data of the fertilization experiment at Oberwarmensteinach (Hantschel, 1987, pp. 142, 143).

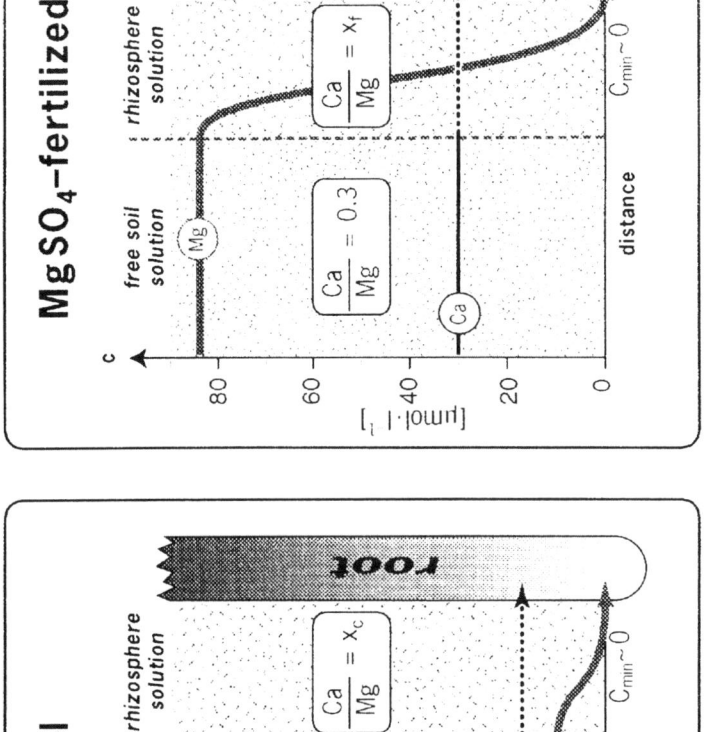

Figure 8. Illustration of a hypothesis to explain how $MgSO_4$ application could increase Ca (and K) supply. Soil solution data from the fertilization trial at Oberwarmensteinach, Fichtelgebirge; Hantschel, 1987, pp. 143, 144

In the long run, however, K and Ca will be depleted from soil by advanced leaching loss. Since K is less tightly absorbed by the cation-exchange complex than Ca, the beneficial effect of $MgSO_4$ application is less lasting. This hypothesis is in accordance with the pattern of the K and Ca ratios shown in Figure 7.

Although there are several general papers available on tree nutrition and fertilization, e.g. *Nutrition and Fertilization of Spruce* (Zöttl, 1990), hardly any comment on the rate of Mg application is offered. Most of the above-cited experiments use Mg fertilizer rates which are high compared with the Mg requirement of forests. No trials are available which have aimed to determine the exact quantity of Mg fertilizer needed.

Based on the experience derived from diagnostic fertilization trials, Hüttl (1989) recommended application rates of $5-10\,dt\,ha^{-1}$ of easily soluble Mg and K fertilizers for practical use. Kaupenjohann and Zech (1987) proposed $3-8\,dt\,ha^{-1}$ which should depend on site characteristics and be based on needle and soil analytical verification.

These recommended Mg fertilizer quantities are more than one order of magnitude higher than the annual requirement of Mg which ranges from about 10 to $30\,kg\,ha^{-1}\,y^{-1}$. Needs for wood production, and thus net Mg fixation, are again about one order of magnitude less than the annual requirements. Thus, the net use of fertilizer Mg for wood production is in the range of 1% of the recommended fertilization rate.

With application of a nutrient by soil fertilization, however, it must be assumed that only a fraction of the applied quantity will be available to the plant. Some will be leached out of the soil: this is dependent upon the fertilizer formulation. Some may be biologically fixed by, for example, the ground vegetation of the forest. Furthermore, only a part of the fertilized nutrients is readily available since plant roots are probably not as evenly distributed in the soil as the fertilizer. Thus, the Mg fertilizer requirement is generally greater than the quantity of Mg needed for tree tissue production.

A simple $MgSO_4$ adsorption experiment was carried out using six different forest soils to evaluate the recommended Mg fertilizer rates (Zapf, 1995). Liu and Trüby (1989) showed that $2\,meq\,Mg\,kg^{-1}$ of soil should be enough for sufficient Mg supply. Thus, the study of Zapf (1995) was designed to define the amount of $MgSO_4$ needed to supply the soil with $2\,meq\,Mg\,kg^{-1}$. Dependent upon the upper $40\,cm$ of the forest soils, $MgSO_4$ rates of 30, 320, 490, 700, 770, and $810\,kg\,ha^{-1}$ were necessary; this is close to the quantities recommended by Hüttl (1989) and Kaupenjohann and Zech (1987).

8.5. Summary and conclusions

This review summarizes the results of Mg fertilization trials in forests with respect to nutritional effects and aims to highlight general observations. In order to evaluate the results, the Mg requirements of forests and the means of meeting these Mg requirements are defined first.

Magnesium nutrition of trees is achieved by root uptake which can be described by the theory of mass flow and diffusion. Thus, environmental factors which restricted root growth of a forest will increasingly reduce Mg supply on low-fertility soils where a major part of Mg is supplied to the tree root via diffusion.

Besides root geometry, Mg uptake is also affected by antagonistic ions; this effect is often evaluated in terms of ion ratios (e.g. Mg/Al, K/Mg ratios). Interpretation of such data, however, should consider that root uptake of nutrients may significantly alter the ion ratios at the root surface compared with the bulk soil solution (which is generally analysed).

The Mg requirements of forests, which are defined as the quantity of Mg used for annual production of all plant parts, including substitution of Mg-canopy-leaching loss, range from about 3 to 30 kg Mg ha^{-1} y^{-1}, which is between about 10 and 20% of the total Mg pool in the biomass of forest trees. Net wood production requires from 0.1 to 6.8 kg Mg ha^{-1} y^{-1}. Most of the requirement is provided via litter cycling (the biogeochemical cycle). In old coniferous stands, atmospheric deposition (the geochemical cycle) adds considerably to Mg supply. The internal retranslocation of Mg (the biochemical cycle) accounts for about 10% of the Mg requirement.

Magnesium fertilization has been conducted to bridge a shortage of natural Mg supply. The evaluation of older lime and fertilizer trials, which did not usually address Mg, showed that, whenever Mg was contained in the applied fertilizer, Mg deficiency and 'new-type forest decline' were not as severe as on unfertilized or only limed stands. Specific Mg fertilization trials, which have been conducted since the beginning of the eighties, have shown a very rapid nutritional response of forest trees to highly water-soluble Mg fertilizers, while response to dolomite and especially silicate-bound Mg was retarded. The current results suggest that the effects of all Mg fertilizers are lasting. Thus it is concluded that Mg fertilization strategies, as proposed in the literature, present valuable tools for rapid and sustainable recovery of Mg nutrition in Mg-deficient forest ecosystems.

8.6. References

Ames RN, Reid CPP, Porter LK, Cambardella C. *et al.* 1983. Hyphal uptake and transport of nitrogen from two ^{15}N-labelled sources by Glomus mosseae, a vesicular–arbuscular mycorrhizal fungus. New Phytol. 95, 381–396.

Arthur MA, Fahey ThJ. 1992. Biomass and nutrients in an Engelmann spruce–subalpine fir forest in north central Colorado: pools, annual production, and internal cycling. Can. J. For. Res. 22, 315–325.

Baldwin JP, Nye PH, Tinker PB. 1973. Uptake of solutes by multiple root systems from soil. III. A model for calculating the solute uptake by a randomly dispersed root system developing in a finite volume of soil. Plant Soil 38, 621–635.

Barber SA. 1984. Soil Nutrient Bioavailability – A Mechanistic Approach. J Wiley & Sons, New York, 398 pp.

Bazilevich NI, Shitikova TY. 1989. Biogeochemistry of certain forested landscapes of different temperature regions. Pochvovedeniye. 7, 11–23.

Blair JM. 1988. Nutrient release from decomposing foliar litter of three tree species with special reference to calcium, magnesium and potassium dynamics. Plant Soil. 110, 49–55.

Bockheim JD, Leide JE. 1991. Foliar nutrient dynamics and nutrient-use efficiency of oak and pine on a low-fertility soil in Wisconsin. Can. J. For. Res. 21, 925–934.

Bosch Ch. 1986. Standorts- und ernährungskundliche Untersuchungen zu den Erkrankungen der Fichte (Picea abies (L.) Karst.) in höheren Gebirgslagen. Schriftenreihe der forstwissenschaftlichen Fakultät

294

der Universität München und der Bayerischen Forstlichen Versuchs- und Forschungsanstalt 75, München, 241 pp.

Bussler W, Marschner H. 1975. Das Fachgebiet Pflanzenernährung. Z. Pflanzenernähr. Bodenk. 138, 367.

Clarkson DT. 1985. Factors affecting mineral nutrient acquisition by plants. Ann. Rev. Plant Physiol. 36, 77–115.

Crow TR, Mroz GD, Gale MR. 1991. Regrowth and nutrient accumulations following whole-tree harvesting of a maple–oak forest. Can. J. For. Res. 21, 1305–1315.

Dreyer E, Fichter J, Bonneau M. 1994. Nutrient content and photosynthesis of young yellowing Norway spruce trees (Picea abies (L.) Karst.) following calcium and magnesium fertilization. Plant Soil. 160, 67–78.

Edwards GS, Friend AL, O'Neil EG, Tomlinson PT. 1992. Seasonal patterns of biomass accumulation and carbon allocation in Pinus taeda seedlings exposed to ozone, acidic precipitation, and reduced soil Mg. Can. J. For. Res. 22, 640–646.

Feger KH, Raspe S, Schmid M, Zöttl HW. 1991. Verteilung der Elementvorräte in einem schlechtwüchsigen 100jährigen. Fichtenbestand auf Buntsandstein. Forstw. Cbl. 110, 248–262.

Foster NW, Mitchel MS, Morrison IK, Shepard JP. 1992. Cycling of acid and base cations in deciduous stands of Huntington Forest, New York and Turkey Lakes, Ontario. Can. J. For. Res. 22, 167–174.

Gillespie AR, Pope PE. 1990. Rhizosphere acidification increases phosphorous recovery of black locust. II. Model predictions and measured recovery. Soil Sci. Soc. Am. J. 54, 538–541.

Göranson A, Eldhuset TD. 1987. Effects of aluminium on growth and nutrient uptake of Betula pendula seedlings. Physiol. Plantarum. 69, 193–199.

Grenzius R. 1984. Starke Versauerung der Waldböden Berlins. Forstw. Cbl. 103, 131–139.

Grimme H. 1983. Aluminium induced magnesium deficiency in oats. Z. Pflanzenernähr. Bodenk. 146, 666–676.

Gussone HA, Zöttl HW. 1975. Die Wirkung jahreszeitlich verschiedener Düngung auf junge Fichten. Forstw. Cbl. 94, 334–343.

Hantschel R. 1987. Wasser- und Elementbilanz von geschädigten, gedüngten Fichtenökosystemen im Fichtelgebirge unter Berücksichtigung von physikalischer und chemischer Bodenheterogenität. Bayreuther Bodenkundl. Ber. 3, Bayreuth, 219 pp.

Hantschel R, Kaupenjohann M, Horn R, Gradl J, Zech W. 1988. Ecologically important differences between equilibrium and percolation soil extracts. Bavaria. Geoderma. 43, 213–227.

Hildebrand EE. 1986. Zustand und Entwicklung der Austauschereigenschaften von Mineralböden aus Standorten mit erkrankten Waldbeständen. Forstw. Cbl. 105, 60–76.

Horn R. 1987. The role of structure for nutrient sorptivity of soils. Z. Pflanzenernähr. Bodenk. 150, 13–16.

Hüttl RF. 1985. 'Neuartige' Waldschäden und Nährelementversorgung von Fichtenbeständen (Picea abies (L.) Karst.) in Südwestdeutschland. Freiburger Bodenkundl. Abh. 16, Freiburg, 195 pp.

Hüttl RF. 1988. Forest decline and nutritional disturbances. In: DW Cole, SP Gessel, Eds. Forest Site Evaluation and Long-term Productivity. University of Washington Press. 180–186.

Hüttl RF, Fink S. 1988. Diagnostische Düngungsversuche zur Revitalisierung geschädigter Fichtenbestände (Picea abies (L.) Karst.) in Südwestdeutschland. Forstw. Cbl. 107, 173–183.

Hüttl RF. 1989. Neuartige Waldschäden aus dem Blickwinkel der Waldernährungslehre. Kali-Briefe. 19, 367–389.

Hüttl RF, Zöttl HW. 1993. Liming as mitigation tool in Germany's declining forests – reviewing results from former and recent trials. For. Ecol. Manage. 61, 325–338.

Ilvesniemi H. 1992. The combined effect of mineral nutrition and soluble aluminium on Pinus sylvestris and Picea abies seedlings. For. Ecol. Manage. 51, 227–238.

Ingestad T. 1970. A definition of optimum nutrient requirements in birch seedlings. I. Physiol. Plant. 23, 1127–1138.

Ingestad T. 1971. A definition of optimum nutrient requirements in birch seedlings. II. Physiol. Plant. 24, 118–125.

Ingestad T. 1974. Towards optimum fertilization. Ambio. 3, 49–54.

Ingestad T. 1976. Nitrogen and cation nutrition of three ecologically different plant species. Physiol. Plant. 38, 29–34.

Ingestad T. 1979. Mineral nutrient requirements of Pinus sylvestris and Picea abies seedlings. Physiol. Plant. 45, 373–380.

Jia H, Ingestad T. 1984. Nutrient requirements and stress response of Populus simonii and Paulownia tomentosa. Physiol. Plant. 62, 117–124.

Johnson DW, Van Miegroet H, Lindberg SE, Todd DE, Harrison RB. 1991. Nutrient cycling in red spruce forests of the Great Smoky Mountains. Can. J. For. Res. 21, 769–787.

Johnson DW, Kelly JM, Swank WT, *et al.*, 1988. The effects of leaching and whole-tree harvesting on cation budgets of several forests. J. Environ. Qual. 17, 418–424.

Hecht-Buchholz Ch. 1985. Aluminiuminduzierter Magnesium- und Calciummangel im Laborversuch bei Fichtensämlingen. Allg. Forstz. 46, 1248–1252.

Junga U. 1984. Sterilkultur als Modellsystem zur Untersuchung des Mechanismus der Aluminium-Toxizität bei Fichtenkeimlingen (Picea abies (L.) Karst.). Ber. des Forschungszentrums Waldöko-systeme/Waldsterben 5, Göttingen, 173 pp.

Kaupenjohann M, Zech W. 1987. Walddüngung und neuartige Waldschäden: Ergebnisse aus Düngungs-und Kalkungsversuchen. In G Glatzel, Ed. Möglichkeiten und Grenzen der Sanierung immissionsgeschädigter Waldökosysteme. pp. 82–98. FIW, Universität für Bodenkultur, Wien.

Kaupenjohann M, Zech W, Hantschel R, Horn R. 1987. Ergebnisse von Düngungsversuchen mit Magnesium an vermutlich immissionsgeschädigten Fichten (Picea abies (L.) Karst.) im Fichtelgebirge. Forstw. Cbl. 106, 78–84.

Kaupenjohann M. 1989. Chemischer Bodenzustand und Nährelementversorgung immissionsbelasteter Fichtenbestände in NO-Bayern. Bayreuther Bodenkundliche Berichte 11, Bayreuth, 202 pp.

Kaupenjohann M, Forster JC, Bäumler R. 1994. Düngung von Weihnachtsbaum- und Schmuckreisigbe-ständen. AFZ. 14, 797–800.

Kayser AT, Wilcke W, Kaupenjohann M, Joslin JD. 1994. Small scale heterogeneity of soil chemical properties I. A technique for rapid aggregate fractionation. Z. Pflanzenernähr. Bodenk. 157, 453–458.

Kelly JM, Barber SA. 1991. Magnesium uptake kinetics in loblolly pine seedlings. Plant Soil. 134, 227–232.

Kenk G, Unfried P, Evers FH, Hildebrand EE. 1984. Düngung zur Minderung der neuartigen Waldschäden – Auswertung eines alten Düngungsversuchs zu Fichte im Bundsandstein-Odenwald. Forstw. Cbl. 103, 307–320.

Knoepp JD, Swank WT. 1994. Long-term soil chemistry changes in aggrading forest ecosystems. Soil Sci. Soc. Am. J. 58, 325–331.

Koide RT. 1991. Nutrient supply, nutrient demand and plant response to mycorrhizal infection. New Phytol. 117, 365–389.

Le Goaster S, Dambrine E, Ranger J. 1990. Mineral supply of healthy and declining trees of a young spruce stand. Water Air Soil Pollut. 54, 269–280.

Leonardi S, Flückiger W. 1989. Effects of cation leaching on mineral cycling and transpiration: investi-gations with beech seedlings, Fagus sylvatica L. New Phytol. 111, 173–179.

Liu J-Ch. 1988. Ernährungskundliche Auswertung von diagnostischen Düngungsversuchen in Fichtenbeständen (Picea abies (L.) Karst.) Südwestdeutschlands. Freiburger Bodenkundl. Abh. 21, Freiburg, 193 pp.

Liu J-Ch, Trüby P. 1989. Bodenanalytische Diagnose von K- und Mg-Mangel in Fichtenbeständen (Picea abies (L.) Karst.). Z. Pflanzenernähr. Bodenk. 152, 307–311.

Marschner H, Römheld V, Horst WJ, Martin P. 1986. Root-induced changes in the rhizosphere: importance for the mineral nutrition of plants. Z. Pflanzenernähr. Bodenk. 149, 441–456.

Marschner H, Römheld V, Cakmak I. 1986. Root-induced changes of nutrient availability in the rhizosphere. J. Plant Nutr. 10, 1175–1184.

Mengel K, Kirkby EA. 1987. Principles of Plant Nutrition. International Potash Institute Bern, Worblaufen-Bern. 687 pp.

Mengel K, Lutz H-J, Breininger M. 1987. Auswaschung von Nährstoffen durch sauren Nebel aus jungen intakten Fichten (Picea abies (L.) Karst.). Z. Pflanzenernähr. Bodenk. 15, 61–68.

Mengel K, Breininger M Th, Lutz HJ. 1990. Effect of simulated acidic fog on carbohydrate leaching, CO_2 assimilation and development of damage symptoms in young spruce trees (Picea abies (L.) Karst.). Environ. Exp. Bot. 30, 165–173.

Miller HG. 1981. Nutrient cycles in forest plantations, their change with age and the consequence for fertilizer practice. In: Productivity and Perpetuity. Proceedings of the Australian Forest Nutrition Workshop. pp. 187–199. Canberra.

Monk CD, Day FP. 1985. Vegetation analysis, primary production and selected nutrient budgets for a southern Appalachian oak forest: a synthesis of IBP studies at Coweeta. For. Ecol. Managm. 10, 87–113.

Murach D. 1984. Die Reaktion der Feinwurzeln von Fichten (Picea abies (L.) Karst.) auf zunehmende Bodenversauerung. Göttinger Bodenkundliche Berichte 77, Göttingen, 126 pp.

Nye PH, Tinker PB. 1977. Solute Movement in the Soil-Root System. Blackwell, Oxford, 342 pp.

296

Oren R, Schulze E-D, Werk KS, Meyer J. 1988. Performance of two Picea abies (L.) Karst. stands at different stages of decline. VII. Nutrient relations and growth. Oecologia. 77, 163–173.

Pape Th, van Breemen N, van Oeveren H. 1989. Calcium cycling in an oak–birch woodland on soils of varying CaCO₃ content. Plant Soil. 120, 253–261.

Pehl ChE, Tuttle ChL, Houser JN, Moehring DM. 1984. Total biomass and nutrients of 25-year-old loblolly pines (Pinus taeda L.). For. Ecol. Manage. 9, 155–160.

Potter CS. 1990. Nutrient leaching from Acer rubrum leaves by experimental acid rainfall. Can. J. For. Res. 21, 222–229.

Sauter U. 1989. Düngung zur Vitalisierung neuartig geschädigter Waldbestände in Bayern. Kali-Briefe (Büntehof). 19 (6), 443–459.

Sauter U. 1991. Versuche zur Wirkung von sulfatisch, carbonatisch und silikatisch gebundenem Magnesium auf Ernährungszustand und Wachstum junger Fichten, chemischen Bodenzustand und Sickerwasserbefrachtung. Forstl. Forschungssber. München. 114, München, 442 pp.

Schaaf W. 1992. Elementbilanz eines stark geschädigten Fichtenökosystems und deren Beeinflussung durch neuartige basische Magnesiumdünger. Bayreuther Bodenkundliche Berichte 23, Bayreuth, 169 pp.

Schulze E-D, Oren R, Lange OL. 1989. Nutrient relations of trees in healthy and declining Norway spruce stands. In E-D Schulze, OL Lange, R Oren, Eds. Forest Decline and Air Pollution. Ecol. Stud. 77, 393–417.

Seggewiss B, Jungk A. 1988. Einfluß der Kaliumdynamik im wurzelnahen Boden auf die Magnesiumaufnahme von Pflanzen. Z. Pflanzenernähr. Bodenk. 151, 91–96.

Stevens PA, Hornung M, Hughes S. 1989. Solute concentrations, fluxes and major nutrient cycles in a mature Sitka-spruce plantation in Beddgelert Forest, North Wales. For. Ecol. Manage. 27, 1–20.

Stienen H, Bauch J. 1988. Element content in tissues of spruce seedlings from hydroponic cultures simulating acidification and deacidification. Plant Soil. 106, 231–238.

St. John TV. 1980. Root size, root hairs and mycorrhizal infection: A re-examination of Baylis's hypothesis with tropical trees. New Phytol. 84, 483–487.

Switzer GL, Nelson LE. 1972. Nutrient accumulation and cycling in loblolly pine (Pinus taeda L.) plantation ecosystems: The first twenty years. Soil Sci. Soc. Am. Proc. 36, 143–147.

Ulrich B. 1989. Effects of acid deposition on forest ecosystems in Europe. Adv. Environ. Sci. 2, 189–272.

Van den Driessche R. 1984. Nutrient Storage, Retranslocation and Relationship of Plantation Forests. Academic Press, London, 181–209.

Van Rees KCJ, Comerford NB, McFee WW. 1990. Modeling potassium uptake by slash pine seedlings from low potassium supplying soils of the southeastern coastal plain. Soil Sci. Soc. Am. J. 54, 1413–1421.

Zapf R. 1995. Kieseritdüngung auf sauren Waldböden. MgSO₄-Sorption und Auswirkungen auf den Bodenlösungschemismus. Mitteilgn. Deutsch. Bodenkundl. Gesellsch. 176/II, 975–978.

Zech W, Popp E, 1983. Magnesiummangel, einer der Gründe für das Fichten- und Tannensterben in NO-Bayern. Forstw. Cbl. 102, 50–55.

Zöttl HW. 1990. Ernährung und Düngung der Fichte. Forstw. Cbl. 109, 130–137.

9
Structural aspects of magnesium deficiency

S. FINK

9.1. Introduction

Structural changes of cells and tissues in herbaceous agricultural plants caused by nutrient deficiencies have already been studied extensively at the levels of both light and electron microscopy, and have partially proved to be of considerable diagnostic value (Thomson and Weier, 1962; Thomson *et al.*, 1964; Bussler, 1964, 1981; Whatley, 1971; Hecht-Buchholz, 1972, 1983; Pissarek, 1973, 1979; Bangerth, 1979). In trees, however, similar studies have been carried out only in a few cases, and, so far, the conifers especially have received little attention. There has been some limited research on the effects of Ca deficiency (Davis, 1949) and B deficiency (Blaser *et al.*, 1967; Raitio, 1979) upon the structure of needles, stems, and roots of *Pinus* spp. and *Thuja* spp. Recent interest in the general effects of mineral deficiencies upon the structure and function of conifers (especially *Picea abies* and *Pinus silvestris*) arose with the occurrence of 'New-Type Forest Decline', which proved to be associated in many instances with Mg deficiency at high elevations on acid soils (Type I Spruce Decline; FBW, 1986; Roberts *et al.*, 1989) or, to a lesser extent, K deficiency on calcareous soils (Type IV Spruce Decline; FBW, 1986). In the light of initial discussions on the causes of this damage (primarily soil changes by increased input of acidity or primarily direct effects of air pollutants upon the foliage), anatomical investigations of conifer needles subjected to either mineral deficiencies or various gaseous pollutants have proved to be a valuable tool for a differential diagnosis, if compared with structural changes found in affected needles from declining trees in the field (Fink, 1988, 1989, 1991, 1993; Holopainen and Nygren, 1989; Barsig *et al.*, 1990; Holopainen *et al.*, 1992). In this context, some of the most pronounced structural changes caused by Mg deficiency in Norway spruce needles will be described here, and discussed with regard to their implications on the growth of affected trees. Furthermore, structural reorganization of cells and tissues after application of Mg fertilizer will be considered.

9.2. Normal structure of a spruce needle

In cross-section, a healthy green spruce needle is characterized by an intact epidermis with stomata, turgescent mesophyll cells, and a vascular bundle with open xylem, cambium, and phloem cells which are surrounded by transfusion cells and a closed circular bundle sheath (Figures 1 and 3). Furthermore, resin cavities lined by

R. F. Hüttl & W. Schaaf (eds): Magnesium Deficiency in Forest Ecosystems, 297–307.
© 1997 *Kluwer Academic Publishers. Printed in Great Britain.*

298

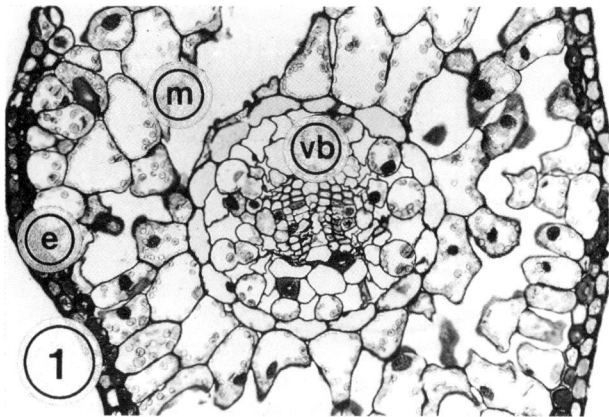

Figure 1. Cross-section through a healthy green 2-year-old spruce needle. e = epidermis, m = mesophyll, vb = vascular bundle; ×180

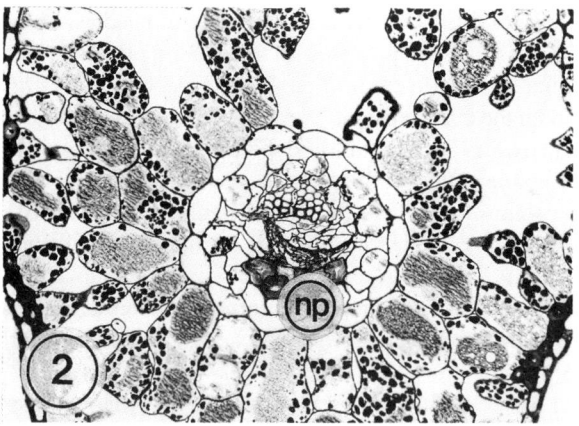

Figure 2. Cross-section through a corresponding yellow spruce needle suffering from Mg deficiency. np = necrotic phloem; ×160

epithelial cells may be found on one or both lateral sides of a needle (see also Marco, 1939).

During normal ageing processes, the principal changes appear in the phloem part of the vascular bundle. After each vegetation period, the active sieve cells collapse and are substituted in the following spring by new sieve cells from the active cambium. Consequently, in older needles a large number of obliterated phloem cells can be found. Xylem cells, on the contrary, stay functional during the lifetime of the needle and no new tracheids are normally differentiated from the cambium.

With regard to the main physiological functions of these structures, the main

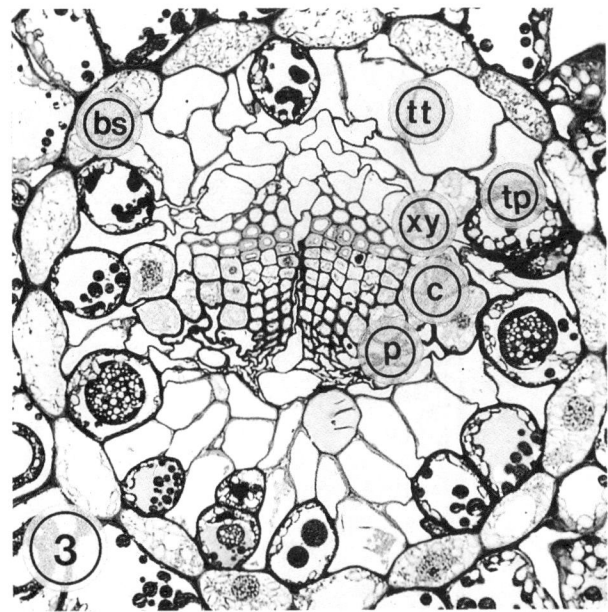

Figure 3. Detailed view of the vascular bundle of a healthy spruce needle. bs=bundle sheath, tt=transfusion tracheids, tp=transfusion parenchyma, xy=xylem, c=cambium, p=phloem; ×370

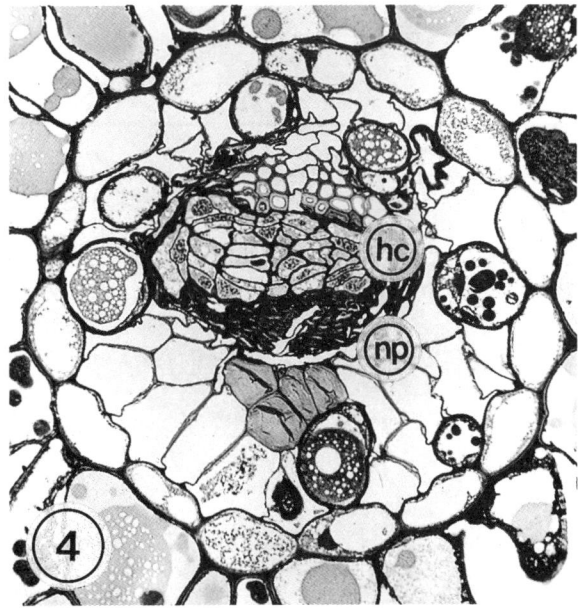

Figure 4. Vascular bundle from a corresponding yellow needle with Mg deficiency showing hyperplastic, disorganized cambium (hc) and necrotic phloem (np); ×340

300

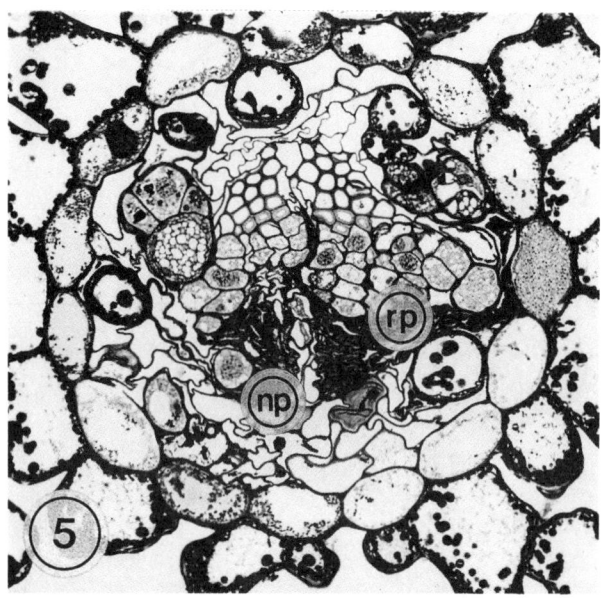

Figure 5. Vascular bundle of a 2-year-old spruce needle which had been suffering from Mg deficiency but regreened within a few months after fertilization with Mg sulfate; necrotic phloem (np) is still evident but regenerated functional phloem (rp) is beginning to develop from the reorganized cambium; ×330

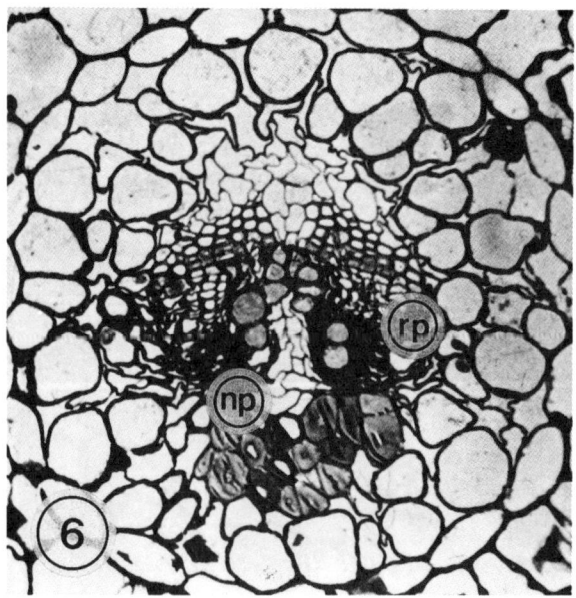

Figure 6. Vascular bundle of a 3-year-old spruce needle which regreened after Mg fertilization; regenerated phloem (rp) has substituted the old necrotic phloem (np); ×280

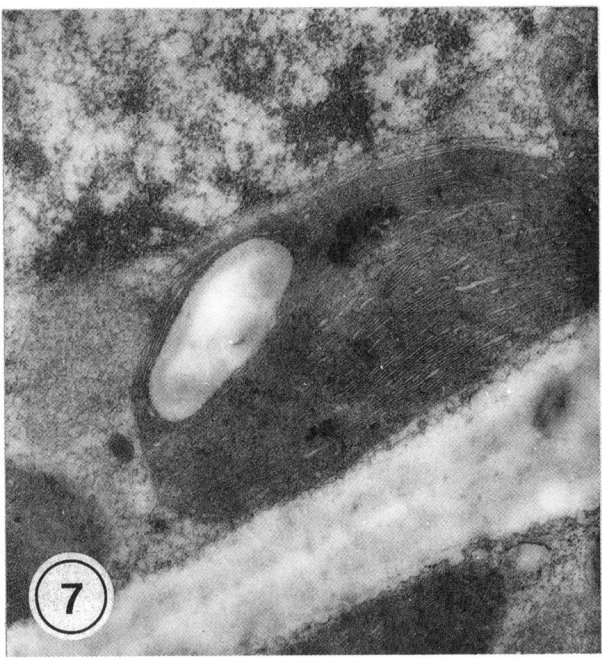

Figure 7.   Chloroplast from a healthy green 2-year-old spruce needle with very few plastoglobuli and a small starch grain; ×26000

pathway of water and minerals into the needle is via the vascular tracheids and subsequently the transfusion tracheids up to the bundle sheath. Since here apoplastic transport seems to be at least partially blocked by the heavy lignin incrustations into the radial walls (similar to the suberin incrustations of the Casparian strips in root endodermis), water and dissolved minerals are forced into the symplast, and thus come under the active control of the cells for further transport into the mesophyll (see Scholz and Bauch, 1973). Here, water evaporates from the cell walls into the intercellular spaces and the water vapour finally leaves through the stomata. In the other direction, assimilated sugars from the mesophyll cells pass the bundle sheath and are transported to the phloem by means of the transfusion parenchyma cells. The loading of the sieve cells with these sugars is then accomplished by the 'Strasburger cells', specialized cells of the phloem rays which control and regulate the activities of the sieve cells (Sauter and Braun, 1968). Phloem-mobile minerals, such as magnesium or potassium, may take the same route to leave older needles, and are transported into younger needles when there is strong demand and insufficient supply. Phloem-immobile minerals, such as calcium and manganese, cannot be transported out of the needle in this way and consequently accumulate with age.

On the ultrastructural level, the chloroplasts in the mesophyll cells are the most conspicuous organelles. In healthy needles, they exhibit a well-developed thylakoid

302

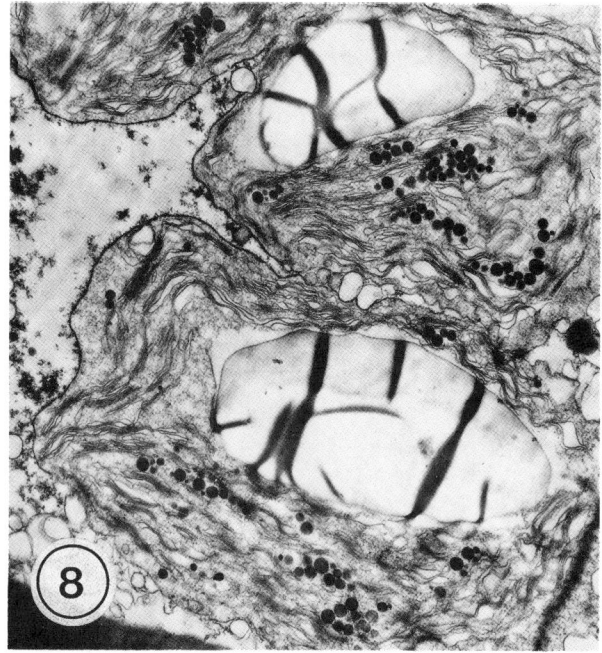

Figure 8. Chloroplast from a yellow 3-year-old spruce needle with Mg deficiency; the thylakoid system is swollen and disorganized and large starch grains are included; ×24000

system, few plastoglobuli, and no, or only small, starch grains (depending upon the time of sampling; Figure 7). During the normal ageing processes, the thylakoid membranes slowly become degraded and the degradation products accumulate as lipids in plastoglobuli, so that the amount of membranes decreases and the amount of plastoglobuli increases.

9.3. Structural changes due to Mg deficiency

A corresponding cross-section from a Norway spruce needle suffering from severe Mg deficiency shows marked changes, especially in the vascular bundle where nearly all sieve cells in the phloem may have collapsed, whereas the adjoining parenchyma and cambial cells are partially enlarged (Figures 2 and 4). The mesophyll cells stay turgescent but are filled with many starch grains (black dots in Figure 2). This is either the consequence of the phloem collapse, since the carbohydrates produced by still-ongoing photosynthesis can no longer be transported out of the needle, or the cause of the phloem collapse, since supply of sugars to maintain the high turgor within active sieve cells may be lacking if translocation of sugars to the phloem is already inhibited within the mesophyll cells. These patterns of alterations in chloroplast structures and phloem integrity were observed early in yellow

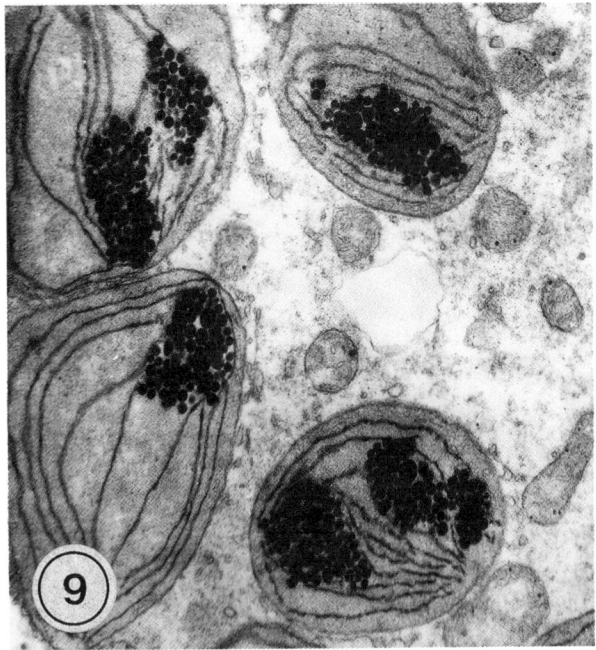

Figure 9. Other chloroplasts from a similar needle showing a much-reduced membrane system and numerous plastoglobuli; ×15 000

needles of declining trees from the field (e.g. in high regions of the Black Forest) in current research on 'New-Type Forest Decline', though they were originally thought to be possibly caused by the direct impact of gaseous air pollutants (Fink, 1983; Parameswaran *et al.*, 1985; Sutinen, 1987). Later controlled studies showed, however, that these gaseous pollutants cause quite different structural symptoms, which are mainly characterized by necrosis of mesophyll cells (especially close to stomata where the pollutants enter), whereas the vascular bundles are much less affected (Fink, 1988, 1989, 1991, 1993; Holopainen *et al.*, 1992). Identical structural changes in yellow needles from the field were found in needles from culture experiments with Mg or K deficiency, as described in the present paper. The effects upon the structure of the phloem cells caused by mineral deficiencies are so pronounced that they may even become visible before any alterations in chloroplast structure in the mesophyll cells are obvious (Schmitt *et al.*, 1986). Thus, these first changes in vascular bundles can be used as a microscopic symptom for the diagnosis of mineral deficiencies in the needles.

The peculiar phloem collapse in Mg-deficient needles, on the other hand, apparently has not yet been described in deficient angiospermous plants, where collapse and necrosis seem to be restricted to mesophyll cells in the leaves and

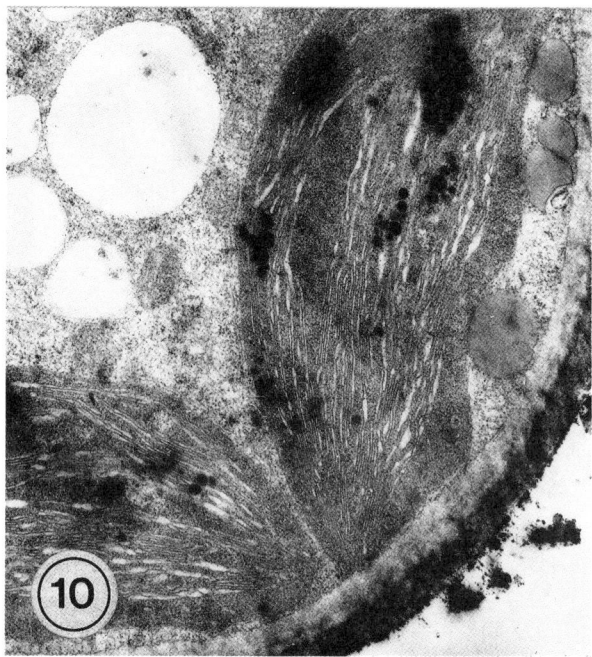

Figure 10. Chloroplasts from a 3-year-old needle which originally suffered from Mg deficiency but which regreened within a year after Mg fertilization; a slight swelling of thylakoids and certain number of plastoglobuli can still be discerned but, in general, the structure has changed back to the normal functional appearance of chloroplasts; ×23 000

cortical cells in the petioles and stems (Bussler, 1964; Pissarek, 1973). The different reaction in conifers may perhaps be due to the specific symplastic mechanisms involved in phloem loading in conifer needles (Blechschmidt-Schneider, 1989). From the physiological point of view, Mg and K are essential for the activation of ATPases involved in phloem loading. Furthermore, K is the most important inorganic osmoticum for maintaining the high turgor of sieve cells. The exact physiological mechanisms for the induction of phloem necrosis under mineral deficiencies have not yet been conclusively evaluated on an experimental basis, however. The observed starch accumulation has been found in a number of declining spruce stands (Parameswaran *et al.*, 1985; Fink, 1988, 1993; Jung and Wild, 1988; Forschner *et al.*, 1989; Hasemann and Wild, 1990) and also in herbaceous plants with Mg deficiency (Marschner, 1986; Fischer and Bussler, 1988). This inhibition of the translocation of carbohydrates to stem and roots may have negative influences, especially upon root development, which then in turn will further decrease the capacity for cation absorption. There is circumstantial evidence that, in declining spruce stands, root development is primarily impeded by reduced assimilate production and transport from the shoots to the roots (Roberts *et al.*, 1989). Also controlled culture experiments have recently shown that, in the case of Mg or K deficiency, less carbohydrates are allocated to the roots (Ericsson, 1989).

At the level of cellular ultrastructure, Mg and K deficiencies lead to a reduction and swelling of the thylakoids in chloroplasts, which are frequently also distorted because of the occurrence of large starch grains; reduced and irregularly organized grana compartments and separation of thylakoid membranes which form electron-translucent gaps are typical, as well as an increased amount of plastoglobuli (Figures 8 and 9). Such changes have been reported in a number of tree species and in herbaceous plants, though they cannot be regarded as specific for Mg deficiency (Whatley, 1971; Hamzah and Gomez, 1979; Fink, 1989, 1993; Holopainen and Nygren, 1989; Holopainen et al., 1992). These structural changes, together with the decrease in the amount of chlorophyl in the yellow needles, point to a reduction of photosynthetic capacity, though assimilation is still continuing to an extent such that appreciable quantities of carbohydrates can accumulate as starch in these cells. These structural changes are essentially the same in the Mg-deficient needles from the field and from culture experiments with defined induction of deficiencies (Fink, 1988, 1991, 1993). It must be stressed that, in spite of the partial disorganization of the organelles, the mesophyll cells in yellow needles may remain alive and turgescent for 2–4 years.

9.4. Structural reorganization after fertilization

Macroscopically visible regreening of yellow needles has been observed sometimes only a few months after application of easily soluble Mg salts as fertilizers (Huettl and Fink, 1988; Huettl, 1991). Microscopically, this regreening process is accompanied by a structural reorganization of the needles. In the vascular bundle, the still functional, though somewhat disorganized, cambium starts to differentiate new functional sieve cells again so that the translocation capacity is restored and starch accumulation disappears. On the ultrastructural level, chloroplasts repair their damaged membrane system resp. become 'digested' in lytic vacuoles (see Matile, 1975) and new, intact chloroplasts are formed by division of existing ones (Figure 10). In this way, the visible increase in chlorophyll content is paralleled by a reorganization of the chloroplast structures (Huettl and Fink, 1988).

The main prerequisite for this type of quick and successful recovery is the fact that the mesophyll cells and the cambial cells are able to stay turgescent and alive for several years even though they are heavily disorganized. This is in contrast to the effect of other mineral deficiencies (e.g. K deficiency) or the direct impact of gaseous pollutants, which lead to a partial necrosis, especially of mesophyll cells (Fink, 1988, 189, 1993; Holopainen et al., 1992). Dead cells, however, cannot become revitalized even when growth conditions become much more favorable. In this respect, the apparent ability of the cells in the conifer needles to survive Mg deficiency for some years is a peculiar phenomenon, which helps in the success of quick revitalization measures.

In general, evergreen conifers are likely to exhibit such positive effects since they have a high inherent regeneration capability. Each year, during winter, the sieve cells routinely collapse and the photosynthetic apparatus suffers from considerable damage, so that each spring the sieve cells have to be formed again and the damaged chloroplasts have to be repaired (Senser *et al.*, 1975; Senser and Beck, 1979). Thus, in a way, the above described reorganization of Mg-deficient needles after Mg fertilization is at least partially analogous to the springtime recovery and surely takes advantage of the inherent self-repair mechanisms of the long-living evergreen needles.

9.5. References

Bangerth F. 1979. Calcium-related physiological disorders of plants. Ann. Rev. Phytopathol. 17, 97–122.

Barsig M, Endler W, Weese G, Hafner L. 1990. Cytomorphologische Untersuchungen an Nadeln von Kiefern (*Pinus silvestris* L.) eines Ballungsgebietes. II. Das Spektrum feinstruktureller Schadphänomene. Angew Bot. 64, 303–315.

Blaser HW, Marr C, Takahashi D. 1967. Anatomy of boron-deficient *Thuja plicata*. Am. J. Bot. 54, 1107–1113.

Blechschmidt-Schneider S. 1989. Phloembeladung bei *Picea abies* (L.) Karst. Physiologische Betrachtungen. Kali-Briefe. 19, 467–489.

Bussler W. 1964. Comparative examinations of plants suffering from potash deficiency. Weinheim: Verlag Chemie.

Bussler W. 1981. Microscopical possibilities for the diagnosis of trace element stress in plants. J. Plant Nutr. 3, 115–128.

Davies DE. 1949. Some effects of calcium deficiency on the anatomy of *Pinus taeda*. Am. J. Bot. 36, 276–282.

Ericsson T. 1989. Dry matter partitioning in birch seedlings – a balance between internal nitrogen and carbon fluxes (Abstr). Proceedings International Conference on Nitrogen-fixing Trees and Fast-growing Trees, Marburg, 8–12.10.1989.

FBW (Forschungsbeirat Waldschäden/Luftverunreinigungen). 1986. 2. Bericht. KfK Karlsruhe, 229 pp.

Fink S. 1983. Histologische und histochemische Untersuchungen an Nadeln erkrankter Tannen und Fichten im Südschwarzwald. Allg. Forstz. 38, 660–663.

Fink S. 1988. Histological and cytological changes caused by air pollutants and other abiotic factors. In: Schulte-Hostede S, Darrall NM, Blank LW, Wellburn AR, eds. Air Pollution and Plant Metabolism. London, New York: Elsevier Appl. Sci. Publ. pp. 36–54.

Fink S. 1989. Pathological anatomy of conifer needles subjected to gaseous pollutants or mineral deficiencies. Aquilo Ser. Bot. 27, 1–6.

Fink S. 1991. Structural changes in conifer needles due to Mg and K deficiency. Fertil. Res. 27, 23–27.

Fink S. 1993. Microscopical criteria for the diagnosis of abiotic injuries to conifer needles. In: Huettl RF, Mueller-Dombois D, Eds. Forest Decline in the Atlantic and Pacific Region, Springer-Verlag, Berlin, Heidelberg, New York, pp. 175–188.

Fischer ES, Bussler W. 1988. Effects of magnesium deficiency on carbohydrates in *Phaseolus vulgaris*. Z. Pflanzenernähr Bodenk. 151, 295–298.

Forschner W, Schmitt V, Wild A. 1989. Investigations on the starch content and ultrastructure of spruce needles relative to the occurrence of novel forest decline. Botanica Acta. 102, 208–221.

Hamzah S, Gomez JB. 1979. Ultrastructure of mineral deficient leaves of *Hevea*. I. Effects of macronutrient deficiencies. J. Rubber Res. Inst. Malaysia. 27, 132–142.

Hasaemann G, Wild A. 1990. The loss of structural integrity in damaged spruce needles from locations exposed to air pollution. I. Mesophyll and central cylinder. J. Phytopathol. 128, 15–32.

Hecht-Buchholz C. 1972. Wirkung der Mineralstoffernährung auf die Feinstruktur der Pflanzenzelle. Z. Pflanzenernähr Bodenk. 132, 45–69.

Hecht-Buchholz C. 1983. Light and electron microscopic investigations of the reactions of various genotypes to nutritional disorders. Plant Soil. 72, 151–165.

Holopainen T, Nygren P. 1989. Effects of potassium deficiency and simulated acid rain, alone and in combination, on the ultrastructure of Scots pine needles. Can. J. For. Res. 19, 1402–1411.

Holopainen T, Anttonen S, Wulff A, Palomäki V, Kärenlampi L. 1992. Comparative evaluation of the effects of gaseous pollutants, acidic deposition and mineral deficiencies: structural changes in the cells of forest plants. Agric. Ecosyst. Environ. 42, 365–398.

Huettl RF. 1991. Die Nährelementversorgung geschädigter Wälder in Europa und Nordamerika. Freib. Bodenkundl. Abhandl. 28, 440.

Huettl RF, Fink S. 1988. Diagnostische Düngungsversuche zur Revitalisierung geschädigter Fichtenbestände (*Picea abies* Karst.) in Südwestdeutschland. Forstw. Cbl. 107, 173–183.

Jung G, Wild A. 1988. Electron microscopic studies of spruce needles in connection with the occurrence of novel forest decline. I. Investigations of the mesophyll. J. Phytopathol. 122, 1–12.

Marco HF. 1939. The anatomy of spruce needles. J. Agric. Res. 58, 357–368.

Marschner H. 1986. The Mineral Nutrition of Higher Plants. New York, London: Academic Press.

Matile P. 1975. The lytic compartment of plant cells. Cell Biol. Monogr. 1, Springer-Verlag, Wien, 183.

Parameswaran N, Fink S, Liese W. 1985. Feinstrukturelle Untersuchungen an Nadeln geschädigter Tannen und Fichten aus Waldschadensgebieten im Schwarzwald. Eur. J. For. Pathol. 15, 168–182.

Pissarek H-P. 1973. Zur Entwicklung der Kalium-Mangelsymptome an Sommerraps. Z. Pflanzenernähr Bodenk. 136, 1–19.

Pissarek H-P. 1979. Einfluß von Ca-Mangel auf den Aufbau des Blatt- und Stegelgewebes von Mais und Sonnenblumen. Angew Botanik. 53, 215–224.

Raitio H. 1979. Growth disturbances of Scots pine caused by boron deficiency on an afforested abandoned peatland field. Description and interpretation of symptoms. Folia Forestalia. 412, 1–16.

Roberts TM, Skeffington RA, Blank LW. 1989. Causes of type I spruce decline in Europe. Forestry. 62, 179–222.

Sauter JJ, Braun HJ. 1969. Histologische und zytochemische Untersuchungen zur Funktion der Baststrahlen von *Larix decidua* Mill., unter besonderer Berücksichtigung der Strasburger-Zellen. Z. Pflanzenphysiol. 59, 420–438.

Schmitt U, Liese W, Ruetze M. 1986. Ultrastrukturelle Veeränderungen in grünen Nadeln geschädigter Fichten. Angew Botanik. 60, 441–450.

Scholz F, Bauch J. 1973. Anatomische und physiologische Untersuchungen zur Wasserbewegung in Kiefernnadeln. Planta. 109, 105–119.

Senser M, Beck E. 1979. Kälteresistenz der Fichte. II. Einfluß von Photoperiode und Temperatur auf die Struktur und photochemischen Reaktionen von Chloroplasten. Ber. Dt. Bot. Ges. 92, 243–259.

Senser M, Schoetz F, Beck E. 1975. Seasonal changes in structure and function of spruce chloroplasts. Planta. 126, 1–10.

Sutinen S. 1987. Cytology of Norway spruce needles. II. Changes in yellowing spruces from the Taunus Mountains, West Germany. Eur. J. For. Pathol. 17, 74–85.

Thomson WW, Weier TE. 1962. The fine structure of chloroplasts from mineral-deficient leaves of *Phaseolus vulgaris*. Am. J. Bot. 49, 1047–1055.

Thomson WW, Weier TE, Drever H. 1964. Electron-microscopic studies on chloroplasts from phosphorus-deficient plants. Am. J. Bot. 51, 933–939.

Whatley JM. 1971. Ultrastructural changes in chloroplasts of *Phaseolus vulgaris* during development under conditions of nutrient deficiency. New Phytol. 70, 725–742.

10
Fine-root development

S. RASPE

The root system represents the connecting link between the soil and the above-ground part of the plant. The roots of forest trees have to fulfil different functions. They anchor the trees in the mineral soil and take up water and nutrients from the soil. Making the soil accessible, they change its physical and chemical properties. Moreover, they also store water and organic reserves, such as starch and lipids. Lastly, they are a 'laboratory for biochemical synthesis' (Kern et al., 1961) by producing phytohormones, such as cytokinine.

The functions of the root system are divided within different diameter classes. Stronger roots ($\phi > 2$ mm) hold and support the tree as well as providing storage and transport of nutrients. They constitute most of the below-ground biomass (Kodrik, 1991; Raspe et al., 1989; Raspe, 1992). The uptake of water and nutrients, however, takes place through non-lignified parts of finest ($\phi < 1$ mm) and fine roots ($\phi < 2$ mm) and their mycorrhizae. Also, the biosynthetic functions of the roots occur in the tips of fine roots. High turnover rates of fine roots enable the root system to react in a sensible way to different conditions and provide organic material to the soil.

Root systems of tree species are determined genetically (Gruber, 1992). However, the external form is modified by conditions. The physical, chemical, and biological properties of soils have profound effects on the rate of root growth and development. Therefore, the habit and intensity of root systems of established trees are determined by the soil conditions to a considerable extent (Pritchett and Fisher, 1987). Variations in root systems among individuals of the same species grown in different soils, consequently, may be as great as those among different species on the same soil (Wiedemann, 1927). Thus a root system can act as an indicator of the soil conditions on which it was developed (Weller, 1965). Besides direct influences in the soil, environmental influences affect the root system indirectly through the above-ground parts of the plant.

Fertilization directly affects the nutrient supply in the soil and the mobilization vs. demobilization of potentially toxic ions. Fertilization can also provoke changes in the spectrum of soil organisms (meso- and macrofauna, fungi, micro-organisms). Soil temperature, texture, aeration and moisture are, however, only slightly influenced.

The factors influencing root growth indirectly are closely related to the amount of assimilates produced in the plant. These factors are mainly light, temperature and humidity (Göttsche, 1972a), but also the CO_2 concentration and gases, like ozone, which damage the assimilation organs (Prinz et al., 1982) as well as SO_2 (Keller,

309

R. F. Hüttl & W. Schaaf (eds): Magnesium Deficiency in Forest Ecosystems, 309–332.
© 1997 Kluwer Academic Publishers. Printed in Great Britain.

1979). The quantity of photosynthesis has a great influence on root growth. Huss (1977) showed that reduced photosynthesis inhibits root growth considerably.

The direct and indirect environmental factors are combined in the nutrient supply (Feger and Raspe, 1992). The pattern of distribution of assimilates to the above- and below-ground biomass can be changed by nutrient deficiency. Fine-root growth is reduced in the case of Mg deficiency due to insufficient assimilate supply (Hüttl, 1991).

Therefore, optimal root development presumes a sufficient Mg supply in the soil. At Mg-deficient sites, the fine-root growth can be improved by special Mg additions. Therefore, a chemotropic impulse for fine-root growth originates from the dissolution of Mg fertilizer, dependent on its depth. Owing to the higher Mg uptake, root growth is further stimulated indirectly by a higher photosynthetic production and translocation of assimilates from the needles. The direct and indirect effects of fertilization on the fine-root system should therefore always be considered together.

Although dolomitic liming and Mg fertilization are currently commonly used to restabilize damaged forests, the effects on the root systems of trees have seldom been examined so far. Before a summary of these investigations can be presented, the importance of Mg supply for fine-root development should be demonstrated.

10.1. Soil chemical conditions, root development and Mg uptake

A general estimate of the total Norway spruce root dry matter growth rate is presented in section 4.5.8. of Chapter 4. Nutrient availability in the soil must affect the growth rate, distribution and nutrition of roots. By nutrient solution experiments with birch seedlings, Ericsson (1990) was able to show that the assimilation supply of roots can be changed by nutrient deficiency (Figure 1). In the case of N, P or S deficiency, root growth is stimulated relative to shoot growth by intensified translocation of carbohydrates. In the case of Mg deficiency, however, the transfer of carbohydrates into roots is clearly inhibited. Therefore, fine-root development is restricted by nutrition and fine-root biomass is closely correlated to the Mg supply in the soil (Hüttl, 1991). Mg deficiency leads to a reduced content of chlorophyl and carotinoids (Beyschlag et al., 1987) as well as changes in enzyme activity of chlorophyl oxidation (Fink, 1983). This leads to a reduction of the capacity for photosynthesis (Mehne, 1989). In addition, the translocation of assimilates from needles into roots is restrained by a collapse of the phloem (Fink, 1988, 1992). Furthermore, inhibition of root growth reinforces the reduction of Mg, Ca and water uptake as these uptakes are especially high in apical zones of growing roots (Häußling et al., 1988); in this connection, the number of active root tips is particularly important.

Gonzalez Cascon et al. (1988) proved in pot experiments with different soils, that the growth of fine roots of immature silver fir depends on the Mg supply. Under field conditions, there is also a close relationship between the amount of fine roots and the Mg supply in the soil. This was shown by Raspe (1992) for a 50-year-old Norway spruce stand grown on a podzol which had originated from granite, and which is extremely poor in base cations. The strongest correlation was found

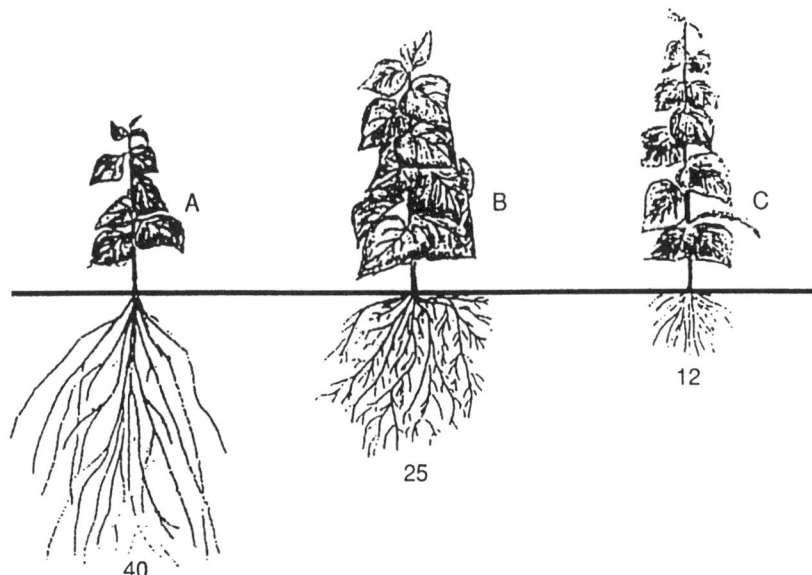

Figure 1. Root distribution and percentage of below-ground biomass from total dry matter of birch seedlings cultivated in nutrient solution with N deficiency (A), sufficient nutrition (B) and Mg deficiency (C). (From Ericsson, 1989 in Hüttl, 1991)

between the fine-root intensity and the exchangeable Mg^{2+} content of the soil, if depth intervals were stratified (Table 1). Also the Mg content of fine roots is closely correlated with the Mg supply in solid soil (Figure 2) and soil solution (Table 2). Moreover, Schneider and Zech, 1989 as well as Eichhorn, 1987 observed close correlations between fine-root density and Mg content of both living fine roots and needles. The density of living fine roots (Figure 3) and the level of mycorrhizal infection (Figure 4) were also positively correlated with the Mg concentration in the soil solution (Meyer *et al.*, 1988). Although the statistical evaluation of these studies is problematic, due to the high spatial variability of fine-root parameters, nevertheless diagrams indicate close relationships between fine-root growth and Mg supply.

The rate of Mg uptake can be strongly depressed by other cations, such as K^+, NH_4^+, Ca^{2+}, and Mn^{2+}, as well as by H^+ (Marschner *et al.*, 1986). In acidified forest soils, concentrations of Al^{3+} and partly of Mn^{2+} are usually higher in the soil solution. Al inhibits principally Ca uptake whereas Mg uptake is mainly reduced by elevated Mn concentrations (Stienen and Bauch, 1988). These additional inhibitory effects reflect the importance of Ca^{2+} and Mg^{2+} loading at the cell walls in the apoplast of the root bark for their uptake into the roots (Marschner, 1992).

Cations like Ca^{2+} and Mg^{2+} are bound exchangeably to the carboxy groups in the cell walls. Due to this, the concentration of these ions is higher in the apoplast of the root bark than in the soil solution. Therefore their uptake into the cytoplasm is favoured. Above all, Mg^{2+} can be exchanged by other cations, like H^+, Mn^{2+} and

Table 1. Pearson coefficient of correlation (r) between fine-root intensity of a 50-year-old spruce stand and contents of exchangeable cations of an acidic granite site at the Black Forest

		Depth (cm)			
	Forest floor	0–10	10–20	20–30	30–40
K⁺	0.266	0.490*	0.244	0.174	0.379
Ca²⁺	−0.202	0.236	0.162	0.084	0.081
Mg²⁺	0.382	0.695**	0.576**	0.585**	0.142
Mn²⁺	−0.072	−0.175	−0.515*	−0.254	0.336
Na⁺	−0.026	0.191	−0.040	−0.119	0.110
Fe³⁺	−0.334	0.138	0.319	0.419	−0.007
Al³⁺	−0.268	−0.246	−0.17	0.160	0.301
H⁺	−0.121	0.522*	0.477*	0.447*	0.384
CEC	0.017	0.379	0.190	0.260	0.305
BS	0.193	0.494*	0.372	0.153	−0.013
n	36	36	36	36	36

CEC_e = effective cation exchange capacity NH_4Cl-extract; BS = base saturation
**$p < 0.001$; *$0.001 < p < 0.01$; *n* = number of samples

Table 2. Significant correlation (*r*-values) between the Ca and Mg contents in needles and roots and the element concentrations in the water extract

	Concentration in water extract		
	Ca	Mg	Al
Calcium			
Needles	+0.57⁺⁺		−0.21
Roots	+0.82⁺⁺⁺		−0.48⁺
Magnesium			
Needles		+0.83⁺⁺⁺	−0.07
Roots		+0.58⁺⁺	−0.34

⁺$\alpha = 0.05$; ⁺⁺$\alpha = 0.01$; ⁺⁺⁺$\alpha = 0.001$; *n* = 45. Based on Gonzalez Cascon *et al.*, 1989

especially monomeric Al species (Al^{n+}) like Al^{3+}. In the case of low Mg supply, antagonistic Al uptake can induce Mg deficiency easily, whereas high Mg supply rarely provokes growth depression (Marschner, 1988). Therefore stabilization of the root system by Mg fertilization is often essential for recovery from Mg deficiency on acid forest soils.

10.2. The objective of Mg fertilization on root development

Mg is taken up by the fine roots of trees. Recovery from diagnosed Mg deficiency is therefore closely connected with the growth and development of fine roots. For restabilization of the whole ecosystem, application of lime and fertilizer aims to improve the soil chemistry of the main rooting horizons (Hüttl, 1991). A successful fertilizer application will therefore stabilize the fine-root system in the mineral soil. Besides the whole fine-root biomass, the ratio of living to dead fine roots as well as

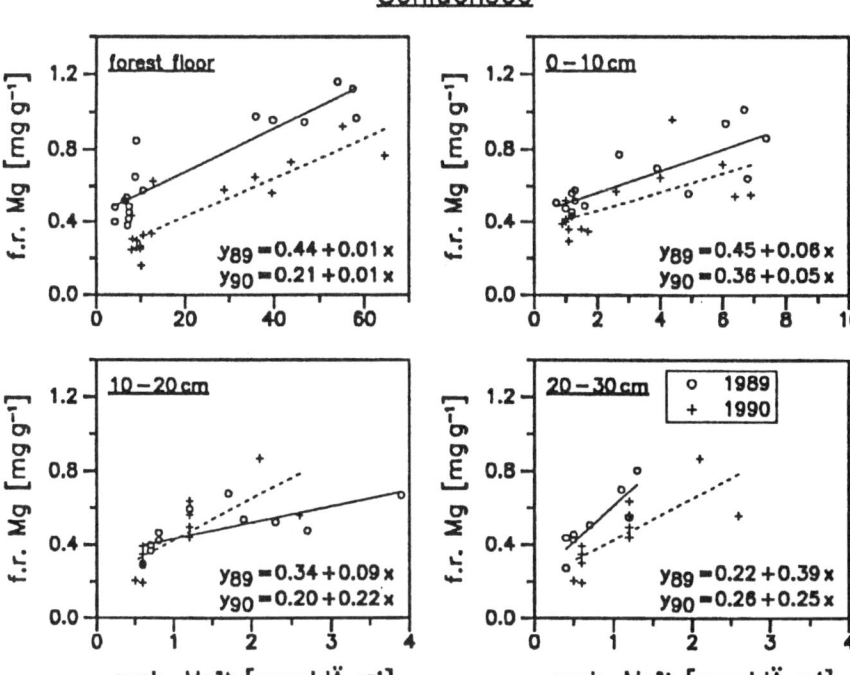

Figure 2. Correlation (linear regression) between Mg content of fine roots (f.r. Mg [mg/g⁻¹]) and exchangeable Mg content in bulk soil (exch. Mg²⁺ [μmol IÄ g⁻¹]) in different soil depths at the Schluchsee in the Black Forest

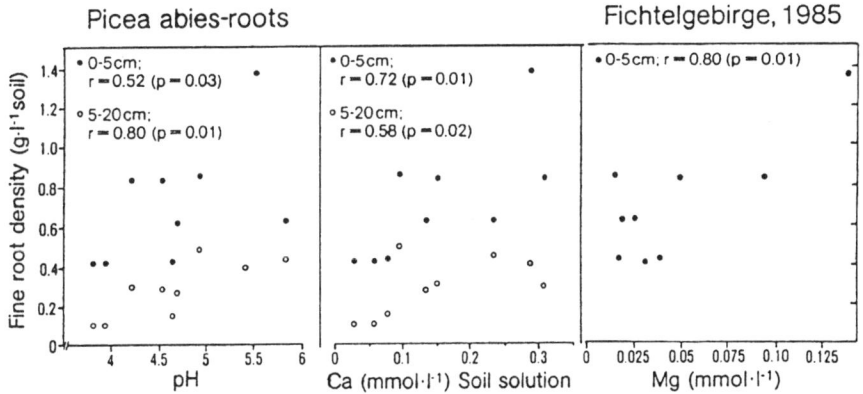

Figure 3. Correlation (linear regression) between fine-root biomass in two mineral soil horizons and pH, Ca and Mg concentrations of soil solution from suction cups ($n=9$). Solution from the suction cups was collected at 20 cm soil depth by means of a negative pressure, which slightly exceeds the natural water tension in the soil. The best correlation was observed between fine-root density and Mg concentration. (From Schneider *et al.*, 1989)

314

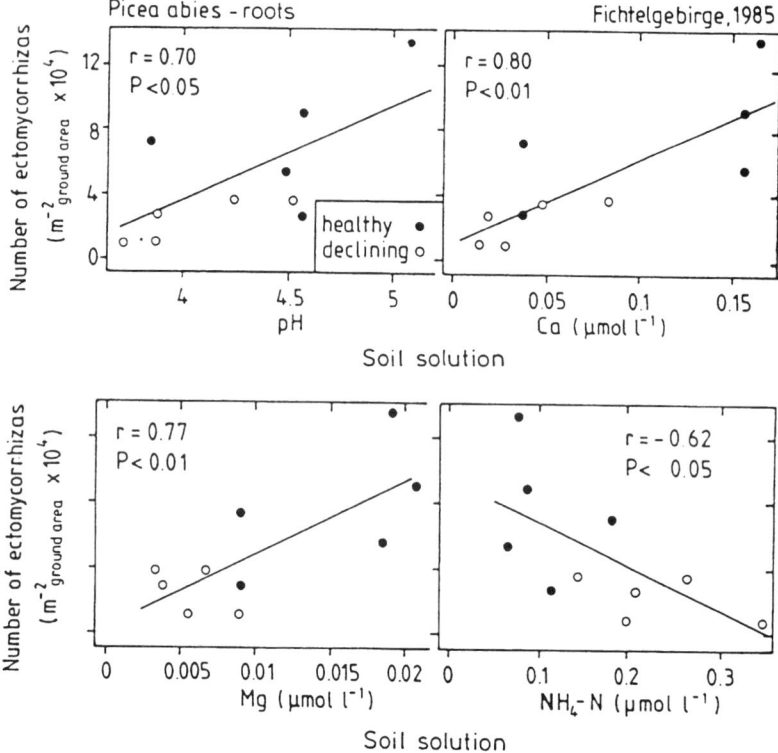

Figure 4. Correlation (linear regression) between number of ectomycorrhizas in two mineral soil horizons (combined) and pH, Ca, Mg and NH_4 concentrations of free-drainage soil solution ($n=10$) in young healthy and declining spruce stands (Meyer *et al.*, 1988). The free-drainage soil solution was collected just below the forest floor. (From Schneider *et al.*, 1989)

the numbers of active root tips and their distribution are the usual parameters for characterization of fertilizer effects on root systems.

Spruce at acid sites, in particular, build up an extremely shallow fine-root system (e.g. Murach, 1984; Raspe *et al.*, 1989; Raspe, 1992). The objective of Mg fertilization is to counteract the dangers resulting from increased dry stress during summer droughts, loss of nutrients due to eluviation into scarcely rooted or non-rooted horizons as well as the long-term danger of wind throw. Therefore, fertilization has to promote fine-root growth in deeper soil horizons. For that purpose, readily soluble mineral salts like $MgSO_4$ (Kieserite) have proved to be very useful (Raspe, 1992).

In order to counteract the effects of acid precipitation and corresponding soil acidification, the application of dolomitic limestone was proposed (Gussone, 1984; Ulrich, 1972, 1986). Therefore, a deeper fine-root distribution can be expected by increasing the pH level of the soil. FWL (1989) also recommends liming to deepen rooting of forest trees. It is however necessary to add Mg at the same time.

It is always the intention to obtain a better Mg nutrition of both the roots and the whole tree to improve stabilization of the stand (Huber, 1991). Dependent on the intensity of the fertilizer solution, the soil solution will be enriched with Mg^{2+} ions (Schüler, 1992; Feger et al., 1990). This affects the replacement of exchangeable Fe, Mn and Al ions in the soil against Mg^{2+}. According to Ebben (1989), Jorns and Hecht-Buchholz (1985) and Rost-Siebert (1983), a more favourable chemical milieu for the plant roots is obtained by extended Mg/Al relations. The predominant objective of previous forest fertilization, i.e. increase in productivity, is now replaced by revitalization of damaged forest stands and long-term preservation of an intact root system.

10.3. Effects of Mg fertilization on fine-root growth and distribution

Due to the close relationship between Mg supply in the soil and the fine-root biomass (section 10.1), an increase in fine-root mass is to be expected if Mg-deficient sites are ferti-lized with Mg. The reaction of roots to Mg addition depends on the amount of adsorbed fertilizer Mg in the mineral soil. The length of the fine-root reaction depends, however, on the fertilizer remaining in the soil.

10.3.1. $MgSO_4$

Readily soluble 'neutral salts', such as $MgSO_4$, have proved to quickly increase the Mg nutrition of Mg-deficient stands (Hüttl, 1989). Most of the applied Mg is stored in plant-available form at the exchange-complex in the mineral soil. This likely provides a chemotropic stimulus to deeper fine-root development. On the other hand, the input of unbuffered fertilizer salts leads initially to a decrease in soil solution pH and mobilization of inorganic Al (Feger et al., 1991) which is potentially toxic to roots. This might result in reduced growth of fine roots in the mineral soil (Ulrich, 1988).

A significant increase in fine-root biomass after $MgSO_4$ fertilization was observed in a 40-year-old Norway spruce stand at a typical Mg-deficient site in the Black Forest (Raspe and Feger, 1992; Raspe, 1992). Three growing seasons after application, a high proportion, approximately 70%, of the fertilizer Mg remained mainly exchangeable in the intensively rooted soil horizons (Feger, 1992). This led to a significant increase in fine-root intensity (expressed as g dry matter per dm^3 of soil volume) in the mineral soil up to a depth of 30 cm (Figure 5). Thus the fine-root pool was increased by more than 100%. In the mineral soil, in particular, the amount of fine roots increased from 170% to more than 300%.

The significance of the particular depth interval for the development of fine roots after Kiserite fertilization can be seen clearly by the relative distribution of fine roots in depth (RDFD) in the soil profile. This quotient is defined as the percentage of a depth interval of the whole root pool of the site (Raspe and Haug, 1992). The differences of RDFD between a $MgSO_4$-fertilized site and an untreated control site

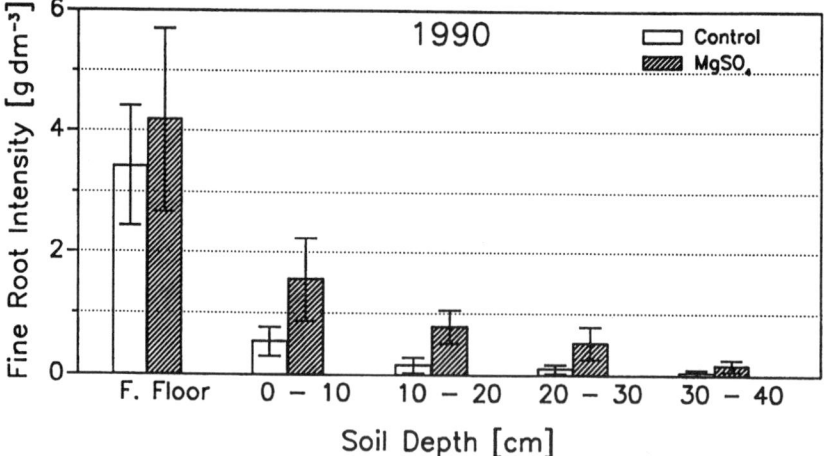

Figure 5. Fine-root intensity three growing seasons after MgSO₄ fertilization (750 kg/ha) of a typical Mg-deficient site in the Black Forest in comparison with an untreated control plot (lines show standard deviation)

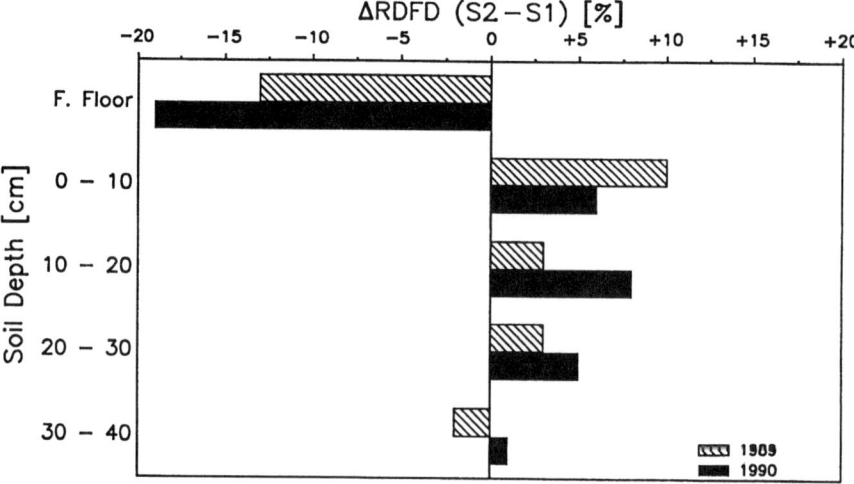

Figure 6. Changes in the relative distribution of fine roots with depth (RDFD) at the MgSO₄-treated plot (S2) calculated as the percentage of fine-root biomass at certain depths of the whole fine-root biomass at the MgSO₄-treated site minus the difference of those at the control site two (1989) and three (1990) growing seasons after fertilizer application

are shown in Figure 6. Negative signs in the forest floor indicate an elevated importance of this layer for the fine-root distribution at the control plot. Compared to the control plot, a much better fine-root development is noted in the mineral soil at the MgSO₄ plot. From the second to the third vegetation period, fine-root displacement into deeper mineral soil levels is noticeable. After five growing seasons, this trend still continues. The roots seem to follow a chemotropic stimulus of fertilizer Mg percolating into the soil.

Ulrich (1988) feared that the application of 'neutral salts', such as $MgSO_4$, would provoke Al stress or other forms of acidic toxicity. Up to now, this has not been verified by field experiments. Although, directly after Kiserite application, very high concentrations of potentially toxic Al^{3+} have been measured in the soil solution of the intensively rooted upper soil (Feger et al., 1990; Prietzel and Feger, 1992), no retreat of fine roots from the mineral soil into the forest floor has been observed.

10.3.2. Dolomite

Comparing $MgSO_4$ fertilization with dolomitic liming, Schaaf and Zech (1991) showed that the depth of fine-root reaction depends on the solubility of the products used. They observed a better rooting of the mineral soil up to a depth of 30 cm after $MgSO_4$ application. At the limed plot, the effect on fine-root development was limited to the forest floor and the uppermost part of the mineral soil (Figure 7).

Schüler and Zwick (1992) obtained similar results at pine stands of the Palatinate. In a comparison of different liming variants, they decided that the vertical distribution of fine and weak ($2\,mm < \phi < 5\,mm$) roots is in exact accordance with the improvement of the nutrient supply, especially of Mg. This comparison included 4 treatments. One plot was treated with 3 t/ha granulated dolomite (0–2 mm), another with an extreme dose of 15 t/ha fine-ground dolomite (0–0.09 mm with 1.5 t/ha Hyperphos). The third plot was fertilized with a dolomite suspension of 3 t dry matter per hectar including 3% dicalcium phosphate. One plot remained untreated. Three years after application of the dolomite suspension, both the fine-root intensity and the vitality of the roots in the deeper Bvhs horizon (Figure 8) increased significantly. Comparable tendencies also occurred in the 15 t/ha plot, though they were not so pronounced. After conventional dolomite application, a further concentration of fine and weak root distribution to the forest floor was observed.

Long-term effects of liming on fine-root growth were seldom examined. The oldest research was done by Zöttl (1964). Twenty years after carbonate fertilization with the most macronutrients, he observed, at mature spruce stands in Bavaria, a reduction of fine-root mass in the mineral soil as well as an increase in the forest floor. Murach and Schünemann (1985) also observed more shallow fine-root systems of adult spruce stands in the Solling which had been limed 5, 10 and 12 years before. Hölzer (1992), however, reported an increase in fine-root vitality in the mineral soil at a 9-year-old spruce stand in the Northern Eifel 8 years after liming with a dolomitic suspension. The vital fine-root mass increased at a depth of 10–15 cm from $0.4\,g/dm^3$ on average on the control plot to $1.8\,g/dm^3$ on the fertilized plot. In addition, fine-root necromass decreased significantly. This is probably due to the improved depth effect of dissolved lime suspension compared with coarsely ground or granulated lime.

Persson and Ahlström (1991a) observed a strongly dose-linked reaction between

318

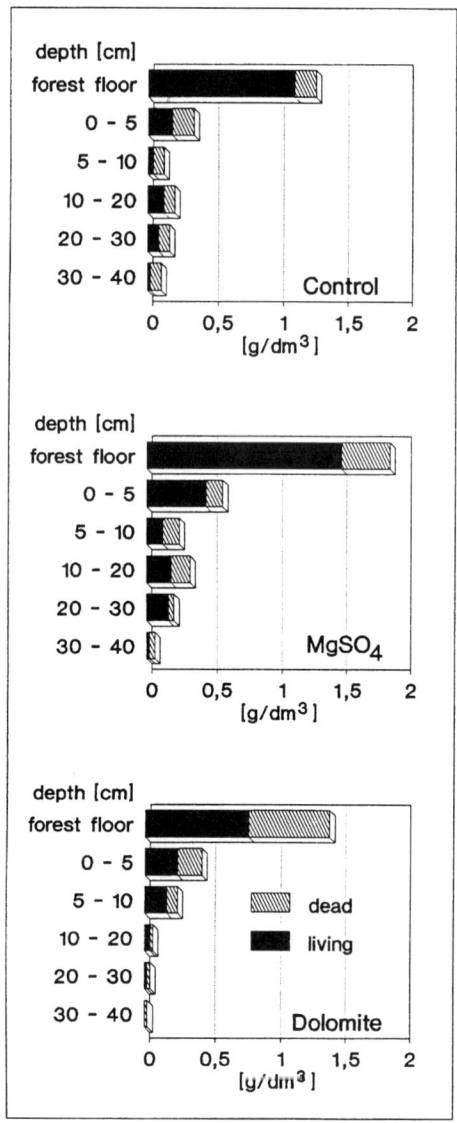

Figure 7. Fine-root distribution (< 2 mm) at three experimental plots (control; MgSO₄; dolomite) at the spruce site Oberwarmensteinbach in the Fichtelgebirge four years after fertilizer application. (Redrawn from Schaaf and Zech, 1991)

fine-root system and dolomite addition in Norway spruce and Scots pine stands in South and Central Sweden. Data from ingrowth core experiments indicate a positive effect of liming at a low dose of crushed dolomite (1.550 kg/ha). With higher doses of dolomite (3.5 and 8.6 t ha, respectively) no effects on fine-root growth were to be

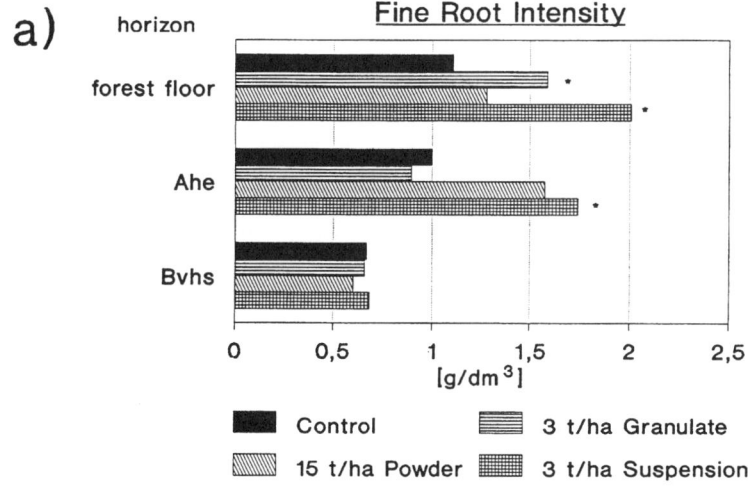

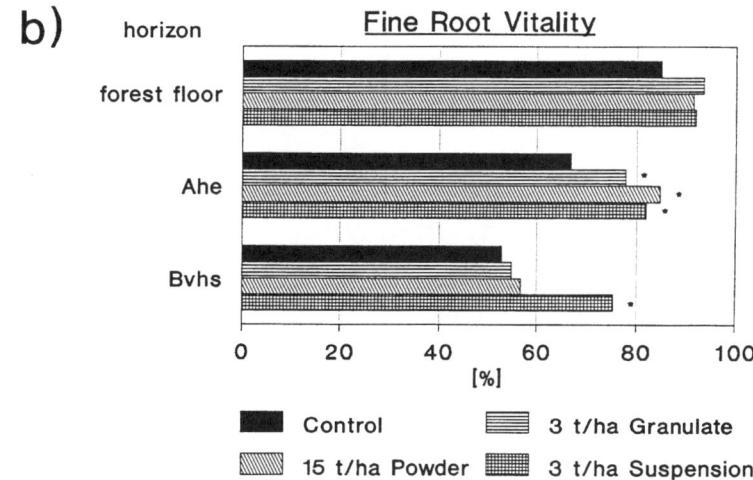

Figure 8. Fine-root density (a) and percentage of vital root tips (b) in three soil horizons after different lime treatments at a pine stand in the Palatinate. * = significantly different from the control ($p \leq 0.05$, *t*-test). (Data from Schüler and Zwick, 1992)

found. Furthermore, liming in combination with N fertilization has a strong negative effect on the development of fine roots. It is therefore likely that lime application at high levels may cause a decrease in fine-root growth, in particular in areas with high N deposition. In order to minimize the risk of negative effects on fine root growth arising from lime application, the rate of dolomite lime application should be low and, if possible, in repeated doses (Persson and Ahlström, 1991b).

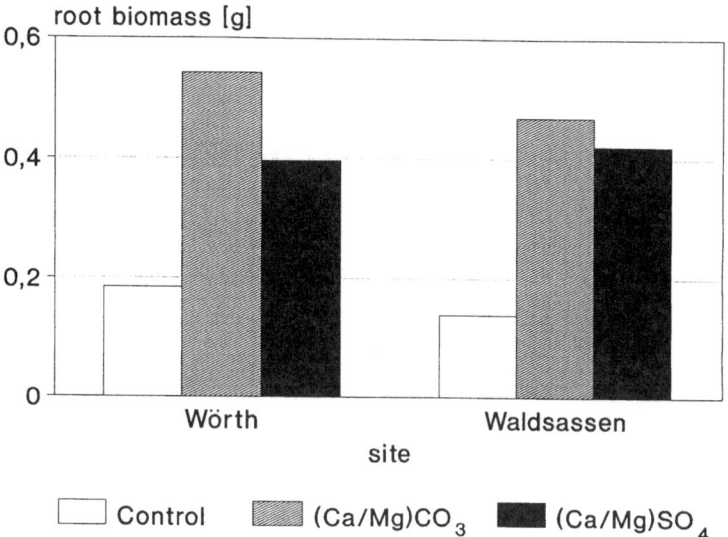

Figure 9. Average of root weights (g fresh weight per seedling) in pot experiments with different treatments (control; (Ca/Mg)CO$_3$; (Ca/Mg)SO$_4$) and soils from two sites (Wörth and Waldsassen) in Bavaria. (Data from Gonzalez Cascon *et al.*, 1989)

10.3.3. Pot experiments

The reactions of fine roots to Mg addition are also confirmed under standardized conditions in the laboratory. In container experiments with young silver fir trees growing on very acidic and poorly based soils rich in Al, Gonzalez Cascon *et al.* (1989) observed an increase in mineral soil rooting after Mg and Ca sulfate fertilization. If the fertilizers were given in the correct doses, the fine-root reaction was similar to that after addition of carbonates (Figure 9). In contrast, Sauter (1991) did not find any increase in the root growth of young spruce trees after fertilization with Mg sulfate. However, the apprehensions of Ulrich (1988), concerning fine-root damage in mineral soil due to Mg sulfate fertilization were not confirmed. Neither the molar Ca/Al and Mg/Al ratios in the soil solution nor the rates of nutrient uptake and growth showed any evidence for MgSO$_4$-induced acid toxicity. In this study, however, an improvement in root growth was only found after fertilization with Mg hydroxycarbonate. The general application of such container experiments can only be proved by similar field studies representing realistic field conditions.

The direct transfer of frequently performed pot experiments with nonmycorrhizal seedlings under hydrocultural conditions (e.g. Jorns and Hecht-Buchholz, 1985; Rost-Siebert, 1983; Schlegel and Hüttermann, 1990) seems to be doubtful (Rehfeuss, 1988; Zöttl, 1983). Therefore, we will not discuss these experiments in this connection.

10.4. Effects of Mg fertilization on fine-root nutrition

10.4.1. Magnesium

Changes in soil chemistry affect modifications of fine-root nutrition. Mg content of fine roots parallels the increase of fine-root density in those soil horizons where the exchangeable Mg^{2+} content increases after fertilization. Therefore, at the above-mentioned spruce site in the Black Forest (section 10.3.), the Mg content of fine roots increased significantly up to a soil depth of 40 cm after fertilization with $MgSO_4$, whereas liming with dolomite only resulted in an increase in the forest floor (Figure 10). All in all, both fertilizers improved the preliminary deficient Mg supply of this site to the level of dolomitic sites in the Alps (Sandhage-Hofmann, 1993). The improvement of Mg fine root content correlates with an increase of vital fine-root biomass at the same depth interval (section 10.3.).

At another spruce stand better supplied with Mg in the Solling (Murach and Schünemann, 1985), fine-root biomass did not increase after dolomitic liming, although the Mg content of fine roots did improve. Therefore, Murach and Schünemann (1985) concluded that there was a sufficient fine-root density, even at the control plot, and that no additional fine-root growth was required. Furthermore, due to the improved growth conditions in the forest floor, fine-root biomass shifted to some extent into the humus layer. Obviously, fine-root distribution at this site was less influenced by Mg supply than it was at the above-described site in the Black Forest.

Also Hölzer (1992) determined a parallel development of fine-root growth and their Mg content at spruce sites fertilized with a dolomitic suspension. On the other hand, the uptake of additional Mg in the mineral soil does sometimes seem to be inhibited. Schüler and Zwick (1992) observed, after liming of pine and spruce stands, no modification of Mg content of the fine roots in the B-horizon although exchangeable base cations increased in the soil. Probably, the corresponding anion is responsible for the quantity of Mg uptake. After fertilization with $MgSO_4$, SO_4^{2-} is taken up as the associated anion to a considerable extent (Feger, 1992), whereas lime includes no corresponding anion. Liming often enhances nitrification. But, spruce adapted to acidic site conditions take up NH_4^+ preferentially (Marschner *et al.*, 1991). The considerable importance of the anion associated with the uptake of Mg was also emphasized by Kaupenjohann (1989).

10.4.2. Calcium/aluminium

After fertilization with $MgSO_4$, the increase in Mg nutrition correlates with a decrease in Ca content of fine roots in the whole profile (Figure 11). At the same time, the Al content of fine roots in the mineral soil increases (Figure 12a) due to a short-term mobilization of ionic Al species in the soil solution for a few months after $MgSO_4$ application (Feger *et al.*, 1990). Although soil chemistry normalized after several months, a response of the fine-root system can still be observed up to the fifth vegetation period after fertilization (Raspe *et al.*, 1994). The molar ratio of

322

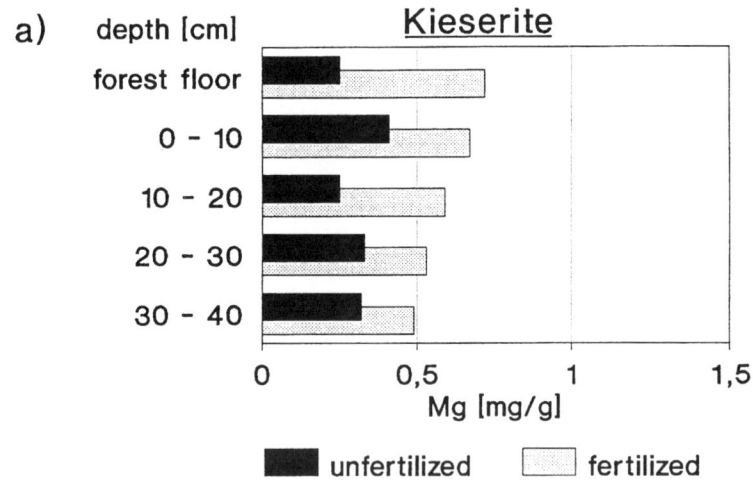

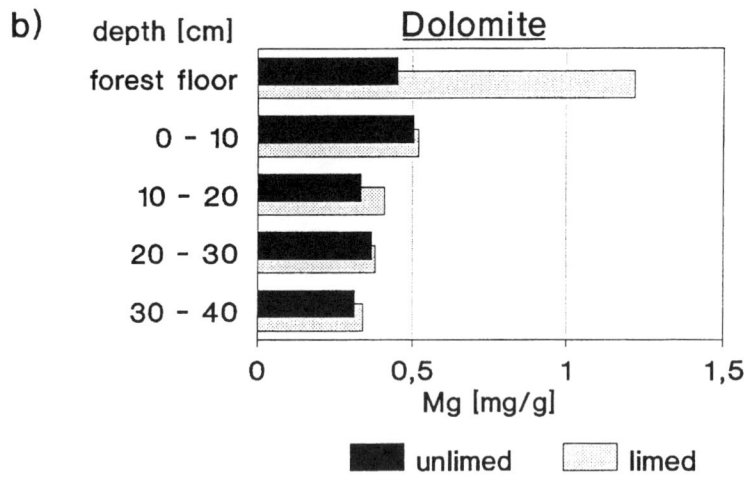

Figure 10. Mg contents of fine roots in different soil depth three growing seasons after application of 700 kg/ha MgSO₄ (Kieserite) (a) and 4 t/ha dolomite (b) at a Mg-deficient spruce site in the Black Forest

Ca to Al decreased significantly (Figure 13a). In spite of extremely low Ca/Al ratios, no fine-root damage was observed. Fine-root intensity, on the contrary, increased mainly in the mineral soil (section 10.3).

On the other hand, Mg addition by liming is expected to increase the Ca/Al ratio by increasing the Ca and decreasing the Al contents of fine roots. This was verified by, for example, Murach and Schünemann (1985) at older liming plots in the Solling

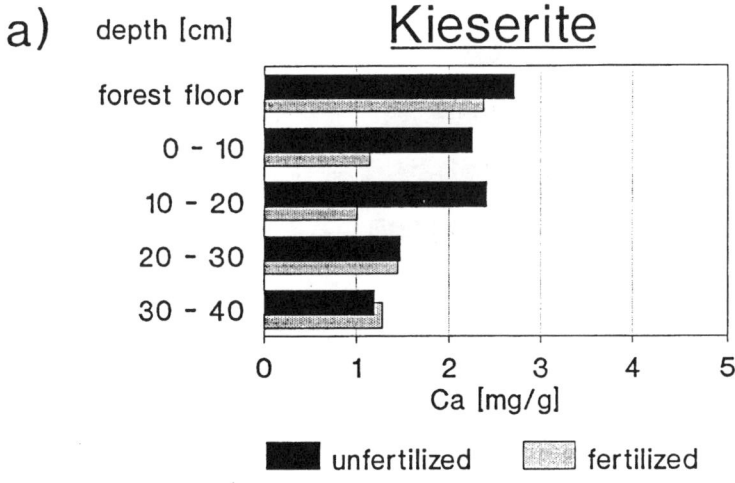

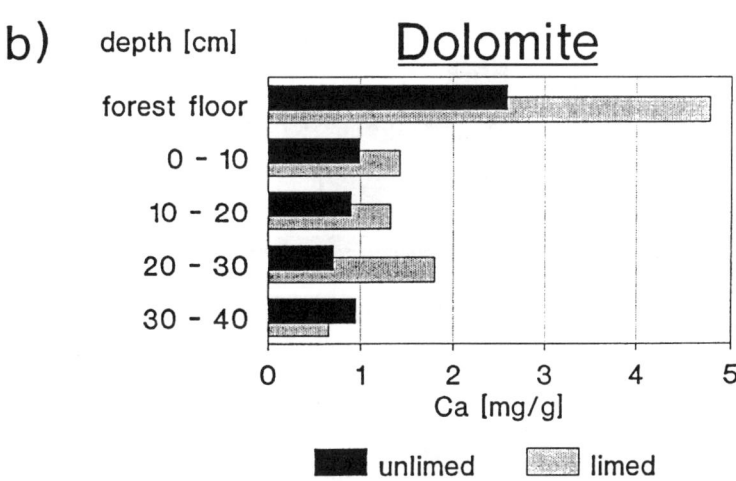

Figure 11. Ca contents of fine roots at different soil depths three growing seasons after application of 700 kg/ha MgSO$_4$ (Kieserite) (a) and 4 t/ha dolomite (b) at a Mg-deficient spruce site in the Black Forest

and by Wiedey (1991) at the Hils, as well as by Schüler and Zwick (1992) for conventional lime doses in spruce and pine stands of the Palatinate. But the increase in Ca/Al ratio was less than that expected from the ratio in soil solution. This is due to selective binding of Al to acidic groups located in cell walls of the root bark (Junga, 1984; Rost-Siebert, 1985). In all cases, fine-root growth increased after liming only in the upper soil horizons (forest floor and uppermost 10 cm of the mineral soil). Obviously the increased Ca/Al ratio is not sufficient to cause

324

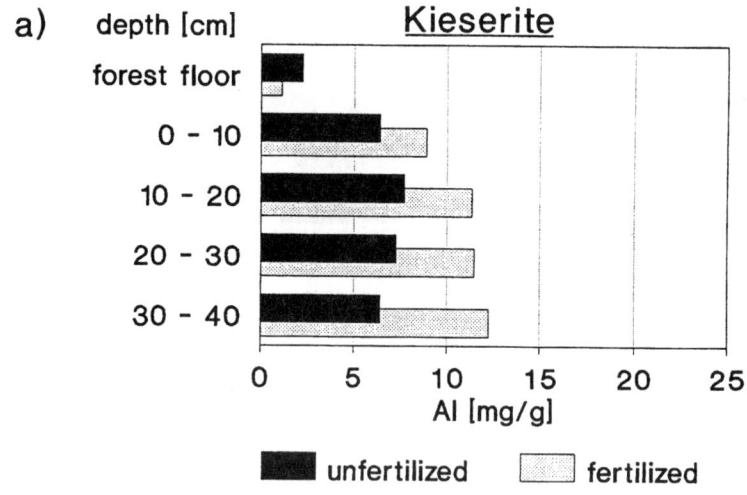

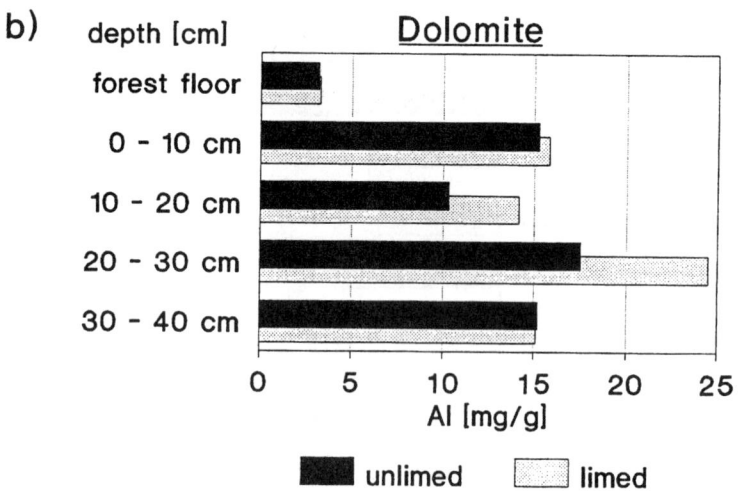

Figure 12. Al contents of fine roots at different soil depths three growing seasons after application of 700 kg/ha MgSO$_4$ (Kieserite) (a) and 4 t/ha dolomite (b) at a Mg-deficient spruce site in the Black Forest

proliferation of fine-root biomass of conifers in deeper mineral soil horizons.

Usually, Al content of fine roots in both hardwood (Safford, 1974) and coniferous stands (Bauch *et al.*, 1985; Raspe *et al.*, 1994; Stienen, 1986) increases even during the initial phase after liming. Increasing nitrification after liming (Wölfelschneider, 1994) yields protons for additional mobilization of Al^{3+}. Therefore, not only the Ca and Mg contents of spruce fine roots, but also the Al content had increased three

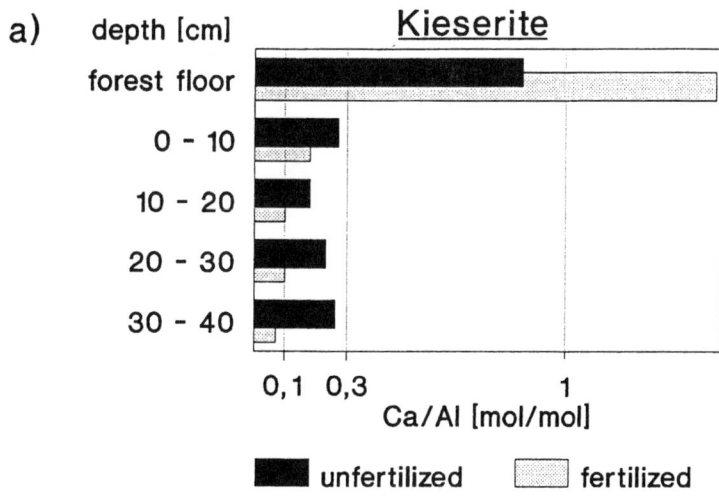

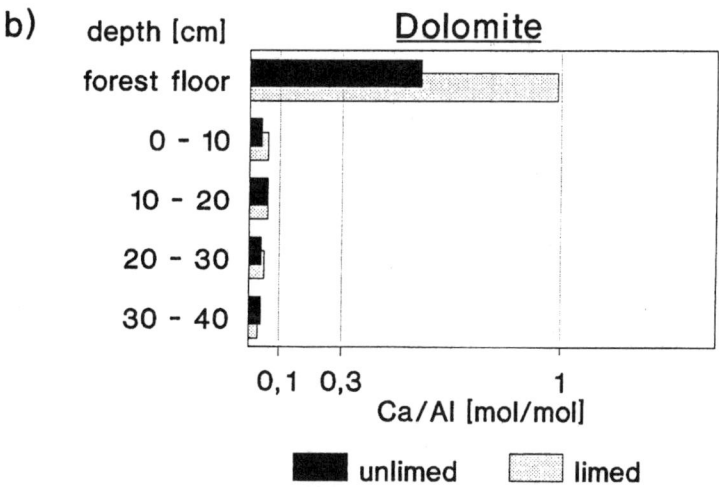

Figure 13. Ca/Al ratio in fine roots at different soil depths three growing seasons after application of 700 kg/ha MgSO$_4$ (Kieserite) (a) and 4 t/ha dolomite (b) at a Mg-deficient spruce site in the Black Forest

growing seasons after application of 4 t/ha granulated dolomite to an acidic granite site in the Black Forest (Figure 12b). The Ca/Al ratios of fine roots in the mineral soil remained constant (Figure 13b). After liming a considerable inhibition of cation uptake via competition of Al^{3+} ions is to be excluded (Stienen, 1986). Therefore the Mg and Ca contents of fine roots increased.

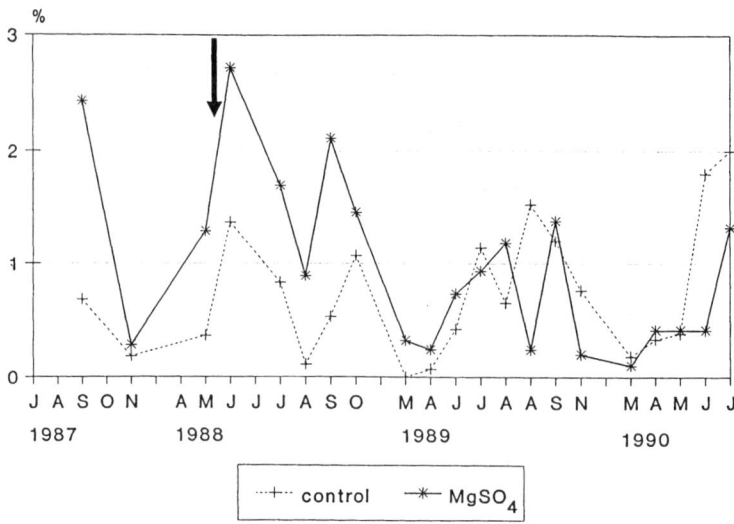

Figure 14. Temporal variation of the percentage of nonmycorrhizal root tips (PNRT) in the top layer (O/A horizon) from a control and a MgSO$_4$-fertilized (700 kg/ha) plot in the Black Forest. (Redrawn from Haug *et al.*, 1992)

10.5. Effects of Mg fertilization on mycorrhizae

Tree roots are usually colonized by mycorrhizae fungi (Frank, 1885). The mycorrhizae are known to influence nutrient uptake, resistance to drought and disease, and growth of forest trees. Diverse species of fungi build up typical ectomycorrhizae. The composition of mycorrhizae species depends on the soil conditions (Göttsche, 1972b; Kottke and Agerer, 1983; Meyer, 1962; Zhao, 1989). Therefore changes in soil chemistry caused by fertilization probably affect the mycorrhizae. Hence the reaction of mycorrhizae to Mg fertilization will be mentioned briefly.

Due to a strong increase in soil pH, liming can considerably reduce both mycorrhization and the number of species which make up the mycorrhizae. This was shown by Fiedler and Hunger (1963) 10 years after application of 161.3 dt/ha dolomite to a 70-year-old spruce stand at the Erzgebirge. Müller and Oberwinkler (1989) observed a reduction in mushroom density after liming connected with an enlarged species variety. Hence, after liming, fructification of mycorrhizae fungi and branching of spruce fine roots decrease because of the reduction in humus and increased decomposition of organic matter. The impact of soil pH on micorrhization of beech stands was also documented by Zhao (1989). The reduction of mycorrhization does not in general lead to a decrease in tree growth (Fiedler and Hunger, 1963). Obviously trees are able to abandon mycorrhizae as a tool for nutrient uptake if Mg supply is sufficient. However, they do lose protection against parasitic fungi (Trappe and Fogel, 1977).

On the other hand, the mycorrhizae are less influenced by addition of MgSO$_4$. No modification of relative mycorrhizae abundance in the forest floor was observed over a period of three years after application of 750 kg/ha $MgSO_4$ at an acidic spruce site in the Black Forest (Raspe and Haug, 1992). However, directly after fertilizer application, the percentage of non-mycorrhizal root tips slightly increased (Figure 14). Nevertheless, one year later, this deviation had disappeared. Therefore, this was only a short-term effect of Kiserite application. The appearance and disappearance of species of mycorrhizae was not observed (Haug et al., 1992). Damage to the mycorrhizae caused by elevated Al^{3+} concentration in soil solution was also excluded (Haug and Feger, 1991). The long-term effects of $MgSO_4$ fertilization on mycorrhizae have not been studied up to now. Therefore, no final report of the effects of $MgSO_4$ fertilization on mycorrhizae can be given.

10.6. Influence of site conditions

The studies described above mainly relate to typical Mg-deficient sites. However, if Mg is added at sufficiently supplied sites, in general, no reaction of fine roots occurs. This was shown by Schneider and Zech (1991) at three different spruce sites of the Fichtelgebirge. Two years after fertilization with $MgSO_4$ and half-burned dolomite, the fine-root density in the forest floor of a Mg-deficient 40-year-old spruce stand was significantly increased compared with a non-fertilized spruce stand. No such changes occurred at a second site where the nutrients were sufficient. At a third site with low K availability but high Ca and Mg saturation of the bulk soil, half-burned dolomite caused an increase in root growth due to a reinforced antagonism between Ca and Mg uptake on the one hand and K uptake on the other hand.

This study also shows a possible delay in the fine-root reaction after fertilization with $MgSO_4$. Two years after application of $MgSO_4$, Schneider and Zech (1989) observed no increased rooting in the mineral soil. Another two years later, Schaaf and Zech (1991) detected a significant increase in fine-root growth in the mineral soil at the Kiserite-fertilized plot on the same site. However, Raspe (1992) reported a rapid reaction of spruce fine roots to $MgSO_4$ application at a well-drained granite site in the Black Forest with an annual precipitation of 2000 mm. Two years after fertilization a significant depth effect on fine-root growth was detectable (section 10.3.).

At another site in the Black Forest, no reaction of fine-root growth to Mg fertilization was observed (Raspe, 1992). This site was a 100-year-old spruce stand growing on acidic brown forest soils ('Sauerbraunerden') derived from mesozoic quartz-sandstone ('Buntsandstein'). Vertical flow in the soil was restricted due to compact subsoil horizons with a higher clay content. As a consequence, periodical water logging occurred, possibly critically affecting the oxygen supply to the roots (Raspe and Feger, 1990). Therefore, the positive effect of Mg fertilization on root distribution is limited if superficial rootage is due not only to a generally insufficient supply of Mg but also to soil physical conditions. The importance of soil moisture regime and soil aeration to rooting depth was also emphasized by Köstler et al. (1968) and Kreutzer (1961).

10.7. Conclusions and practical applications

In general, stimulation of fine-root growth by fertilization of Mg-deficient sites is restricted on soil horizons with increased Mg supply. The stabilizing effect of the Mg addition to the ecosystem depends on the amount and type of adsorbed fertilizer Mg in the mineral soil. Fertilizing with $MgSO_4$ rapidly stimulates fine-root growth in mineral soil horizons at well-drained sites and improves Mg nutrition of all tree components. The effects of liming are usually restricted to the forest floor.

When a readily soluble Mg fertilizer, such as $MgSO_4$ is used, a greater volume of soil is utilized by the roots, providing access to more moisture and nutrients in addition to a possible improvement in windfirmness. Additional organic matter and mineral nutrients contained in the roots will be released and improve the fertility of the lower mineral horizons when the present mature stand is harvested. Decay and release of increased nutrient levels from roots within the forest floor and especially in deeper soil horizons should enrich the soil for a substantial period of time through recycling of nutrients within the tree–soil system. This enrichment should benefit root development and consequently the stand development of future trees on this soil. However, the positive effect of Mg fertilization on root distribution is limited when superficial rootage is also due to soil physical conditions. Also, no growth reaction was observed when spruce was already sufficiently supplied with Mg.

Due to the slow solubility of lime, application of dolomite usually promotes shallow rooting of spruce at Mg-deficient sites. Therefore trees become more susceptible to drought when precipitation is periodically insufficient. Moreover, useage of nutrient pools in deeper soil horizons decreases and loss of nutrients from the rooted soil horizons occur. Therefore, addition of slowly soluble carbonates increases the danger of further destabilization of the root systems of forest trees. However, application of dissolved MgO solution may induce fine-root reactions similar to those after sulfate fertilization if the added Mg moves rapidly into the mineral soil.

In general, there is a close relationship between root growth in the mineral soil and the Mg content in the fine roots. The reaction of the fine-root system therefore depends on the solubility of the added fertilizer. From the point of view of fine-root development, Mg fertilization with mineral salts seems to have a lower risk than liming with dolomite.

10.8. References

Bauch J, Stienen H, Ulrich B, Matzner E. 1985. Einfluß einer Kalkung bzw. Düngung auf den Elementgehalt in Feinwurzeln und das Dickenwachstum von Fichten aus Waldschadensgebieten. Allg. Forstz. 40, 1148–1150.
Beyschlag W, Wedler M, Lange OL, Heber U. 1987. Einfluß einer Magnesiumdüngung auf Photosynthese und Transpiration von Fichten an einem Magnesium-Mangelstandort im Fichtelgebirge. Allg. Forstz. 43, 738–741.
Ebben U. 1989. Die toxische Wirkung von Aluminium auf das Wurzelsystem der Buche. Allg. Forstz. 44, 781–783.

Eichhorn J. 1987. Vergleichende Untersuchungen von Feinwurzelsystemen bei unterschiedlich geschädigten Altfichten (*Picea abies* Karst.) PhD Thesis, Faculty of Forestry, University of Göttingen, Germany, 154 p.

Ericsson T. 1989. Dry matter partitioning in birch seedlings – a balance between internal nitrogen and carbon fluxes (Abstr.) In: Wener D, Müller P, Eds. Nitrogen Fixing Trees and Fast-growing Trees. pp. 232–233. Gustav Fischer Verlag, Stuttgart, New York.

Feger KH. 1992. Bilanzierung von Stofflüssen in magnesiumgedüngten Fichtenökosystemen im Schwarzwald (Projekt ARINUS). In Glatzel G, Jandl R, Sieghardt M, Hager H, Eds. Magnesiummangel in Mitteleuropäischen Waldökosystemen. pp. 88–101. Forstl. Schriftenreihe Univ. f. Bodenkultur Wien 5. Austrian Society for Forest Ecosystem Research and Experimental Tree Research, University of Agriculture, Vienna, Austria.

Feger KH, Raspe S. 1992. Ernährungszustand von Fichtennadeln und -wurzeln in Abhängigkeit vom Nährstoffangebot im Boden. Forstw. Cbl. 111, 73–86.

Feger KH, Zöttl HW, Brahmer G. 1990. Projekt ARINUS: IV. Auswirkungen der Kieseritdüngung. KfK-PEF-Berichte. 35(1), 27–38.

Feger KH, Zöttl HW, Brahmer G. 1991. Assessment of the ecological effects of forest fertilization using an experimental watershed approach. Fertil. Res. 27, 49–61.

Fiedler H-J, Hunger W. 1963. Über den Einfluß einer Kalkdüngung auf Vorkommen, Wachstum und Nährelementgehalt höherer Pilze im Fichtenbestand. Archiv f. Forstwesen 12, 936–962.

Fink S. 1983. Histologische und histochemische Untersuchungen an Nadeln erkrankter Tannen und Fichten im Südschwarzwald. Allg. Forstz. 38, 660–663.

Fink S. 1988. Histologische und histochemische Untersuchungen zur Nährstoffdynamik in Waldbäumen in Hinblick auf die 'Neuartigen Waldschäden'. KfK-PEF-Berichte. 35(1), 209–243.

Fink S. 1992. Physiologische und strukturelle Veränderungen an Bäumen unter Magnesiummangel. In: Glatzel G, Jandl R, Sieghardt M, Hager H, Eds. pp. 16–26. Forstl. Schriftenreihe Univ. f. Bodenkultur Wien 5. Austrian Society for Forest Ecosystem Research and Experimental Tree Research, University of Agriculture, Vienna, Austria.

Frank AB. 1885. Über die auf Wurzelsymbiose beruhende Ernährung gewisser Bäume durch unterirdische Pilze. Ber. Dtsch. Bot. Ges. 3, 128–145.

FWL (Forschungsbeirat Waldschäden/Luftverunreinigungen). 1989. 3. Bericht. Kernforschungszentrum Karlsruhe, 611 S.

Gonzalez Cascon MR, Alcubilla M, Rehfeuss KE. 1988. Entwicklung von Tannensämlingen (*Abies alba* Mill.) in Abhängigkeit von der Basensättigung natürlicher Substrate. Allg. Forst-u. Jagdz. 160, 233–241.

González Cascón MR, Alcubilla M, Rehfeuess KE. 1989. Wirkung von Magnesium- und Calcium-Sulfate und Carbonat auf Sproß- und Wurzelentwicklung junger Weißtannen (*Abies alba* Mill.) im Topfversuch mit sauren Böden. Allg. Forst-u. Jagdz. 161, 21–28.

Göttsche D. 1972a. Verteilung von Feinwurzeln und Mykorrhizen im Bodenprofil eines Buchen- und Fichtenbestandes im Solling. Mitt. Bundesforschungsanstalt für Forst und Holzwirtschaft. 88, 102 S.

Göttsche D. 1972b. Distribution of fine roots and mycorrhizae in the soil profiles of one beech and one spruce stand in the Solling range. Mit Bundesforschungsanstalt f. Forst und Holzwirtschaft. 88, 102.

Gruber F. 1992. Dynamik und Regeneration der Gehölze. Berichte des Forschungzentrums Waldökosysteme, A 86/1, Göttingen, Germany, 420 p.

Gussone H-A. 1984. Empfehlungen zur Kompensationsdüngung. Forst- u. Holzw. 39, 154–160.

Haug I, Feger KH. 1991. Effects of fertilization with $MgSO_4$ and $(NH_4)_2SO_4$ on soil solution chemistry, mycorrhiza and nutrient content of fine roots in a Norway spruce stand. Water, Air Soil Pollut. 54, 453–467.

Haug I, Pritsch K, Oberwinkler F. 1992. Der Einfluß von Düngung auf Feinwurzeln und Mykorrhizen im Kulturversuch und im Freiland. KfK-PEF-Berichte. 97, 159.

Häußling M, Jorns CA, Lembecker G, Hecht-Buchholz Ch, Marschner H. 1988. Ion and water uptake in relation to root development in Norway spruce (*Picea abies* (L.) Karst.). J. Plant Physiol. 133, 486–491.

Hölzer G. 1992. Auswirkungen einer Suspensionsdüngung auf den chemischen Zustand von Boden, Wurzeln und Nadeln sowie auf die Feinwurzelvitalität eines Fichtenbestandes in der Nordeifel. Diploma Thesis, Faculty of Agriculture, University of Bonn, Germany.

Huber B. 1991. Sektion Waldernährung tagte in Freiburg: Salzdüngung oder Kalkung in geschädigten Waldbeständen? Allg. Forstz. 46, 80–85.

Huss J. 1977. Vergleichende ökologische Untersuchungen über die Reaktion junger Fichten auf

Lichtentzug und Düngung im Freigelände und in Beschattungskästen. Göttinger Bodenkundl. Berichte 51, University of Göttingen, Germany.

Hüttl RF. 1989. Liming and fertilization as migation tools in declining forest ecosystems. Water Air Soil Pollut. 44, 93–118.

Hüttl RF. 1991. Die Nährelementversorgung geschädigter Wälder in Europa und Nordamerika. Freiburger Bodenkundl. Abh. 28, University of Freiburg, Germany, 440 p.

Jorns A, Hecht-Buchholz Ch. 1985. Aluminium induzierter Magnesium- und Calciummangel im Laborversuch bei Fichtensämlingen. Allg. Forstz. 40, 1248–1252.

Junga U. 1984. Sterilkulture as Modellsystem zur Untersuchung des Mechanismus der Aluminiumtoxizität bei Fichtenkeimlingen (*Picea abies* Karst.). Berichte des Forschungszentrums Waldökosysteme/ Waldsterben 5, Göttingen, Germany, 263 p.

Kaupenjohann M. 1989. Chemischer Bodenzustand und Nährelementversorgung immissionsbelasteter Fichtenbestände in NO-Bayern. Bayreuther Bodenkundl. Ber. 11, Bayreuth, Germany, 202 pp.

Keller T. 1979. Der Einfluß langdauernder SO_2-Begasung auf das Wurzelwachstum der Fichte. Schweizerische Zeitschrift für Forstwesen. 130, 429–435.

Kern KG, Moll W, Braun HJ. 1961. Wurzeluntersuchungen in Rein- und Mischbeständen des Hochschwarzwaldes (Vfl. Todtmoos 2/I-IV). Allg. Forst- u. Jagdz. 132, 241–260.

Kodrik M. 1991. Root biomass distribution of Norway spruce under different air pollution regimes. In Kodrik M, Ed. Forest Ecosystems in Slovakia. p. 20. Excursion informations of III-rd International Society of Root Research Symposium, Self publishers, Verein für Wurzelforschung, Klagenfurt, Austria.

Köstler JN, Brückner E, Bibelriether H. 1968. Die Wurzeln der Waldbäume. Published by Paul Parey Hamburg u. Berlin, Germany, 285 p.

Kottke I, Agerer R. 1983. Untersuchungen zur Bedeutung der Mykorrhiza in älteren Laub- und Nadelwaldbeständen des Südwestdeutschen Keuperberglandes. Mitteilungen des Vereins für Forstliche Standortskunde und Forstpflanzenzüchtung. 33, 30–39.

Kreutzer K. 1961. Wurzelbildung junger Waldbäume auf Pseudogleyböden. Forstw. Cbl. 80, 356–392.

Marschner H. 1988. Mineral Nutrition of Higher Plants. Academic Press, 674 S.

Marschner H. 1992. Bodenversauerung und Magnesiumernährung der Pflanzen. In: Glatzel G, Jandl R, Sieghardt M, Hager H, Ed. Magnesiummangel in Mitteleuropäischen Waldökosystemen. pp. 1–15. Forstl. Schriftenreihe Univ. f. Bodenkultur Wien 5. Austrian Society for Forest Ecosystem Research and Experimental Tree Research, University of Agriculture, Vienna, Austria.

Marschner H, Häussling M, Leisen E. 1986. Rhizosphere pH of Norway spruce trees grown under both controlled and field conditions. In: Indirect Effects of Air Pollution on Forest Trees-Root-Rhizosphere Interactions. pp. 113–118. Proceedings of a workshop organized by the Commission of the European Communities and the KFA Jülich GmbH, 5–6 December, 1985.

Marschner H, Häussling M, George E. 1991. Ammonium and nitrate uptake rates and rhizosphere pH in nonmycorrhizal roots of Norway spruce [*Picea abies* (L.) Karst.] Trees. 5, 14–21.

Mehne BM. 1989. Physiologische Untersuchungen an Fichten mit unterschiedlicher Magnesiumversorgung. Allg. Forstz. 44, 1248.

Meyer FH. 1962. Die Buchen und Fichtenmykorrhiza in verschiedenen Bodentypen, ihre Beeinflussung durch Mineraldünger sowie für die Mykorrhizabildung wichtige Faktoren. Mittl. d. Bundesforschungsanstalt f. Forst- und Holzwirtschaft Hamburg 54, 1–73.

Meyer J, Schneider BU, Werk KS, Oren R, Schulze ED. 1988. Performance of two *Picea abies* (L,) Karst. stands of different stages of decline. V. Root tip and ectomycorrhiza development and their relation to aboveground and soil nutrients. Oecologia. 77, 7–13.

Müller C, Oberwinkler F. 1989. Einfluß von Düngung im Wald auf die Fruktifikation von Mykorrhizapilzen und kleinstandörtliche Beeinflussung der Stickstoffmineralisation durch Fruchtkörper. KfK-PEF-Berichte. 55, 97–105.

Murach D. 1984. Die Reaktion der Feinwurzeln von Fichten (*Picea abies* Karst.) auf zunehmende Bodenversauerung. Göttinger Bodenkundl. Berichte, University of Göttingen, Germany, 127 S.

Murach D, Schünemann E. 1985. Reaktion der Feinwurzeln von Fichten auf Kalkungsmaßnahmen. Allg. Forstz. 40, 1151–1154.

Persson H, Ahlström K. 1991a. The effects of forest liming on fertilization on fine-root growth. Water Air Soil Pollut. 54, 365–375.

Persson H, Ahlström K. 1991b. The effects of alkalising compounds on fine-root growth in forest stands. In: Hübl E, Kutschera-Mitter L, Lichtenegger E, Sobotik M, Eds. Root Ecology and its Practical Application – Abstracts. ISRR 3th Int. Symposium, Vienna. p. 110. Self publishers, Verein für

Wurzelforschung, Klagenfurt, Austria.

Prietzel J, Feger KH. 1992. Dynamics of aqueous aluminium species in a podzol affected by experimental $MgSO_4$ and $(NH_4)_2SO_4$ treatments. Water Air Soil Pollut. 65, 153–173.

Prinz B, Krause GHM, Stratmann H. 1982. Waldschäden in der Bundesrepublik Deutschland. LIS-Berichte 28 der Landesanstalt für Immissionsschutz des Landes NRS, 154 p.

Pritchett WL, Fisher RF. 1987. Properties and Management of Forest Soils. 2nd edition, John Wiley & Sons, New York, Chichester, Brisbane, Toronto, Singapore, 148–164.

Raspe S. 1992. Biomasse und Mineralstoffgehalte der Wurzeln von Fichtenbeständen (Picea abies Karst.) des Schwarzwaldes und Veränderungen nach Düngung. Freiburger Bodenkundl. Abh. 29, University of Freiburg, Germany, 197 p.

Raspe S, Feger KH. 1990. Element distribution in roots of two contrasting Norway spruce stands in the Black Forest/Germany. In: Persson H, Ed. Above and Below-ground Interactions in Forest Trees in Acidified Soils. pp. 137–146. Commission of the European Communities, Air Pollution Research Report 32.

Raspe S, Feger KH. 1992. Distribution and nutritional status of Norway spruce (Picea abies Karst.) fine roots subjected to fertilization with $MgSO_4$ and $(NH_4)_2SO_4$. In: Kutschera L, Hübl E, Lichtenegger E, Persson H, Sobotik M, Eds. Root Ecologie and its Practical Application. pp. 487–490. Self publishers, Verein für Wurzelforschung, Klagenfurt, Austria.

Raspe S, Haug I. 1992. Wirkung einer Magnesiumsulfat-Düngung im Schwarzwald auf Fichtenwurzeln und Mykorrhizen. In: Glatzel G, Jandl R, Sieghardt M, Hager H, Eds. Magnesiummangel in Mitteleuropäischen Waldökosystemen. pp. 102–109. Forstl. Schriftenreihe Univ. f. Bodenkultur Wien 5, Austrian Society for Forest Ecosystem Research and Experimental Tree Research, University of Agriculture, Vienna, Austria.

Raspe S, Feger KH, Zöttl HW. 1989. Erfassung der Elementvorräte in der Wurzelbiomasse eines 100jährigen Fichtenbestandes (Picea abies Karst.) im Schwarzwald. Angew. Botanik. 63, 145–163.

Raspe S, Feger KH, Zöttl HW. 1994. Projekt ARINUS: VIII. Feinwurzelverteilung und -ernährung nach experimenteller Düngung. KfK-PEF-Berichte. 117, 13–27.

Rehfuess KE. 1988. Übersicht über die bodenkundliche Forschung in Zusammenhang mit den neuartigen Waldschäden. KfK-PEF-Bericht. 35(1), 1–26.

Rost-Siebert K. 1983. Aluminium-Toxizität und -Toleranz an Keimpflanzen von Fichte (Picea abies Karst.) und Buche (Fagus sylvatica L.). Allg. Forstz. 38, 686–689.

Rost-Siebert K. 1985. Untersuchungen zur H- und Al-Ionentoxizität an Keimpflanzen von Fichte und Buche in Lösungskultur. Ber. d. Forschungszentr. Waldökostst./Waldsterben, University of Göttingen, Germany, 12, 219 p.

Safford LO. 1974. Effect of fertilization on biomass and nutrient content of fine roots in a beech–birch–maple stand. Plant Soil. 40, 359–363.

Sandhage-Hofmann A. 1993. Feinwurzeldynamik unterschiedlich geschädigter kalkalpiner Fichtenbestände (Wank-Massiv). Bayreuther Bodenkundl. Ber. 33, University of Bayreuth, Germany, 243 p.

Sauter U. 1991. Versuche zur Wirkung von sulfatisch, carbonatisch und silikatisch gebundenem Magnesium auf Ernährungszustand und Wachstum junger Fichten, chemischen Bodenzustand und Sickerwasserbefrachtung. -Forstl. Forschungsber. München 114, Faculty of Forestry University of Munich, Germany, 442 S.

Schaaf W, Zech W. 1991. Bodenchmie, Wurzelwachstum und Ernährungszustand: Einfluß unterschiedlicher Löslichkeit von Düngern. Allg. Forstz. 46, 760–768.

Schlegel H, Hüttermann A. 1990. Identification of ion stress in roots of forest trees. In: Persson H, Ed. Above and Below-ground Interactions in Forest Trees in Acidified Soils. pp. 110–118. Commission of the European Communities, Air Pollution Research Report 32.

Schneider U, Zech W. 1989. Über den Einfluß Mg-haltiger Dünger auf das Wachstum und die Elementgehalte von Feinwurzeln immissionsgeschädigter Fichten. IMA-Querschnittseminar 'Düngung geschädigter Waldbestände', KfK-PEF-Berichte. 55, 107–118.

Schneider BU, Zech W. 1991. The influence of Mg fertilization on growth and mineral content of fine roots in (Picea abies (Karst) L.) stands at different stages of decline in NE-Bavaria. Water Air Soil Pollut. 54, 469–476.

Schneider BU, Meyer J, Schulze E-D, Zech W. 1989. Root and mycorrhizal development in healthy and declining Norway spruce stands. In: Schulze ED, Lange OL, Oren R, Eds. Forest Decline and Air Pollution, a Study of Spruce (Picea abies) on Acid Soils. pp. 370–391. Ecological Studies 77, Springer Verlag.

Schüler G. 1992. Erste Auswirkungen der Bodenschutzkalkung auf den Sickerwasserchemismus in versauerten Waldökosystemen. Mitteilungen aus der Forstlichen Versuchsanstalt Rheinland-Pfalz 21/92, 27–67.

Schüler G, Zwick N. 1992. Die Beeinflussung von Feinwurzelmasse und -vitalität eines Kiefernbestandes (*Pinus sylvestris* L.) mit unterständiger Buche (*Fagus sylvatica* L.) durch pflanzenverfügbare Elemente, sowie Veränderungen nach unterschiedlichen Kalkungsmaßnahmen. Mitteilungen aus der Forstlichen Versuchsanstalt Rheinland-Pfalz 21/92, 69–98.

Stienen H. 1986. Nährelementgehalte in den Feinwurzeln der Fichte nach saurer Beregnung und Kalkung. Forstw. Cbl. 105, 321–324.

Stienen H, Bauch J. 1988. Element content in tissues of spruce seedlings from hydroponic cultures simulating acidification and deacidification. Plant Soil. 106, 294–298.

Trappe JM, Fogel R. 1977. Ecosystematic functions of mycorrhizae. In: Marschall JK, Ed. The Belowground Ecosystem: A synthesis of Plant-associated Processes. pp. 205–214. Range Science Department, Science Series 26, Colorado State University, Fort Collins, USA.

Ulrich B. 1972. Forstdüngung und Umweltschutz. Allg. Forstz. 27, 147–148.

Ulrich B. 1986. Die Rolle der Bodenversauerung beim Waldsterben: langfristige Konsequenzen und forstliche Möglichkeiten. Forstw. Cbl. 105, 421–435.

Ulrich B. 1988. Bodenkundliche Forschung in Zusammenhang mit den neuartigen Waldschäden; Stellungnahme zu einem Artikel von K.-E. Rehfuess. Allg. Forstz. 43, 1171–1173.

Weller F. 1965. Die Ausbreitung der Pflanzenwurzeln im Boden in Abhängigkeit von genetischen und ökologischen Faktoren. Arbeiten der Landwirtschaftlichen Hochschule Hohenheim 32, Eugen Ulmer Stuttgart, Germany.

Wiedemann E. 1927. Der Wurzelbau älterer Waldbäume. Forstarchiv. 229–233.

Wiedey G-A. 1991. Ökosystemare Untersuchungen in zwei unterschiedlich exponierten Fichtenaltbeständen und in einem Kalkungs- und Düngungsversuch im Hils. Berichte des Forschungszentrums Waldökosysteme A 63, 205.

Wölfelschneider A. 1994. Einflußgrößen der N- und S-Mineralisierung auf unterschiedlich behandelten Fichtenstandorten im Südschwarzwald. PhD Thesis Faculty of Forestry University of Freiburg.

Zhao Z. 1989. Untersuchungen über Zusammenhänge zwischen chemischen Eigenschaften von Waldböden und Buchenmykorrhizen im Wienerwald. Dissertation Univ. f. Bodenkultur Wien, 35, 98 S.

Zöttl H. 1964. Düngung und Feinwurzelverteilung in Fichtenbeständen. Mitt. Bayer. Staatsforstverw. 34, 333–342.

Zöttl HW. 1983. Zur Frage der toxischen Wirkung von Aluminium auf Pflanzen. Allg. Forstz. 38(8), 206–208.

11
Evaluation of different magnesium fertilization strategies

W. SCHAAF

11.1. Strategies of Mg fertilization

The evaluation of Mg fertilization treatments in forest ecosystems should be made against the background of the specific aims and objectives of the application, the effects at different levels (plant cell, whole tree, stand, soil, whole ecosystem, environment), and the consequences of taking no measures.

Mg fertilization measures in forest ecosystems have been intensively carried out and investigated during the last 15 years, mainly in the context of the so-called 'new forest decline' symptoms, such as needle/leaf yellowing and foliage reductions, both at stand or watershed levels and on broad-scale practical application. A large number of different fertilizer forms, singly or in combinations, and amounts in many different ecosystems have been used with a number of aims to be reached and questions to be addressed. This results in considerable difficulty in evaluating the positive and negative effects of these fertilization strategies.

Another question of interest is, what criteria are appropriate for an evaluation of the effects of forest fertilization measures? According to the points mentioned above, possible criteria at the different levels could be: tree nutrition (element contents in foliage, ratios of nutrients, damage symptoms), tree vitality (a frequently used but still scientifically undefined term), stand growth/production, soil chemical status (e.g. pH, base saturation, exchangeable nutrient stores, acid/base neutralizing capacity), turnover processes, seepage quality and element budgets.

Many different terms and concepts have been used to specify the objectives of forest ecosystem fertilization. Compensation, restoration, amelioration, revitalization, and regeneration are the most frequent concepts for fertilizer applications (see Andersson and Persson, 1988). These terms refer to different fields, such as atmospheric deposition, nutrition, soil status, and forest management. Consequently, the aims of these fertilization strategies vary considerably (Evers and Huettl, 1990; Feger, 1993; Hanisch, 1989; Huettl, 1989; Jenkins et al., 1991; Zoettl, 1990).

The most important aims of recent strategies in forest fertilization are:

- To mitigate acute deficiency symptoms,
- To ameliorate the soil chemical status,
- To neutralize acid atmospheric deposition,
- To avoid further soil acidification by increasing the buffering capacity,
- To protect surface- and ground-waters from acidification and pollution,
- To restore a desired soil quality,

R. F. Hüttl & W. Schaaf (eds): Magnesium Deficiency in Forest Ecosystems, 333–355.
© 1997 *Kluwer Academic Publishers. Printed in Great Britain.*

- To facilitate management strategies, like changes of tree species, and
- To maintain or increase productivity.

However, the types or forms of fertilizers to be applied is still controversial (Huber, 1991). Most experience has been gathered in experiments using dolomitic lime (Huettl and Zoettl, 1993; Feger *et al.*, 1994; Kaupenjohann, 1991; Kreutzer, 1995) or magnesium sulfate (Feger *et al.*, 1991; Kaupenjohann and Zech, 1989). Experiments with rock meal are few (Hildebrand, 1990; Schueler, 1991, 1994a,b). Some reports are available on treatments that tried to combine the effects of different fertilizer forms, e.g. by utilizing more-soluble forms of magnesite (Jandl and Katzensteiner, 1992; Schaaf and Zech, 1993b), granulated magnesium hydroxide (Schaaf, 1995) or by the combination of lime and $MgSO_4$ (Hildebrand, 1990; Wiedey, 1991).

An important factor to be taken into consideration is 'time': different fertilization strategies have or may have different aims with respect to time. For example, neutral salts are usually applied for a short-term correction of deficiencies. Compensatory liming or complete soil restoration treatments rather follow long-term perspectives.

Since almost all trials with different Mg fertilizer forms showed positive and long-lasting effects on tree nutrition with only minor differences due to e.g. fertilizer solubility or tree age (see Chapter 8), these effects will not be discussed here further.

11.2. Fertilizer forms

Forms most frequently used in forest fertilization programs are lime, neutral salts, and, to a lesser extent, rock meals. With respect to Mg fertilization, the most common forms are dolomitic lime, magnesium sulfate, magnesite, and magnesium hydroxide. Their composition, chemical reaction, and dissolution kinetics are discussed in Chapter 7. Lime materials vary considerably with respect to composition and Mg contents. Also, grain size distribution varies with origin, production, and treatment (e.g. pelleting or granulating).

Since many studies show that the effects of fertilization measures are highly dependent on site characteristics, like tree species, stand age and growth, historic land use and management, soil status, humus form, climate, degree of forest dam age, atmospheric and internal input of acids, and level of nitrogen deposition, the best way to follow differences between application forms and amounts would be to carry out the different treatments at one site and in one stand at the same time. But, because forest experiments with an ecosystem approach are usually very time consuming and expensive, there are only very few studies that allow this direct comparison. Laboratory or small-scale studies, like column experiments, can be helpful in testing potential effects.

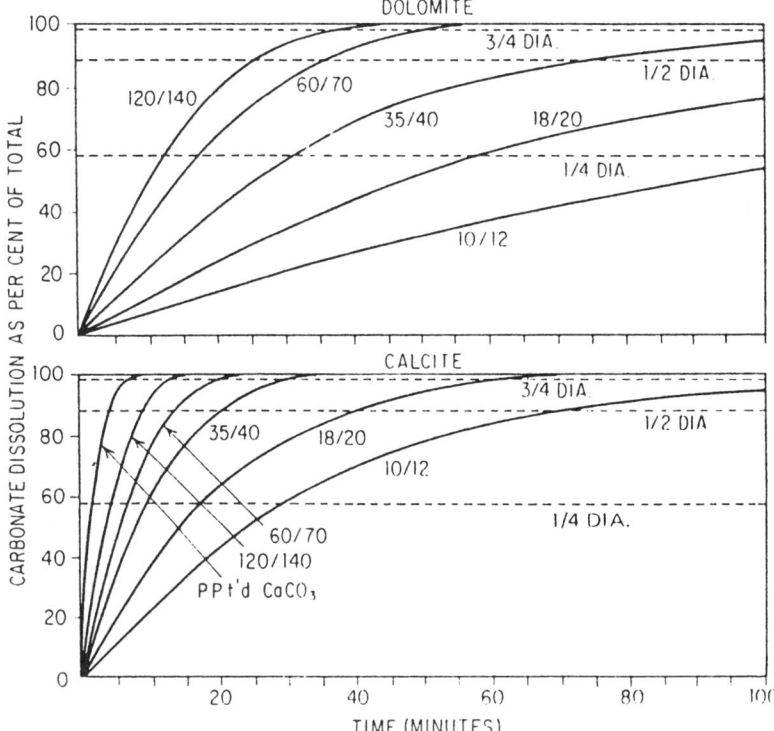

Figure 1. Reaction rates of calcite and dolomite of different sizes in boiling 5 N NH₄Cl solution (from Tisdale *et al.*, 1985)

11.3. Evaluation of dolomitic lime effects

In the controversy about 'lime' and 'neutral-salt' application with respect to Mg fertilization, the two forms mainly discussed are dolomite or dolomitic lime and $MgSO_4$ (Huber, 1991). Numerous experiments at the laboratory, soil column, pot, field plot and even watershed levels have been carried out. But, in most cases, no difference was found between different 'lime' forms when discussing the results.

Calcitic lime ($CaCO_3$) has a higher solubility rate than dolomite ($CaMg(CO_3)_2$; see Andersson and Persson, 1988, and Figure 1). Also, the differences in grain size, preparation and composition of dolomitic material used in many studies are to be considered when interpreting the experimental results of fertilizer effects. Figure 1 shows the reactivities and the effects of varying grain size. Due to these different reactivities, liming with dolomite results in smaller pH increases than with $CaCO_3$. Lyngstad (1993) reported the effects of calcite and dolomite of varying grain size on soil pH over a period of ten years. pH increases were the greater with smaller particle size and time spans to reach the same pH effects varied from one to six years for the same fertilizer amounts of different grain sizes. In all cases, dolomite reacted

Table 1. Effect of grain size on reactivity and acid neutralization capacity (ANC) of dolomite fertilizers (data from Schueler, 1991)

	0–0.09 mm	0.1–2 mm
%CaO	31.8	29.5
% MgO	19.2	19.8
% reactivity at pH 2.0	24.0	6.0
% reactivity at pH 3.5	14.0	5.0
kmol(c) ha^{-1} ANC at pH 2.0	14.4	3.6
kmol(c) ha^{-1} ANC at pH 3.5	8.4	3.0

more slowly than calcite. It is important to keep this difference in mind when discussing 'liming' experiments with other types of Mg amendment. Half-calcined dolomite ($CaCO_3 \cdot MgO$) contains magnesium in the form of MgO which has a higher solubility than $MgCO_3$.

A comparison of results from two dolomite fertilization experiments reported by Schueler (1991) demonstrates the influence of grain size. Two Norway spruce stands were treated with 3 t ha^{-1} of dolomite: at the Linz site, the grain size was 0–0.09 mm and at the Pruem site it was 0.1–2 mm. While the contents of CaO and MgO were almost the same, the reactivity, as tested in short-term studies with HCl, was three to four times higher for the fine-ground dolomite (Table 1). Correspondingly, the acid neutralization capacity (ANC) of the amounts applied to the forest soil were different. Two years after application of the two fertilizers, the pH values of the organic layers were increased by 1.5 and 0.3 units in Linz and Pruem, respectively, compared with the control. Base saturation in the upper mineral soil was increased from 6% to 12% in Linz and from 3% to 5.5% in Pruem. Similar results were obtained by Hildebrand and Schack-Kirchner (1990) in soil column studies.

Therefore, for a proper comparison, it is problematic that in many cases these characterizations of the fertilizer material is not reported in detail.

In general pH increases are correlated with the applied amount of dolomite (Figure 2). In the Hoeglwald experiment, the application of 4000 kg ha^{-1} dolomitic lime increased pH in the organic surface layer from values of 2.8–3.5 to 3.5–6.5 (Kreutzer et al., 1991).

Positive effects of dolomitic liming have also been shown with respect to increasing exchangeable Mg stores and base saturation (Aldinger, 1987), although there seems to be a need for an appropriate determination method of exchangeable Mg and CEC if the soil sample contains undissolved lime material. Base saturation was increased from 50–80% to 100% in the organic surface layer, and, in the upper mineral soil, from 6–10% to 17–33% in the Hoeglwald experiment (Kreutzer, 1995). An application of 4 t ha^{-1} lime plus 2 t ha^{-1} dolomitic lime increased the base saturation from 5.0–6.0% to 18.5–13.0% in the upper 20 cm of the profile (Wiedey, 1991).

Almost all reports on surface liming experiments agree that these main positive effects on soil chemistry are restricted to a few cm depth of the profile, in most cases

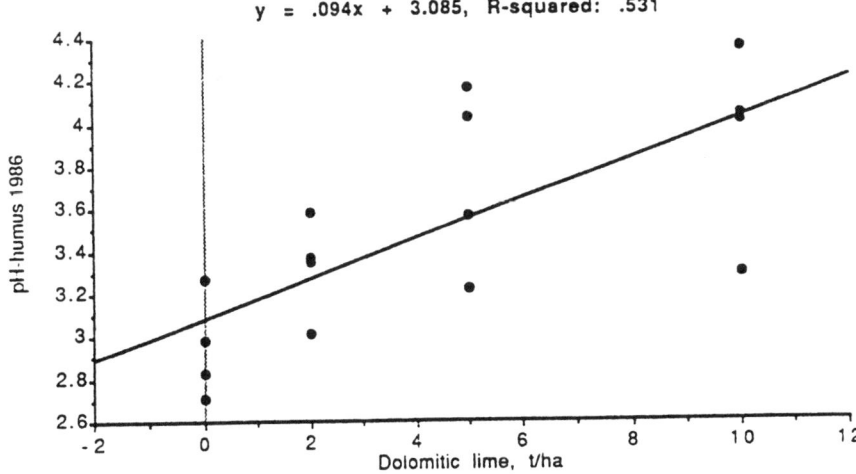

Figure 2. Relationship between pH(KCl) in the humus layer and dolomite applications from different beech forests in South Sweden (from Andersson and Persson, 1988)

to the organic surface layer or the upper few cm of the top mineral soil (Kreutzer *et al.*, 1991; Matzner and Meiwes, 1991). Only with very high dosages or after a long time are effects in deeper soil layers reported (Kreutzer, 1995; Marschner, 1990; Matzner *et al.*, 1985; Reiter *et al.*, 1986; Wenzel, 1989).

In some cases, enhanced acidification of the subsoil due to base-cation losses as a result of NO_3^- leaching following enhanced nitrification have been reported after surface liming (Kaupenjohann, 1989).

These effects may lead to an even more shallow distribution of the fine-root system described by several authors (Murach, 1988; Raspe, 1992; Schaaf and Zech, 1991; see Chapter 10, section 10.3).

Effects of surface liming on mineralization and NO_3^- leaching seem to be highly dependent on the N-status of the site. The C/N ratio of the O-horizon is thought to be an important indicator (Persson *et al.*, 1989). Many studies report increased soil microbial activity and leaching of NO_3^- and DOC from the organic surface layer after liming (Feger *et al.*, 1994; Kreutzer, 1995; Marschner *et al.*, 1989; Schierl *et al.*, 1986). This is thought to be related to the increases in soil pH, nutrient availability, and, thus, better soil chemical conditions for micro-organisms. Effects on soil biology seem to include a shift from fungi to bacteria as the dominating micro-organisms in organic matter decomposition (Badalucco *et al.*, 1992; Woelfelschneider, 1994). Considering the differences in dissolution kinetics between calcitic and dolomitic lime, these effects should be less pronounced with dolomite, when comparable amounts are applied. However, too little data are available up to now.

On the other hand, there are reports, mainly from N-limited sites, of an initial depression in tree growth after liming due to reduced N-mineralization rates

(Persson *et al.*, 1991). These results were obtained at sites where abiotic ammonia immobilization in soil organic matter and reduced nitrification were found (Persson, 1994). It was shown in laboratory studies that dolomitic liming increased potential nitrification in the upper soil, but, under field conditions, enhanced nitrification was observed only at NH_4 concentrations higher than $50\,mg\,g^{-1}$ in the humus layer (Persson, 1994).

11.4. Evaluation of $MgSO_4$ effects

The main reason for the application of Mg as a neutral salt is its quick supply in cases of clear deficiencies indicated by needle or leaf yellowing (Zech and Popp, 1983). Initial experiments with young spruce and beech trees showed a very fast response and a regreening of deficient trees (e.g. Ende and Zoettl, 1991; Huettl, 1991; Kaupenjohann, 1989; Liu, 1989). An important argument against the application of neutral salts emphasizes that they have no positive effect on the soil chemical status (e.g. soil pH) and that they carry the risk of enhanced leaching losses of base cations from the rhizosphere, especially of Mg and Al, due to the introduction of a mobile anion resulting in further soil and seepage acidification and the risk of root damage (Huber, 1991). However, experiments at the stand scale revealed that, after an initial concentration peak (mainly Mg^{2+}, Al^{n+} and SO_4^{2-}), no negative effects on tree nutrition and root vitality could be observed (Haug and Feger, 1990). In these studies, the absence of increased losses of SO_4 was attributed to the obviously high capacity of the mineral soils for SO_4 immobilization by anion sorption or the formation of new mineral phases like $AlOHSO_4$. However, no clear scientific proof for this formation has been achieved under field conditions. The same is true for the predicted toxicity of increased Al^{3+} concentrations on roots and mycorrhizae. Experiments in the Black Forest, for example, showed no negative effects on the root development despite high initial Al^{3+} concentrations in the soil solution after $MgSO_4$ application (see Chapter 10).

Experiments with $MgSO_4$ showed that SO_4^{2-} can be retained within the forest ecosystem in considerable amounts despite its high solubility and mobility and despite high atmospheric S inputs at some of the sites under investigation.

Hantschel (1987) reported an increase in S output from 72.6 to $146.8\,kg\,S\,ha^{-1}$ during the first two years after the application of $1000\,kg\ MgSO_4\,ha^{-1}$. Thus, the application of $232\,kg\,S\,ha^{-1}$ in form of $MgSO_4$ resulted in an output surplus of $74\,kg\,S\,ha^{-1}$ or 32% of the applied amounts. These data imply that there is storage of large quantities of S within the mineral soil, corresponding to the complete S deposition ($42\,kg\,S\,ha^{-1}\,y^{-1}$) of approximately four years.

From a watershed study in the Black Forest, Feger *et al.* (1991) reported an output at the weir of only 12% of $170\,kg\,S\,ha^{-1}$ applied as $MgSO_4$ within the first 1.5 y; $85\,kg\,S\,ha^{-1}$ were retained in the mineral soil (0–80 cm depth). The 'missing' $65\,kg\,ha^{-1}$ were attributed to the debris zone and the aquifer. In this case, a study area with low atmospheric input, the storage of $85\,kg\,S\,ha^{-1}$ in the mineral soil equals the

S deposition rates of eight years and amounts to more than 10% of the total S pool of the mineral soil.

These unexpected results are in contrast to the high mobility of the SO_4^{2-} anion found in laboratory experiments (Hildebrand, 1990) and also explain the high retention rates of fertilized Mg observed in these systems.

However, information is still lacking about the long-term fate and behaviour of these increased S stores in forest soils. In a study on reversibility of soil acidification, Alewell and Matzner (1992) stressed that the binding form of soil S is an important factor controlling the release of stored S. The authors concluded that organically bound S is less important than inorganic processes for long-term S storage (see Gustafsson and Jacks, 1993; Singh, 1984), and that high S stores in the soil, especially if Jurbanit ($AlOHSO_4$) is the controlling mineral phase, can maintain high soil acidity over long periods (Hauhs, 1988). On the other hand, sorption of SO_4^{2-} in soils may increase CEC and, thus, provide exchange sites for the counter-cation Mg^{2+} (Mulder and Cresser, 1994; see also Chapter 7, section 7.3.3.). For example, Raspe (1992) reported statistically significant increases in CEC between 11% and 30% for different soil layers 2.5 y after application of 750 kg $MgSO_4 ha^{-1}$. Kaupenjohann (1992) found increases in actual CEC from 112 to 159 $mmol_c kg^{-1}$ in 0–5 cm depth of the mineral soil after $MgSO_4$ fertilization. In contrast, dolomitic liming decreased actual CEC (77 $mmol_c kg^{-1}$). Effects on potential CEC were even more pronounced.

11.5. Evaluation of $Mg(OH)_2$ effects

In order to evaluate dolomitic lime and highly soluble mineral fertilizer, Schaaf (1992) studied the effects of granulated $Mg(OH)_2$ which was thought to combine the positive effects of both fertilizer forms.

The experimental site of Hohe Matzen is located in the eastern part of the Fichtel Mountains in NE Bavaria (FRG). The soils were mainly typical Dystrochrepts derived from granite. The 60-year-old Norway spruce stand showed severe needle yellowing typical of Mg deficiency.

The site was characterized by high atmospheric inputs with deposition rates of 1.25 kg H^+, 42 kg S, and 32 kg N per ha per year. The soil was acidic down to a great depth. The pH values in soil solutions of the organic surface layer and the upper mineral soil were around 3.5. Concentrations of Al^{n+} (4–8 mg L^{-1}), SO_4^{2-} (25–30 mg L^{-1}), and especially NO_3^- (30–40 mg L^{-1}) and DOC (12–55 mg L^{-1}) were very high. The element budget indicated a significant influence of N inputs and processes of N turnover on the chemical status of the soil and probably on tree nutrition (Schaaf and Zech, 1993a). Nitrification in the upper mineral soil had led to a transformation of a major part of NH_4^+ into NO_3^-, which was quantitatively leached, resulting in an ecosystem-internal H^+ production of 1.8 kmol(c) $ha^{-1} y^{-1}$. NO_3^- and SO_4^{2-} dominated the seepage output from the ecosystem.

Before application of the fertilizer, the buffer capacity and reactivity was tested in pH-stat.-titrations using 0.1 g of fertilizer material and 0.05 mol/L H_2SO_4 at six

340

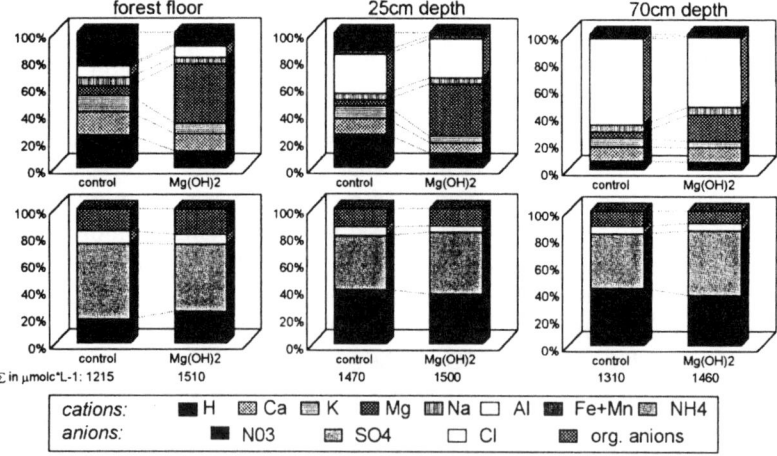

Figure 3. Ionic composition of soil solutions at three soil depths of control and Mg(OH)$_2$-fertilized plots at Hohe Matzen, Fichtel Mountains/FRG (mean values over 2.5 y after application)

pH levels. At pH 7, within two hours, 70% of the total acid neutralization capacity (ANC) of the granules was exhausted. With decreasing pH, almost all of the buffering capacity was released within two hours. These results characterized the fertilizer as quickly soluble with a total ANC of 33 mol kg^{-1} (Schaaf and Zech, 1993b).

The application was carried out in May 1988: 2650 kg Mg(OH)$_2$ ha^{-1} corresponding to 1040 kg Mg ha^{-1} was applied to experimental plots with five replications each.

Mg(OH)$_2$ fertilization resulted in manifold increased Mg^{2+} concentrations in the soil solution down to 70 cm soil depth and to a significant increase of pH down to 25 cm mineral soil depth. NO$_3^-$ concentrations were elevated after fertilization but decreased within 15 months to below the levels of the control plots. As a mean, over the whole experimental period of three years, N output was not increased by fertilization. Figure 3 presents a summary of the fertilization effects on the composition of the soil solution at three soil depths. Molar ratios of Mg/Al$_{total}$ and Ca/H clearly increased after fertilization, reducing the potential risk of Al^{3+} and H$^+$ toxicity and thus creating more favourable conditions for fine-root growth even in deeper soil horizons.

Figure 4 shows that large amounts of fertilizer Mg remained within the system in exchangeable form, especially in the organic surface layer, but also in the mineral soil. Two and a half years after application, only 10% of the Mg fertilizer had leached from the humus layer into the mineral soil and only about 4% had left the ecosystem with seepage. Despite a higher internal H$^+$ production rate due to nitrification processes, the acid-buffering capacity of the soil was clearly increased without any indication of Al mobilization due to the treatment.

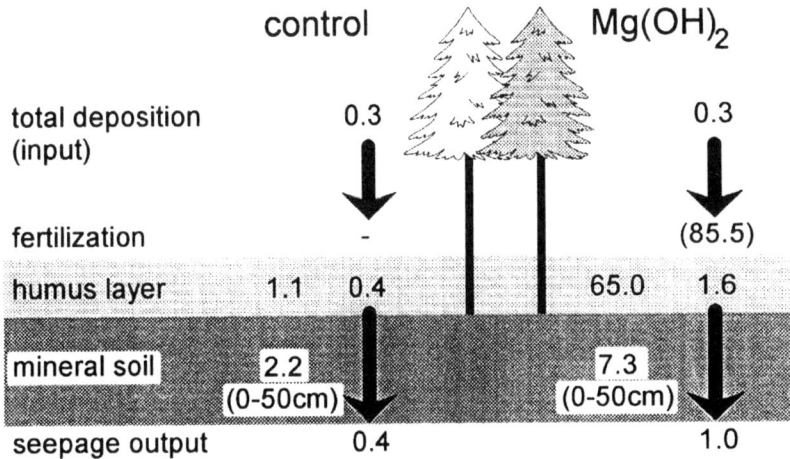

Figure 4. Magnesium budget and exchangeable soil stores of Mg of fertilized and control plots at Hohe Matzen, Fichtel Mountains/FRG (mean annual fluxes and exchangeable stores 2.5 y after fertilization in kmol$_c$ ha^{-1})

The results presented indicate that Mg(OH)$_2$ is a suitable fertilizer to improve the chemical status of acidified forest soils. Due to its high solubility, Mg(OH)$_2$ is more effective at depth than surface application of dolomite lime (see Huettl and Zoettl, 1993).

11.6. Comparison of element budgets from liming and fertilizer studies at the ecosystem level

Element budgets of forested ecosystems, at a plot or watershed scale, are very useful to study the various effects of additional Mg inputs via liming/fertilization. If continued over sufficiently long periods, i.e. at least several years after the treatment, these budget data can be used for an evaluation of the measures at the ecosystem level. This includes all the processes at the different compartment levels and allows the integration of factors influencing the treatment itself, like management and deposition history as well as site conditions (soil, climate etc.). The concept of element budgets, introduced by Ulrich *et al.* (1979), is used in many studies and is regarded as a very useful tool to characterize the status of an ecosystem and its stability and elasticity against internal and external impacts (Abrahamsen *et al.*, 1989; Bergkvist and Folkeson, 1993; Brahmer, 1990; Bredemeier, 1987; de Vries, 1988; Driscoll 1992; Einsele, 1986; Feger, 1993; Hambuckers and Remacle, 1990; Hantschel *et al.*, 1990; Hauhs, 1989; Johnson and Lindberg, 1989; Kreutzer, 1989; Marschner, 1990; Matzner, 1989; Meiwes and Beese, 1988; Moldan and Cerny, 1994; Rasmussen, 1990; Schulze, 1989; Tamm, 1989; Tuerk, 1992; Ulrich, 1989; van Breemen *et al.*, 1989; Weissen *et al.*, 1990).

In the following, some budget studies of fertilized and unfertilized forest

Table 2. Experimental sites, treatments, and references of element budget studies

Reference	Location	Treatment	Amount	Tree species	Age
Meyer, 1992	Wingst, N Germany	Dolomite Dolomite Dolomite + $K_2SO_4 \cdot MgSO_4$	$3\,t\,ha^{-1}$ $6\,t\,ha^{-1}$ $6\,t\,ha^{-1}$ + $550\,kg\,ha^{-1}$	Norway spruce	93
Wiedey, 1991	Hils, NE Germany	Ca_2SiO_4 + dolomite + $K_2SO_4 \cdot MgSO_4$	$4\,t\,ha^{-1}$ + $2\,t\,ha^{-1}$ + $1\,t\,ha^{-1}$	Norway spruce	69
Marschner, 1990	Berlin, Germany	Dolomite + $K_2SO_4 \cdot MgSO_4$	$6.1\,t\,ha^{-1}$ + $145\,kg\,ha^{-1}$	Scots pine	40
Feger *et al.*, 1994	Black Forest, SW Germany	Dolomite	$4\,t\,ha^{-1}$	Norway spruce	40–80
Feger *et al.*, 1991	Black Forest, SW Germany	$MgSO_4$	$750\,kg\,ha^{-1}$	Norway spruce	45
Hantschel, 1987	Fichtel Mountains, SE Germany	$MgSO_4$	$1\,t\,ha^{-1}$	Norway spruce	45
Tuerk, 1992	Fichtel Mountains, SE Germany	Dolomite	$10\,t\,ha^{-1}$	Norway spruce	45
Schaaf, 1992	Fichtel Mountains, SE Germany	$Mg(OH)_2$	$2.6\,t\,ha^{-1}$	Norway spruce	60–70

ecosystems are compared with special respect to losses of fertilizer Mg and other elements. Table 2 gives an overview of the experimental sites, treatments, and references.

Meyer (1992) investigated the effects of two increasing doses of dolomitic lime in Northern Germany. The soils were characterized as typical haplorthods derived from sand. The following data show the mean annual element budgets for the two years following application.

The control plot showed slightly positive budgets for the base cations, Ca^{2+} and Mg^{2+} (Figure 5). For aluminium, a strongly negative budget of $2.6\,kmol_c\,ha^{-1}\,y^{-1}$ was calculated. Also, sulfate and nitrate output exceeded total deposition. The application of $3\,t\,ha^{-1}$ dolomitic lime resulted in an increased NO_3-N output. Ca and Mg outputs were also increased but were still below the inputs (not including the fertilizer inputs) and S output was reduced. After treatment with $6\,t\,ha^{-1}$ dolomitic lime, the budgets of Mg and especially Ca became negative while Al output was almost zero. Also, sulfate output was significantly reduced. Addition of Kalimagnesia ($K_2SO_4 * MgSO_4$) also resulted in high Al and S outputs, but reduced Ca and Mg outputs. Within the first two years, only 0.2–4.2% Ca and Mg in the fertilizer were lost from the soil.

A combined fertilization with Ca_2SiO_4, dolomite, and Kalimagnesia increased the outputs of almost all elements, as shown in Figure 6, especially Al and sulfate (Wiedey, 1991). Nitrate was only slightly affected. Ca and Mg outputs were elevated

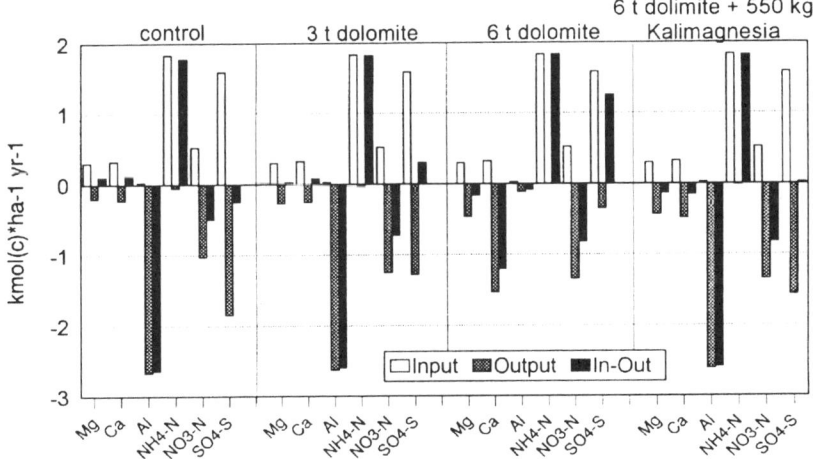

Figure 5. Element budgets of control and fertilized plots at Wingst, NW Germany (data from Meyer, 1992)

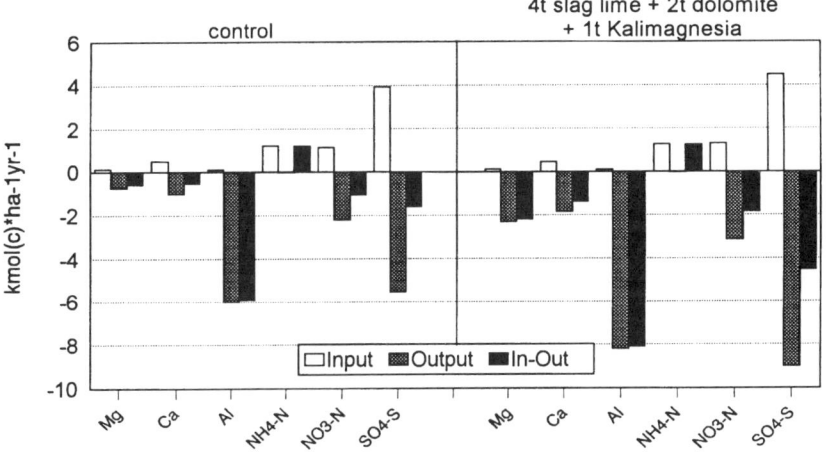

Figure 6. Element budgets of control and fertilized plots at the Hils Mountains, Germany (data from Wiedey, 1991)

compared to the control, which already showed negative budgets, but small compared to the fertilizer amounts.

Furthermore, the treatment resulted in an increased canopy buffering, from 19% to 30% of the deposited protons, due to increased leaching of base cations, especially K^+ and Ca^{2+}, from 1.6 to 2.1 kmol(c) ha^{-1} y^{-1} (see Table 3).

Forty-year-old Scots pine ecosystems on dystric cambisols in Berlin (FRG) were studied by Marschner (1990). Element budgets were calculated over a period of

Table 3. Canopy leaching and H⁺ buffering in kmol(c) ha⁻¹ y⁻¹ (data from Wiedey, 1991)

	Control	Fertilized
Ca leaching	0.85	1.10
K leaching	0.48	0.65
Mg leaching	0.18	0.20
Mn leaching	0.12	0.13
Cation leaching	1.63	2.08
Canopy buffering of H⁺	0.40	0.77

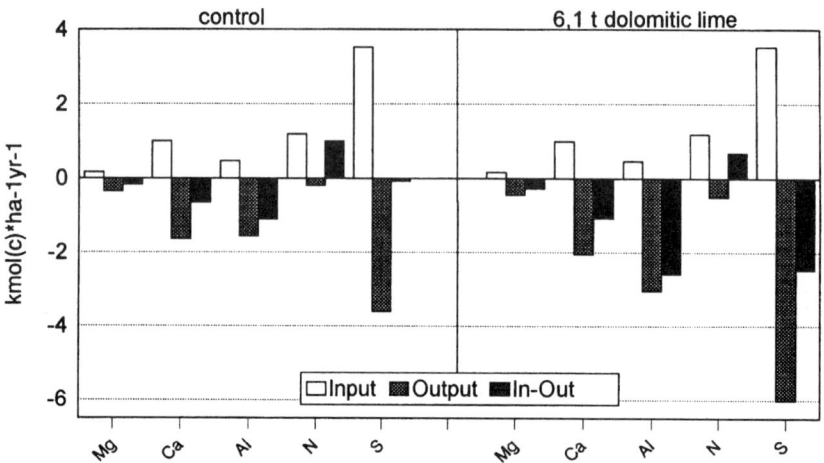

Figure 7. Element budgets of control and fertilized plots in Berlin, Germany (data from Marschner, 1990)

3.5 y on a control plot and a plot treated with dolomitic lime plus Kalimagnesia. The total amounts in the fertilizer were 2450 kg Ca, 144 kg Mg, 34 kg K, and 20 kg S. Except for N, all budgets on the control plot were negative (Figure 7). The outputs of all fertilized M_b cations were increased, and outputs of aluminium and sulfate were almost doubled, resulting in strongly negative budgets.

From a watershed study in the Black Forest (FRG), Feger *et al* (1994) reported element budgets of Norway spruce stands growing on iron–humus podzols. One of the watersheds was limed with dolomite. During the observation period of three years after application, the treatment caused an increase in Ca and Mg outputs from the catchment (Figure 8). The budgets of both elements were negative on the control. Outputs of sulfate and nitrate were slightly decreased and increased, respectively. Al output showed no change after the treatment.

In a parallel study by Feger *et al.* (1991), the effects of $MgSO_4$ fertilization were studied in two watersheds in the Black Forest. The soils were comparable to the site with the dolomite treatment. The application resulted in increased outputs of the fertilizer Mg^{2+} and SO_4^{2-} one year after the treatment (Figure 9). Both elements also showed negative budgets on the control watershed.

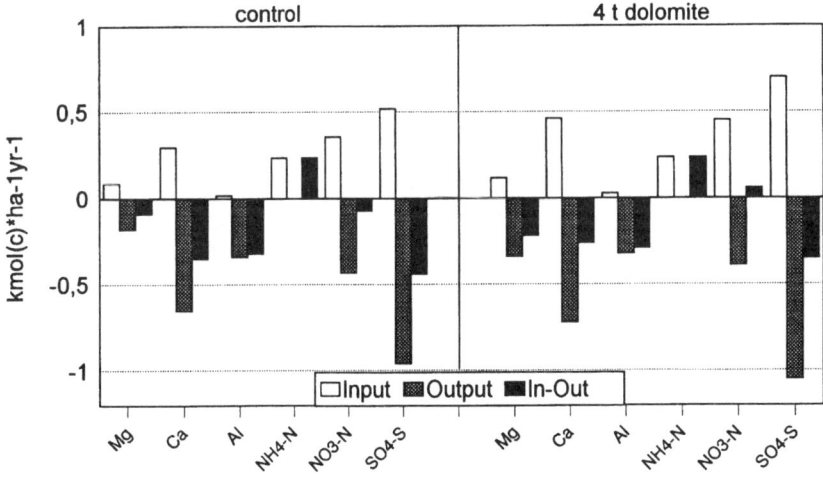

Figure 8. Element budgets of control and limed watersheds in the Black Forest, Germany (data from Feger *et al.*, 1994)

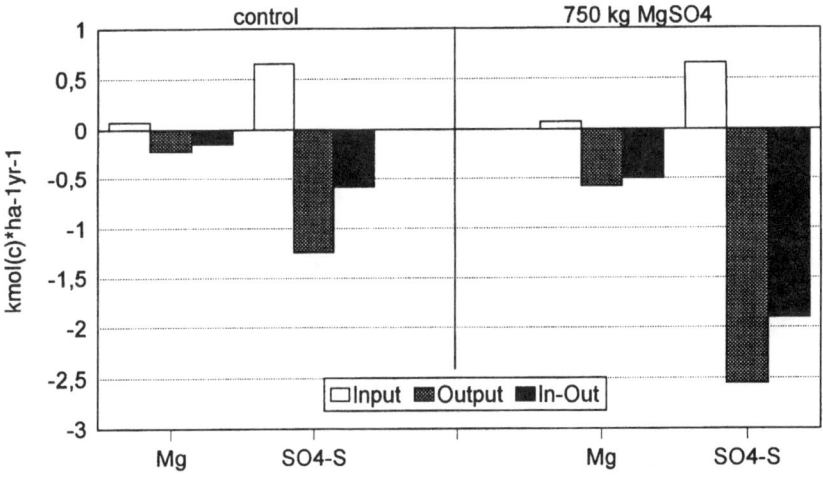

Figure 9. Element budgets of control and fertilized watersheds in the Black Forest, Germany (data from Feger *et al.*, 1991)

In the Fichtel Mountains, several studies were carried out in one Mg-deficient damaged and in one healthy Norway spruce stand on podzols derived from phyllite (Hantschel, 1987; Tuerk, 1992). At both sites, fertilization experiments included plots with an application of half-calcined dolomite and of $MgSO_4$, respectively. Tuerk (1992) reported budget data for the control and the limed plots over a period of six years for both sites. The very large amount of 10 t dolomite, corresponding to 1715 kg Mg ha^{-1} increased the Mg output, but not the Ca output at the damaged site

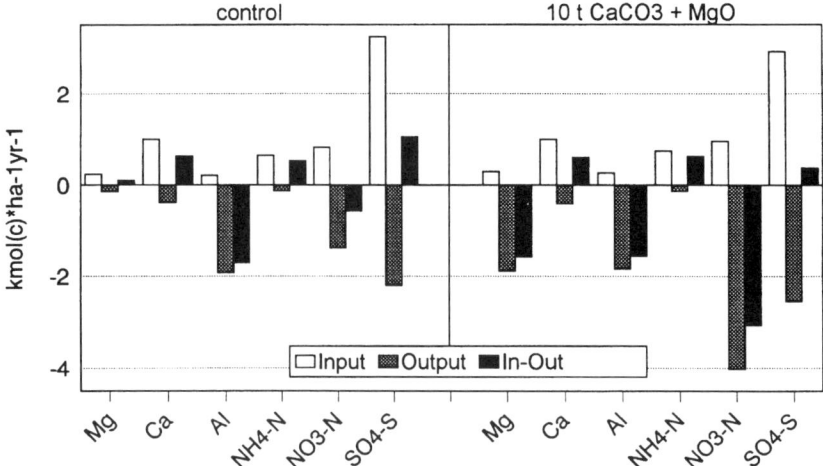

Figure 10. Element budgets of control and fertilized plots at Oberwarmensteinach in the Fichtel mountains, Germany (data from Tuerk, 1992)

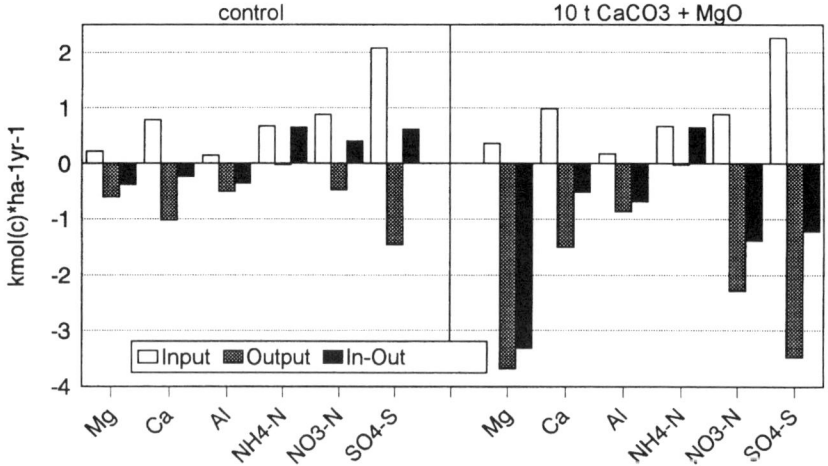

Figure 11. Element budgets of control and fertilized plots at Wuelfersreuth in the Fichtel Mountains, Germany (data from Tuerk, 1992)

within the period of six years after treatment (Figure 10). There were only minimal increases in Al and S outputs; however, nitrate output was tripled.

In contrast, the control at the undamaged site already showed negative budgets for Mg and Ca which were enhanced by the treatment (Figure 11). Outputs of these two fertilizer elements were much higher at the healthy site than at the damaged site. Also Al and especially S outputs were clearly increased. Nitrate losses from the system were more than four times higher compared to the control. Whereas N and

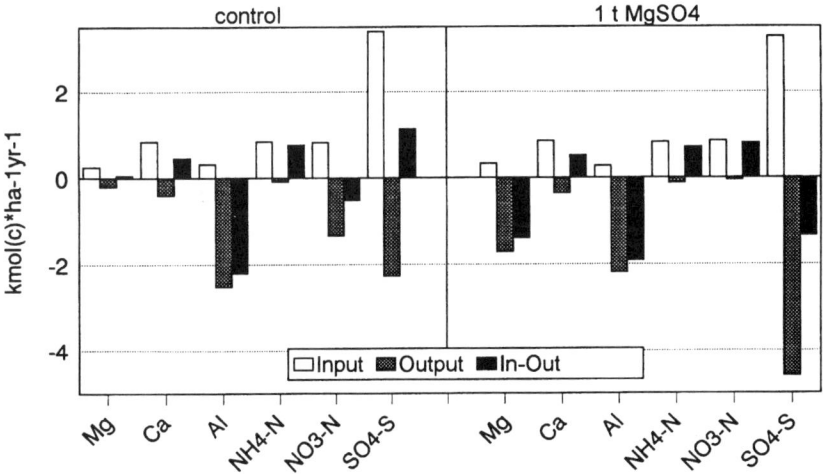

Figure 12. Element budgets of control and fertilized plots at Oberwarmensteinach in the Fichtel Mountains, Germany (data from Hantschel, 1987)

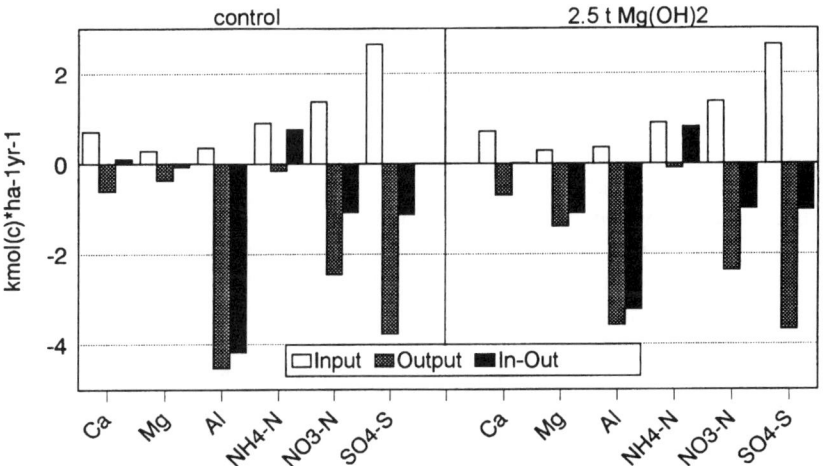

Figure 13. Element budgets of control and fertilized plots at Hohe Matzen in the Fichtel Mountains, Germany (data from Schaaf, 1992)

S showed positive budgets on the untreated plot, the limed plot had strongly negative budgets for all elements except NH$_4$.

For the MgSO$_4$-fertilized plots at the damaged site, Hantschel (1987) calculated input–output budgets for a 2-year period following application. The applied Mg amounts were only about 10% of the dolomite plot (174 kg Mg ha^{-1}). The fertilizer elements, Mg and S, showed clear increases in output and a change in budget from

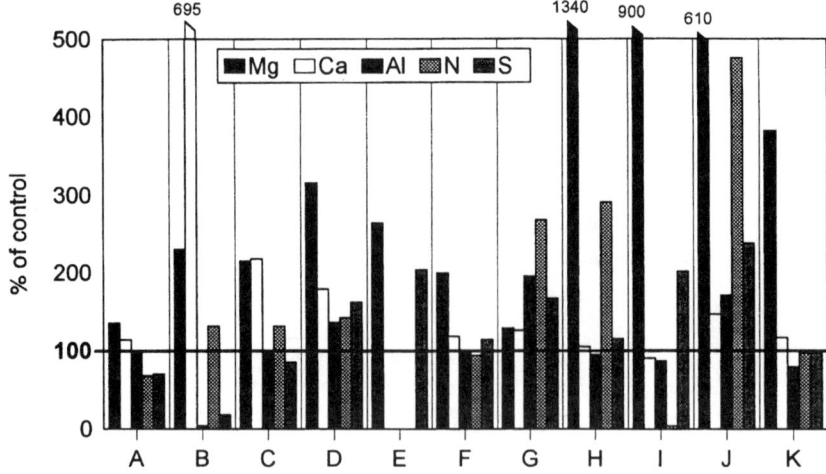

Figure 14. Effects of different Mg fertilizer forms and amounts on element budgets from 12 studies: budgets (throughfall deposition – seepage/watershed output) are given in % of the control plot/watershed (see Figures 5–13 for data sources) **A** = 3 t ha⁻¹ dolomite; **B** = 6 t ha⁻¹ dolomite; **C** = 6 t ha⁻¹ dolomite + 550 kg Kalimagnesia (K₂SO₄*MgSO₄) ha⁻¹; **D** = 4 t ha⁻¹ Ca silicate + 2 t ha⁻¹ dolomite + 1 t ha⁻¹ Kalimagnesia, **E** = 750 kg MgSO₄ ha⁻¹; **F** = 4 t ha⁻¹ dolomite; **G** = 6.1 t ha⁻¹ dolomite + 145 kg ha⁻¹ Kalimagnesia; **H** = 10 t ha⁻¹ dolomite; **I** = 750 kg MgSO₄ ha⁻¹; **J** = 10 t ha⁻¹ dolomite; **K** = 2.5 t Mg(OH)₂ ha⁻¹

positive to negative (Figure 12). N output was reduced to almost zero, whereas Ca and Al were hardly affected.

The absolute amounts of Mg lost from the ecosystem were similar for both treatments, but, in relation to the amounts of Mg applied, the losses after fertilization with salt were 10 times higher compared with the dolomite treatment.

Treatment with Mg(OH)₂ increased Mg-losses as described in section 11.5, but reduced Al output and hardly affected N and S budgets on average over a period of 2.5 y after application (Figure 13).

These examples, summarized in Figure 14, elucidate the statement that there is no uniform reaction of forest ecosystems or forest soils to the various fertilization treatments. The general assumption that application of quickly soluble neutral salts results in acidification and enhanced risks of Al leaching is contradicted, for example, by the studies in the Fichtel Mountains (Hantschel, 1987) and in the Black Forest (Feger *et al.*, 1991). Although the MgSO₄ applications result in enhanced SO_4^{2-} leaching, the reported losses are relatively small compared to the fertilizer amounts, leading to the assumption that these soils have a considerable potential for SO_4^{2-} sorption (see section 11.4.).

On the other hand, liming, especially with dolomite, does not necessarily result in increased nitrate losses, as shown for example by Feger *et al.* (1994) in their catchment study. One of the major keys to understanding the effects of liming on the N budget of a site seems to be its N status with respect to 'nitrogen saturation' (Kreutzer, 1995), including atmospheric deposition as well as site and management

Table 4. Correlation coefficients between element budget data and soil parameters (only values significant at $p < 0.05$)

	Mg_f	$N_{in/out}$	BS_{sub}	CEC	Al_{top}	Al_{sub}	N_{out}	S_{out}
Mg_{out}	0.5089		0.6587		0.6910			
Al_{out}		0.7364					0.5868	0.7442
N_{out}	0.7008	0.8091						
S_{out}			0.5273	0.5200	0.6265	−0.6729		

(Mg_{out}, Al_{out}, N_{out}, S_{out} = output in % of control; Mg_f = Mg amount fertilized in kg ha^{-1}; $N_{in/out}$ = N input/N output ratio of control; BS_{top} = base saturation in mineral subsoil (<50 cm); BS_{sub} = base saturation in mineral subsoil (>50 cm); CEC = cation exchange capacity of the mineral soil; Al_{top} = Al saturation of CEC in the top soil; Al_{sub} = Al saturation of CEC in the subsoil)

history. This finding is underlined by the correlation between the N output of fertilized plots and the input/output regime of the control plots (Table 4) found in the data from the 11 studies.

Furthermore, the examples reveal that liming may lead to enhanced Al leaching from the soil (see Marschner, 1990), although the results of Meyer (1992) may suggest that this could be true only for 'insufficient' amounts of lime. But no explanation is given for the reported differences between Al budgets of the 3 t and 6 t treatments. The comparison of applications of 6 t ha^{-1} dolomite with and without Kalimagnesia seems to underline the effects of neutral salts on Al^{n+} and SO_4^{2-} leaching risks. But the differences between the two lime treatments and the comparison with the control are still unexplained.

Statistical analysis of the dataset furthermore revealed significant correlations between the Al output on the one hand and N and S outputs on the other hand. The leaching losses of Al are obviously controlled by the 'carrier' anions in the output, not only by SO_4^{2-}, but also by NO_3^-. Thus, Al output is also dependent on the N status of the ecosystem. S output correlates with CEC, base saturation of the subsoil, and Al saturation of the exchange complex, but with different signs for the topsoil and the subsoil (Table 4). This may indicate the importance of exchange site characteristics, mainly of the subsoil, for SO_4^{2-} leaching. High base and low Al saturation in the deeper soil horizons together with higher pH seem to be more unfavourable for SO_4^{2-} sorption and/or storage in form of $AlOHSO_4$.

Most of the reports cited covered only a short period after the application so that long-term effects cannot be evaluated from the reported data. This holds true especially for the question of long-term effects of surface liming with slowly soluble material and for the assessment of reported sulfate storage after application of sulfate fertilizers and the possibilities and conditions for a remobilization and its subsequent effects on the chemical soil status.

11.7. Conclusions

This chapter discusses the differences between lime and fertilizer application with respect to Mg in forest ecosystems. The main topics are:

- Effects on the chemical soil status (e.g. pH, base saturation, CEC),
- Effects on the soil solution (e.g. Al mobilization),
- Effects on the plant availability of Mg,
- Effects on mineralization (e.g. NO_3^- leaching),
- Effects on the storage of fertilizer elements,
- Effects on element budgets, and
- Differences in effect with soil depth and time.

These differences can be explained by solubility, reaction kinetics, contents and composition of the material, and the accompanying or chemically produced anion. Furthermore, the same treatment at different sites or different treatments at the same site may lead to quite different results due to the influence of many other factors, like site history, management, and environmental conditions. This finally leads to the conclusion that it is hardly possible to predict the effects of lime/fertilizer applications at sites where these factors are not known and it explains the variety of different findings of many experiments and studies.

It is important to keep in mind the time factor for an evaluation of treatment effects as discussed in section 11.6. Most of these studies were carried out only over a few years after the application and, thus, cover only the initial effects. Since dolomitic lime and $MgSO_4$ or other quickly soluble fertilizers may show considerable differences in effects with respect to time, the comparability of the data is problematic. The initial effects of neutral salt application may occur very soon after application whereas liming obviously has more long-term effects.

As has been shown in sections 11.3 and 11.6, the effects of dolomitic liming on mineralization seem to depend on the N status of the site. The fate of mineralized and/or nitrified N is, beside the factors mentioned already, also dependent on the N demand of the vegetation and the capacity of the stand to use this N supply – this varies with stand age, tree species and understory vegetation. Figure 15 shows a good example of the temporal dynamics of the N demand and the consequences for the N budget of a forest ecosystem (Bredemeier and Ulrich, 1992). From this figure, it can be seen how important it is to take these temporal changes into account when planning forest liming measures. Furthermore, an increase in nitrification (H^+ production) and subsequent NO_3^- leaching may counteract the increase in buffer capacity (H^+ consumption) of liming. The influence of tree species is illustrated by the results of Matzner and Meiwes (1991), who reported significant NO_3^- losses through leaching after liming one Norway spruce stand in the Solling/Germany, whereas, in a neighbouring beech stand, no increase in NO_3^- leaching could be detected. This result may be explained by the deeper root system of the beech stand.

Another problem to deal with is the abundance of aims or objectives of forest ecosystem fertilization discussed in section 11.1 that may influence the evaluation of the 'success' of different treatments. There is a considerable number of indications that there is not one single strategy or treatment that leads to the desired result. For example, the quick elevation of the Mg^{2+} concentration in soil solutions of Mg-deficient sites alone may not be sufficient to improve the nutritional status of

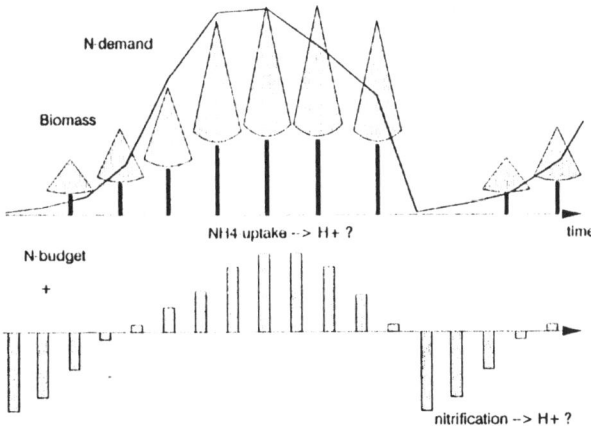

Figure 15. Temporal dynamics of N demand and N budget of a Norway spruce stand (from Bredemeier and Ulrich, 1992)

the stand. This was found mainly in older stands, where Mg contents in foliage did not change for several years despite a quick supply (Feger *et al.*, 1991; Kaupenjohann, 1989; Schaaf, 1995). Possible reasons for this delay may be: the overall slower reactivity of older trees; the greater significance of internal nutrient allocation; the different structure of the root system compared to young trees; or soil chemical conditions may restrict Mg uptake. Schaaf (1995) concluded, from his study of tree nutrition over a 5-year period following $Mg(OH)_2$ application in a 70-year-old Norway spruce stand, that under unfavourable soil conditions an increase in Mg supply alone (e.g. by fertilization with $MgSO_4$) without amelioration of the chemical soil status may not be sufficient to mitigate tip yellowing, i.e. Mg deficiency.

On the other hand, a quick supply of Mg has been shown to correct deficiency symptoms within only a few weeks in younger stands and even to repair the phloem collapse phenomenon as found by Fink (Chapter 9). This reaction may also explain the positive effects of $MgSO_4$ fertilization on root development described by Raspe (1992) and Schaaf and Zech (1991). Besides the positive effects on water and nutrient uptake and tree stability, an increase in rooting depth and root biomass may also decrease the leaching losses. Furthermore, the direct input of carbon into the soil from the increased allocation (from the foliage) to the root, rather than input via litter fall, may have implications on the long-term storage of carbon in forest soils with respect to the elevated atmospheric CO_2 concentrations expected in the future.

The most promising strategy for the correction of Mg deficiencies in forest eco-systems influenced by atmospheric deposition of N and/or acidity seems to be the application of a combination of dolomitic lime and neutral salt on the basis of a thorough evaluation of the actual site conditions. A possible alternative could be the application of quickly soluble materials like $Mg(OH)_2$.

11.8. References

Abrahamsen G, Seip HM, Semb A. 1989. Long-term acidic precipitation studies in Norway. In: Adriano DC, Havas M, Eds. Acid Precipitation, Vol. 1: Biological and Ecological Effects. New York, Springer, 137–179.

Aldinger E. 1987. Elementgehalte in Boden und Nadeln verschieden stark geschädigter Fichten-Tannen-Bestände auf Praxiskalkungsflächen im Buntsandstein. Freiburger Bodenkundl. Abh. 19, 266pp.

Alewell C, Matzner E. 1992. Reversibility of soil solution acidity and of sulfate retention in acid forest soils. Water Air Soil Pollut. 71, 155–165.

Andersson F, Persson T, Eds. 1988. Liming as a measure to improve soil and tree condition in areas affected by air pollution. National Swedish Environmental Protection Board, Report 3518, 131pp.

Badalucco L, Grego S, Dell'Orco S, Nannipieri P. 1992. Effect of liming on some chemical, biochemical, and microbiological properties of acid soils under spruce (*Picea abies* L.). Biol. Fertil. Soils. 14, 76–83.

Bergkvist B, Folkeson L. 1993. Soil acidification and element fluxes of a Fagus sylvatica forest as influenced by simulated nitrogen deposition. Water Air Soil Pollut. 65, 111–122.

Brahmer G. 1990. Wasser- und Stoffbilanzen bewaldeter Einzugsgebiete im Schwarzwald unter besonderer Berücksichtigung naturräumlicher Ausstattung und atmogener Einträge. Freiburger Bodenkundl. Abh. 25, 295pp.

Bredemeier M. 1987. Stoffbilanzen, interne Protonenproduktion und Gesamtsäurebelastung des Bodens in verschiedenen Waldökosystemen Norddeutschlands. Ber. Forschungszentrum Waldökosysteme. A33, 1–138.

Bredemeier M, Ulrich B. 1992. Input/output-analysis of ions in forest ecosystems. In: Teller A, Mathy P, Jeffers JNR, Eds. Responses of Forest Ecosystems to Environmental Changes. London, Elsevier, 229–243.

de Vries W. 1988. Critical deposition levels for nitrogen and sulphur on Dutch forest ecosystems. Water Air Soil Pollut. 42, 221–239.

Driscoll CT. 1992. Patterns in the biogeochemistry at the Hubbard Brook experimental forest, New Hampshire, USA. In: Teller A, Mathy P, Jeffers JNR, Eds. Responses of Forest Ecosystems to Environmental Changes. Elsevier Applied Science, London. 244–255.

Einsele G, Ed. Das landschaftsökologische Forschungsprojekt Naturpark Schönbuch. DFG-Forschungsbericht. Weinheim (VCH), 1–636.

Ende HP, Zoettl HW. 1991. Effects of magnesium fertilizer on the vitality and nutrition of a European beech (*Fagus sylvatica* L.) stand in the southern Black Forest of West Germany. Water Air Soil Pollut. 54, 561–566.

Evers FH, Huettl RF. 1990. A new fertilization strategy in declining forests. Water Air Soil Pollut. 54, 495–508.

Feger KH. 1993. Boden- und Wasserschutz in forstlich genutzten Ökosystemen. In: Alfred-Wegener-Stiftung, Ed. Die benutzte Erde – Ökosysteme, Rohstoffgewinnung, Herausforderungen. Berlin, 315–329.

Feger KH, Zöttl HW, Brahmer G. 1991. Assessment of the ecological effects of forest fertilization using an experimental watershed approach. Fertil. Res. 27, 49–61.

Feger KH, Armbruster M, Raspe S, Woelfelschneider A, Brahmer G. 1994. Dolomitic limestone application to a forested catchment – chemical and biological responses. In: Jenkins A, Ed. Experimental Manipulations of Biota and Biogeochemical Cycling in Ecosystems. Commission of the European Communities Ecosystems Research Report.

Gustafsson JP, Jacks G. 1993. Sulphur status in some Swedish podzols as influenced by acidic deposition and extractable organic carbon. Environ. Pollut. 81, 185–191.

Hambuckers A, Remacle J. 1990. A six-year nutrient balance for a coniferous watershed receiving acidic rain inputs. In: Harrison AF, Ineson P, Heal OW, Eds. Nutrient Cycling in Terrestrial Ecosystems. London, Elsevier, 130–138.

Hanisch B, Ed. 1989. IMA-Querschnittseminar 'Düngung geschädigter Waldbestände'. KfK-PEF Bericht 55, Kernforschungszentrum Karlsruhe, 343pp.

Hantschel R. 1987. Wasser- und Elementbilanz von geschädigten, gedüngten Fichtenökosystemen im Fichtelgebirge unter besonderer Berücksichtigung von physikalischer und chemischer Bodenheterogenität. Bayreuther Bodenkundl. Ber. 3, 219pp.

Hantschel R, Kaupenjohann M, Horn R, Zech W. 1990. Water, nutrient and pollutant budgets in damaged Norway spruce stands in NE-Bavaria (FRG) and their changes after different fertilization treatments. Water Air Soil Pollut. 49, 273–297.

Haug I, Feger KH. 1990. Effects of fertilization with and on soil solution chemistry, mycorrhiza and nutrient content of fine roots in a Norway spruce stand. Water Air Soil Pollut. 65, 153–173.

Hauhs M. 1988. Reversibility of acidification. In: Mathy P, Ed. Air Pollution and Ecosystems. Dordrecht, Kluwer, 407–417.

Hauhs M. 1989. Lange Brahmke: An ecosystem study of a forested catchment. In: Adriano DC, Havas M, Eds. Acid Precipitation, Vol. 1: Biological and Ecological Effects. New York, Springer, 275–305.

Hildebrand EE. 1990. Der Einfluß von Forstdüngungen auf die Lösungsfracht des Makroporenwassers. Allg. Forst. 45, 604–607.

Hildebrand EE, Schack-Kirchner H. 1990. Der Einfluß der Korngröße oberflächlich ausgebrachter Dolomite auf Lösungsverhalten und vertikale Wirkungstiefe. Forst- und Holzwirt. 45, 139–142.

Huber B. 1991. Sektion Waldernährung tagte in Freiburg: Salzdüngung oder Kalkung in geschädigten Waldbeständen. Allg. Forst. 465, 80–85.

Huettl RF. 1989. Liming and fertilization as mitigation tools in declining forest ecosystems. Water Air Soil Pollut. 44, 93–118.

Huettl RF. 1991. Die Nährelementversorgung geschädigter Wälder in Europa und Nordamerika. Freiburger Bodenk. Abh. 289, 440pp.

Huettl RF, Zöettl HW. 1993. Liming as a mitigation tool in Germany's declining forests – reviewing results from former and recent trials. Forest Ecol. Manage. 61, 325–338.

Jandl R, Katzensteiner K. 1992. Düngungsversuche mit Magnesitdüngern zu Fichte im Magnesium-mangelgebiet Schöneben. In: Glatzel G, Jandl R, Eds. Magnesiummangel in mitteleuropäischen Waldökosystemen. Forstl. Schriftenreihe Universität für Bodenkultur, Wien, Band. 5, 152–161.

Jenkins A, Waters D, Donald A. 1991. An assessment of terrestrial liming strategies in upland Wales. J. Hydrol. 124, 243–261.

Johnson DW, Lindberg SE. 1989. Acidic deposition on Walker Branch watershed. In: Adriano DC, Havas M, Eds. Acid Precipitation, Vol. 1: Biological and Ecological Effects. New York, Springer, 1–38.

Kaupenjohann M. 1989. Chemischer Bodenzustand und Nährelementversorgung NO-Bayerischer Fichtenbestände. Bayreuther Bodenkundl. Ber. 9. 202pp.

Kaupenjohann M. 1991. Magnesiummangel im Wald. Kali Briefe (Büntehof). 20, 561–577.

Kaupenjohann M. 1992. Mehrjährige Erfahrungen mit der Magnesiumdüngung in Waldökosystemen des Fichtelgebirges. In: Flatzel G, Jandl R, Eds. Magnesiummangel in miteleuropäischen Waldöko-systemen. Forstl. Schriftenreihe Universität für Bodenkultur, Wien, Band. 5, 122–131.

Kaupenjohann M, Zech W. 1989. Waldschäden und Düngung. Allg. Forstz. 37, 1002–1008.

Kreutzer K. 1989. The effects of acid irrigation and compensative liming on soil and trees in a mature Norway spruce stand (Höglwald project). In: Ulrich B, Ed. International Congress on Forest Decline Research: State of Knowledge and Perspectives. Kernforschungszentrum Karlsruhe GmbH, Vol. II, 667–690.

Kreutzer K, Göttlein A, Pröbstle 1991. Dynamik und chemische Auswirkungen der Auflösung von Dolomitkalk unter Fichte (Picea abies [L.] Karst.). In: Ökosystemforschung Höglwald. Kreutzer K, Göttlein A, Eds. Parey, Hamburg, 186–204..

Kreutzer K. 1995. Effects of liming on soil processes. Plant Soil. 168–169, 447–470.

Liu JC. 1989. Ernährungskundliche Auswertung von diagnostischen Düngungsversuchen in Fichten-beständen (Picea abies Karst.) Südwestdeutschlands. Freiburger Bodenk. Abh. 21, 193pp.

Lyngstad I. 1993. The effect of particle size of calcite and dolomitic limestones on pH in soil (in Swedish). Norsk landbruksforskning. Suppl. No. 16, 142–154.

Marschner B. 1990. Elementumsätze in einem Kiefernforstökosystem auf Rostbraunerde unter dem Einfluß einer Kalkung/Düngung. Ber. des Forschungsz. Waldökosysteme. A60, 1–192.

Marschner B, Stahr K, Renger M. 1989. Potential hazards of lime application in a damaged pine forest ecosystem in Berlin, Germany. Water Air Soil Pollut. 48, 45–57.

Matzner E. 1989. Acidic precipitation: Case study Solling. In: Adriano DC, Havas M, Eds. Acid Precipitation, Vol. 1: Biological and Ecological Effects. New York, Springer, 39–83.

Matzner E, Meiwes KJ. 1991. Effects of liming and fertilization on soil solution chemistry. Water Air Soil Pollut. 54, 377–389.

Matzner E, Khanna PK, Meiwes KJ, Ulrich B. 1985. Effects of fertilization and liming on the chemical soil conditions and element distribution in forest soils. Plant Soil. 87, 405–415.

Meiwes KJ, Beese F. 1988. Ergebnisse der Untersuchung des Stoffhaushaltes eines Buchenwaldöko-systems auf Kalkgestein. Ber. Forschungszentrum Waldökosysteme, B9, 1–141.

Meyer M. 1992. Untersuchungen zur Restabilisierung geschädigter Waldökosysteme im norddeutschen Küstenraum (Fallstudie Wingst II). Ber. Forschungszentrums Waldökosysteme, A94, 306.

354

Moldan B, Cerny J, Eds. 1994. Biogeochemistry of Small Catchments. SCOPE Report 51, Chichester, Wiley & Sons, 419pp.

Mulder J, Cresser MS. 1994. Soil and Soil Solution Chemistry. In: Moldan B, Cerny J, Eds. Biogeochemistry of Small Catchments. SCOPE Report 51. Chichester, Wiley & Sons, 107–131.

Murach D. 1988. Judgement of the applicability of liming to restabilise forest stands – with special consideration of root ecological aspects. In: Mathy P, Ed. Air Pollution and Ecosystems. Dordrecht, Kluwer, 445–451.

Persson T. 1994. Effects of liming on soil organisms and soil biological processes. Aktuelt fra Skogforsk 14 (Norwegian Forest Research Institute), 13–16.

Persson T, Lundkvist H, Wiren A, Hyvönen R, Wessen B. 1989. Effects of acidification and liming on carbon and nitrogen mineralization and soil organisms in mor humus. Water Air Soil Pollut. 45, 77–96.

Persson T, Wiren A, Andersson S. 1991. Effects of liming on carbon and nitrogen mineralization in coniferous forests. Water Air Soil Pollut. 54, 351–364.

Rasmussen L, Ed. 1990. Study on acid deposition effects by manipulating forest ecosystems (EXMAN). Air Pollution Research Report 24. Commission of the European Communities, Brüssel, 1–44.

Raspe S. 1992. Biomasse und Mineralstoffgehalte der Wurzeln von Fichtenbeständen (Picea abies Karst.) des Schwarzwaldes und Veränderungen nach Düngung. Freiburger Bodenkundl. Abh. 29, 197.

Reiter H, Bittersohl J, Schierl R, Kreutzer K. 1986. Einfluß von saurer Beregnung und Kalkung auf austauschbare und gelöste Ionen im Boden. Forstw. Cbl. 105, 300–309.

Schaaf W. 1992. Elementbilanz eines stark geschädigten Fichtenökosystems und deren Beeinflussung durch neuartige basische Magnesiumdünger. Bayreuther Bodenkundl. Ber. 23, 169.

Schaaf W. 1995. Effects of $Mg(OH)_2$-fertilization on nutrient cycling in a heavily damaged Norway spruce ecosystem (NE Bavaria/FRG). Plant Soil. 168–169, 505–511.

Schaaf W, Zech W. 1991. Einfluß unterschiedlicher Löslichkeit von Düngern. Allg. Forstz. 15, 766–768.

Schaaf W, Zech W. 1993a. Element turnover in a heavily damaged Norway spruce ecosystem influenced by N deposition. Plant Soil. 152, 277–285.

Schaaf W, Zech W. 1993b. Düngung mit gebranntem Magnesit und Magnesiumhydroxid zur Standortmelioration in einem stark geschädigten Fichtenökosystem. Z. Pflanzenernähr. Bodenk. 156, 357–364.

Schierl R, Göttlein A, Hohmann E, Trübenbach E, Kreutzer K. 1986. Einfluß von saurer Beregnung und Kalkung auf Humusstoffe sowie Aluminium- und Schwermetalldynamik in wässrigen Bodenextrakten. Forstw. Cbl. 105, 309–313.

Schueler G. 1991. Initial compensation of acidic deposition in forest ecosystems by different rock meals. Water Air Soil Pollut. 54, 435–444.

Schueler G. 1994a. Forstliche Stabilisierungsmaßnahmen bei Luftschadstoffeinträgen durch silikatische Gesteinsmehle im Vergleich zur Bodenschutzkalkung (Teil 1). Die Naturstein-Industrie. 1, 18–23.

Schueler G. 1994b. Forstliche Stabilisierungsmaßnahmen bei Luftschadstoffeinträgen durch silikatische Gesteinsmehle im Vergleich zur Bodenschutzkalkung (Teil 2). Die Naturstein-Industrie. 2, 36–40.

Schulze ED. 1989. The ecosystem balance of *Picea abies* stands in the Fichtelgebirge. In: Ulrich B, Ed. International Congress on Forest Decline Research: State of Knowledge and Perspectives. Kernforschungszentrum Karlsruhe GmbH, Vol II, 643–648.

Singh BR. 1984. Sulfate sorption by acid forest soils: 3. Desorption of sulfate from adsorbed surfaces as a function of time, desorbing ion, pH, and amount of adsorption. Soil Sci. 138 (5), 346–353.

Tamm CO. 1989. Comparative and experimental approaches to the study of acid deposition effects on soils as substrate of forest growth. Ambio. 18 (3), 184–191.

Tisdale SL, Nelson WL, Beaton JD. 1985. Soil Fertility and Fertilizers. New York, Macmillan, 754.

Tuerk T. 1992. Die Wasser- und Stoffdynamik in zwei unterschiedlich geschädigten Fichtenstandorten im Fichtelbirge. Bayreuther Bodenkundl. Ber. 22, 203.

Ulrich B. 1989. Effects of acid precipitation on forest ecosystems in Europe. In: Adriano DC, Johnson AH, Eds. Acid Precipitation, Vol. 2: Biological and Ecological Effects. New York, Springer, 189–272.

Ulrich B, Mayer R, Khanna PK. 1979. Deposition von Luftverunreinigungen und ihre Auswirkungen in Waldökosystemen im Solling. Schriften aus der Forstlichen Fakultät der Universität Göttingen und der Niedersächsischen Forstlichen Versuchsanstalt, Band 58. Frankfurt (Sauerländer), 1–291.

van Breemen N, Boderie PMA, Booltink HWG. 1989. Influence of airborn ammonium sulfate on soils of an oak woodland ecosystem in the Netherlands: seasonal dynamics of solute fluxes. In: Adriano DC, Johnson AH, Eds. Acid Precipitation, Vol. 1: Biological and Ecological Effects. New York, Springer, 209–236.

Weissen F, Hambuckers A, Praag HJ van, Ramacle J. 1990. A decennial control of N-cycle in the Belgian Ardenne forest ecosystems. Plant Soil. 128, 59–66.

Wenzel B. 1989. Kalkungs- und Meliorationsexperimente im Solling: Initialeffekte auf Boden, Sickerwasser und Vegetation. Ber. Forschungszentrum Waldökosysteme. A51, 1–274.

Wiedey GA. 1991. Ökosystemare Untersuchungen in zwei unterschiedlich exponierten Fichtenaltbeständen und in einemn Kalkungs- und Düngungsversuch im Hils. Ber. d. Forschungszentrums Waldökosysteme. A63, 205.

Woelfelschneider A. 1994. Einflußgrößen der Stickstoff- und Schwefel-Mineralisierung auf unterschiedlich behandelten Fichtenstandorten im Südschwarzwald. Freiburger Bodenkundl. Abh. 34, 191.

Zech W, Popp E. 1983. Magnesiummangel, einer der Gründe für Fichten- und Tannensterben in NO-Bayern. Forstw. Cbl. 102, 50–55.

Zoettl HW. 1990. Ernährung und Düngung der Fichte. Forstw. Cbl. 109, 130–137.

Concluding remarks

R. F. HÜTTL

For as long as the so-called 'new type forest damage' has been discussed, nutritional disturbances have been indicated as symptomatic findings. Meanwhile the complex and numerous relationships between nutrient supply and new type forest damage in Europe, North America and elsewhere are well established.

To characterize the extent of the temporary development and the spatial distribution of the new type forest damage, annual forest damage inventories are made. To a large extent, these surveys are based on unspecific foliar losses which do not allow adequate estimates related to the actual health or vitality status of forest trees and/or stands. Many of the damage symptoms characterized as new have in fact been known for a long time or can be explained by natural causes. New or new-type damage phenomena indeed appear to be exclusively related to nutritional disturbances. The most prevailing nutrient disorder with regard to the new type forest damage is Mg deficiency.

For macroscopic diagnosis of nutrition-related forest damage, specific discoloration symptoms can be utilized. The contribution by Ende and Evers (Chapter 1) discusses the visual deficiency symptoms in coniferous and deciduous trees. In addition the relevant threshold values for foliar and soil analyses are presented. Hence, for exact evaluation of the nutritional status of forest trees and stands, foliar analysis has been shown to be the most appropriate tool. For a number of tree species, sufficiently assured threshold values are available. Element ratios allow insight into balanced and unbalanced nutrient supplies.

Landmann, Hunter and Hendershot (Chapter 2) detail the temporal and spatial development of Mg deficiency in forest stands of Europe, North America and New Zealand. It has become evident that the Mg deficiency symptoms, which have been observed since the mid-1970s particularly in Norway spruce forests at higher altitudes in middle-range mountains of central Europe, spread relatively fast over larger areas in the early 1980s. Notably, in Europe this sudden and widespread occurrence can currently best be explained by extreme weather conditions related to a number of dry periods at the end of the 1970s and the beginning of the 1980s. The observed stagnation of discoloration as well as the phenomenon of natural regeneration of tip-yellowed spruce trees coincides with vegetation periods with more favorable precipitation regimes documented since the mid-1980s.

In this context also, elevated nitrogen deposition rates stimulating growth and eventually causing nutrient imbalances must be considered.

In general, with the Mg-deficiency phenomenon, healthy trees can be found right next to yellowed individuals. This pattern appears to be related to small-scale differences in soil water supply and Mg availability as well as to genetic differences

R. F. Hüttl & W. Schaaf (eds): Magnesium Deficiency in Forest Ecosystems, 357–362.

between the trees with regard to their Mg-absorption capacities. Magnesium deficiency can be observed in Norway spruce only when Mg deficiency prevails in the rooted solum. In cases of insufficient Mg supply, high Al^{3+} as well as NH_4^+ contents can additionally reduce Mg uptake. Mg deficiency prevails in forest stands that are characterized by very low atmospheric Mg input loads and base-poor, acidic parent materials.

Most of the earlier observations of Mg deficiency are clearly linked to former land use practices. In Europe at upper and middle elevations, Mg deficiency often coincides with former agricultural land use, including above all litter raking. The role of forestry in promoting Mg deficiency is most clearly demonstrated in New Zealand, where the first symptoms were linked to the introduction of nutrient-demanding species. Also, in Europe, the large-scale anthropogenic introduction of Norway spruce and to some extent also of Scots pine on sites with acidic nutrient-poor soils formerly covered by insufficiently growing beech or beech–oak stands and thus more recent nutrient-demanding forestry practices presumably played an important role in bringing about Mg deficiency.

As indicated above, monitoring forest health has confirmed that climate is the synchronizing factor for short-term dynamics of Mg-deficiency situations. In areas only episodically affected by Mg deficiency, such as in southern Scandinavia, climate may be considered the cause of Mg deficiency. However, it may be argued that climate only reveals a problem caused by other factors. Summer drought was found to be the key climatic feature triggering increases in Mg deficiency.

The biogeochemistry of Mg in forest ecosystems is discussed by Feger (Chapter 3). As an example, man-made coniferous forest stands which have been managed over longer time periods illustrate that Mg soil supply, notably on Mg-poor sites, is clearly related to the rate of Mg mobilization from rock weathering as well as from mineralization of organic matter. Clearly, uncoupling of weathering and uptake processes with regard to soil depth represents an important factor for forest ecosystem degradation, particularly increasing Mg losses from the ecosystem via soil leaching.

Also, from a physiological aspect, Mg plays a key role in tree nutrition. Magnesium is catalytically involved in most essential biochemical and regulatory processes of the plant metabolism. Slovik (Chapter 4) states that our knowledge of the physiology of Mg would seem to be sufficient to explain the Mg deficiency symptoms in the field. Magnesium is mobile in the xylem and the phloem tissue of forest trees. Magnesium appears to recycle several times per year within the canopy. In the spring, the young flush of conifers is supplied with Mg that originates mainly from the previous needle-age classes and from the sapwood. The pool of transiently depleted Mg in sapwood and older needles is substituted during the subsequent summer by net uptake of Mg from the soil. The varying Mg contents in different tree organs and tissues depend on the supply rate of Mg from the soil as well as on the tree species. Slovik further argues that, based on complete information of Mg cycling rates and balances, the impact of man-made air pollutants on Mg nutrition and tree vitality can eventually be quantified.

Investigations into the effects of Mg deficiency on the growth of trees are mainly carried out as pot experiments with young trees, different soil substrates, fertilizer forms, and doses. However, some field trials also exist. From these trials, Makkonen-Spiecker and Spiecker (Chapter 5) conclude that, with regard to Mg, neutral or positive growth effects can be demonstrated. Varying results can be related to differences in tree age and species or to different methodological approaches. It is also learned that only very low foliar Mg concentrations lead to reduced shoot growth. In conifers, Mg deficiency can also provoke needle losses. Hence, positive effects of Mg fertilization on tree growth occur only when foliar Mg contents are extremely low. Whenever Mg deficiency restricts tree development, root growth appears to be affected first.

To explain the far-reaching changes in Mg nutrition of many forest ecosystems in Europe, North America and elsewhere, various hypotheses have been formulated. These can be separated into direct and indirect cause–effect relationships.

Direct impacts of gaseous air pollutants, such as SO_2, NO_x, NH_x and O_3, as single factors or in combination with additive or synergistic effects, were relatively soon excluded as primary causes.

However, enhanced H^+ and NH_4^+/NH_3 input rates can increase crown leaching of mobile nutrient elements, such as Mg, with or without damage to foliar tissue. But it appears unlikely that this acid-leaching mechanism can actually provoke Mg deficiency.

There is no doubt that indirect effects, i.e. mechanisms and processes that lead to changes in substrate conditions in forest ecosystems, may cause Mg deficiency. In this context, the acid rain/Al toxicity hypothesis must be mentioned. As soon as this hypothesis was stated, it was controversial. Nevertheless, it appears plausible that, due to enhanced soil acidification related to larger man-made acid deposition input and other influences, acid base-poor soils may be degraded of their Mg supply to the point that the forest stands they carry can no longer be supplied sufficiently with Mg and acute Mg-deficiency symptoms may appear. On the other hand, it was found that most forest tree species impacted by Mg deficiency are relatively tolerant against Al toxicity. Apparently these tree species are well adapted to acid site conditions and have developed adequate protection strategies. So far, Al toxicity postulated for Norway spruce and European beech based on hydroponic culture studies has not been verified under field conditions.

As already indicated, increased N input into forest ecosystems has to be considered as an important factor. Improved NH_4^+ supply and uptake may lead to reduced Mg absorption as well as to lower pH values in the rhizosphere. However, anion uptake might be favored by this process. Better NO_3^- supply acts in the opposite direction. In areas with high atmospheric N input, N uptake may also occur through the crown directly. In general, better N supply increases tree and stand growth. From this, dilution effects and nutrient imbalances can result. High N deposition may also enhance soil acidification and lead to a lower availability of base cations. However, because of strong regional and local variations in N deposition rates and forms, the ammonium hypothesis can be disregarded as an overall explanation model. N deposition affects different forests in different but specific ways. The expected effects can be either positive or negative.

It is now clear that the human impact on the atmosphere and the resulting element deposition has to be seen as yet another site factor, whereby emissions must be differentiated according to their type, form, amount, and ratios. Consequently, an explanation model was formulated and tested successfully, in which Mg-related forest damage was considered to be caused by site- and stand-specific combinations of multiple stress factors. In this Mg-deficiency concept, among the various anthropogenic influences, site history plays a decisive role. Externally accelerated soil acidification, higher N input rates, reduced Mg deposition and larger Mg crown leaching losses are ecosystematic processes which locally add to this situation and can be characterized as accompanying factors. Extreme weather conditions, particularly drought periods during the growth period are viewed as initiating as well as accompanying factors. Biotic diseases clearly have a secondary effect. As the constellation of causal factors is generally complex and as the various influences are interrelated as well as site and stand dependent, the specific damage causes can only be determined at the site itself. Global explanation approaches are thus not useful.

Fertilizer trials represent a classic method to diagnose nutritional disturbances in forests. Through the application of fast-soluble fertilizers, it can be tested within a short time period whether deficiency symptoms can be reduced or completely mitigated. With regard to Mg deficiency, this is also possible by applying Mg-containing lime materials. Since the early 1980s, in forest areas exhibiting Mg deficiency damage, numerous diagnostic Mg fertilization and liming trials have been initiated. They prove that Mg deficiency in Norway spruce, Silver fir, Douglas fir, European beech and Scots pine can be overcome by appropriate Mg application.

Using fast-soluble Mg fertilizers, a remarkable revitalization associated with a reduction in or complete resolution of discoloration symptoms was found in young trees within only a few months. In older trees and stands, these effects occurred after 1–3 years. In fact, a sustainable fertilization effect can be obtained which has now been demonstrated over a period of almost 15 years. Similar positive effects were obtained from the application of Mg-containing lime materials. However, depending on particle size, Mg content and application dosage, these effects were generally retarded.

Even extremely high exchangeable Al^{3+} contents of acid soils as well as the related unfavorable Mg/Al or Ca/Al ratios did not impede enhanced Mg uptake when Mg supply was improved through Mg fertilization. In acid base-poor substrates, Mg salt applications did not cause fine-root or micorrhiza damage, even when, after the application of e.g. Kieserite ($MgSO_4$), exchange processes brought about an increase in exchangeable Al contents in the soil solution combined with a pH decrease. This short-term process of enhanced Al contents and reduced pH values, however, did not cause a long-term acidification of the top- or subsoil. Fertilized trees, because of the enhanced Mg supply, exhibited improved fine-root growth in the mineral soil. This important and desired effect is well substantiated by the contribution of Raspe (Chapter 10).

Because of slower dissolution kinetics, lime applications, particularly using larger particle size fractions, led to enhanced fine-root growth mainly in the organic layer. Notably in shallow-rooting tree species, such as Norway spruce, this effect may increase the danger of frost, drought and storm impact. Further ecological risks of liming are enhanced NO_3^- production and displacement; also, mobilization of heavy metals and Al were found after undifferentiated liming treatments. On K-poor substrates, the application of lime can further decrease K uptake and even induce K deficiency. In order to ensure the desired positive, and to prevent potential negative, effects, a thorough evaluation is necessary for all sites to be considered for liming. Besides soil pH values, humus type and content, site history, atmospheric N input and N uptake capacity of the stand, heavy metal contamination of the soil and soil vegetation along with several other factors have to be considered. Schaaf (Chapter 11) further evaluates the different Mg fertilization strategies that are detailed by Augustin, Mindrup and Meiwes (Chapter 7), and also by the contributions of Feger (Chapter 3), Raspe (Chapter 10), Kaupenjohann (Chapter 8) as well as Katzensteiner and Glatzel (Chapter 6).

Reduced fine-root growth observed in Mg-deficient conifers is apparently related to changes in assimilate allocation. Fink (Chapter 9) indicates that this phenomenon might be caused by a collapse of the phloem tissue within the vascular bundle of Mg-deficient tip-yellowed foliage, consequently impeding translocation of assimilates. Mg deficiency also reduces photosynthesis rates. When Mg deficiency was alleviated through appropriate fertilization, a regeneration of the phloem leading to a normal carbon allocation pattern was detected in revitalized foliar tissue. This remarkable effect explains the improved fine-root system, height and diameter growth of fertilized trees in comparison with unfertilized Mg-deficient individuals. The sustainable elimination of Mg deficiency is hence a decisive precondition in order to improve the health status and growth of yellowed trees on a long-term basis and to prevent irreversible damage. As, in conifers, acute Mg deficiency leads to specific phloem damage, histological needle analysis can be used as a diagnostic instrument to differentiate between direct and indirect damage pathways; because phytotoxic concentrations of gaseous air pollutants cause damage in the substomatal mesophyll tissue, while the phloem tissue stays intact.

Experiments to induce Mg deficiency showed that the application of NH_4-N on Mg-poor soils may further enhance or even initiate Mg deficiency. These experimental results can be transferred to forest ecosystems with acid base-poor soils and high NH_4-N deposition rates. However, numerous fertilizer trials also indicate that coniferous and deciduous forest stands can tolerate relatively high N input loads over long time periods without damage and even use the improved N supply to enhance biomass production when the availability of the other essential nutrients and water is adequate.

Finally, through analysis of previous Mg fertilizer trials, it was found that the application of the missing nutrient in the form of a neutral salt had in fact improved the resistance of trees remarkably and impeded the appearance of acute Mg-deficiency symptoms during the time period when new type forest damage symptoms developed. This was also true for areas that had received moderate H, S and N input loads during recent decades.

Also, former Mg liming improved the Mg supply of spruce and fir stands on acid base-poor substrates sustainably. Former $CaCO_3$ application without any Mg content caused soil chemical changes, such as significantly reduced Al^{3+} concentrations, better base saturation and improved cation-exchange capacity, but the development of Mg deficiency or related symptoms of the new type forest damage could not be prevented through this measure.

Today as in the past, besides liming as a soil amelioration measure, site- and stand-specific fertilization to improve the nutritional status and the resistance of tree stands to any kind of negative impact can be recommended. Even though aspects of soil and water protection have to be considered when silvicultural practices, such as liming or fertilization are applied, the tree and/or stand as an important ecosystem compartment can not be neglected. It would be an unsustainable approach to try to establish healthy forest ecosystems, including a healthy hydro- and pedosphere, without trying to optimize the health, i.e. the nutritional status, of the trees.

As the degradation of forest ecosystems is often related to a complex set of causal factors, regeneration measures in many cases must exceed soil amelioration or fertilization measures. These practices have to be integrated into an overall forest management concept in which nutritional and soil-related aspects, but also site history, silvicultural, ecological and landscape management factors as well as growth parameters, must be considered. In any case, all measures have to be adapted to the site- and stand-specific requirements and thus have to be determined case by case.